MEDICAL
INTELLIGENCE
UNIT

Molecular Pathogenesis of Cholestasis

Michael Trauner, M.D.

Department of Medicine
Division of Gastroenterology and Hepatology
Medical University Graz
Graz, Austria

Peter L.M. Jansen, M.D., Ph.D.

Department of Gastroenterology and Hepatology
Academic Medical Center
Amsterdam, The Netherlands

SPRINGER SCIENCE+BUSINESS MEDIA, LLC

MOLECULAR PATHOGENESIS OF CHOLESTASIS

Medical Intelligence Unit

http://www.wkap.nl/

Please address all inquiries to the Publishers:
Eurekah.com / Landes Bioscience, 810 South Church Street
Georgetown, Texas, U.S.A. 78626
Phone: 512/ 863 7762; FAX: 512/ 863 0081
www.Eurekah.com
www.landesbioscience.com

ISBN 978-1-4613-4767-5

Molecular Pathogenesis of Cholestasis edited by Michael Trauner and Peter L.M. Jansen, Landes / Kluwer dual imprint / Landes series: Medical Intelligence Unit

While the authors, editors and publisher believe that drug selection and dosage and the specifications and usage of equipment and devices, as set forth in this book, are in accord with current recommendations and practice at the time of publication, they make no warranty, expressed or implied, with respect to material described in this book. In view of the ongoing research, equipment development, changes in governmental regulations and the rapid accumulation of information relating to the biomedical sciences, the reader is urged to carefully review and evaluate the information provided herein.

Library of Congress Cataloging-in-Publication Data

Molecular pathogenesis of cholestasis / [edited by] Michael Trauner,
Peter L.M. Jansen.
    p. ; cm.
Includes bibliographical references and index.
ISBN 978-1-4613-4767-5   ISBN 978-1-4419-9034-1 (eBook)
DOI 10.1007/978-1-4419-9034-1
1. Cholestasis. 2. Cholestasis--Molecular aspects. 3.
Cholestasis--Pathogenesis. 4. Molecular biology. I. Trauner, Michael.
II. Jansen, Peter L. M.
    [DNLM: 1. Cholestasis--genetics. 2. Cholestasis--metabolism. 3.
Cholestasis--physiopathology. 4. Molecular Biology--methods. WI 703
M718 2003]
RC854.C45M64 2003
616.3'6--dc22   2003022028

CONTENTS

EDITORS

Michael Trauner, M.D.
Department of Medicine
Division of Gastroenterology and Hepatology
Medical University Graz
Graz, Austria
Chapter 20

Peter L.M. Jansen, M.D., Ph.D.
Department of Gastroenterology and Hepatology
Academic Medical Center
Amsterdam, The Netherlands
Chapter 13

CONTRIBUTORS

Gianfranco Alpini
Department of Internal Medicine
and Medical Physiology
Scott & White Hospital
Texas A & M University System
Health Science Center
Temple, Texas, U.S.A.
Chapter 6

M. Sawkat Anwer
Departments of Biomedical
and Clinical Sciences
Tufts University School
of Veterinary Medicine
North Grafton, Massachusetts, U.S.A.
Chapter 9

Irwin M. Arias
Department of Physiology
Tufts University School of Medicine
Boston, Massachusetts, U.S.A.
Chapter 5

Margaret F. Bassendine
Centre for Liver Research
The Medical School
Framlington Place
Newcastle Upon Tyne, U.K.
Chapter 17

Ulrich Beuers
Department of Internal Medicine II
University of Münich
Münich, Germany
Chapter 24

Einar Björnsson
Department of Internal Medicine
Section of Gastroenterology
and Hepatology
Sahlgrenska University Hospital
Gothenburg, Sweden
Chapter 18

James L. Boyer
Department of Medicine
and Liver Center
Yale University School of Medicine
New Haven, Connecticut, U.S.A.
Chapter 1

Laura Bull
UCSF Liver Center Laboratory
San Francisco General Hospital
San Francisco, California, U.S.A.
Chapter 7

Roger W. Chapman
Department of Gastroenterology
John Radcliffe Hospital
Oxford, U.K.
Chapter 18

James M. Crawford
Department of Pathology
University of Florida College of Medicine
Gainesville, Florida, U.S.A.
Chapter 12

Yunhai Cui
Division of Tumor Biochemistry
Deutsches Krebsforschungszentrum
Heidelberg, Germany
Chapter 15

Harald Dobnig
Department of Internal Medicine
Karl Franzens University
Graz, Austria
Chapter 22

Wihelma Echevarría
Yale University School of Medicine
New Haven, Connecticut, U.S.A.
Chapter 4

Ronald Oude Elferink
Department of Experimental Hepatology
Academic Medical Center
Amsterdam, The Netherlands
Chapter 14

Astrid Fahrleitner
Department of Internal Medicine
Karl Franzens University Hospital
Graz, Austria
Chapter 22

Peter Fickert
Department of Internal Medicine
Karl-Franzens University
Graz, Austria
Chapter 20

Heather Francis
Department of Internal Medicine
Scott & White Hospital
Texas A & M University System
   Health Science Center
Temple, Texas, U.S.A.
Chapter 6

Siddhartha S. Ghosh
Departments of Medicine
   and Molecular Genetics
Albert Einstein College of Medicine
New York, New York, U.S.A.
Chapter 25

Jonathan D. Gitlin
Edward Mallinckrodt Department
   of Pediatrics
Washington University School
   of Medicine
St. Louis, Missouri, U.S.A.
Chapter 16

Shannon Glaser
Department of Internal Medicine
Scott & White Hospital
Texas A & M University System
   Health Science Center
Temple, Texas, U.S.A.
Chapter 6

Gregory J. Gores
Division of Gastroenterology
   and Hepatology
Mayo Medical School, Clinic,
   and Foundation
Rochester, Minnesota, U.S.A.
Chapter 10

Chandan Guha
Departments of Medicine
   and Molecular Genetics
Albert Einstein College of Medicine
New York, New York, U.S.A.
Chapter 25

Bruno Hagenbuch
Department of Medicine
University Hospital
Zurich, Switzerland
Chapter 2

Iqbal Hamza
Edward Mallinckrodt Department
  of Pediatrics
Washington University School
  of Medicine
St. Louis, Missouri, U.S.A.
Chapter 16

Hajime Higuchi
Division of Gastroenterology
  and Hepatology
Mayo Medical School, Clinic,
  and Foundation
Rochester, Minnesota, U.S.A.
Chapter 10

Guido J.E.J. Hooiveld
Division of Human Nutrition
University Wageningen
Wageningen, The Netherlands
Chapter 3

Saul J. Karpen
Department of Pediatrics/GI
  and Nutrition
Baylor College of Medicine
Houston, Texas, U.S.A.
Chapter 8

Dietrich Keppler
Division of Tumor Biochemistry
Deutsches Krebsforschungszentrum
Heidelberg, Germany
Chapter 15

Helmut Kipp
Department of Physiology
Tufts University School of Medicine
Boston, Massachusetts, U.S.A.
Chapter 5

Jörg König
Division of Tumor Biochemistry
Deutsches Krebsforschungszentrum
Heidelberg, Germany
Chapter 15

Folkert Kuipers
Center for Liver, Digestive
  and Metabolic Diseases
Academic Hospital
Groningen, The Netherlands
Chapter 23

Gerd A. Kullak-Ublick
Department of Internal Medicine
University Hospital
Zürich, Switzerland
Chapter 19

Gene LeSage
Department of Internal Medicine
Scott & White Hospital
Texas A & M University System
  Health Science Center
Temple, Texas, U.S.A.
Chapter 6

Sung W. Lee
Departments of Medicine
  and Molecular Genetics
Albert Einstein College of Medicine
New York, New York, U.S.A.
Chapter 25

Marco Marzioni
Department of Internal Medicine
Scott & White Hospital
Texas A & M University System
  Health Science Center
Temple, Texas, U.S.A.
Chapter 6

Peter J. Meier
Department of Medicine
University Hospital
Zürich, Switzerland
Chapter 2

Michael Müller
Division of Epidemiology and Nutrition
University Wageningen
Wageningen, The Netherlands
Chapters 3, 13

Michael H. Nathanson
Department of Digestive Diseases
Yale University School of Medicine
New Haven, Connecticut, U.S.A.
Chapter 4

Anne T. Nies
Division of Tumor Biochemistry
Deutsches Krebsforschungszentrum
Heidelberg, Germany
Chapter 15

Jayanta Roy-Chowdhury
Departments of Medicine
   and Molecular Genetics
Albert Einstein College of Medicine
New York, New York, U.S.A.
Chapter 25

Namita Roy-Chowdhury
Departments of Medicine
   and Molecular Genetics
Albert Einstein College of Medicine
New York, New York, U.S.A.
Chapter 25

Christian Rust
Department of Internal Medicine II
University of Münich
Münich, Germany
Chapter 24

Ekkehard Sturm
Department of Pediatrics
University Hospital
Groningen, The Netherlands
Chapter 13

Yuichi Sugiyama
School of Pharmaceutical Sciences
The University of Tokyo
Tokyo, Japan
Chapter 11

Hiroshi Suzuki
School of Pharmaceutical Sciences
The University of Tokyo
Tokyo, Japan
Chapter 11

Mark G. Swain
Health Sciences Center
University of Calgary
Calgary, Alberta, Canada
Chapter 21

Henkjan J. Verkade
Center for Liver, Digestive
   and Metabolic Diseases
Academic Hospital
Groningen, The Netherlands
Chapter 23

Cynthia R.L. Webster
Departments of Biomedical
   and Clinical Sciences
Tufts University School of Veterinary
   Medicine
North Grafton, Massachusetts, U.S.A.
Chapter 9

Anniek Werner
Center for Liver, Digestive
   and Metabolic Diseases
Academic Hospital
Groningen, The Netherlands
Chapter 23

Gernot Zollner
Department of Internal Medicine
Karl-Franzens University
Graz, Austria
Chapter 20

PREFACE

The aim of this book was to provide a cutting-edge overview on the molecular pathogenesis of cholestasis. For this purpose we asked the leading experts in their fields to cover the key aspects of bile formation and cholestatic liver diseases. The topics of this book reach from basic mechanisms of hepatobiliary transport and its regulation, over general molecular and cellular concepts of cholestatic liver injury, to specific molecular mechanisms of hereditary and acquired cholestatic liver diseases, their complications and treatment. This is an attempt to link basic research to clinical hepatology and related areas of clinical medicine. From these top-class reviews, it will become evident to the reader that recent advances in molecular biology now permit a better understanding of cholestatic liver diseases and that such new insights are beginning to impact the clinical management of these patients.

We hope that basic research and physician scientists in the areas of hepatology, genetics, molecular and cell biology, pharmacology, pathology, gastroenterology, endocrinology, and students will find this book instructive, insightful and stimulating for their own research.

Knowledge about hepatobiliary transport in health and disease is rapidly increasing since this is a very active area of research. Therefore, one new aspect of this book is to provide an internet version in addition to the traditional hardcopy, to facilitate easy access and encourage continuous updating of individual chapters.

We want to thank all authors for their excellent contributions which reflect the high level of their own research. We extend our thanks to Cynthia Conomos and Ron Landes from Landes Bioscience for their support and excellent co-operation in publishing this book.

Michael Trauner, M.D.
Graz, Austria
Peter Jansen, M.D., Ph.D.
Amsterdam, The Netherlands

CHAPTER 1

Mechanisms of Bile Formation: An Introduction

James L. Boyer

Introduction

The formation of bile is a unique and vital function of the liver. Failure to form bile results in progressive cholestatic liver injury and death. Knowledge of the mechanism of bile formation has progressed rapidly in the past decade and this introduction provides a background and historical perspective for the subsequent reviews. Bile is a complex aqueous secretion composed of ~95% water and endogenous solid constituents consisting of bile salts, phospholipid and cholesterol, amino acids, steroids, enzymes, porphyrins, vitamins, and heavy metals as well as exogenous drugs, xenobiotics and toxins.[1] Bile salts are the major organic solute in bile and are necessary for the emulsification and digestive absorption of dietary lipids. Bile serves to eliminate potentially harmful organic lipophilic substances, including xenobiotics and toxins as well as endogenous substrates such as bilirubin and bile salts, not easily excreted by the kidney. Bile is also the major route of excretion for cholesterol. Disorders that impair the production of bile result in the syndrome known as cholestasis.[2,3]

The primary secretion of bile is generated by osmotic gradients that are formed within the bile canaliculus of the hepatocytes by energy dependent mechanisms. The creation of these osmotic gradients depends upon the function of a number of transporting polypeptides that are located on basolateral and apical plasma membrane domains. This primary secretion of bile is subsequently modified by the bile duct epithelium where transport systems in the luminal membranes of cholangiocytes both secrete and absorb certain biliary constituents.[4]

Historical Aspects

Prior to the middle of the 20th century, little was known about the fundamental mechanisms that produced this vital secretion. Knowledge lagged far behind the understanding of the formation of urine, primarily because bile was a "hidden" secretion whose product could only be examined after a surgical laparotomy and cannulation of the external bile ducts. The primary secretion formed by the hepatocytes enters minute bile canaliculi whose microscopic dimensions of ·1 um are best observed by electron microscopy. Thus there is no hope of sampling primary hepatocyte bile so that the early scientific literature was largely confined to reports of chemical analyses from bile collected surgically or following placement of an indwelling biliary cannula.[5,6]

A mechanistic understanding of biliary secretion entered the modern era when Ivar Sperber, a Swedish physiologist working at the Royal College of Agriculture in Uppsala, first formulated the "osmotic theory of bile formation" in a classic review, entitled "Secretion of organic anions in the formation of urine and bile", published in Pharmacological Reviews in 1959.[7] This report was heavily influenced by the work of many different investigators and served to complement Sperber's own observations of the secretion of phenol red in urine and bile

Molecular Pathogenesis of Cholestasis, edited by Michael Trauner and Peter L.M. Jansen.
©2004 Eurekah.com and Kluwer Academic / Plenum Publishers.

of the anesthetized chicken. Sperber realized that organic solutes, bile acids and other "cholephiles" when injected intravenously, were taken up by the liver and subsequently concentrated in bile, resulting in a choleresis. He reasoned, perhaps influenced by the fundamental observations of Pappenheimer (who proposed that water would flow across a semi-permeable membrane as a result of osmotic gradients) that the concentrative transport of solutes in bile created osmotic gradients that then stimulated the passive diffusion of water (and electrolytes) into bile thereby promoting its secretion.[8] Thus a paradigm was established upon which all further studies have been based.

In the early 1950's, Ralph Brauer, a physiologist working for the US Navy, demonstrated that bile was not formed by hydrostatic filtration as was urine, but was secreted against pressures that exceeded the vascular perfusion pressure of the isolated perfused rat liver.[9] This seminal study clearly demonstrated that bile production was an energy dependent process, findings that were observed by others using metabolic inhibitors that resulted in inhibition of bile production.[10,11]

Subsequent investigators endeavored to more clearly define the basic mechanisms that generated this secretion. The next advance came from the observations of Henry Wheeler and his colleagues who adopted the technique of solute clearances from renal physiologists.[12] By measuring the biliary clearance of radiolabled inert solutes such as erythritol or mannitol that entered the bile at the level of the bile canaliculus. Wheeler and subsequent investigators were able to distinguish canalicular bile production from fluid secretions formed by the bile duct epithelium.[13] These studies defined "bile salt dependent" and "bile salt independent" (BSDF and BSIF) components of canalicular bile flow and also led to the recognition that considerable species differences existed regarding the relative contribution of the bile duct epithelium to bile flow.[1] These studies were first performed in dogs and rats with biliary fistulae, but were soon used to measure these components in patients following routine cholecystectomy.[14] These studies demonstrated that man produced ~ 750 ml of bile daily with ~75% of bile formed at the canaliculus in the adult. The source of canalicular bile in man was reasonably evenly divided between BSDF and BSIF while a variable fraction (~25%) of the daily bile production originated from the bile ducts in response to meal induced release of secretin.

The next important advance came from the demonstration that the sodium pump, Na^+, K^+-ATPase, was localized to the basolateral membrane or the hepatocyte.[15-17] Prior to this finding, most biliary physiologists believed that bile salt independent bile flow was generated by the active extrusion of sodium ions into the canalicular lumen. The finding that the sodium pump was localized to the basolateral membrane, exactly as demonstrated for classic epithelia, indicated that the liver, whose apical membrane surrounded the hepatocyte like a belt, was physiologically more similar to other polarized epithelia with respect to the polarized location of specific transport proteins.[18] Thus, the sodium gradient, which is generated by the sodium pump, is directed inwardly from plasma into the cell, and can be utilized as a driving force when coupled to other solutes, as initially demonstrated for the uptake of conjugated bile salts.[19,20] Thus, as in all cells, the sodium gradient can be coupled to move solutes "up-hill" energetically from plasma to the cell interior. Studies performed in isolated hepatocytes and hepatocyte couplets (a novel in-vitro model that enabled studies of bile secretion to be made without the confounding effects of blood flow and pressure and independent of the contribution of the bile ducts) confirmed these concepts and demonstrated that this process was also electrogenic.[20,21] Thus the sodium pump, together with potassium channels in the basolateral membrane, were shown to generate both chemical and electrical driving forces that could be used for transmembrane transport of organic solutes.

The recognition that the sodium pump was localized to the basolateral membrane of the liver cells led to its use as a biochemical marker of this membrane that then facilitated the isolation of purified canalicular membranes.[22] Since the apical canalicular membrane represents only 3-5 % of the surface membrane of hepatocytes, it was a significant advance to be able to isolate purified canalicular membrane subfractions that were relatively free of basolateral membrane contamination and thus could then be used to study transport function when prepared as membrane vesicles. These approaches defined a number of transport functions in the plasma membranes of hepatocytes and ultimately led to the recognition that bile salts and other solutes were transported into bile largely by ATP dependent transport mechanisms.[23-25]

The advent of molecular cloning techniques and cellular expression systems rapidly accelerated progress in this field and led to the molecular characterization of most of the major membrane transport proteins that determine both the hepatic uptake of organic solutes as well as bile salt dependent and bile salt independent canalicular excretion.

Overview of the Molecular Mechanisms of Bile Formation

The major transporters that determine bile formation are illustrated in Figure 1 and are discussed in greater detail in the subsequent chapters. The enterohepatic circulation of bile salts maintains bile salt dependent bile flow. Bile salts are excreted into bile and are largely absorbed in the terminal ileum and then efficiently removed from the portal circulation at the basolateral plasma membrane of the hepatocyte. Each of these steps is dependent on the function of bile salt transport proteins.

Hepatocellular Transporters

Transport of conjugated bile salts from plasma into hepatocytes is mediated predominantly by the sodium taurocholate co-transporter (NTCP),[26] gene symbol *SLC10A1*. The driving forces for this coupled uptake process are the inwardly directed sodium gradient and the electronegative cellular membrane potential generated by Na^{+}, K^{+}-ATPase and K^{+}-channels on the basolateral membrane.[27] The sodium pump extrudes 3 sodium ions in exchange for 2 potassium ions resulting in an intracellular negative potential that is further enhanced by the outward diffusion of K^{+} through channels in the plasma membrane. A form of epoxide hydrolase is located on the basolateral membrane and has also been shown to mediate sodium dependent uptake of unconjugated bile salts in in-vitro studies.[28] However the physiologic significance of this transporter remains to be determined. Unconjugated bile salts, as well as many organic anions such as bromosulfophthalein (BSP) and numerous other lipophilic albumin-bound compounds are transported from plasma into hepatocytes by Na^{+}-independent transporters primarily by a family of organic anion transporting polypeptides known as OATPs[29,30] (Fig. 1). The OATP's function as electroneutral anion exchangers where the uptake of the organic solutes appears to be driven by the outwardly directed gradients of intracellular anions such as glutathione and possible bicarbonate.[31-33]

Small organic cations are taken up into hepatocytes by members of the solute carrrier protein family OCT (*SLC22A*) by sodium independent mechanisms. Members of the organic anion transporter, OAT (*SLC22A*) family are also located at the basolateral membrane of hepatocytes and determine the uptake of organic anions like para-aminohippuric acid in exchange for intracellular dicarboxylic acids. OCTs and OATs were originally cloned from kidney but are also expressed in hepatocytes and transport a variety of organic solutes including drugs.[30]

Once within the hepatocyte, bile salts and other cholephiles bind to cytosolic binding proteins and rapidly diffuse toward the apical side of the cell where they are extruded by export pumps located at the canalicular membrane. Active solute transport across the canalicular membrane of hepatocytes represents the rate-limiting step in bile formation. This unidirectional concentrative transport step is driven by an array of ATP-dependent export pumps that belong to the ATP-Binding-Cassette, (ABC) superfamily of membrane transporters (Fig. 1).[34] The first canalicular ATP dependent transporter to be cloned was the multidrug resistance 1 P-glycoprotein, MDR1 (*ABCB1*). Originally described in tumor cell lines, MDR1 mediates the biliary excretion of bulky lipophilic cations, including the chemotherapeutic agents, daunomycin and adriomycin and other drugs such as verapamil and cyclosporin A.[35] MDR1 is not highly expressed in the liver, and its endogenous substrates are not known. Thus the physiologic role of MDR1 in the overall production of bile remains unclear. In contrast, a closely

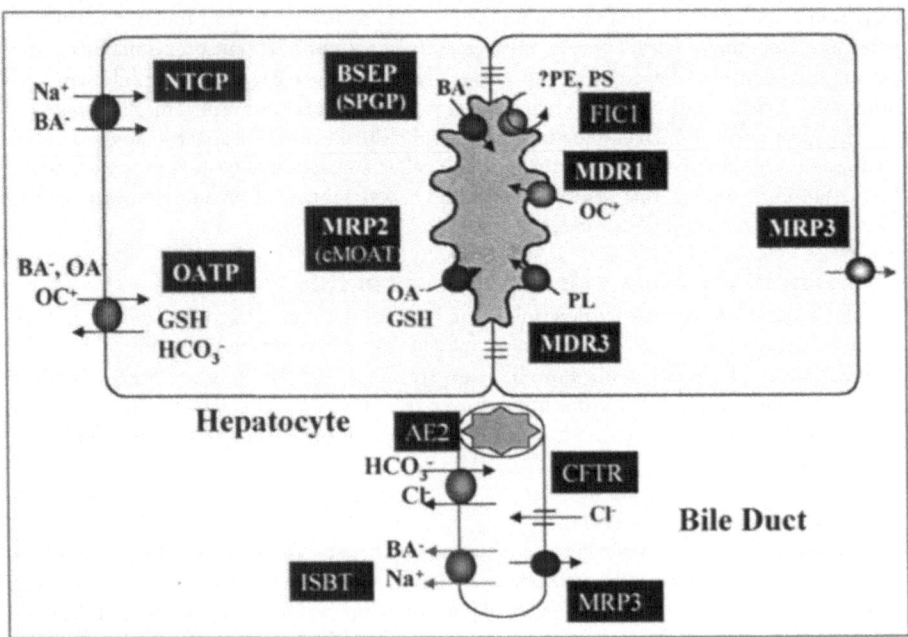

Figure 1. Major hepatobiliary transport systems.

related family member, the multidrug resistance 3 P-glycoprotein, MDR3 (*ABCB4*), Mdr2 in rodent liver,[36] is an important determinant of phospholipid excretion in bile. Knock-out Mdr2 (-/-) mice and studies in transfected yeast, demonstrate that this P-glycoprotein functions not as a drug export pump but as a phospholipid flippase that translocates phosphatidylcholine from the inner to the outer bilayer of the canalicular membrane. Within the lumen of the bile canaliculus, bile salts are thought to selectively extract this phospholipid from the canalicular membrane resulting in the formation of mixed micelles.[37]

A third important bile export pump, MRP2 (*ABCC2*) is a member of the multidrug resistance associated protein family, MRP, (*ABCC*) and is also known as the canalicular multispecific organic anion transporter (cMOAT).[37a,37b] MRP2 functions as a conjugate export pump and mediates ATP-dependent canalicular excretion of a wide range of amphipathic anionic substrates including leukotriene C_4, glutathione-S-conjugates, glucuronides (e.g., bilirubin diglucuronide, estradiol-17ß- glucuronide), sulfate conjugates and heavy metal conjugates.[38] MRP2 also exports glutathione into bile[39,40] resulting in the generation of the major component of canalicular bile salt independent bile flow.[41] MRP2 is an important target for drug-induced cholestasis.

In contrast, bile salt dependent bile flow (originally believed to be driven by the intracellular negative electrical potential) is generated largely if not entirely by the transport activity of the ATP dependent bile salt export pump, BSEP (*ABCB11*). BSEP was previously known as the sister of P-glycoprotein, due to close homology to MDR1.[42] However once the full-length cDNA was cloned and functional expression of the rat homologue was achieved in Sf9 insect cells, its role as the bile salt export pump became established.[43] BSEP is highly conserved in vertebrate evolution.[44] Multiple mutations in Human BSEP have been described in children with progressive familial intrahepatic cholestatic liver disorders known as PFIC type 2, firmly establishing the importance of this ABC transporter as the major if not sole determinant of bile salt dependent bile flow.[45]

Other members of the MRP family, MRP1, 3, 4 and 6 (*ABCC1,3,4,6*) have been localized to the basolateral membrane of the hepatocyte. Mrp1 and 3 are weakly expressed in normal liver and thus their physiologic significance in the normal liver is not known. However, Mrp3 is markedly up regulated in cholestatic liver injury where it is speculated (based on in-vitro studies in membrane vesicles)[46] to function as an adaptive mechanism for the extrusion of bile salt conjugates (sulfates and glucuronides).[46-48] MRP6 is found on the lateral membrane of the hepatocyte and can transport the endothelial receptor antagonist BQ-123.[49]

In addition to the activity of membrane transport proteins at the basolateral and canalicular plasma membrane of hepatocytes, the final composition and magnitude of canalicular bile flow depends on several other cellular mechanisms These include canalicular exocytosis of transcytotic and sub-canalicular vesicles, cytoskeletal structures including a pericanalicular actin web that surrounds the bile canalicular membrane and results in periodic canalicular membrane contractions via actin/myosin interactions, thereby facilitating the forward movement of bile through the network of canaliculi within the hepatic lobule. In addition various electrolyte transporters and ion channels in the bile ductular epithelial cells modify the primary secretion by both active secretory and absorptive mechanisms.[4,50,51]

Cholangiocyte Transporters

Bile formation at the level of the biliary epithelium is regulated by receptors on the basolateral membrane of cholangiocytes that respond to circulating hormones such as secretin, VIP and bombesin. These hormones stimulate the secretion of a bicarbonate enriched fluid from the luminal surface of cholangiocytes that line the bile duct lumen. While details of this secretory process can be found elsewhere[4,51-53] and in following chapters, the process is mediated by a Cl^-/HCO_3^- anion exchanger isoform (AE2) at the apical membrane of the bile duct epithelial cells in concert with a chloride channel, the cystic fibrosis transmembrane regulator (CFTR), also expressed on the luminal membrane of cholangiocytes (Fig. 1). Secretin stimulates the net secretion of bicarbonate anions via cyclic AMP dependent opening of this chloride channel which in turn results in the electrogenic outward movement of chloride ions which are recycled to the hepatocyte via AE2, resulting in net secretion of bicarbonate. Whether bicarbonate is excreted electrogenically is not known. The signal transduction pathways for VIP and bombesin stimulated secretion are also not yet fully known although the process is also dependent on chloride and bicarbonate anions.

Sodium and chloride ions and water are reabsorbed by the bile duct epithelia via the coupled activity of sodium/hydrogen exchange isoforms 2 and 3 and the Cl^-/HCO_3^- anion exchanger isoform (AE2). Cyclic AMP dependent inhibition of the sodium hydrogen exchangers also contributes to net increases in cholangiocyte bile production.[54]

Transporters for the re-absorption of solutes such as amino acids, glucose and bile salts are also located at the luminal membrane of cholangiocytes. Isbt, the ileal sodium dependent bile salt transporter in rodents (*Slc10a2*), also called the apical bile salt transporter (Asbt) is capable of transporting bile salts from the biliary space into the bile duct epithelium. The physiologic significance of this process is not yet clear. However during cholestasis, bile duct epithelial cells proliferate and Isbt might then facilitate removal of bile salts from the cholestatic biliary lumen. MRP3, another member of the MRP family is normally located on the basolateral membrane of cholangiocytes and may function to remove bile salts and other toxic substances back to the systemic circulation thereby protecting this epithelium from further injury. A truncated form of Asbt (t-Asbt) is also located on the basolateral membrane of cholangiocytes and has been reported to contribute to this process.

Finally, purigenic receptors have been identified that line the biliary epithelium, and release of nucleotides from hepatocytes and cholangiocytes may function in a paracrine manner to regulate secretion from this epithelium.[55] Details of the mechanism of many of these bile secretory processes are discussed in following chapters.

Bile Salt Transporters in Intestine and Kidney

The majority of the bile salt pool is re-circulated following absorption in the terminal ileum by the ileal sodium dependent bile salt transporter, *ISBT (SLC10A2)*. ISBT is associated with the ileal bile acid binding protein (I-BABP) but the functional relationship is not yet known. Oatp3 is located in the small intestine and could also function in the uptake of bile salts (*Slc21a7*).[56] Bile salts are extruded into mesenteric vessels from the terminal ileum by uncertain mechanisms although Mrp3 and t-Asbt are located on the basolateral domain of intestinal cells and Mrp3 is highly expressed in the terminal ileum and colon.[57] Bile salts that are filtered at the glomerulus are re-absorbed by Isbt localized on the luminal membrane of the proximal tubule.

Regulation of the Expression of Bile Transporters

The functional expression of the transporters that regulate bile secretion can be modified by both transcriptional and post-transcriptional mechanisms with the latter involving changes in RNA stability, targeting and phosphorylation/dephosphorylation reactions.[3] Short-term regulation is generally achieved by post-transcriptional events while long term modifications are determined by changes in nuclear transcription.[58,59] For example, Ntcp, Mrp2 and Bsep can each be rapidly inserted or retrieved from their functional locations in the plasma membrane by cyclic AMP mediated phosphorylation/dephosphorylation reactions.[60] Changes in osmolarity or infusions of bile salts can stimulate the insertion of transport proteins including Mrp2 and Bsep into the apical canalicular membrane.[3]

Regulation at the level of gene transcription provides for more chronic responses to increases in substrate load as well as adaptive responses that minimize tissue accumulation of bile salts during liver injury, particularly during cholestatic conditions. Both the metabolism of bile salts from cholesterol as well as the expression of bile salts transporters in liver, intestine and kidney are exquisitely regulated by the nuclear receptor, farnesoid X receptor (FXR). As bile salts in tissue rise, they bind to FXR in the nucleus, promoting the formation of heterodimers with the retinoic X receptor (RXR-α) which then is able to interact with a specific response element in the promoter regions of several bile salt transport proteins as well as another nuclear receptor, the short heterodimeric protein (Shp-1) that inhibits both the expression of Cyp7α that converts cholesterol to bile salts and the transcription of Ntcp. These nuclear receptor gene response element interactions result in decreased bile salt synthesis, increased expression of Bsep, promoting the hepatic excretion of bile salts and inhibition of Ntcp expression (via down regulation by Shp-1) resulting in diminished hepatic bile salt uptake. Down-regulation of ileal bile salt uptake also occurs. This brief example serves to illustrate the highly integrated and complex feedback regulation that governs the expression of the major determinants of bile salt synthesis and their enterohepatic transport. Subsequent chapters will expand on these basic concepts.

References

1. Boyer JL, Nathanson MH. Bile formation. In: Schiff ER, Sorrell MF, Maddrey WC, eds. Schiff's Diseases of the Liver. Lippincott-Raven, 1999:119-146.
2. Trauner M, Meier PJ, Boyer JL. Molecular pathogenesis of cholestasis. N Eng J Med 1998; 339:1217-1227.
3. Trauner M, Boyer JL. Bile salt Transporters: Molecular Characterization, Function and Regulation. Physiol Rev 2003; 83: in press.
4. Kanno N, LeSage G, Glaser S et al. Regulation of cholangiocyte bicarbonate secretion. Am J Physiol Gastrointest Liver Physiol 2001; 281:G612-G625
5. Sobotka H. Physiological chemistry of the bile. Baltimore: Williams & Wilkins, 1937.
6. Hazelwood GAD. Recent developments in our knowledge of bile salts. Physiol Rev 1955; 35:178-196.
7. Sperber I. Secretion of organic anions in the formation of urine and bile. Pharmacol Rev 1959; 11:109-134.
8. Boyer JL. Milestones in Liver Disease—A commentary. J Hepatol 2002; 36:4-7.

9. Brauer RW, Leong GF, Holloway RJ. Mechanics of bile secretion: effect of perfusion pressure and temperature on bile flow and secretion pressure. Am J Physiol 1954; 177:103-112.

10. Bizard G. Enzyme inhibitors and biliary secretion. In: Taylor W, ed. The Biliary System. Oxford: Blackwell, 1965:315-324.

11. Boyer JL. Canalicular bile formation in the isolated perfused rat liver. Am J Physiol 1971; 221:1156-1163.

12. Wheeler HO. Canalicular bile production in dogs. Amer J Physiol 1968; 214:866-874.

13. Wheeler HO. Secretion of bile acids by the liver and their role in the formation of hepatic bile. Arch Intern Med 1972; 130:533-541.

14. Boyer JL, Bloomer JR. Canalicular bile secretion in man: studies utilizing the biliary clearance of (14C) mannitol. J Clin Invest 1974; 54:773-781.

15. Blitzer BL, Boyer JL. Cytochemical localization of Na+,K+-ATPase in the rat hepatocyte. J Clin Invest 1978; 62:1104-1108.

16. Latham PS, Kashgarian M. The ultrastructural localization of transport ATPase in the rat liver at nonbile canalicular plasma membranes. Gastroenterology 1979; 76:988-996.

17. Sztul ES, Biemersderfer D, Caplan MJ et al. Localization of Na$^+$,K$^+$-ATPase a-subunit to the sinusoidal and lateral but not canalicular membranes of rat hepatocytes. J Cell Biol 1987; 104:1239-1248.

18. Boyer JL. New concepts of mechanisms of hepatocyte bile formation. Physiol Rev 1980; 60:303-326.

19. Anwer MS, Hegner D. Effect of Na on bile acid uptake by isolated rat hepatocytes. Evidence for a heterogeneous system. Hoppe-Seyler's Z Physiol Chem 1978; 359:181-192.

20. Van Dyke RW, Stephens JE, Scharschmidt BF. Effect of ion substitution on bile acid-dependent and bile acid-independent bile formation by the isolated perfused rat liver. J Clin Invest 1982; 70:505-517.

21. Graf J, Henderson RM, Krumpholz B et al. Cell membrane and transepithelial voltages and resistances in isolated rat hepatocyte couplets. J Membrane Biol 1987; 95:241-254.

22. Meier PJ, Boyer JL. Preparation of basolateral (sinusoidal) and canalicular plasma membrane vesicles for the study of hepatic transport processes. In: Fleischer S, Fleischer B, eds. Methods in Enzymology. New York: Academic Press, Inc., 1990:534-545.

23. Nishida T, Gatmaitan Z, Che MX et al. Rat liver canalicular membrane vesicles containing an ATP-dependent bile acid transport system. Proc Natl Acad Sci 1991; 88:6590-6594.

24. Muller M, Ishikawa T, Berger U et al. ATP-dependent transport of taurocholate across the hepatocyte canalicular membrane mediated by a 110-kDa glycoprotein binding ATP and bile salt. J Biol Chem 1991; 266:18920-18926.

25. Stieger B, O'Neill B, Meier PJ. ATP-dependent bile-salt transport in canalicular rat liver plasma-membrane vesicles. Biochem J 1992; 284:67-74.

26. Hagenbuch B, Meier PJ. Molecular cloning, chromosomal localization, and functional characterization of a human liver Na+/bile acid cotransporter. J Clin Invest 1994; 93:1326-1331.

27. Moseley RH, Boyer JL. Mechanisms of electrolyte transport in the liver and their functional significance. In: Popper H, ed. Seminars in Liver Disease. New York: Thieme-Stratton Inc., 1985:122-135.

28. Vondippe P, Amoui M, Stellwagen RH et al. The Functional Expression of Sodium-dependent Bile acid Transport in Madin-Darby Canine Kidney Cells Transfected with the cDNA for Microsomal Epoxide Hydrolase. J Biol Chem 1996; 271:18176-18180.

29. Kullak-Ublick GA, Ismair MG, Stieger B et al. Organic anion-transporting polypeptide B (OATP-B) and its functional comparison with three other OATPs of human liver. Gastroenterology 2001; 120:525-533.

30. Suzuki H, Sugiyama Y. Transport of drugs across the hepatic sinusoidal membrane: sinusoidal drug influx and efflux in the liver. Semin Liver Dis 2000; 20:251-263.

31. Li L, Lee TK, Meier PJ et al. Identification of glutathione as a driving force and leukotriene C4 as a substrate for oatp1, the hepatic sinusoidal organic solute transporter. J Biol Chem 1998; 273:16184-16191.

32. Satlin LM, Amin V, Wolkoff A. Organic anion transporting polypeptide mediates organic anion/HCO$_3^-$ exchange. Journal Biological Chemistry 1997; 272:26340-26345.

33. Li L, Meier PJ, Ballatori N. Oatp2 mediates bidirectional organic solute transport: a role for intracellular glutathione. Mol Pharmacol 2000; 58:335-340.

34. Keppler D, Arias IM. Introduction: Transport across the hepatocyte canalicular membrane. FASEB J 1997; 11:15-18.

35. Thiebaut F, Tsuruo T, Hamada H et al. Cellular localization of the multidrug-resistance gene product P-glycoprotein in normal human tissues. Proc Nat Acad Sci USA 1987; 84:7735-7738.

36. Oude Elferink RPJ, Groen AK. Mechanisms of biliary lipid secretion and their role in lipid homeostasis. Sem Liv Dis 2000; 20:293-305.

37. Oude Elferink RPJ, Tytgat GNJ, Groen AK. The role of mdr2 P-glycoprotein in hepatobiliary lipid transport. FASEB J 1997; 11:19-28.
37a. Paulusma CC, Bosma PJ, Zaman GJR et al. Congenital jaundice in rats with a mutation in a multi-drug resistance associated protein gene. Science 1996; 271:1126-1128.
37b. Buchler M, Konig J, Brem M et al. cDNA cloning of the hepatocyte canalicular isoform of the multi-drug resistance protein, cMrp, reveals a novel conjugate export pump deficient in hyperbilirubinemic mutant rats. J Biol Chem 1996; 271:15091-15098.
38. Konig J, Nies AT, Cui Y et al. Conjugate export pumps of the multidrug resistance protein (MRP) family: localization, substrate specificity, and MRP2-mediated drug resistance. Biochim Biophys Acta 1999; 1461:377-394.
39. Paulusma C, van Geer M, Evers R et al. Canalicular multispecific organic anion transporter/multidrug resistance protein 2 mediates low-affinity transport of reduced glutathione. Biochem J 1999; 338:393-401.
40. Rebbeor JF, Connolly GC, Henson JH et al. ATP-dependent GSH and glutathione S-conjugate transport in skate liver: role of an mrp functional homologue. Am J Physiol Gastrointest Liver Physiol 2000; 279:G417-G425
41. Ballatori N, Truong AT. Glutathione as a primary osmotic driving force in hepatic bile formation. Amer J Physiol 1992; 263:G617-G624
42. Childs S, Yeh RL, Georges E et al. Identification of a sister gene to p-glycoprotein. Cancer Res 1995; 55:2029-2034.
43. Gerloff T, Stieger B, Hagenbuch B et al. The sister P-glycoprotein represents the canalicular bile salt export pump of mammalian liver. J Biol Chem 1998; 273:10046-10050.
44. Cai S-Y, Wang L, Ballatori N et al. Bile salt export pump is highly conserved during vertebrate evolution and its expression is inhibited by PFIC type II mutations. Am J Physiol 2001; 281:G316-G322
45. Strautnieks SS, Bull L, Knisely AS et al. A gene encoding a liver-specific ABC transporter is mutated in progressive familial intrahepatic cholestasis. Nat Genet 1998; 20:233-238.
46. Hirohashi T, Suzuki H, Takikawa H et al. ATP-dependent transport of bile salts by rat multidrug resistance-associated protein 3 (Mrp3). J Biol Chem 2000; 275:2905-2910.
47. Soroka CJ, Lee JM, Azzaroli F et al. Cellular localization and up-regulation of multidrug resistance-associated protein 3 in hepatocytes and cholangiocytes during obstructive cholestasis in rat liver. Hepatology 2001; 33:783-791.
48. Donner MG, Keppler D. Up-regulation of basolateral multidrug resistance protein 3 (Mrp3) in cholestatic rat liver. Hepatology 2001; 34:351-359.
49. Madon J, Hagenbuch B, Landmann L et al. Transport function and hepatocellular localization of mrp6 in rat liver. Mol Pharmacol 2000; 57:634-641.
50. Boyer JL. Bile duct epithelium: frontiers in transport physiology. Am J Physiol 1996; 270:G1-G5
51. Baiocchi L, LeSage G, Glaser S et al. Regulation of cholangiocyte bile secretion. J Hepatol 1999; 31:179-191.
52. Roberts SK, LaRusso N. Pathobiology of biliary epithelia. Current Opinion in Gastroenterology 1994; 10:526-533.
53. Strazzabosco M. Biliary tract physiology. Curr Opin in Gastro 1998; 14:395-401.
54. Mennone A, Biemersderfer D, Negoianu D et al. Role of sodium/hydrogen exchanger isoform NHE3 in fluid secretion and absorption in mouse and rat cholangiocytes. Am J Physiol Gastrointest.Liver Physiol 2001; 280:G247-G254
55. Feranchak AP, Fitz JG. Adenosine Triphosphate Release and Purinergic Regulation of Cholangiocyte Transport. Sem Liv Dis 2002; 22:251-262.
56. Walters HC, Craddock AL, Fusegawa H et al. Expression, transport properties, and chromosomal location of organic anion transporter subtype 3. Am J Physiol 2000; 279:G1188-G1200
57. Rost D, Mahner S, Sugiyama Y et al. Expression and localization of the multidrug resistance-associated protein 3 in rat small and large intestine. Am J Physiol 2002; 282:G720-G726
58. Chiang JYL. Bile acid Regulation of Gene Expression: Roles of Nuclear Hormone Receptors. Endocr Rev 2002; 23:443-463.
59. Karpen SJ. Nuclear receptor regulation of hepatic function. J Hepatol 2002; 36:832-850.
60. Kipp H, Arias IM. Intracellular trafficking and regulation of canalicular ATP-binding cassette transporters. Semin Liver Dis 2000;20:339-351.

CHAPTER 2

Hepatocellular Transport Systems: Basolateral Membrane

Bruno Hagenbuch and Peter J. Meier

Summary

The basolateral membrane of hepatocytes is equipped with efficient transport systems for uptake of bile salts (Ntcp/NTCP) and xenobiotics (Oatps/OATPs). Under physiological conditions these transporters are important for ongoing bile formation and for efficient hepatic detoxification. Especially Oatps/OATPs seem to play an important role in the hepatic first-pass clearance of xenobiotics and can influence the bioavailability of certain drugs. Under cholestatic conditions these uptake systems are down-regulated to prevent an overflow of hepatocytes with potentially toxic compounds. Furthermore an efficient efflux system is expressed (Mrp3/MRP3) at the basolateral membrane that helps to protect liver cells from intracellular accummulation of toxic bile salts and biotransformation products. Hence, the correct interplay of regulatory mechanisms that control the expression of basolateral transporters is vital for the adaptation of efficient hepatocellular functions to various physiological and pathophysiological conditions.

Introduction

Hepatocytes are polarized cells. Their sinusoidal (basolateral) membrane faces the portal blood and contains several transport systems that are important for uptake of bile salts and xenobiotics while the canalicular membrane harbours mainly ATP-dependent transporters that are involved in their secretion into bile (see chapter by Guido Hooiveld and Michael Müller in this book).

The most important basolateral transport systems for bile salts and xenobiotics are the sodium dependent Na^+/taurocholate cotransporting polypeptide (Ntcp/NTCP), sodium independent organic anion transporting polypeptides (Oatps/OATPs) and multidrug resistance associated proteins (Mrps/MRPs). Here we summarize their molecular characteristics, substrate specificities and expression under normal and cholestatic conditions.

The Na⁺/Taurocholate Cotransporting Polypeptide (Ntcp/NTCP)

Uptake of conjugated bile salts into hepatocytes is mainly a sodium dependent process and accounts for more than 80% of taurocholate uptake. The out-to-in sodium gradient that is maintained by the Na^+/K^+-ATPase provides the energy for this secondary active transport step, which is mediated by the Na^+/taurocholate cotransporting polypeptide (rodents: Ntcp; man: NTCP; gene symbols: rodents: *Slc10a1*; man: *SLC10A1*). This Na^+-dependent bile salt uptake system has been characterized in detail using different experimental models such as the perfused rat liver,[1,2] isolated hepatocytes[3,4] and isolated sinusoidal membrane vesicles.[5]

So far Na^+/taurocholate cotransporting polypeptides have been cloned and characterized from rat (Ntcp),[6] mouse (Ntcp1, Ntcp2),[7] rabbit[8] and human liver (NTCP).[9]

Molecular Pathogenesis of Cholestasis, edited by Michael Trauner and Peter L.M. Jansen.
©2004 Eurekah.com and Kluwer Academic / Plenum Publishers.

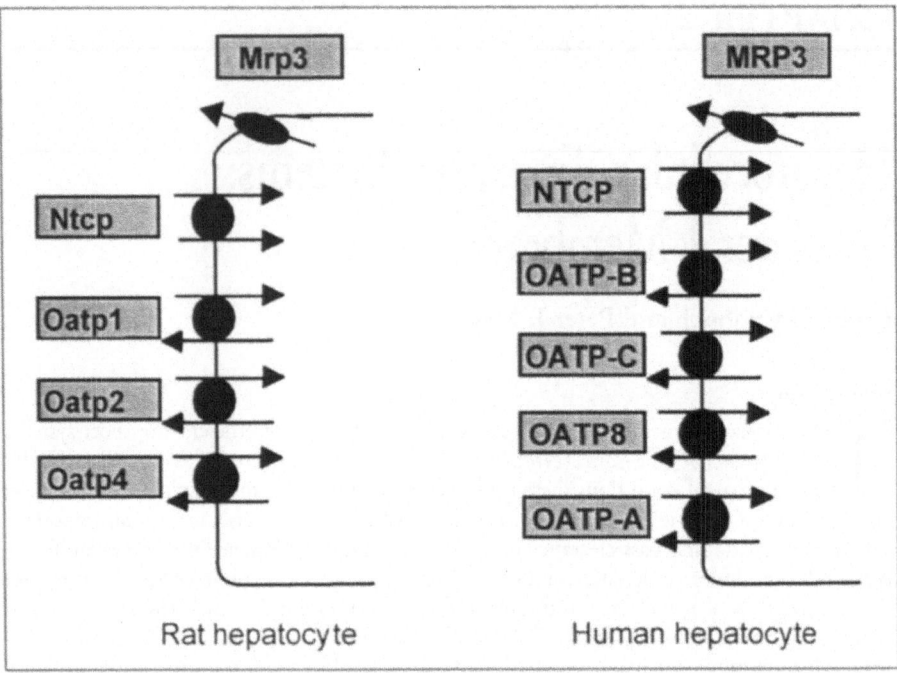

Figure 1. Bile salt and organic anion transporters expressed at the basolateral membrane of rat and human hepatocytes.

The rat Ntcp (*Slc10a1*) was the first sodium dependent bile salt transporter isolated by expression cloning using the *Xenopus laevis* expression system.[6] It is a glycoprotein of 362 amino acids with an apparent molecular mass of 51 kDa.[10,11] Ntcp is expressed at the basolateral membrane of hepatocytes (Fig. 1)[10] and at the luminal membrane of pancreatic acinar cells.[12] Computer programs predicted 7 transmembrane domains, however, recent experimental evidence suggest rather a 9 transmembrane topology.[13] The rat *Ntcp*-gene consists of 5 exons, spans approximately 16 kb of genomic sequence[14] and is located on rat chromosome 6q24,[15] which is syntenic to the location of the mouse *Ntcp*-gene on chromosome 12[16] and the human *NTCP*-gene on 14q24.1.[17]

Expression of sodium dependent taurocholate transport in *Xenopus laevis* oocytes was almost completely blocked when rat liver mRNA was treated with Ntcp-specific antisense oligonucleotides.[18] This suggests that Ntcp is the major if not the only sodium dependent taurocholate uptake system in rat liver. Besides *Xenopus laevis* oocytes, Ntcp was expressed and functionally characterized in several mammalian expression systems including COS-7,[19] CHO,[20] HPCT-1E3[21] and rat hepatoma cells.[22] The results demonstrated that Ntcp could account for most Na^+-dependent bile salt transport characteristics that have been previously determined in hepatocytes.[23] It preferentially transports conjugated bile salts and has an apparent Km value between 11 and 34 µM for taurocholate (depending on the expression systems used).[6,19,20,22] Transport is electrogenic with a sodium:taurocholate stoichiometry of 2:1[24,25] and the only non-bile salt substrate identified so far is estrone-3-sulfate, which is transported with an apparent Km value of 27 µM.[20] Thus, Ntcp is a rather specific bile salt transporter, which is important for the vectorial uptake of bile salts from portal blood into hepatocytes.

Ntcp expression is extensively regulated under different physiological and pathophysiological conditions. During development Ntcp protein expression and Na^+-dependent bile salt uptake can be detected simultaneously around days 18-20 of gestation.[26,27] In primary cultured rat hepatocytes a parallel decrease of Ntcp protein expression and Na^+-dependent taurocholate transport was observed.[28,29] Also, Ntcp downregulation occurs in various rat models of cholestasis including partial hepatectomy,[30] bile duct ligation[31] and ethinyl estradiol,[32] endotoxin[33,34] and cytokine[35] administration, indicating that Ntcp downregulation serves as a protective mechanism agains bile salt overload of hepatocytes.

The molecular mechanisms of *Ntcp* gene regulation have been the focus of extensive investigations, and several important hepatocyte-enriched transcription factors and ligand-activated nuclear receptors[14,36] that interact with the *Ntcp*-gene have been identified and are summarized in detail elsewhere (chapters by Saul Karpen and Michael Trauner et al in this book).

In conclusion, Ntcp is the major if not the only Na^+-dependent bile salt uptake system of hepatocytes that is downregulated under cholestatic conditions to reduce hepatocellular damage due to intracellular bile salt accumulation.

The human NTCP (*SLC10A1*) is a 349 amino acid glycoprotein with an apparent molecular mass of 56 kDa that is expressed at the basolateral membrane of human hepatocytes (Fig. 1).[9,37] It mediates Na^+-dependent bile salt uptake, however with higher affinities than the rat Ntcp (taurocholate Km ~6μM, taurochenodeoxycholate Km ~2μM).[9,38] These high affinities allow the human hepatocytes to efficiently extract bile salts from portal blood to keep plasma bile salt concentrations at a minimum. Expression of NTCP protein was studied in 2 patients with hypercholanemia and found to be normal.[37] Another study were NTCP expression was studied in biopsies of 37 patients demonstrated, similar to the cholestatic animal models, that NTCP downregulation occurred in livers from patients with inflammatory cholestasis caused by cholestatic alcoholic hepatitis, cholestatic autoimmune hepatitis and drug-induced and obstructive cholestasis while NTCP expression was normal in patients with primary billiary cirrhosis and chronic hepatitis.[39]

The Organic Anion Transporting Polypeptides (Oatps/OATPS)

Members of of the organic anion transporting polypeptide superfamily (rodents: Oatps; human: OATPs; gene symbols rodents: *Slc21a*; man: *SLC21A*) (Table 1) mediate the Na^+-independent uptake of bile acids and other organic solutes including steroid hormones and drugs into hepatocytes (for an extensive review see 40). Of all the different Oatps/OATPs (Fig. 2) only Oatp1 (*Slc21a1*),[41] Oatp2 (*Slc21a5*)[42] and Oatp4 (*Slc21a10*)[43] have been identified in rat liver, while in human liver OATP-A (*SLC21A3*),[44] OATP-B (*SLC21A9*),[45] OATP-C (*SLC21A6*)[46-48] and OATP8 (*SLC21A8*)[49] are expressed to various degrees (Fig. 1).

The Organic Anion Transporting Polypeptide 1 (Oatp1; Slc21a1)

Oatp1 (previously also called Oatp) was the first organic anion transporting polypeptide isolated from rat liver.[41] It is a 12 transmembrane domain glycoprotein with 670 amino acids and an apparent molecular mass of approximately 80 kD.[50,51] Oatp1 is strongest expressed in the liver, kidney and brain with minor reactive mRNA species in lung, skeletal muscle and proximal colon.[41] In rat liver, Oatp1 is exclusively expressed at the basolateral plasma membrane of hepatocytes where it mediates Na^+-independent transport of a variety of organic solutes.

Originally, Oatp1 has been cloned as a sodium-independent bromosulfophthalein (BSP) uptake system of rat liver.[41,52] However, later it has been shown that Oatp1 can mediate transmembrane transport of a wide range of amphipathic organic compounds including bile salts, glutathione, steroids and their conjugates, thyroid hormones, leukotriene C4, linear and cyclic peptides, organic cations, the mycotoxin ochratoxin A, and numerous drugs including the angiotensin-converting enzyme inhibitors enalapril and temocaprilat, the HMG-CoA reductase inhibitor pravastatin and the antihistamine fexofenadine (Table 2).

The driving force for Oatp1-mediated organic anion transport appears to be exchange against intracellular anions such as bicarbonate, BSP[52,53] and glutathione (GSH).[54]

In the liver, basolateral expression of Oatp1 is down-regulated after cholate feeding,[55] in cholestatic liver and following partial hepatectomy[56] and in primary cultured rat hepatocytes.[28]

Table 1. *Molecular characteristics of the organic anion transporting polypeptides (Oatp/OATPs). Only human OATP-A, OATP-B, OATP-C and OATP8 as well as the rat Oatp1, Oatp2 and Oatp4 have been identified in the liver on the protein level*

Human Proteins	Gene Symbols	Rat Proteins	Gene Symbols	Amino Acids	Main Location
		Oatp1	*Slc21a1*	670	Liver, kidney
HPGT	*SLC21A2*	PGT	*Slc21a2*	643	Ubiquitous
OATP-A	*SLC21A3*			670	Brain, liver
		OAT-K1	*Slc21a4*	669	Kidney
		Oatp2	*Slc21a5*	661	Liver, BBB, choroid plexus, retina
OATP-C	*SLC21A6*			691	Liver
		Oatp3	*Slc21a7*	670	Jejunum, brain
OATP8	*SLC21A8*			702	Liver
OATP-B	*SLC21A9*	Oatp9	*Slc21a9*	682	Liver, placenta, brain
		Oatp4	*Slc21a10*	687	Liver
OATP-D	*SLC21A11*	Oatp11	*Slc21a11*	710	Ubiquitous
OATP-E	*SLC21A12*	Oatp12	*Slc21a12*	722	Ubiquitous
		Oatp5	*Slc21a13*	670	Kidney
OATP-F	*SLC21A14*	Oatp14	*Slc21a14*	716	Brain

Similar to other hepatocellular membrane transporters, *Oatp1* gene transcription is controlled by a hierarchical network of hepatocyte-enriched transcription factors and ligand-activated nuclear receptors.[57,58] Details of Oatp/OATP regulation can be found elswhere in this book (chapters by Saul Karpen and Michael Trauner and his colleagues).

On the protein level, functional down-regulation of Oatp1 occurs via serine phosphorylation by extracellular ATP.[59] Similarly, protein kinase C activation leads to decreased transport of estrone-3-sulfate in Oatp1 expressing *X. laevis* oocytes.[60]

The Organic Anion Transporting Polypeptide 2 (Oatp2; Slc21a5)

Oatp2 was independently isolated from rat brain[42] and rat retina.[61] The 661 amino acid protein was immunolocalized at the blood-brain barrier in endothelial cells and at the basolateral plasma membrane of the choroid plexus epithelial cells.[62] In rat liver, Oatp2 is expressed at the sinusoidal membrane of hepatocytes and has an apparent molecular mass of 76-92 kDa.[63,64] Oatp2-mRNA was also detected in kidney[42] and retina,[61] and the protein was fine localized to the apical region of rat retinal pigment epithelium.[65]

Although it is also a multispecific transport protein with a broad substrate specificity, a unique feature of Oatp2 is its high affinity transport activity for the cardiac glycoside digoxin (apparent Km value ~0.24μM).[42] Other transport substrates of Oatp2 include bile salts, steroid conjugates, thyroid hormones, cyclic and linear peptides, ouabain, organic cations and drugs like fexofenadine and pravastatin (Tabe 2).

Similar to Oatp1, the driving force for Oatp2 mediated transport could be GSH since taurocholate uptake was stimulated by high intracellular GSH concentrations.[66]

Regulation of Oatp2 expression has been demonstrated on transcriptional and posttranscriptional levels. In primary cultured rat hepatocytes Oatp2 mRNA is down-regulated rapidly by more than 70% during the first 24 hours of culturing time. However, Oatp2 protein is kept around 50% of initial expression during 3 days of culturing.[28] Similar discrepancies between expression of Oatp2 on the mRNA and protein levels have also been shown in pregnant and postpartum female rats.[67] Whether these differences are caused by posttranslational modifications

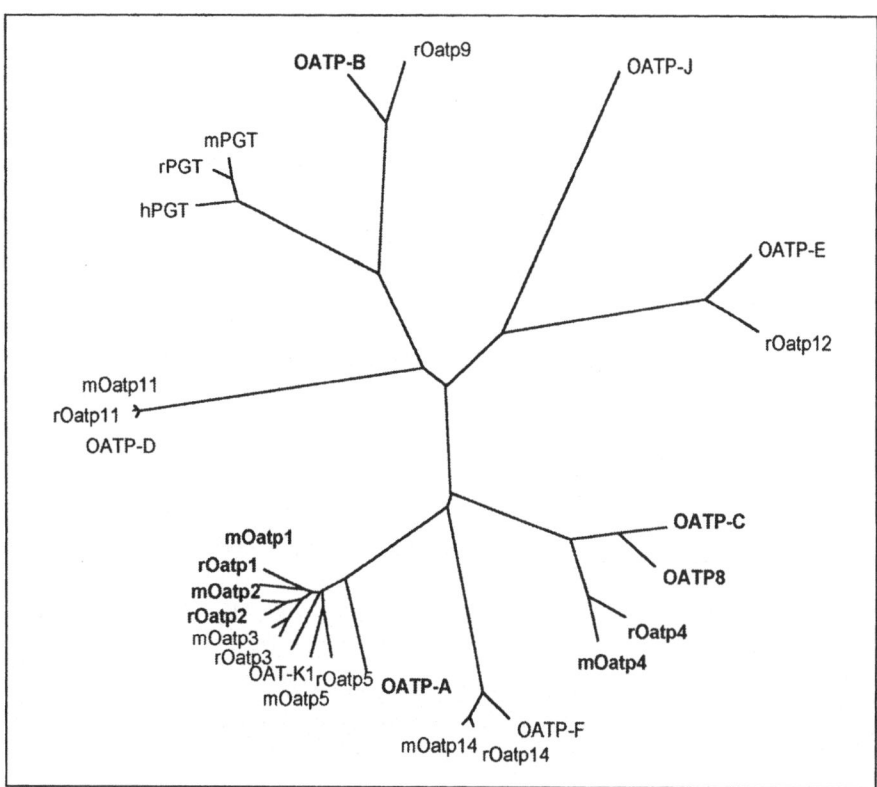

Figure 2. Phylogenetic tree of the Oatp/OATP superfamily. Members in bold are expressed in the liver and discussed in the text.

of the protein such as for example via phosphorylation by protein kinase C[60] remains to be investigated under various physiologic and pathophysiologic conditions.

Similar to Oatp1, *Oatp2* gene expression is down regulated in HNF1α knockout mice.[68] However, unlike Oatp1, *Oatp2* gene expression is induced by phenobarbital[69] and pregnenolone-16α-carbonitrile (PCN) treatment[70,71] through a pregnane X receptor (PXR) mechanism.[72]

The Organic Anion Transporting Polypeptide 4 (Oatp4; Slc21a10)

Oatp4 has been isolated from rat and mouse livers,[43,73-75] and is a 687 amino acid protein with the 12 transmembrane domain topology that is common to all other Oatps/OATPs.[43,74] It is selectively expressed at the basolateral (sinusoidal) plasma membrane of hepatocytes and has an apparent molecular mass of 85 kDa.[43]

The transport functions of Oatp4 have been characterized in *X. laevis* oocytes. While a short splice variant, rlst-1, transported only taurocholate,[73] the full length Oatp4 is a multispecific organic anion transporter. Its spectrum of transport substrates includes taurocholate, BSP, steroid conjugates, thyroid hormones, eicosanoids, cyclic peptides such as BQ-123, DPDPE and CCK-8 as well as the toxins microcystin and phalloidin (Table 2).

Although the regulation of *Oatp4* gene expression has not been studied in detail, recent findings indicate that HNF-1α might be positively involved,[68,76] thus supporting the concept that HNF-1α is essential for bile salt uptake across the basolateral hepatocyte plasma membrane.[68] Which transcription factors are responsible for the reported down-regulation of Oatp4 expression during cholestasis [73] is not yet known.

Table 2. Substrate specificity of the liver expressed members of the OATP superfamily

Transporter	Substrates (Km value)	References
Rat Oatps		
Oatp1 (Slc21a1)	Cholate (54 µM), glycocholate (54 µM), taurocholate (19-50 µM), TCDCA (7 µM), TUDCA (13 µM), sulfotaurolithocholate (6 µM), Aldosterone (15nM), cortisol (13 µM), DHEAS (5 µM), $E_2 17\beta G$ (3-20 µM), E-3-S (5-12 µM), T_3, rT_3, T_4, LTC_4 (270 nM), BQ-123 (600 µM), CRC220 (30-57 µM), deltorphin II (137 µM), DPDPE (48 µM), GSH, Dexamethasone, enalapril (214 µM), fexofenadine (32 µM), gadoxetate (3.3 mM), ouabain (1.7-3 mM), pravastatin (30 µM), temocaprilat (47 µM), Monoglucuronosyl bilirubin, BSP (1-3 µM), BSP- DNP-SG (408 µM), E3040 glucuronide, APD-ajmalinium, N-methylquinidine, rocuronium, Ochratoxin A (17-29 µM)	41, 48, 51-54, 63, 81, 93-105
Oatp2 (Slc21a5)	Cholate (46 µM), glycocholate (40 µM), taurocholate (35 µM), TCDCA (12 µM), TUDCA (17 µM), DHEAS (17 µM), $E_2 17\beta G$ (3 µM), E-3-S (11 µM), T_3 (6 µM), T_4 (7 µM), BQ-123 (30 µM), DPDPE (19 µM), Leu-enkephalin, Biotin, digoxin (240 nM), fexofenadine (6 µM), ouabain (470 µM), pravastatin (38 µM), APD-ajmalinium, rocuronium	42, 61, 63, 64, 81, 102, 104, 106
Oatp4 (Slc21a10)	Taurocholate (27 µM), DHEAS (5 µM), $E_2 17bG$ (32 µM), E-3-S (37 µM), T_3, T_4, LTC_4 (7 µM), prostaglandin E_2 (13 µM), BSP (1 µM), mycrocystin, phalloidin[1]	43, 86, 107, 108
Human OATPs		
OATP-A (SLC21A3)	Cholate (93 µM), glycocholate, taurocholate (60 µM), TCDCA, TUDCA (19 µM), DHEAS (7 µM), $E_2 17\beta G$, E-3-S (59 µM), T_3 (7 µM), rT_3, T_4 (8 µM), Prostaglandine E_2, BQ-123, CRC220, deltorphin II (330 µM), DPDPE (202 µM), Chlorambuciltaurocholate, fexofenadine (6 µM), Gd-B 20790, ouabain (5.5 mM), BSP (20 µM), APD-ajmalinium, N-methylquinine (26 µM), N-methylquinidine (5 µM), rocuronium, Microcystin	38, 45, 80, 81, 102, 104, 108-113
OATP-C (SLC21A6)	Cholate (11 µM), glycocholate, taurocholate (10-34 µM), DHEAS (22 µM), $E_2 17\beta G$ (8-10 µM), E-3-S (13 µM), T_3 (3 µM), T_4 (3 µM), LTC_4, LTE_4, prostaglandine E_2, thromboxane B_2, BQ-123, DPDPE, Benzylpenicillin, methotrexate, pravastatin (14-35 µM), rifampicin (13 µM), Bilirubin, monoglucuronosyl bilirubin (100 nM), bisglucuronosyl bilirubin (300 nM), BSP (100-300 nM), Microcystin, phalloidin[1]	45, 47, 48, 78, 85, 108, 114-116
OATP8 (SLC21A8)	Glycocholate, taurocholate (6 µM), DHEAS, $E_2 17\beta G$ (5 µM), E-3-S, T_3 (6 µM), T_4, LTC_4, BQ-123, CCK-8 (11 µM), deltorphin II, DPDPE, Digoxin, methotrexate (25 µM), ouabain, rifampicin (2 µM), Monoglucuronosyl bilirubin (500 nM), BSP (0.4-3 µM), Microcystin, phalloidin[1]	45, 49, 85, 86, 108, 114, 116
OATP-B (SLC21A9)	E-3-S (6 µM), DHEAS, benzylpenicillin, BSP (0.7 µM)	45, 78

1) unpublished own observation.
Abbreviations: APD-ajmalinium: N-(4,4-azo-n-pentyl)-21-deoxyajmalinium; BSP: bromosulfophthalein; BSP-SG: gluthathione-conjugated BSP; DNP-SG: dinitrophenyl-glutathione; CCK8: cholecystokinin-8; DHEAS: dehydroepiandrosterone-sulfate; DAMGO: D-Ala[2],N-Me-Phe[4],Gly[5]-ol]-Enkephalin; DPDPE: [D-Pen[2,5]]-Enkephalin; E-3-S: estrone-3-sulfate; $E_2 17\beta G$: estradiol-17β-glucuronide; GCDCA, glycochenodeoxycholate; Gd-B 20790, gadolinium derivative; GUDCA, glycoursodeoxycholate; TCDCA, taurochenodeoxycholate; TUDCA, tauroursodeoxycholate.

The Human Organic Anion Transporting Polypeptide A (OATP-A; SLC21A3)

OATP-A (previously also called OATP or OATP1) was the first human OATP isolated from human liver.[38] Its gene maps to chromosome 12p12[77] and the protein contains 670 amino acids. Although OATP-A was originally cloned from a human liver cDNA library, its strongest expression is in brain followed by the kidney, liver, lung and testis.[38,78] In addition, OATP-A is expressed in the human hepatoma cell line HepG2,[79] in colon adenocarcinoma (GI-112) cells[78] and in hepacellular carcinoma.[80] In human liver, OATP-A exhibits an apparent molecular mass of approximately 85 kDa,[80] while in brain capillary endothelial cells (blood brain barrier) its apparent molecular mass is ~60kDa because of incomplete glycosylation.[81]

Transport substrates of OATP-A include bile salts and BSP, steroid conjugates, thyroid hormones, prostaglandin E2, ouabain, organic cations, the cyanobacterial toxin microcystin and several drugs including the endothelin receptor antagonist BQ-123, the thrombin inhibitor CRC-220, the opioid receptor agonists [D- penicillamine 2,5]enkephalin (DPDPE) and deltorphin II, fexofenadine and certain magnetic resonance imaging contrast agents (Table 2). Thus, human OATP-A has a similar wide spectrum of transport substrates as rat Oatp1, and it transports the largest number of amphipathic substrates of all human OATPs.[45]

Expression of OATP-A is maintained or even moderately increased in patients with cholestasis due to primary sclerosing cholangitis.[80] However, transcriptional and posttranscriptional regulation of OATP-A expression remain to be investigated.

The Human Organic Anion Transporting Polypeptide B (OATP-B; SLC21A9)

OATP-B, the gene of which was mapped to chromosome 11[78] was originally isolated from human brain as a 709 amino acid protein.[82] However, its strongest expression is in the liver, followed by the spleen, placenta, lung, kidney, heart, ovary, small intestine, and brain.[45,78,83] In human liver, OATP-B is exclusively expressed at the basolateral (sinusoidal) plasma membrane of hepatocytes and exhibits an apparent molecular mass of 85 kDa.[45]

Functionally, OATP-B has been studied in *X. laevis* oocytes and in transiently transfected HEK293 cells. Compared to other OATPs it has a rather narrow substrate specificity and transports only BSP, estrone-3-sulfate and DHEAS.[45] For other compounds such as PGE2 and estradiol-17β-glucuronide controversial results have been presented[45,78] (Table 2).

The Human Organic Anion Transporting Polypeptide C (OATP-C; SLC21A6)

OATP-C (also called LST-1, OATP2 or OATP6) was cloned from human liver by several groups.[46-48,78] Its gene is located within the *OATP* gene cluster on chromosome 12p12 and encodes a 691 amino acid glycoprotein with an apparent molecular mass of 84 kDa.[46] Expression of OATP-C appears to be restricted to the liver, where it is confined to the basolateral (sinusoidal) plasma membrane.[46-48,78] Its exclusive expression in human liver suggests that OATP-C plays a crucial role in the hepatic clearance of albumin-bound amphipathic organic compounds.

The transport functions of OATP-C have been characterized in *X. laevis* oocytes[45,47] and in stably transfected HEK-293 cells.[46,48,78] The transport substrates include bile salts, conjugated and unconjugated bilirubin, BSP, steroid conjugates, thyroid hormones, eicosanoids, cyclic peptides, drugs like benzylpenicillin, methotrexate, pravastatin and rifampicin and the natural toxins microcystin and phalloidin (Table 2). Thus, OATP-C exhibits also a wide substrate spectrum and, most importantly, also transports unconjugated bilirubin, which could not be shown so far to be transported by any other Oatp/OATP.

Transcriptional expression of OATP-C appears to be decreased in primary sclerosing cholangitis since these patients experience a decrease of mRNA levels to about 50% as compared to healthy control livers [84]. In addition, similar to Oatp1 and Oatp2, basal expression of human OATP-C is dependent on the liver-enriched transcription factor HNF-1α.[76]

The Human Organic Anion Transporting Polypeptide 8 (OATP8; SLC21A8)

OATP8 (also called LST-2) was also cloned from human liver.[49,85] Its gene belongs to the same cluster on chromosome 12p12 as OATP-A and OATP-C. It encodes a 702 amino acid glycoprotein that has an apparent molecular mass of 120 kDa in human liver.[49] Under normal physiological conditions, OATP8 seems to be predominantly, if not exclusively, expressed at the basolateral plasma membrane of hepatocytes.[49,85] However, OATP8 is also expressed in several different cancer cell lines derived from gastric, colon, pancreas, gallbladder, lung and brain cancers,[85] but the significance of this expression remains to be investigated.

The transport functions of OATP8 have been characterized in *X. laevis* oocytes and in stably transfected HEK-293 cells. The substrates include bile salts, monoglucuronosyl bilirubin, BSP, steroid conjugates, thyroid hormones, leukotriene C_4, linear and cyclic peptides, the cardiac glycosides digoxin and ouabain, the antineoplastic organic anion methothexate, the antibiotic rifampicin and the natural toxins microcystin and phalloidin (Table 2). Thus, OATP8 is a second liver specific organic anion transporter with a broad substrate specificity. However, as compared to OATP-C, OATP8 has unique transport properties for the intestinal peptide cholecystokinin 8 (CCK-8),[86] the opioid peptide deltorphin II[45] and the cardiac glycosides digoxin and ouabain[45] (Table 2).

Hepatic expression of OATP8 appears to be dependent on HNF-1α[76] as well as on the bile acid nuclear receptor FXR/BAR.[87] The latter findings indicate that induction of *OATP8* gene expression by bile acids could serve to maintain hepatic extraction of xenobiotics and peptides under cholestatic conditions.

The Multidrug Resistance Associated Protein 3 (Mrp3/MRP3)

Mrp3 (gene symbol: *Abcc3*) is a 1527 amino acid protein that belongs to the multidrug resistance associated protein family and is expressed at the basolateral membrane of liver (Fig. 1), adrenal, pancreas, kidney and gut epithelial cells.[88] It is an ATP-dependent organic anion transporter that prefers glucuronidated substrates over GSH conjugates and also transports bile salts.[88,89] Under normal physiological conditions, Mrp3 is hardly detectable at the basolateral membrane of hepatocytes. However, under cholestatic conditions it is upregulated and can function as basolateral "bile salt export pump" that compensates at least in part for the disrupted canalicular bile salt secretion.[90] In Mrp2-deficient rats[91] and in livers of Dubin-Johnson patients[92] Mrp3/MRP3 expression is also upregulated. Whether Oatps/OATPs also contribute to basolateral bile salt efflux remains to be investigated.

Acknowledgements

The authors were supported by the Swiss National Science Foundation (grants 31-59204.99 to B.H. and 31-64140.00 to P.J.M.)

References

1. Reichen J, Paumgartner G. Uptake of bile acids by perfused rat liver. Am J Physiol 1976; 231(3):734-742.
2. Dietmaier A, Gasser R, Graf J et al. Investigations on the sodium dependence of bile acid fluxes in the isolated perfused rat liver. Biochim Biophys Acta 1976; 443(1):81-91.
3. Anwer MS, Hegner D. Effect of Na⁺ on bile acid uptake by isolated rat hepatocytes. Hoppe-Seyler's Z Physiol Chem 1978; 359:181-192.
4. Van Dyke RW, Stephens JE, Scharschmidt BF. Bile acid transport in cultured rat hepatocytes. Am J Physiol 1982; 243:G484-G492.
5. Zimmerli B, Valantinas J, Meier PJ. Multispecificity of Na⁺-dependent taurocholate uptake in basolateral (sinusoidal) rat liver plasma membrane vesicles. J Pharmacol Exp Ther 1989; 250(1):301-308.
6. Hagenbuch B, Stieger B, Foguet M et al. Functional expression cloning and characterization of the hepatocyte Na⁺/bile acid cotransport system. Proc Natl Acad Sci USA 1991; 88:10629-10633.
7. Cattori V, Eckhardt U, Hagenbuch B. Molecular cloning and functional characterization of two alternatively spliced Ntcp isoforms from mouse liver. Biochim Biophys Acta 1999; 1445:154-159.

8. Kramer W, Stengelin S, Baringhaus KH et al. Substrate specificity of the ileal and the hepatic Na(+)/bile acid cotransporters of the rabbit. I. Transport studies with membrane vesicles and cell lines expressing the cloned transporters. J Lipid Res 1999; 40(9):1604-17.
9. Hagenbuch B, Meier PJ. Molecular Cloning, Chromosomal Localization, and Functional Characterization of a Human Liver Na⁺ Bile acid Cotransporter. J Clin Invest 1994; 93(3):1326-1331.
10. Stieger B, Hagenbuch B, Landmann L et al. In situ localization of the hepatocytic Na⁺/taurocholate cotransporting polypeptide (Ntcp) in rat liver. Gastroenterology 1994; 107:1781-1787.
11. Ananthanarayanan M, Ng OC, Boyer JL et al. Characterization of cloned rat liver Na⁺-bile acid cotransporter using peptide and fusion protein antibodies. Am J Physiol 1994; 30(4):G637-G643.
12. Kim JY, Kim KH, Lee JA et al. Transporter-mediated bile acid uptake causes Ca2⁺-dependent cell death in rat pancreatic acinar cells. Gastroenterology 2002; 122(7):1941-53.
13. Hallen S, Mareninova O, Branden M et al. Organization of the membrane domain of the human liver sodium/bile acid cotransporter. Biochemistry 2002; 41(23):7253-66.
14. Karpen SJ, Sun AQ, Kudish B et al. Multiple factors regulate the rat liver basolateral sodium-dependent bile acid cotransporter gene promoter. J Biol Chem 1996; 271(25):15211-15221.
15. Cohn MA, Rounds DJ, Karpen SJ et al. Assignment of a rat liver Na⁺ bile acid cotransporter gene to chromosome 6q24. Mammalian Genome 1995; 6(1):60.
16. Green RM, Ananthanarayanan M, Suchy FJ et al. Genetic mapping of the Na⁺-taurocholate cotransporting polypeptide to mouse chromosome 12. Mammalian Genome 1998; 9(7):598-598.
17. Clark RF, Cruts M, Korenblat KM et al. A yeast artificial chromosome contig from human chromosome 14q24 spanning the Alzheimer's disease locus AD3. Human Molecular Genetics 1995; 4(8):1347-1354.
18. Hagenbuch B, Scharschmidt BF, Meier PJ. Effect of antisense oligonucleotides on the expression of hepatocellular bile acid and organic anion uptake systems in Xenopus laevis oocytes. Biochem J 1996; 316:901-904.
19. Boyer JL, Ng OC, Ananthanarayanan M et al. Expression and Characterization of a Functional Rat Liver Na⁺ Bile acid Cotransport System in COS-7 Cells. Am J Physiol 1994; 266(3 Part 1):G382-G387.
20. Schroeder A, Eckhardt U, Stieger B et al. Substrate specificity of the rat liver Na⁺/bile salt cotransporter in Xenopus laevis oocytes and in CHO cells. Am J Physiol 1998; 274:G370-G375.
21. Platte HD, Honscha W, Schuh K et al. Functional characterization of the hepatic sodium-dependent taurocholate transporter stably transfected into an immortalized liver-derived cell line and V79 fibroblasts. Eur J Cell Biol 1996; 70(1):54-60.
22. Torchia EC, Shapiro RJ, Agellon LB. Reconstitution of bile acid transport in the rat hepatoma McArdle RH-7777 cell line. Hepatology 1996; 24(1):206-211.
23. Meier PJ. Molecular mechanisms of hepatic bile salt transport from sinusoidal blood into bile. Am J Physiol 1995; 269:G801-G812.
24. Hagenbuch B, Meier PJ. Sinusoidal (basolateral) bile salt uptake systems of hepatocytes. Semin Liver Dis 1996; 16(2):129-136.
25. Weinman SA. Electrogenicity of Na⁺-coupled bile acid transporters. Yale Journal of Biology and Medicine 1998; 70(4):331-340.
26. Hardikar W, Ananthanarayanan M, Suchy FJ. Differential ontogenic regulation of basolateral and canalicular bile acid transport proteins in rat liver. J Biol Chem 1995; 270(35):20841-20846.
27. Suchy FJ, Bucuvalas JC, Goodrich AL et al. Taurocholate transport and Na⁺-K⁺-ATPase activity in fetal and neonatal rat liver plasma membrane vesicles. Am J Physiol 1986; 251:G665-G673.
28. Rippin SJ, Hagenbuch B, Meier PJ et al. Cholestatic expression pattern of sinusoidal and canalicular organic anion transport systems in primary cultured rat hepatocytes. Hepatology 2001; 33:776-782.
29. Liang D, Hagenbuch B, Stieger B et al. Parallel Decrease of Na⁺ -Taurocholate Cotransport and Its Encoding Messenger RNA in Primary Cultures of Rat Hepatocytes. Hepatology 1993; 18(5):1162-1166.
30. Green RM, Gollan JL, Hagenbuch B et al. Regulation of hepatocyte bile salt transporters during hepatic regeneration. Am J Physiol 1997; 273:G621-G627.
31. Gartung C, Ananthanarayanan M, Rahman MA et al. Down-regulation of expression and function of the rat liver Na⁺/bile acid cotransporter in extrahepatic cholestasis. Gastroenterology 1996; 110(1):199-209.
32. Simon FR, Fortune J, Iwahashi M et al. Ethinyl Estradiol Cholestasis Involves Alterations In Expression Of Liver Sinusoidal Transporters. Am J Physiol 1996; 34(6):G1043-G1052.
33. Green RM, Beier D, Gollan JL. Regulation of hepatocyte bile salt transporters by endotoxin and inflammatory cytokines in rodents. Gastroenterology 1996; 111(1):193-198.
34. Trauner M, Arrese M, Lee H et al. Endotoxin downregulates rat hepatic ntcp gene expression via decreased activity of critical transcription factors. J Clin Invest 1998; 101(10):2092-2100.
35. Denson LA, Auld KL, Schiek DS et al. Interleukin-1beta suppresses retinoid transactivation of two hepatic transporter genes involved in bile formation. J Biol Chem 2000; 275(12):8835-43.

36. Karpen SJ. Nuclear receptor regulation of hepatic function. J Hepatol 2002; 36:832-850.
37. Shneider BL, Fox VL, Schwarz KB et al. Hepatic Basolateral Sodiu-Dependent-Bile acid Transporter Expression in Two Unusual Cases of Hypercholanemia and in Extrahepatic biliary atresia. Hepatology 1997; 25:1176-1183.
38. Kullak-Ublick GA, Hagenbuch B, Stieger B et al. Molecular and functional characterization of an organic anion transporting polypeptide cloned from human liver. Gastroenterology 1995; 109(4):1274-1282.
39. Zollner G, Fickert P, Zenz R et al. Hepatobiliary transporter expression in percutaneous liver biopsies of patients with cholestatic liver diseases. Hepatology 2001; 33(3):633-46.
40. Hagenbuch B, Meier PJ. The Superfamily of Organic anion transporting Polypeptides (OATPs). Biochim Biophys Acta 2003; 1609:1-18.
41. Jacquemin E, Hagenbuch B, Stieger B et al. Expression Cloning of a Rat Liver Na⁺-Independent Organic anion transporter. Proc Natl Acad Sci USA 1994; 91(1):133-137.
42. Noé B, Hagenbuch B, Stieger B et al. Isolation of a multispecific organic anion and cardiac glycoside transporter from rat brain. Proc Natl Acad Sci USA 1997; 94(19):10346-10350.
43. Cattori V, Hagenbuch B, Hagenbuch N et al. Identification of organic anion transporting polypeptide 4 (Oatp4) as a major full-length isoform of the liver-specific transporter-1 (rlst-1) in rat liver. FEBS Lett 2000; 474:242-245.
44. Kullak-Ublick GA, Beuers U, Paumgartner G. Molecular and functional characterization of bile acid transport in human hepatoblastoma HepG2 cells. Hepatology 1996; 23(5):1053-1060.
45. Kullak-Ublick GA, Ismair MG, Stieger B et al. Organic anion-transporting polypeptide B (OATP-B) and its functional comparison with three other OATPs of human liver. Gastroenterology 2001; 120:525-533.
46. König J, Cui Y, Nies AT et al. A novel human organic anion transporting polypeptide localized to the basolateral hepatocyte membrane. Am J Physiol 2000; 278:G156-G164.
47. Abe T, Kakyo M, Tokui T et al. Identification of a novel gene family encoding human liver-specific organic anion transporter LST-1. J Biol Chem 1999; 274:17159-17163.
48. Hsiang B, Zhu Y, Wang Z et al. A novel human hepatic organic anion transporting polypeptide (OATP2). J Biol Chem 1999; 274:37161-37168.
49. König J, Cui Y, Nies AT et al. Localization and genomic organization of a new hepatocellular organic anion transporting polypeptide. J Biol Chem 2000; 275:23161-23168.
50. Bergwerk AJ, Shi XY, Ford AC et al. Immunologic distribution of an organic anion transport protein in rat liver and kidney. Am J Physiol 1996; 271(2):G231-G238.
51. Eckhardt U, Schroeder A, Stieger B et al. Polyspecific substrate uptake by the hepatic organic anion transporter oatp1 in stably transfected CHO cells. Am J Physiol 1999; 276:G1037-G1042.
52. Shi XY, Bai S, Ford AC et al. Stable inducible expression of a functional rat liver organic anion transport protein in HeLa cells. J Biol Chem 1995; 270(43):25591-25595.
53. Satlin LM, Amin V, Wolkoff AW. Organic anion transporting polypeptide mediates organic anion/HCO3- exchange. J Biol Chem 1997; 272(42):26340-26345.
54. Li LQ, Lee TK, Meier PJ et al. Identification of glutathione as a driving force and leukotriene C-4 as a substrate for oatp1, the hepatic sinusoidal organic solute transporter. J Biol Chem 1998; 273(26):16184-16191.
55. Fickert P, Zollner G, Fuchsbichler A et al. Effects of ursodeoxycholic and cholic acid feeding on hepatocellular transporter expression in mouse liver. Gastroenterology 2001; 121:170-183.
56. Lee JM, Trauner M, Soroka CJ et al. Expression of the bile salt export pump is maintained after chronic cholestasis in the rat. Gastroenterology 2000; 118:163-172.
57. Müller M. The transcriptional control of hepatocanalicular transporter gene expression. Semin Liver Dis 2000; 20:223-235.
58. Meier PJ, Stieger B. Bile salt transporters. Annu Rev Physiol 2002; 64:635-661.
59. Glavy JS, Wu SM, Wang PJ et al. Down-regulation by extracellular ATP of rat hepatocyte organic anion transport is mediated by serine phosophorylation of Oatp1. J Biol Chem 2000; 275:1479-1484.
60. Guo GL, Klaassen CD. Protein kinase C suppresses rat organic anion transporting polypeptide 1- and 2-mediated uptake. J Pharmacol Exp Ther 2001; 299:551-557.
61. Abe T, Kakyo M, Sakagami H et al. Molecular characterization and tissue distribution of a new organic anion transporter subtype (oatp3) that transports thyroid hormones and taurocholate and comparison with oatp2. J Biol Chem 1998; 273(35):22395-401.
62. Gao B, Stieger B, Noè B et al. Localization of the organic anion transporting polypeptide 2 (Oatp2) in capillary endothelium and choroid plexus epithelium of rat brain. J Histochem Cytochem 1999; 47:1255-1264.
63. Reichel C, Gao B, van Montfoort J et al. Localization and function of the organic anion-transporting polypeptide Oatp2 in rat liver. Gastroenterology 1999; 117:688-695.
64. Kakyo M, Sakagami H, Nishio T et al. Immunohistochemical distribution and functional characterization of an organic anion transporting polypeptide 2 (oatp2). FEBS Lett 1999; 445:343-346.

65. Gao B, Wenzel A, Grimm C et al. Localization of organic anion transport protein 2 in the apical region of rat retinal pigment epithelium. Invest Ophthalmol Vis Sci 2002; 43:510-514.
66. Li L, Meier PJ, Ballatori N. Oatp2 mediates bidirectional organic solute transport: a role for intracellular glutathione. Mol Pharmacol 2000; 58:335-340.
67. Cao J, Huang L, Liu Y et al. Differential regulation of hepatic bile salt and organic anion transporters in pregnant and postpartum rats and the role of prolactin. Hepatology 2001; 33:140-147.
68. Shih DQ, Bussen M, Sehayek E et al. Hepatocyte nuclear factor-1alpha is an essential regulator of bile acid and plasma cholesterol metabolism. Nature Genet 2001; 27:375-382.
69. Hagenbuch N, Reichel C, Stieger B et al. Effect of phenobarbital on the expression of bile salt and organic anion transporters of rat liver. J Hepatol 2001; 34:881-887.
70. Rausch-Derra LC, Hartley DP, Meier PJ et al. Differential effects of microsomal enzyme-inducing chemicals on the hepatic expression of rat organic anion transporters, OATP1 and OATP2. Hepatology 2001; 33:1469-1478.
71. Staudinger JL, Goodwin B, Jones SA et al. The nuclear receptor PXR is a lithocholic acid sensor that protects against liver toxicity. Proc Natl Acad Sci USA 2001; 98:3369-3374.
72. Guo GL, Staudinger J, Ogura K et al. Induction of rat organic anion tarnsporting polypeptide 2 by pregnenolone-16a-carbonitrile is via interaction with pregnane X receptor. Mol Endocrinol 2002; 61:832-839.
73. Kakyo M, Unno M, Tokui T et al. Molecular characterization and functional regulation of a novel rat liver-specific organic anion transporter rlst-1. Gastroenterology 1999; 117:770-775.
74. Choudhuri S, Ogura K, Klaassen CD. Cloning of the full-length coding sequence of rat liver-specific organic anion transporter-1 (rlst-1) and a splice variant and partial characterization of the rat lst-1 gene. Biochem Biophys Res Commun 2000; 274:79-86.
75. Ogura K, Choudhuri S, Klaassen CD. Full-length cDNA cloning and genomic organization of the mouse liver-specific organic anion transporter-1 (lst-1). Biochem Biophys Res Commun 2000; 272(563-570).
76. Jung D, Hagenbuch B, Gresh L et al. Characterization of the human OATP-C (SLC21A6) gene promoter and regulation of liver-specific OATP genes by hepatocyte nuclear factor 1 alpha. J Biol Chem 2001; 276:37206-37214.
77. Kullak-Ublick GA, Beuers U, Meier PJ et al. Assignement of the human organic anion transporting polypeptide (OATP) gene to chromosome 12p12 by fluorescence in situ hybridizyrion. J Hepatol 1996; 25:985-987.
78. Tamai I, Nezu J, Uchino H et al. Molecular identification and characterization of novel members of the human organic anion transporter (OATP) family. Biochem Biophys Res Commun 2000; 273:251-260.
79. Lee TK, Hammond CL, Ballatori N. Intracellular glutathione regulates taurocholate transport in HepG2 cells. Toxicol Appl Pharmacol 2001; 174:207-215.
80. Kullak-Ublick GA, Glasa J, Boker C et al. Chlorambucil-taurocholate is transported by bile acid carriers expressed in human hepatocellular carcinomas. Gastroenterology 1997; 113(4):1295-1305.
81. Gao B, Hagenbuch B, Kullak-Ublick GA et al. Organic anion-transporting polypeptides mediate transport of opioid peptides across blood-brain barrier. J Pharmacol Exp Ther 2000; 294:73-79.
82. Nagase T, Ishikawa K, Suyama M et al. Prediction of the coding sequences of unidentified human genes. XII. The complete sequences of 100 new cDNA clones from brain which code for large proteins in vitro. DNA Res 1998; 5:355-364.
83. St-Pierre MV, Ugele B, Hagenbuch B et al. Characterization of an organic anion transporting polypeptide (OATP-B) in human placenta. J Clin Endocrin Met 2002; 87:1856-1863.
84. Oswald M, Kullak-Ublick GA, Paumgartner G et al. Expression of hepatic transporters OATP-C and MRP2 in primary sclerosing cholangitis. Liver 2001; 21:247-253.
85. Abe T, Unno M, Onogawa T et al. LST-2, a human liver-specific organic anion transporter, determines methotrexate sensitivity in gastrointestinal cancers. Gastroenterology 2001; 120:1689-1699.
86. Ismair MG, Stieger B, Cattori V et al. Hepatic uptake of cholecystikinin octapeptide by organic anion-transporting polypeptides OATP4 and OATP8 of rat and human liver. Gastroenterology 2001; 121:1185-1190.
87. Jung D, Podvinec M, Meyer UA et al. Human organic anion transporting polypeptide OATP8 (SLC21A8) promoter is transactivated by the farnesoid X receptor/bile acid receptor. Gastroenterology 2002; 122:1954-1966.
88. Borst P, Evers R, Kool M et al. A family of drug transporters: the multidrug resistance-associated proteins. J Natl Cancer Inst 2000; 92(16):1295-302.
89. Hirohashi T, Suzuki H, Takikawa H et al. ATP-dependent transport of bile salts by rat multidrug resistance-associated protein 3 (Mrp3). J Biol Chem 2000; 275:2905-2910.
90. Soroka CJ, Lee JM, Azzaroli F et al. Cellular localization and up-regulation of multidrug resistance- associated protein 3 in hepatocytes and cholangiocytes during obstructive cholestasis in rat liver. Hepatology 2001; 33(4):783-91.

91. Hirohashi T, Suzuki H, Ito K et al. Hepatic expression of multidrug resistance-associated protein-like proteins maintained in Eisai hyperbilirubinemic rats. Mol Pharmacol 1998; 53(6):1068-75.
92. Konig J, Rost D, Cui Y et al. Characterization of the human multidrug resistance protein isoform MRP3 localized to the basolateral hepatocyte membrane. Hepatology 1999; 29(4):1156-63.
93. Kanai N, Lu R, Bao Y et al. Estradiol 17 beta-D-glucuronide is a high-affinity substrate for oatp organic anion transporter. Am J Physiol 1996; 270(2):F326-F331.
94. Kullak-Ublick G-A, Hagenbuch B, Stieger B et al. Functional characterization of the basolateral rat liver organic anion transporting polypeptide. Hepatology 1994; 20:411-416.
95. Pang KS, Wang PJ, Chung A et al. The modified dipeptide, enalapril, an angiotensin-converting enzyme inhibitor, is transported by the rat liver organic anion transport protein. Hepatology 1998; 28:1341-1346.
96. Kouzuki H, Suzuki H, Ito K et al. Contribution of organic anion transporting polypeptide to uptake of its possible substrates into rat hepatocytes. J Pharmacol Exp Ther 1999; 288(2):627-634.
97. Bossuyt X, Muller M, Hagenbuch B et al. Polyspecific drug and steroid clearance by an organic anion transporter of mammalian liver. J Pharmacol Exp Ther 1996; 276(3):891-896.
98. Kouzuki H, Suzuki H, Ito K et al. Contribution of sodium taurocholate co-transporting polypeptide to the uptake of its possible substrates into rat' hepatocytes. J Pharmacol Exp Ther 1998; 286(2):1043-1050.
99. Ishizuka H, Konno K, Naganuma H et al. Transport of temocaprilat into rat hepatocytes: role of organic anion transporting polypeptide. J Pharmacol Exp Ther 1998; 287(1):37-42.
100. Friesema EC, Docter R, Moerings EP et al. Identification of thyroid hormone transporters. Biochem Biophys Res Commun 1999; 254(0006-291X TA - Biochem Biophys Res Commun PG - 497-501 SB - M SB - X):497-501.
101. Eckhardt U, Horz JA, Petzinger E et al. The peptide-based thrombin inhibitor CRC 220 is a new substrate of the basolateral rat-liver organic anion-transporting polypeptide. Hepatology 1996; 24(2):380-384.
102. Cvetkovic M, Leake B, Fromm MF et al. Oatp and p-glycoprotein transporters mediate the cellular uptale and excretion of fexofenadine. Drug Metab Dispos 1999; 27:866-871.
103. van Montfoort JE, Stieger B, Meijer DK et al. Hepatic uptake of the magnetic resonance imaging contrast agent gadoxetate by the organic anion transporting polypeptide Oatp1. J Pharmacol Exp Ther 1999; 290:153-157.
104. van Montfoort JE, Hagenbuch B, Fattinger K et al. Polyspecific organic anion transporting polypeptides mediate hepatic uptake of amphipathic type II organic cations. J Pharmacol Exp Ther 1999; 291:147-152.
105. Kontaxi M, Eckhardt U, Hagenbuch B et al. Uptake of the mycotoxin ochratoxin A in liver cells occurs via the cloned organic anion transporting polypeptide. J Pharmacol Exp Ther 1996; 279(3):1507-1513.
106. Tokui T, Nakai D, Nakagomi R et al. Pravastatin, an HMG-CoA reductase inhibitor, is transported by rat organic anion transporting polypeptide, oatp2. Pharm Res 1999; 16:904-908.
107. Cattori V, van Montfoort JE, Stieger B et al. Localization of organic anion transporting polypeptide 4 (Oatp4) in rat liver and comparison of its substrate specificity with Oatp1, Oatp2 and Oatp3. Pflügers Arch 2001; 443:188-195.
108. Fischer WJ, van Montfoort JE, Cattori V et al. Organic anion transporting polypeptides (OATPS) mediate uptake of microcystin into brain and liver. J Toxicol Clin Toxicol 2001; 39:A565.
109. Kullak-Ublick GA, Fisch T, Oswald M et al. Dehydroepiandrosterone sulfate (DHEAS): identification of a carrier protein in human liver and brain. FEBS Lett 1998; 424:173-176.
110. Bossuyt X, Muller M, Meier PJ. Multispecific amphipathic substrate transport by an organic anion transporter of human liver. J Hepatol 1996; 25(5):733-738.
111. Fujiwara K, Adachi H, Nishio T et al. Identification of thyroid hormone transporters in humans: different molecules are involved in a tissue-specific manner. Endocrinology 2001; 142:2005-2012.
112. Pascolo L, Cupelli F, Anelli PL et al. Molecular mechanisms for the hepatic uptake of magnetic resonance imaging contrast agents. Biochem Biophys Res Commun 1999; 257(3):746-752.
113. Meier PJ, Eckhardt U, Schroeder A et al. Substrate specificity of sinusoidal bile acid and organic anion uptake systems in rat and human liver. Hepatology 1997; 26:1667-1677.
114. Cui Y, König J, Leier I et al. Hepatic uptake of bilirubin and its conjugates by the human organic anion transporter SLC21A6. J Biol Chem 2001; 276:9626-9630.
115. Nakai D, Nakagomi R, Furuta Y et al. Human liver-specific organic anion transporter, LST-1, mediates uptake of pravastatin by human hepatocytes. J Pharmacol Exp Ther 2001; 297:861-867.
116. Vavricka SR, van Montfoort J, Ha HR et al. Interactions of rifamycin SV and rifampicin with organic anion uptake systems of human liver. Hepatology 2002; 36:164-172.

CHAPTER 3

The ABC of Canalicular Transport

Guido J.E.J. Hooiveld and Michael Müller

Summary

Bile formation is a regulated process and depends on the coordinated action of a number of transporter proteins in the sinusoidal (basal) and canalicular (apical) domains of the hepatocyte. The secretion of substances in the canalicular lumen is mediated by a set of ATP-dependent transport proteins. Dysfunction of any of these proteins leads to retention of substrates, with e.g., conjugated hyperbilirubinemia or cholestasis as a result. In recent years many of the transport proteins involved in bile formation have been identified, cloned, and functionally characterized. In this chapter we will shortly describe the present knowledge regarding function and regulation of the primary active transport proteins involved in canalicular secretion of biliary components.

Introduction

The liver, the largest single internal organ of the vertebrate body, is uniquely positioned between the digestive tract and the general circulation. Consequently, the liver is continuously exposed to numerous structurally and chemically diverse endogenous and exogenous compounds, such as nutrients and its metabolites, drugs, and other (potentially toxic) xenobiotics. Serving as both a depot and transit station for amino acids, carbohydrates, lipids, and vitamins, the liver efficiently absorbs these substances from the circulation, transforms and metabolizes them, and releases them back into the systemic circulation or into bile. The uptake, biotransformation, and disposition of xenobiotics are other important functions of the liver.

The hepatocyte is a parenchymal cell that constitutes approximately 70 per cent of the total liver cell volume. It is a polarized cell with three functional distinct membrane regions, i.e., the basolateral (sinusoidal), lateral (intercellular), and apical (canalicular) membrane domains. These domains comprise approximately 35, 50, and 15 per cent of the total surface area of hepatocytes[1] (Fig. 1). Hepatocytes are well specialized to transport substances from the portal blood into the liver (and vice versa), and from the liver into bile. Bile, a thick fluid, is formed by a process of osmotic filtration in response to osmotic gradients created within the lumen of the bile canaliculus between adjacent hepatocytes, driving water and some ions from the blood via the thigh-junctions into the canaliculus.[2] This osmotic gradient is established by ongoing active secretion of solutes into the canalicular lumen. Major organic constituents of bile are bile salts, phospholipids, cholesterol, a variety of proteins, small peptides, amino acids, and bilirubin. The main determinant of bile flow is the transport of bile salt, forming the so-called 'bile acid-dependent bile flow'. Secretion of organic anions, especially of reduced glutathione (GSH), further contributes to bile flow and constitutes the 'bile acid-independent bile flow'.[3]

Bile secretion serves different important functions. First, it is one of the main mechanisms for the disposition of many endogenous and exogenous amphipatic compounds, including metabolic waste products, toxins and drugs. Second, it supplies bile salts to the intestine, which is of crucial importance for the emulsification and subsequent digestion and absorption of

Molecular Pathogenesis of Cholestasis, edited by Michael Trauner and Peter L.M. Jansen.
©2004 Eurekah.com and Kluwer Academic / Plenum Publishers.

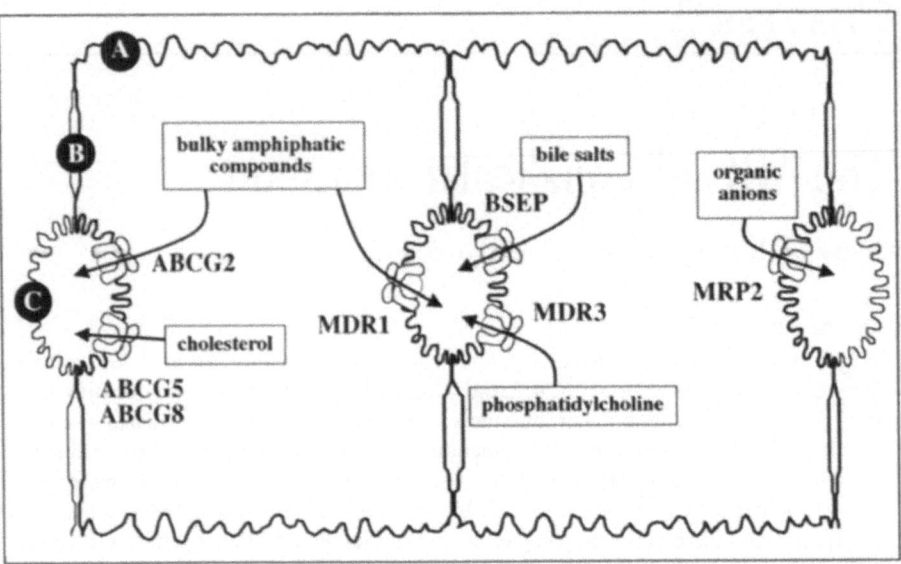

Figure 1. Human canalicular transporter proteins. Hepatocytes are polarized cells with three functional membrane domains, i.e., the sinusoidal (indicated by A), the lateral (indicated by B), and canalicular (indicated by C) membrane domains. Transporter proteins located in the canalicular membrane (represented as spherical structures, with arrows showing direction of transport) are responsible for the biliary secretion of bile salts, PC, cholesterol and anions on the one hand, and for the excretion of potentially toxic compounds on the other hand. These transporter proteins comprise the bile salt transporter BSEP, the phosphatidylcholine translocator MDR3, the cholesterol transporter ABCG5/G8, the anionic conjugate transporter MRP2, and the multidrug transporters MDR1 and ABCG2, respectively. See text for further details.

food, in particular dietary lipids. The inability to produce bile, for instance during cholestatic episodes of benign recurrent intrahepatic cholestasis, is associated with rapid weight loss.

The secretion of solutes into bile depends on the activity of transporter proteins in both the sinusoidal and canalicular membrane domain of hepatocytes. Substances are delivered to the liver via the portal blood. Their uptake into hepatocytes is mediated by a set of transporters that belong to the solute carrier superfamily. These carriers are not directly ATP-dependent for their transport function, but some are driven by a sodium gradient or by anion exchange. Others just 'simply' facilitate diffusion (for recent reviews regarding hepatic uptake systems see refs. 4-9). The final and (under normal physiological conditions) rate-limiting step in hepatobiliary transport is the secretion of substances into the canalicular lumen. For almost all compounds this task is performed by a number of proteins that belong to the large ATP-binding cassette (ABC) transporter superfamily.[4,5,10-12] These proteins are able to pump solutes into bile against a steep concentration gradient.

In this chapter we will focus on ABC proteins involved in secretion of biliary constituents from hepatocytes. We will give a short overview on the molecular structure of ABC transporters and discuss their function in the canalicular membrane. We will also only shortly relate inherited cholestatic syndromes to mutations in transporter genes, and briefly address how the expression of the various transporters is regulated, since the latter two subjects are extensively dealt with in other chapters of this book.

ABC Transporter Proteins

ATP-dependent efflux pumps mediate almost all secretory functions in the canalicular membrane of hepatocytes. These primary active transporter proteins are able to transfer cholephilic compounds from hepatocyte into bile, uphill against a steep (sometimes 100-1000-fold) concentration gradient. This ATP-dependent transport can be considered as the principal driving force of bile flow. In terms of energy bile formation is a costly process: for example, the secretion of one molecule unconjugated bilirubin costs four molecules of ATP equivalents: two molecules of UDP-glucuronic acid for conjugation and two molecules of ATP for canalicular transport.

The ATP-dependent transport proteins belong to the ABC transporter superfamily, one of the largest superfamilies of transmembrane proteins present in prokaryotes and eukaryotes.[13-17] An analysis of the complete human genome sequences revealed that 48 ABC genes are present.[14,18] Proteins are classified as ABC transporters based on the sequence and organization of their ATP-binding domains, also known as nucleotide-binding folds (NBFs). The NBFs contain characteristic motifs, the so-called Walker A and B motifs, separated by approximately 90-120 amino acids, found in all ATP-binding proteins.[17,19] ABC transport proteins also contain an additional element, the signature C motif or 'ABC signature', located just upstream of the Walker B site.[20] The functional protein typically contains two NBFs and two transmembrane (TM) domains[16,21] (Fig. 2). The TM domains contain 6-11 membrane-spanning α-helices and provide the specificity for the substrate. The NBFs are located in the cytoplasm and transfer the energy to transport the substrate across the membrane. ABC pumps are unidirectional. In eukaryotes, most ABC proteins move compounds from the cytoplasm to the outside of the cell, or into an extracellular compartment (endoplasmic reticulum, mitochondria, peroxisome).[14] Most of the known functions of eukaryotic ABC transporters involve the shuttling of compounds either within the cell as part of a metabolic process, or outside the cell for transport to other organs, or secretion from the body. The eukaryotic ABC proteins are organized either as full transporters containing two TMs and two NBFs, or as half transporters[20] (Fig. 2). The latter must form either homodimers or heterodimers to constitute a functional transporter.

Genes encoding for ABC transporters can be divided into subfamilies (A to G) based on similarity in gene structure (half vs. full transporters), order of the domains, and on sequence homology in the nucleotide binding fold and transmembrane domains. To minimize the confusion associated with referring to ABC genes with multiple names, a system of nomenclature for the ABC transporter genes has been introduced,[22] which has been derived from the subdivision made by Allikmets et al.[23] Based on this system, the mammalian ABC gene superfamily can be divided into seven subfamilies, three of which appear to be most important for bile formation: the B, C, and G clusters[10,11,13,14,21] (see Table 1). An extensive overview of all human ABC transporters can be found on http://nutrigene.4t.com/humanabc.htm, and in various recent reviews.[14,16,18,21]

The ABC-A Family

The ABC-A family consists of 13 members. Only two of its members are expressed in liver; ABCA1 and ABCA6.[24,25] Best characterized is ABCA1, which originally has been functionally associated to the process of the engulfment of cells dying by apoptosis.[26] Later studies showed that mutations of the ABCA1 gene are the causative defect in genetic high-density lipoprotein (HDL) deficiency syndromes.[27-29] Affected subjects have a defect in cellular cholesterol removal, which results in the almost complete absence of plasma HDL cholesterol, indicating that ABCA1 has an essential role in cellular lipid efflux. Initially, based on these results it was suggested that ABCA1 is a cholesterol transporter, but recent work has demonstrated this is not the case. Instead, the primary action of ABCA1 likely is acting as a phospholipid translocase.[30-33]

Interestingly, transgenic mice overexpressing *Abca1* have an increased concentration of cholesterol in bile,[34] implying that Abca1 is involved in biliary lipid secretion. This was investigated recently by Groen et al,[35] who measured cholesterol secretion rates into bile in *Abca1* knockout mice. They observed no differences between *Abca1* knockout mice and controls, demonstrating that Abca1 does not contribute to biliary cholesterol efflux. At the same time it was reported that Abca1 is present in the sinusoidal membrane of hepatocytes,[36] which also

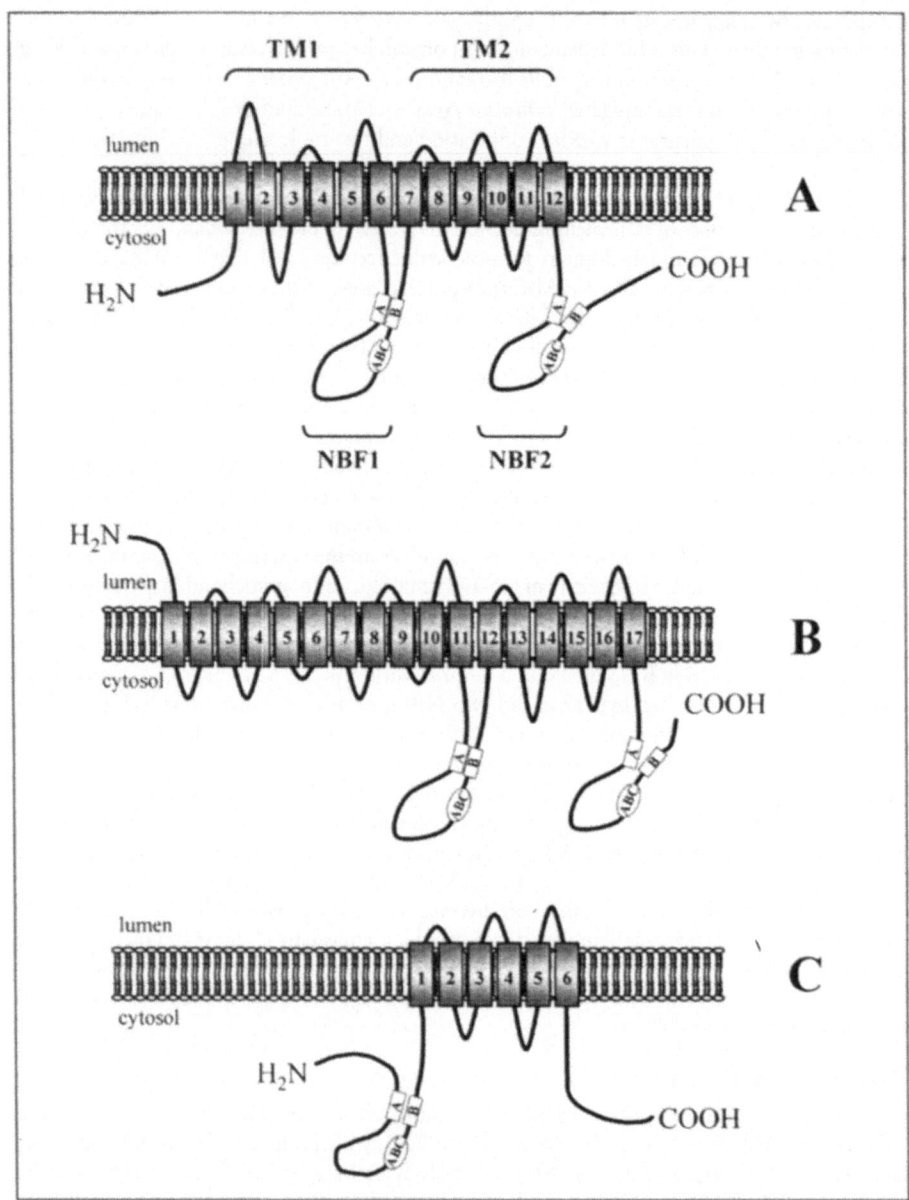

Figure 2. Membrane topology models of ABC transporters. This simplified scheme shows the most plausible membrane topology models for three key ABC transporters, namely MDR1 (ABCB1, panel A), MRP2 (ABCC2, panel B), and ABCG2 (panel C). It is generally accepted that the minimum functional unit requirement for an ABC transporter is the presence of two transmembrane domains (TM), consisting of six or more membrane spanning helices, and two nucleotide binding folds (NBF). These may be present within one polypeptide chain ("full transporters") as is the case for MDR1, or within a membrane-bound homo- or heterodimer of "half transporters" (ABCG2). All ABC proteins contain at least three characteristic peptide sequences: the Walker A and B motifs and the so-called ABC-signature sequence. Whereas the Walker motifs are present in all classes of ATP-binding proteins, the presence of the signature region is a typical feature of ABC proteins that mediate the selective movement of substrates across biological membranes.

Table 1. Human canalicluar transporter proteins

Name	Gene	Chromosome	Size (TMs)	Transport Function	Phenotype
MDR1	*ABCB1*	7q21	1279 (12)	multispecific	mdr
MDR3	*ABCB4*	7q21	1279 (12)	PC	PFIC3
BSEP	*ABCB11*	2q24	1321 (12)	BS	PFIC2
MRP2	*ABCC2*	10q24	1545 (17)	OA	Dubin-Johnson syndrome
ABCG2	*ABCG2*	4q22	655 (6)	multispecifics	mdr
ABCG5	*ABCG5*	2p21	651 (6)	sterols	sitosterolemia
ABCG8	*ABCG8*	2p21	673 (6)	sterols	sitosterolemia

The standard name of the transporters is given in **bold**, the gene name assigned by the nomenclature committee (http://www.gene.ucl.ac.uk/nomenclature/genefamily/abc.html) is given in italics. The size of the protein is given in amino acids. The phenotype associated with overexpression (for MDR1 and ABCG2) or (inherited) defects in the respective transporter genes (for all other genes) is indicated in the last column. See text for further details.

Abbreviations used: TM, most probable number of transmembrane segments; PC, phosphatidylcholine; BS, bile salts; OA, organic anions; mdr, multidrug resistance; PFIC, progressive familial intrahepatic cholestasis.

excludes a primary role of Abca1 in biliary secretion. The role of ABCA1 in hepatocytes is currently unknown, but may involve formation of nascent HDL particles.[34]

Only limited information is available on the regulation of *ABCA1*. Constitutive *ABCA1* gene expression in macrophage and liver cells is mediated by the binding of the transcription factors stimulating protein-1 (Sp1) and –3 (Sp3) to the promoter, while hepatocyte nuclear factor-1α (HNF-1α) is responsible for the hepatocyte-specific expression of *ABCA1*.[37] Expression of *ABCA1* is upregulated in cholesterol-loaded cells as a result of activation by the liver X receptor (LXR),[38-41] a nuclear receptor for oxysterols.[42]

The ABC-B Family

Of the eleven human ABC-B family members, three are expressed in the canalicular membrane of hepatocytes, namely MDR1 (encoded by the *ABCB1* gene), MDR3 (*ABCB4*) and BSEP (*ABCB11*) (Fig. 1). In rodents two Mdr1 proteins fulfill the function of the single human MDR1 protein, which are encoded by the *Abcb1a/1b* genes, respectively. MDR1 and MDR3 (and their mouse and rat orthologues) were the first ABC transporters recognized in canalicular membranes of normal hepatocytes.[43-46] The expression of *ABCB1* in normal human liver is high compared to the expression of *Abcb1a/1b* in laboratory animals.[44,47,48] Various physiological functions of MDR1/Mdr1 have been demonstrated or postulated. These include transport of exogenous and endogenous metabolites or toxins, steroid hormones, hydrophobic peptides or amphiphilic cationic drugs (reviewed in refs. 21,49,50). To establish the physiological function of the *Abcb1* gene products, a set of *Abcb1* gene knockout mice have been generated. These mice grow and develop normally, have no obvious abnormalities under laboratory conditions, and no changes in bile composition become apparent.[51,52] These results indicate that the Mdr1 proteins are not essential for basic physiological functions and bile formation. However, *Abcb1* knockout mice are hypersensitive to many drugs and toxins, due to profoundly reduced clearance of these agents in liver, intestine, and brain.[51-53] From these and other experimental data (reviewed in refs. 49,50), it is evident that MDR1/Mdr1 functions as an excretion transporter for amphipathic basic or cationic compounds. In brain or intestine MDR1/Mdr1 functions in maintenance of the barrier function by continuously removing hydrophobic compounds from the brain or the enterocyte, respectively.[54,55] By analogy, MDR1/Mdr1 likely protects the hepatocyte against hydrophobic toxic components by secreting them into the bile.

Until now, no human disorders are known that are associated with a defect in the *ABCB1* gene. Given the exposure of man to high levels of drugs and environmental toxins, it is surprising that no known inborn hypersensitivity to hydrophobic agents in humans is attributable to absence of MDR1.[21] However, various functional genetic polymorphisms in the *ABCB1* gene have been identified which influence the body distribution and bioavailability of MDR1 substrates (reviewed in refs. 56,57).

The regulation of *ABCB1/Abcb1* expression has been discussed elsewhere in more detail.[47,58,59] Although being unexpectedly complex and far from being fully understood, it is clear that *ABCB1/Abcb1* promoter activation is part of a general stress response in many cells resulting in enhanced cellular resistance.[58,59] Several transcription factors have been identified that regulate the expression of *ABCB1*. These include Sp1 and CCAAT/enhancer binding protein (C/EBP), both necessary for basal expression of *ABCB1*, and various tumor-suppressor and -inducer genes as well as oncogenes, such as p53, nuclear factor-Y (NF-Y), nuclear factor-κB (NF-κB), and T-cell factor-4 (TCF4) (reviewed in ref. 59). Data obtained in vivo on regulation of hepatic *ABCB1* expression are still lacking. In vitro studies demonstrated that the two rodent *Abcb1* genes are differentially regulated.[60-62] This opposite regulation is also observed in vivo: for example, the expression of *Abcb1a* in rat liver is not affected by endotoxin treatment[63] and increases only slightly after bile duct ligation or partial hepatectomy.[64] In contrast, *Abcb1b* expression is markedly enhanced during endotoxin- and bile duct-ligation induced cholestasis, and even more in the remnant liver after partial hepatectomy.[47,63-65] The induction of rat *Abcb1b* expression after exposure to insulin is mediated by binding of the transcription factor NF-κB to a functional NF-κB-like binding site in the promoter.[66] Up-regulation of *Abcb1b* during liver regeneration, after partial hepatectomy, or endotoxin treatment is at least in part tumor necrosis factor-α (TNFα)-dependent. More specifically, activation of the rat *Abcb1b* gene by TNFα is a result of NF-κB signaling.[67] It is thought that hepatic up-regulation of *Abcb1b* provides protection against oxidative-stress induced cell damage.[58]

The human MDR3 and its rodent orthologue Mdr2 are expressed at high levels in canalicular membranes of hepatocytes[46,68,69] (Fig. 1). In contrast to the drug transporting MDR1/Mdr1 proteins, the role of MDR3/Mdr2 seems crucial for basic liver physiology. Its function became apparent when *Abcb4* nullizygous mice were generated.[70] These mice do not secrete phosphatidylcholine (PC) into bile, and biliary cholesterol concentrations are dramatically decreased. Mice heterozygous for the *Abcb4* gene had a 40% decreased PC but normal cholesterol output, demonstrating that absence of PC secretion is the primary effect of *Abcb4* gene disruption.[70,71] From these and other studies, it is now well accepted that Mdr2 and its human counterpart MDR3 acts as a flippase, translocating phospholipids through the canalicular membrane.[72-74] The current working hypotheses for the mechanism of phospholipid secretion are that MDR3/Mdr2 flips PC from the inner to the outer leaflet of the lipid bilayer. Bile salt micelles then selectively solubilize PC. Alternatively, it has been postulated that lipids are concentrated in micro domains in the exoplasmic hemileaflet of the canalicular membrane.[75,76] Bile salts then solubilize PC from these micro domains.

Mutations in the human *ABCB4* gene are the underlying cause of progressive familiar intrahepatic cholestasis type 3 (PFIC3), a syndrome characterized by elevated serum γ-glutamyltransferase levels, decreased biliary phospholipid concentration with normal canalicular secretion of bile salts, resulting in very high bile salt to phospholipid ratios in bile.[77] Patients suffering from PFIC3 have missense mutations or nucleotide deletions in the *ABCB4* gene, resulting in the absence of MDR3 protein.[77]

Studies on regulation of human *ABCB4* gene expression are still missing. Expression of *Abcb4* in rodent liver appears to be unaltered under most conditions of cellular stress.[58] *Abcb4* expression is not affected by endotoxin treatment,[63] and is only slightly enhanced after partial hepatectomy.[64] Hepatic *Abcb4* expression, Mdr2 protein levels, and PC output is increased in mice fed the peroxisome proliferators ciprofibrate or clofibrate,[78,79] suggesting a involvement of the nuclear receptor PPARα[80] in controlling *Abcb4* gene expression. It has been suggested that the type of bile salt in plasma may control the expression level of *Abcb4* and rate of biliary PC secretion in mice. It was observed that the hydrophobic bile salt cholate slightly induced these parameters, whereas a five-fold higher amount of the more hydrophilic bile ursodeoxycholate did not so.[81,82] However, in a similar study these effects could not be observed.[83] Reduction of the bile salt pool by chronic bile diversion in rats decreased *Abcb4* expression level significantly.[84] Other evidence for regulatory functions of hydrophobic bile salts on *Abcb4* expression was provided *in vitro* with isolated rat hepatocytes.[84] When added to the culture medium, taurocholate and taurodeoxycholate both profoundly increased *Abcb4* mRNA levels in a time and concentration-dependent manner. Combined, these studies demonstrate that hydrophobic bile salt induce the expression of *Abcb4*, but also indicate that under normal in vivo conditions the expression level of *Abcb4* is already largely induced. Continuous exposure of rats to simvastatin or pravastatin, two inhibitors of cholesterol biosynthesis, resulted in decreased levels of liver cholesterol and increased biliary PC output.[69,85] This was accompanied by increased levels of *Abcb4* mRNA and Mdr2 protein.[69,85] Similar results were obtained in vitro when another inhibitor was used.[84] In contrast, cholesterol feeding decreased *Abcb4* mRNA considerably.[84] Since the zonal distribution of Mdr2 is very similar to the distribution of 3-hydroxy-3-methylglutaryl-coenzyme A-reductase and -synthase before and after statin treatment, it has been suggested that the factors controlling the expression of these proteins may be similar.[58,69]

Canalicular secretion of bile salts is also an ATP-dependent process.[86-89] The canalicular bile salt export pump BSEP is encoded by the *ABCB11* gene. Until very recently, the only functional data available regarding Bsep-mediated bile salt transport was obtained by expressing the full-length rodent genes.[90-92] This was due to the fact that expression of the human orthologue proved to be very difficult.[93,94] Studies with functional expressed rodent *Abcb11* genes revealed that rat Bsep transports taurocholate in an ATP-dependent fashion with a relative affinity (Km) of 5.3 μM.[90] Rat Bsep also transports the taurine-conjugates of chenodeoxycholate and ursodeoxycholate, but not glycocholate and cholate.[90] Murine Bsep was shown to transport taurocholate, taurochenodeoxycholate and glycocholate.[91,92] Human BSEP functions quite similarly to the rodent Bsep orthologues with respect to the Km for primary and secondary bile salts.[93,94] Also, the Km values for BSEP-mediated taurocholate transport (7.9 μM[93] and 4.3 μM[94]) agrees extremely closely with the Km for taurocholate determined in canalicular membrane vesicles isolated from human liver (4.2 μM[89]). Furthermore, taurochenodeoxycholic acid is the preferred substrate for the human, rat and mouse orthologues. Inhibition of BSEP has been suggested to account for certain forms of drug- and estrogen-induced cholestasis.[94-97]

Several mutations in the *ABCB11* gene have been identified in individuals with progressive familial intrahepatic cholestasis type 2 (PFIC2), an inherited progressive liver disease characterized by high serum bile salt concentrations in conjunction with low γ-glutamyltransferase serum levels.[98] These mutations result in the absence of canalicular BSEP expression and a decrease of biliary bile salts to less than 1% of normal.[99] *Abcb11* knockout mice are cholestatic in the sense that taurocholate accumulates in plasma because its secretion into bile is strongly impaired.[100] However, in contrast to patients, the mice excrete substantial amounts of tauromuricholate into bile as well as a hitherto undefined tetrahydroxy bile salt. Thus, in mice, it appears that significantly different mechanisms of bile salt conjugation are present, which result in bile salt species that are capable of being secreted by other canalicular membrane transporters, for example Mrp2.[100] This escape route prevents severe and progressive cholestasis in *Abcb11* knockout mice and as a consequence the animals have hardly any histopathological signs of liver injury. Because humans are not capable of converting bile salts into muricholate or tetrahydroxy bile salts to any significant extent, this escape route is not present in man.

Interestingly, *Abcb11* nullizygous mice were found to have an increased biliary secretion of phospholipid and cholesterol.[100]

The regulation of rat Bsep has been studied under conditions of endotoxin-treatment, bile duct ligation, and ethinylestradiol-induced cholestasis.[63,101] In the above mentioned cholestatic and stress-models *Abcb11* mRNA and Bsep protein expression levels only slightly decrease compared to levels of the basolateral bile salt uptake carriers or the canalicular transporter Mrp2.[64,101-103] Thus Bsep may continue to secrete bile salts, although at impaired rates. Remarkably, after partial hepatectomy the mRNA level of *Abcb11* is only mildly decreased and the proteins level of Bsep were unaffected in contrast to the bile salt uptake transporters.[64,104] This may explain the fact that after partial hepatectomy the remnant liver is not cholestatic and not damaged by excess bile salts. Expression of *ABCB11/Abcb11* is sensitive to the flux of bile salt through the hepatocyte. Both human[105,106] and rat[107] *ABCB11/Abcb11* promoter regions contain consensus elements that serve as binding sites for the farnesoid X receptor (FXR), a nuclear receptor for bile salts.[108]

The ABC-C Family

The ABC-C family consists of 12 members. Although several members are expressed in liver, only one (MRP2, *ABCC2*) is located in the canalicular membrane of hepatocytes (Fig. 1; Table 1).[109,110] In the liver many endogenous and xenobiotic lipophilic compounds are converted into more hydrophilic anionic conjugates with glutathione, glucuronate, or sulfate. These conjugates are transported across the canalicular membrane into bile by MRP2.

MRP2 was shown to be absent in patients with Dubin-Johnson syndrome,[111,112] an inheritable disorder that is associated with deficient biliary secretion of amphiphilic anionic conjugates, including bilirubin glucuronides. These patients have missense mutations or nucleotide deletions in the ABCC2 gene, resulting in the absence of MRP2 protein.[112,113] Similarly, Mrp2 protein is not expressed in the so-called Transport Deficient/Groningen Yellow (TR⁻/GY) and Eisai hyperbilirubinemic (EHBR) rats.[109,114] These are two different rat strains with a naturally occurring hereditary defect in biliary excretion of non-bile salt organic anions.[115,116] Absence of Mrp2 in these rats is due to mutations that introduce a stop codon in the coding sequence of the *Abcc2* gene.[109,114]

In rats, treatment with endotoxin or ethinylestradiol, or bile duct ligation results in a marked decrease of *Abcc2* expression and Mrp2 protein levels, as well as transport of anionic conjugates into bile.[63,117] The exact molecular mechanism(s) for this reduction is presently unknown, but an important role of the nuclear retinoids receptors (i.e., retinoid X receptor (RXR) and retinoic acid receptor (RAR), reviewed in ref. 118) has been suggested. Exposure to endotoxin and bile duct ligation results in down-regulation of nuclear RXR and RAR protein levels and binding to the *Abcc2* promoter, which in turn likely suppresses *Abcc2* transcription.[119,120] A dose- and time-dependent induction of rat *Abcc2* expression was observed in isolated rat hepatocytes cultured in the presence of xenobiotics, including vincristine, tamoxifen, or rifampicin,[121] indicating that *Abcc2* gene transcription may respond to substrates of Mrp2 itself or of Phase I- and II-enzymes. This response to xenobiotics is also in line with the finding that MRP2 can confer drug resistance in vitro.[122] Sequence analysis of the human *ABCC2* promoter showed a number of putative consensus binding sites for both ubiquitous and liver-enriched transcription factors, including activating protein-1 (Ap1), Sp1, HNF-1α and HNF-3β.[123,124] However, no experimental data is currently available that demonstrate functional interactions of the *ABCC2* promoter with these transcription factors. Recently, regulation of *ABCC2* by three nuclear receptors (i.e., FXR, pregnane X receptor (PXR),[125] and constitutive androstane receptor (CAR)[126]) was demonstrated. These observations suggest that ligands for these nuclear receptors (bile salts for FXR, and xenobiotics and toxins for PXR and CAR) activate transcription of the *ABCC2* gene in order to promote excretion of conjugated toxic agents from the hepatocyte into the bile.[127]

The ABC-G Family

The human ABC-G family consists of five 'half-size' transporters, i.e., proteins consisting of only six transmembrane helices and one nucleotide binding domain as opposed to the classical duplicate motif, and are thought to dimerize to form active membrane transporters (Fig. 2). Except for ABCG2, members of this family are involved in regulation of lipid trafficking.

ABCG1 is ubiquitously expressed, including liver.[128,129] Best characterized is its role in lipid homeostasis in macrophages, but its role in liver is still unclear.[130]

ABCG2 is also known as breast cancer resistance protein (BCRP)[131] and mitoxantrone resistance-associated protein.[132] The encoding gene has been shown to be amplified and overexpressed in human cancer cells and is capable of mediating multidrug resistance, even in the absence of the 'classic' multidrug resistance proteins MDR1 (ABCB1) and MRP1 (ABCC1).[21] Using immunohistochemistry, it was demonstrated that ABCG2 is present in the apical membrane of intestinal epithelial cells and hepatocytes.[133] The latter strongly suggests that ABCG2 is also involved in biliary secretion of unknown hydrophobic agents. Using a specific inhibitor of ABCG2, it was demonstrated that this transporter reduces the oral bioavailability and increases the hepatobiliary elimination of typical ABCG2 substrates.[134] The first data obtained with *Abcg2* knockout mice revealed that these mice are not able to extrude drugs from hemapoietic cells, are hypersensitive to a dietary chlorophyll-breakdown product, and display a previously unknown type of protoporphyria, an inherited disorder of porphyrin-heme metabolism with cutaneous and systemic manifestations.[135,136] No abnormalities were observed in serum levels of cholesterol and phospholipids. Taken together, these data indicate that ABCG2 is, in analogy to MDR1, an important player in the defense system against dietary as well as endogenous toxins.

Recently, exciting progress has been made with respect to ABCG5 and ABCG8. It was already known that mutations in the *ABCG5* and *ABCG8* genes cause sitosterolemia,[137,138] a rare autosomal recessive disorder characterized by accumulation of both plant and animal sterols in blood and tissues.[139,140] Normally, the intestinal absorption of sitosterol and other unwanted plant sterols is dwarfed by the absorption of cholesterol, and the absorbed phytosterols are cleared from the body by highly efficient secretion into bile.[139] However, patients with sitosterolemia display intestinal hyperabsorption of plant sterols and impaired biliary sterol secretion, leading to accumulation of phytosterols in the body.[140] The metabolism of cholesterol is altered in the same manner, but to a lesser extent,[139] rendering many patients hypercholesterolemic and prone to xanthomas and premature atherosclerosis.[139,140] The *ABCG5* and *ABCG8* genes are expressed almost exclusively in liver and small intestine, and are coordinately up-regulated by the nuclear hormone receptor LXR in response to dietary cholesterol.[137,141] High level expression of a human *ABCG5* and *ABCG8* transgene in mice resulted in an approximately 50 per cent reduction in the fractional absorption of cholesterol and a dramatic increase in the biliary secretion of sterols.[142] If coexpressed in the same cells, ABCG5 and ABCG8 are targeted to the apical surface of hepatocytes.[143] Taken together, these data indicate that ABCG5 and ABCG8 form a functional complex that limits the accumulation of dietary sterols by secreting sterols from gut epithelial cells into the lumen and promoting secretion of hepatic sterols into bile. Indeed, simultaneous disruption of the *Abcg5* and *Abcg8* genes in mice resulted in a selective and profound reduction in biliary cholesterol levels and an accumulation of cholesterol in the liver after cholesterol feeding, demonstrating that ABCG5 and ABCG8 are the major hepatobiliary transporter of both dietary and endogenous neutral sterols.[144] Since *Abcg5/Abcg8* knockout mice do display only modest, non-significant reductions in biliary phospholipid and bile acid levels, it is evident that biliary phospholipid and bile acid secretion can proceed in absence of cholesterol secretion, whereas disruption of *Abcb4* prevents biliary secretion of both phospholipids and cholesterol.[70]

Other Hepatic ABC-Transporter Proteins

In addition to the ABC transport proteins discussed above, expression of additional members of the ABC-transporter superfamily has been demonstrated in hepatocytes. These include members of virtually all other clusters. These proteins are not discussed here because of their basolateral localization in hepatocytes (e.g., for MRP3 and MRP6, ABC-C cluster), localization in intracellular organelles such as mitochondria or peroxisomes (e.g., ALDL1 and PXMP1, ABC-D cluster), or simply because of lack of basic information on function and localization (most members of the ABC-E and –F clusters).

Acknowledgements

Preparation of this chapter is supported by the Netherlands Organization for Scientific Research, NWO program grant 902-23-191, NWO project grants 902-23-253, NWO 902-23-257, and NWO 903-39-188.

References

1. Weibel ER, Staubli W, Gnagi HR et al. Correlated morphometric and biochemical studies on the liver cell. I. Morphometric model, stereologic methods, and normal morphometric data for rat liver. J Cell Biol 1969; 42:68-91.
2. Nathanson MH, Boyer JL. Mechanisms and regulation of bile secretion. Hepatology 1991; 14:551-566.
3. Ballatori N, Truong AT. Glutathione as a primary osmotic driving force in hepatic bile formation. Am J Physiol 1992; 263:G617-G624.
4. Meier PJ, Stieger B. Bile salt transporters. Annu Rev Physiol 2002; 64:635-61.
5. Kullak-Ublick GA, Stieger B, Hagenbuch B et al. Hepatic transport of bile salts. Semin Liver Dis 2000; 20:273-292.
6. Kullak-Ublick GA. Regulation of organic anion and drug transporters of the sinusoidal membrane. J Hepatol 1999; 31:563-573.
7. Meier PJ, Eckhardt U, Schroeder A et al. Substrate specificity of sinusoidal bile acid and organic anion uptake systems in rat and human liver. Hepatology 1997; 26:1667-1677.
8. Koepsell H. Organic cation transporters in intestine, kidney, liver, and brain. Annu Rev Physiol 1998; 60:243-66:243-266.
9. Suzuki H, Sugiyama Y. Transport of drugs across the hepatic sinusoidal membrane: sinusoidal drug influx and efflux in the liver. Semin Liver Dis 2000; 20:251-263.
10. Müller M, Jansen PLM. Molecular aspects of hepatobiliary transport. Am J Physiol 1997; 272:G1285-303.
11. Müller M, Jansen PLM. The secretory function of the liver: new aspects of hepatobiliary transport. J Hepatol 1998; 28:344-354.
12. Keppler D, Konig J. Hepatic secretion of conjugated drugs and endogenous substances. Semin Liver Dis 2000; 20:265-272.
13. Dean M, Hamon Y, Chimini G. The human ATP-binding cassette (ABC) transporter superfamily. J Lipid Res 2001; 42:1007-1017.
14. Dean M, Rzhetsky A, Allikmets R. The human ATP-binding cassette (ABC) transporter superfamily. Genome Res 2001; 11:1156-1166.
15. Decottignies A, Goffeau A. Complete inventory of the yeast ABC proteins. Nat Genet 1997; 15:137-145.
16. Klein I, Sarkadi B, Varadi A. An inventory of the human ABC proteins. Biochim Biophys Acta 1999; 1461:237-262.
17. Higgins CF. ABC transporters: from microorganisms to man. Annu Rev Cell Biol 1992; 8:67-113.
18. Dean M, Allikmets R. Complete characterization of the human ABC gene family. J Bioenerg Biomembr 2001; 33:475-479.
19. Walker JE, Saraste M, Runswick MJ et al. Distantly related sequences in the alpha- and beta-subunits of ATP synthase, myosin, kinases and other ATP-requiring enzymes and a common nucleotide binding fold. EMBO J 1982; 1:945-951.
20. Hyde SC, Emsley P, Hartshorn MJ et al. Structural model of ATP-binding proteins associated with cystic fibrosis, multidrug resistance and bacterial transport. Nature 1990; 346:362-365.
21. Borst P, Oude Elferink RPJ. Mammalian ABC transporters in health and disease. Annu Rev Biochem 2002; 71:537-592.

22. ABC Nomenclature Committee. Nomenclature for human ABC-transporter genes. http://www.gene.ucl.ac.uk/nomenclature/genefamily/abc.html. 2002
23. Allikmets R, Gerrard B, Hutchinson A et al. Characterization of the human ABC superfamily: isolation and mapping of 21 new genes using the expressed sequence tags database. Hum Mol Genet 1996; 5:1649-1655.
24. Broccardo C, Luciani M, Chimini G. The ABCA subclass of mammalian transporters. Biochim Biophys Acta 1999; 1461:395-404.
25. Kaminski WE, Wenzel JJ, Piehler A et al. ABCA6, a novel a subclass ABC transporter. Biochem Biophys Res Commun 2001; 285:1295-1301.
26. Luciani MF, Chimini G. The ATP binding cassette transporter ABC1, is required for the engulfment of corpses generated by apoptotic cell death. EMBO J 1996; 15:226-235.
27. Rust S, Rosier M, Funke H et al. Tangier disease is caused by mutations in the gene encoding ATP-binding cassette transporter 1. Nat Genet 1999; 22:352-355.
28. Brooks-Wilson A, Marcil M, Clee SM et al. Mutations in ABC1 in Tangier disease and familial high-density lipoprotein deficiency. Nat Genet 1999; 22:336-345.
29. Bodzioch M, Orso E, Klucken J et al. The gene encoding ATP-binding cassette transporter 1 is mutated in Tangier disease. Nat Genet 1999; 22:347-351.
30. Hamon Y, Broccardo C, Chambenoit O et al. ABC1 promotes engulfment of apoptotic cells and transbilayer redistribution of phosphatidylserine. Nat Cell Biol 2000; 2:399-406.
31. Chambenoit O, Hamon Y, Marguet D et al. Specific docking of apolipoprotein A-I at the cell surface requires a functional ABCA1 transporter. J Biol Chem 2001; 276:9955-9960.
32. Wang N, Silver DL, Thiele C et al. ATP-binding cassette transporter A1 (ABCA1) functions as a cholesterol efflux regulatory protein. J Biol Chem 2001; 276:23742-23747.
33. Tall AR, Costet P, Wang N. Regulation and mechanisms of macrophage cholesterol efflux. J Clin Invest 2002; 110:899-904.
34. Vaisman BL, Lambert G, Amar M et al. ABCA1 overexpression leads to hyperalphalipoproteinemia and increased biliary cholesterol excretion in transgenic mice. J Clin Invest 2001; 108:303-309.
35. Groen AK, Bloks VW, Bandsma RH et al. Hepatobiliary cholesterol transport is not impaired in Abca1-null mice lacking HDL. J Clin Invest 2001; 108:843-850.
36. Neufeld EB, Demosky SJ Jr, Stonik JA et al. The ABCA1 transporter functions on the basolateral surface of hepatocytes. Biochem Biophys Res Commun 2002; 297:974-979.
37. Langmann T, Porsch-Ozcurumez M, Heimerl S et al. Identification of sterol-independent regulatory elements in the human ATP-binding cassette transporter A1 promoter: role of Sp1/3, E-box binding factors, and an oncostatin M-responsive element. J Biol Chem 2002; 277:14443-14450.
38. Costet P, Luo Y, Wang N et al. Sterol-dependent transactivation of the ABC1 promoter by the liver X receptor/retinoid X receptor. J Biol Chem 2000; 275:28240-28245.
39. Schwartz K, Lawn RM, Wade DP. ABC1 gene expression and ApoA-I-mediated cholesterol efflux are regulated by LXR. Biochem Biophys Res Commun 2000; 274:794-802.
40. Repa JJ, Turley SD, Lobaccaro JA et al. Regulation of absorption and ABC1-mediated efflux of cholesterol by RXR heterodimers. Science 2000; 289:1524-1529.
41. Venkateswaran A, Laffitte BA, Joseph SB et al. Control of cellular cholesterol efflux by the nuclear oxysterol receptor LXR alpha. Proc Natl Acad Sci U S A 2000; 97:12097-12102.
42. Peet DJ, Janowski BA, Mangelsdorf DJ. The LXRs: a new class of oxysterol receptors. Curr Opin Genet Dev 1998; 8:571-575.
43. Thiebaut F, Tsuruo T, Hamada H et al. Cellular localization of the multidrug-resistance gene product P- glycoprotein in normal human tissues. Proc Natl Acad Sci U S A 1987; 84:7735-7738.
44. Fojo AT, Ueda K, Slamon DJ et al. Expression of a multidrug-resistance gene in human tumors and tissues. Proc Natl Acad Sci U S A 1987; 84:265-269.
45. Kamimoto Y, Gatmaitan Z, Hsu J et al. The function of Gp170, the multidrug resistance gene product, in rat liver canalicular membrane vesicles. J Biol Chem 1989; 264:11693-11698.
46. Buschman E, Arceci RJ, Croop JM et al. mdr2 encodes P-glycoprotein expressed in the bile canalicular membrane as determined by isoform-specific antibodies. J Biol Chem 1992; 267:18093-18099.
47. Silverman JA, Thorgeirsson SS. Regulation and function of the multidrug resistance genes in liver. Prog Liver Dis 1995; 13:101-123.
48. Silverman JA, Schrenk D. Hepatic canalicular membrane 4: expression of the multidrug resistance genes in the liver. FASEB J 1997; 11:308-313.
49. Ambudkar SV, Dey S, Hrycyna CA et al. Biochemical, cellular, and pharmacological aspects of the multidrug transporter. Annu Rev Pharmacol Toxicol 1999; 39:361-398.
50. Gottesman MM, Pastan I. Biochemistry of multidrug resistance mediated by the multidrug transporter. Annu Rev Biochem 1993; 62:385-427.

51. Schinkel AH, Smit JJM, van Tellingen O et al. Disruption of the mouse mdr1a P-glycoprotein gene leads to a deficiency in the blood-brain barrier and to increased sensitivity to drugs. Cell 1994; 77:491-502.
52. Schinkel AH, Mayer U, Wagenaar E et al. Normal viability and altered pharmacokinetics in mice lacking mdr1-type (drug-transporting) P-glycoproteins. Proc Natl Acad Sci U S A 1997; 94:4028-4033.
53. Schinkel AH, Wagenaar E, van Deemter L et al. Absence of the mdr1a p-glycoprotein in mice affects tissue distribution and pharmacokinetics of dexamethasone, digoxin, and cyclosporin A. J Clin Invest 1995; 96:1698-1705.
54. Schinkel AH. The physiological function of drug-transporting P-glycoproteins. Semin Cancer Biol 1997; 8:161-170.
55. Schinkel AH. Pharmacological insights from P-glycoprotein knockout mice. Int J Clin Pharmacol Ther 1998; 36:9-13.
56. Kerb R, Hoffmeyer S, Brinkmann U. ABC drug transporters: hereditary polymorphisms and pharmacological impact in MDR1, MRP1 and MRP2. Pharmacogenomics 2001; 2:51-64.
57. Brinkmann U, Eichelbaum M. Polymorphisms in the ABC drug transporter gene MDR1. Pharmacogenomics J 2001; 1:59-64.
58. Müller M. Transcriptional control of hepatocanalicular transporter gene expression. Semin Liver Dis 2000; 20:323-337.
59. Labialle S, Gayet L, Marthinet E et al. Transcriptional regulators of the human multidrug resistance 1 gene: recent views. Biochem Pharmacol 2002; 64:943-948.
60. Zhou G, Kuo MT. Wild-type p53-mediated induction of rat mdr1b expression by the anticancer drug daunorubicin. J Biol Chem 1998; 273:15387-15394.
61. Gant TW, Silverman JA, Bisgaard HC et al. Regulation of 2-acetylaminofluorene- and 3-methylcholanthrene-mediated induction of multidrug resistance and cytochrome P450IA gene family expression in primary hepatocyte cultures and rat liver. Mol Carcinog 1991; 4:499-509.
62. Teeter LD, Becker FF, Chisari FV et al. Overexpression of the multidrug resistance gene mdr3 in spontaneous and chemically induced mouse hepatocellular carcinomas. Mol Cell Biol 1990; 10:5728-5735.
63. Vos TA, Hooiveld GJEJ, Koning H et al. Up-regulation of the multidrug resistance genes, mrp1 and mdr1b, and down-regulation of the organic anion transporter, mrp2, and the bile salt transporter, spgp, in endotoxemic rat liver. Hepatology 1998; 28:1637-1644.
64. Vos TA, Ros JE, Havinga R et al. Regulation of hepatic transport systems involved in bile secretion during liver regeneration in rats. Hepatology 1999; 29:1833-1839.
65. Nakatsukasa H, Silverman JA, Gant TW et al. Expression of Multidrug Resistance Genes in Rat Liver During Regeneration and After Carbon Tetrachloride Intoxication. Hepatology 1993; 18:1202-1207.
66. Zhou G, Kuo MT. NF-kappaB-mediated induction of mdr1b expression by insulin in rat hepatoma cells. J Biol Chem 1997; 272:15174-15183.
67. Ros JE, Schuetz JD, Geuken M et al. Induction of Mdr1b expression by tumor necrosis factor-alpha in rat liver cells is independent of p53 but requires NF-kappaB signaling. Hepatology 2001; 33:1425-1431.
68. Smit JJM, Schinkel AH, Mol CAAM et al. Tissue distribution of the human MDR3 p-glycoprotein. Lab Invest 1994; 71:638-649.
69. Hooiveld GJEJ, Vos TA, Scheffer GL et al. 3-Hydroxy-3-methylglutaryl-coenzyme A reductase inhibitors (statins) induce hepatic expression of the phospholipid translocase mdr2 in rats. Gastroenterology 1999; 117:678-687.
70. Smit JJM, Schinkel AH, Oude Elferink RPJ et al. Homogenous disruption of the murine mdr2 P-glycoprotein gene leads to a complete absence of phospholipid from bile and to liver disease. Cell 1993; 75:451-462.
71. Borst P, Schinkel AH. What have we learnt thus far from mice with disrupted P- glycoprotein genes? Eur J Cancer 1996; 32A:985-990.
72. Ruetz S, Gros P. Phosphatidylcholine translocase: A physiological role for the *mdr2* gene. Cell 1994; 77:1071-1081.
73. Oude Elferink RPJ, Groen AK. Mechanisms of biliary lipid secretion and their role in lipid homeostasis. Semin Liver Dis 2000; 20:293-305.
74. van Helvoort A, Smith AJ, Sprong H et al. MDR1 P-glycoprotein is a lipid translocase of broad specificity, while MDR3 P-glycoprotein specifically translocates phosphatidylcholine. Cell 1996; 87:507-517.

75. Crawford JM, Moeckel G-M, Crawford AR et al. Imaging biliary lipid secretion in the rat: ultrastructural evidence for vesiculation of the hepatocyte canalicular membrane. J Lipid Res 1995; 36:2147-2163.

76. Oude Elferink RPJ, Tytgat GN, Groen AK. Hepatic canalicular membrane 1: The role of mdr2 P-glycoprotein in hepatobiliary lipid transport. FASEB J 1997; 11:19-28.

77. de Vree JML, Jacquemin E, Sturm E et al. Mutations in the MDR3 gene cause progressive familial intrahepatic cholestasis. Proc Natl Acad Sci U S A 1998; 95:282-287.

78. Miranda S, Vollrath V, Wielandt AM et al. Overexpression of mdr2 gene by peroxisome proliferators in the mouse liver. J Hepatol 1997; 26:1331-1339.

79. Chianale J, Vollrath V, Wielandt AM et al. Fibrates induce mdr2 gene expression and biliary phospholipid secretion in the mouse. Biochem J 1996; 314:781-786.

80. Desvergne B, Wahli W. Peroxisome proliferator-activated receptors: nuclear control of metabolism. Endocr Rev 1999; 20:649-688.

81. Frijters CMG, Ottenhoff R, van Wijland MJ et al. Influence of bile salts on hepatic mdr2 P-glycoprotein expression. Adv Enzyme Regul 1996; 36:351-63:351-363.

82. Frijters CMG, Ottenhoff R, van Wijland MJ et al. Regulation of mdr2 P-glycoprotein expression by bile salts. Biochem J 1997; 321:389-395.

83. Fickert P, Zollner G, Fuchsbichler A et al. Effects of ursodeoxycholic and cholic acid feeding on hepatocellular transporter expression in mouse liver. Gastroenterology 2001; 121:170-183.

84. Gupta S, Todd SR, Pandak WM et al. Regulation of multidrug resistance 2 P-glycoprotein expression by bile salts in rats and in primary cultures of rat hepatocytes. Hepatology 2000; 32:341-347.

85. Carrella M, Feldman D, Cogoi S et al. Enhancement of mdr2 gene transcription mediates the biliary transfer of phosphatidylcholine supplied by an increased biosynthesis in the pravastatin-treated rat. Hepatology 1999; 29:1825-1832.

86. Nishida T, Gatmaitan Z, Che M et al. Rat liver canalicular membrane vesicles contain an ATP-dependent bile acid transport system. Proc Natl Acad Sci U S A 1991; 88:6590-6594.

87. Müller M, Ishikawa T, Berger U et al. ATP-dependent transport of taurocholate across the hepatocyte canalicular membrane mediated by a 110-kDa glycoprotein binding ATP and bile salt. J Biol Chem 1991; 266:18920-18926.

88. Stieger B, O'Neill B, Meier PJ. ATP-dependent bile-salt transport in canalicular rat liver plasma-membrane vesicles. Biochem J 1992; 284:67-74.

89. Wolters H, Kuipers F, Slooff MJH et al. Adenosine triphosphate-dependent taurocholate transport in human liver plasma membranes. J Clin Invest 1992; 90:2321-2326.

90. Gerloff T, Stieger B, Hagenbuch B et al. The sister of P-glycoprotein represents the canalicular bile salt export pump of mammalian liver. J Biol Chem 1998; 273:10046-10050.

91. Lecureur V, Sun D, Hargrove P et al. Cloning and Expression of Murine Sister of P-Glycoprotein Reveals a More Discriminating Transporter Than MDR1/P-Glycoprotein. Mol Pharmacol 2000; 57:24-35.

92. Green RM, Hoda F, Ward KL. Molecular cloning and characterization of the murine bile salt export pump. Gene 2000; 241:117-123.

93. Noe J, Stieger B, Meier PJ. Functional expression of the canalicular bile salt export pump of human liver. Gastroenterology 2002; 123:1659-1666.

94. Byrne JA, Strautnieks SS, Mieli-Vergani G et al. The human bile salt export pump: characterization of substrate specificity and identification of inhibitors. Gastroenterology 2002; 123:1649-1658.

95. Stieger B, Fattinger K, Madon J et al. Drug- and estrogen-induced cholestasis through inhibition of the hepatocellular bile salt export pump (Bsep) of rat liver. Gastroenterology 2000; 118:422-430.

96. Funk C, Ponelle C, Scheuermann G et al. Cholestatic potential of troglitazone as a possible factor contributing to troglitazone-induced hepatotoxicity: in vivo and in vitro interaction at the canalicular bile salt export pump (Bsep) in the rat. Mol Pharmacol 2001; 59:627-635.

97. Fattinger K, Funk C, Pantze M et al. The endothelin antagonist bosentan inhibits the canalicular bile salt export pump: a potential mechanism for hepatic adverse reactions. Clin Pharmacol Ther 2001; 69:223-231.

98. Strautnieks SS, Bull LN, Knisely AS et al. A gene encoding a liver-specific ABC transporter is mutated in progressive familial intrahepatic cholestasis. Nat Genet 1998; 20:233-238.

99. Jansen PLM, Strautnieks SS, Jacquemin E et al. Hepatocanalicular bile salt export pump deficiency in patients with progressive familial intrahepatic cholestasis. Gastroenterology 1999; 117:1370-1379.

100. Wang R, Salem M, Yousef IM et al. Targeted inactivation of sister of P-glycoprotein gene (spgp) in mice results in nonprogressive but persistent intrahepatic cholestasis. Proc Natl Acad Sci U S A 2001; 98:2011-2016.

101. Lee JM, Trauner M, Soroka CJ et al. Expression of the bile salt export pump is maintained after chronic cholestasis in the rat. Gastroenterology 2000; 118:163-172.

102. Trauner M, Arrese M, Lee H et al. Endotoxin downregulates rat hepatic ntcp gene expression via decreased activity of critical transcription factors. J Clin Invest 1998; 101:2092-2100.
103. Trauner M, Meier PJ, Boyer JL. Molecular pathogenesis of cholestasis. N Engl J Med 1998; 339:1217-1227.
104. Gerloff T, Geier A, Stieger B et al. Differential expression of basolateral and canalicular organic anion transporters during regeneration of rat liver. Gastroenterology 1999; 117:1408-1415.
105. Ananthanarayanan M, Balasubramanian NV, Makishima M et al. Human bile salt export pump (BSEP) promoter is transactivated by the farnesoid X receptor/bile acid receptor (FXR/BAR). J Biol Chem 2001.
106. Plass JRM, Mol O, Heegsma J et al. Farnesoid X receptor and bile salts are involved in transcriptional regulation of the gene encoding the human bile salt export pump. Hepatology 2002; 35:589-596.
107. Gerloff T, Geier A, Roots I et al. Functional analysis of the rat bile salt export pump gene promoter. Eur J Biochem 2002; 269:3495-3503.
108. Chiang JY. Bile acid regulation of gene expression: roles of nuclear hormone receptors. Endocr Rev 2002; 23:443-463.
109. Paulusma CC, Bosma PJ, Zaman GJ et al. Congenital jaundice in rats with a mutation in a multidrug resistance- associated protein gene. Science 1996; 271:1126-1128.
110. Büchler M, König J, Brom M et al. cDNA cloning of the hepatocyte canalicular isoform of the multidrug resistance protein, cMRP, reveals a novel conjugte export pump deficient in hyperbilirubinemic mutant rats. J Biol Chem 1996; 271:15091-15098.
111. Kartenbeck J, Leuschner U, Mayer R et al. Absence of the canalicular isoform of the MRP gene-encoded conjugate export pump from the hepatocytes in Dubin-Johnson syndrome. Hepatology 1996; 23:1061-1066.
112. Paulusma CC, Kool M, Bosma PJ et al. A mutation in the human canalicular multispecific organic anion transporter gene causes the Dubin-Johnson syndrome. Hepatology 1997; 25:1539-1542.
113. Wada M, Toh S, Taniguchi K et al. Mutations in the canilicular multispecific organic anion transporter (cMOAT) gene, a novel ABC transporter, in patients with hyperbilirubinemia II/Dubin-Johnson syndrome. Hum Mol Genet 1998; 7:203-207.
114. Ito K, Suzuki H, Hirohashi T et al. Molecular cloning of canalicular multispecific organic anion transporter defective in EHBR. Am J Physiol 1997; 272:G16-22.
115. Jansen PLM, Groothuis GMM, Peters WHM et al. Selective hepatobiliary transport defect for organic anions and neutral steroids in mutant rats with hereditary conjugated hyperbilirubinemia. Hepatology 1987; 7:71-76.
116. Takikawa H, Sano N, Narita T et al. Biliary excretion of bile acid conjugates in a hyperbilirubinemic mutant sprague-dawley rat. Hepatology 1991; 14:352-360.
117. Trauner M, Arrese M, Soroka CJ et al. The rat canalicular conjugate export pump (Mrp2) is down-regulated in intrahepatic and obstructive cholestasis. Gastroenterology 1997; 113:255-264.
118. Ng KW, Zhou H, Manji S et al. Regulation and regulatory role of the retinoids. Crit Rev Eukaryot Gene Expr 1995; 5:219-253.
119. Denson LA, Auld KL, Schiek DS et al. Interleukin-1β suppresses retinoid transactivation of two hepatic transporter genes involved in bile formation. J Biol Chem 2000; 275:8835-8843.
120. Denson LA, Bohan A, Held MA et al. Organ-specific alterations in RAR alpha:RXR alpha abundance regulate rat Mrp2 (Abcc2) expression in obstructive cholestasis. Gastroenterology 2002; 123:599-607.
121. Kauffmann HM, Keppler D, Kartenbeck J et al. Induction of cMrp/cMoat gene expression by cisplatin, 2- acetylaminofluorene, or cycloheximide in rat hepatocytes. Hepatology 1997; 26:980-985.
122. Cui Y, Konig J, Buchholz JK et al. Drug resistance and ATP-dependent conjugate transport mediated by the apical multidrug resistance protein, MRP2, permanently expressed in human and canine cells. Mol Pharmacol 1999; 55:929-937.
123. Tanaka T, Uchiumi T, Hinoshita E et al. The human multidrug resistance protein 2 gene: Functional characterization of the 5'-flanking region and expression in hepatic cells. Hepatology 1999; 30:1507-1512.
124. Stockel B, Konig J, Nies AT et al. Characterization of the 5'-flanking region of the human multidrug resistance protein 2 (MRP2) gene and its regulation in comparison withthe multidrug resistance protein 3 (MRP3) gene. Eur J Biochem 2000; 267:1347-1358.
125. Dussault I, Forman BM. The nuclear receptor PXR: a master regulator of "homeland" defense. Crit Rev Eukaryot Gene Expr 2002; 12:53-64.
126. Tzameli I, Moore DD. Role reversal: new insights from new ligands for the xenobiotic receptor CAR. Trends Endocrinol Metab 2001; 12:7-10.

127. Kast HR, Goodwin B, Tarr PT et al. Regulation of multidrug resistance-associated protein 2 (MRP2;ABCC2) by the nuclear receptors PXR, FXR, and CAR. J Biol Chem 2001.
128. Chen H, Rossier C, Lalioti MD et al. Cloning of the cDNA for a human homologue of the Drosophila white gene and mapping to chromosome 21q22.3. Am J Hum Genet 1996; 59:66-75.
129. Croop JM, Tiller GE, Fletcher JA et al. Isolation and characterization of a mammalian homolog of the Drosophila white gene. Gene 1997; 185:77-85.
130. Klucken J, Buchler C, Orso E et al. ABCG1 (ABC8), the human homolog of the drosophila white gene, is a regulator of macrophage cholesterol and phospholipid transport. Proc Natl Acad Sci U S A 2000; 97:817-822.
131. Doyle LA, Yang W, Abruzzo LV et al. A multidrug resistance transporter from human MCF-7 breast cancer cells. Proc Natl Acad Sci U S A 1998; 95:15665-15670.
132. Miyake K, Mickley L, Litman T et al. Molecular cloning of cDNAs which are highly overexpressed in mitoxantrone-resistant cells: demonstration of homology to ABC transport genes. Cancer Res 1999; 59:8-13.
133. Maliepaard M, Scheffer GL, Faneyte IF et al. Subcellular localization and distribution of the breast cancer resistance protein transporter in normal human tissues. Cancer Res 2001; 61:3458-3464.
134. Jonker JW, Smit JW, Brinkhuis RF et al. Role of breast cancer resistance protein in the bioavailability and fetal penetration of topotecan. J Natl Cancer Inst 2000; 92:1651-1656.
135. Zhou S, Morris JJ, Barnes Y et al. Bcrp1 gene expression is required for normal numbers of side population stem cells in mice, and confers relative protection to mitoxantrone in hematopoietic cells in vivo. Proc Natl Acad Sci U S A 2002; 99:12339-12344.
136. Jonker JW, Buitelaar M, Wagenaar E et al. The breast cancer resistance protein protects against a major chlorophyll-derived dietary phototoxin and protoporphyria. Proc Natl Acad Sci U S A 2002; 99:15649-15654.
137. Berge KE, Tian H, Graf GA et al. Accumulation of dietary cholesterol in sitosterolemia caused by mutations in adjacent ABC transporters. Science 2000; 290:1771-1775.
138. Lee MH, Lu K, Hazard S et al. Identification of a gene, ABCG5, important in the regulation of dietary cholesterol absorption. Nat Genet 2001; 27:79-83.
139. Salen G, Shefer S, Nguyen L et al. Sitosterolemia. J Lipid Res 1992; 33:945-955.
140. Bjorkhem I, Boberg K, Leitersdorf E: Inborn errors in bile acid biosynthesis and storage of sterols other than cholesterol. Edited by Scriver C, Beaudet A, Sly W, and Valle D. The metabolic and molecular basis of inherited diseases. New York, New York, USA, McGraw-Hill, 2001:2961-2988.
141. Repa JJ, Berge KE, Pomajzl C et al. Regulation of ATP-binding cassette sterol transporters ABCG5 and ABCG8 by the liver X receptors alpha and beta. J Biol Chem 2002; 277:18793-18800.
142. Yu L, Li-Hawkins J, Hammer RE et al. Overexpression of ABCG5 and ABCG8 promotes biliary cholesterol secretion and reduces fractional absorption of dietary cholesterol. J Clin Invest 2002; 110:671-680.
143. Graf GA, Li WP, Gerard RD et al. Coexpression of ATP-binding cassette proteins ABCG5 and ABCG8 permits their transport to the apical surface. J Clin Invest 2002; 110:659-669.
144. Yu L, Hammer RE, Li-Hawkins J et al. Disruption of Abcg5 and Abcg8 in mice reveals their crucial role in biliary cholesterol secretion. Proc Natl Acad Sci U S A 2002; 99:16237-16242.

CHAPTER 4

Gap Junctions in the Liver

Wihelma Echevarría and Michael H. Nathanson

Summary

Gap junctions are hexameric hemichannels that are inserted into the plasma membrane and allow for direct exchange of cytosolic contents among adjacent cells. Connexin 26 and connexin 32 are the specific types of gap junctions found in hepatocytes, and connexin 43 is the predominant gap junction in bile duct epithelia. Intercellular communication of second messengers via gap junctions permits integrated and coordinated responses to hormonal and neural stimuli. This in turn permits regulation of certain liver-specific functions, such as glucose release, bile secretion, hepatic regeneration, and tumor suppression. Alterations in gap junction expression thus may contribute to pathologic conditions such as cholestasis and carcinogenesis.

Introduction

In the liver, cells communicate with each other through various mechanisms. The principal route of communication among hepatocytes is via macula communicans, or gap junctions. These are channels that are formed between adjacent hepatocytes, and allow the direct transfer of molecules from cell to cell. Intercellular communication via gap junctions allows coupled hepatocytes to coordinate their response to extracellular signals. In the past decade much has been learned about the structure and function of gap junctions. Here we will review what is known about the molecular biology and physiology of gap junctions in the liver, and the role of gap junctions in hepatic function in normal and disease states.

Molecular Overview of the Gap Junction

Gap junctions are composed of two paired hemichannels, or connexons, located in the plasma membranes of juxtaposed cells. Connexons from adjacent cells align end-to-end to form intercellular pores, or channels. Each connexon is in turn composed of six subunits, or connexins, that co-align in a cylindrical fashion (Figure 1).[1-3] Connexons can be homomeric (in which all six connexins are identical), or heteromeric (composed of two different kinds of connexins).[4-6] Gap junctions may be either homotypic (in which both connexons are the same), or heterotypic (consisting of two different connexons).[7,8] To date, at least 20 different connexins have been characterized in vertebrates.[9] The most commonly used nomenclature to distinguish among connexins (Cx) is based on their predicted molecular weight.[10] For example, Cx32, the first gap junction protein to be isolated,[11,12] has a predicted molecular weight of 32 kD. Most tissues express more than one kind of connexin, and in the case of the liver, both Cx32[11,12] and Cx26[13] have been identified. The presence of these two connexins in the liver allows for the formation of Cx32/Cx26 heterotypic channels.[14] Structural differences among gap junctions determine their permeability to certain molecules. For example, experiments using HeLa cells transfected with one of several different connexins showed that there are connexin-specific differences in permeability to various tracer dyes.[15] As a result, negatively charged lucifer yellow is transmitted equally well across Cx26 and Cx32, whereas positively charged propidium

Molecular Pathogenesis of Cholestasis, edited by Michael Trauner and Peter L.M. Jansen.
©2004 Eurekah.com and Kluwer Academic / Plenum Publishers.

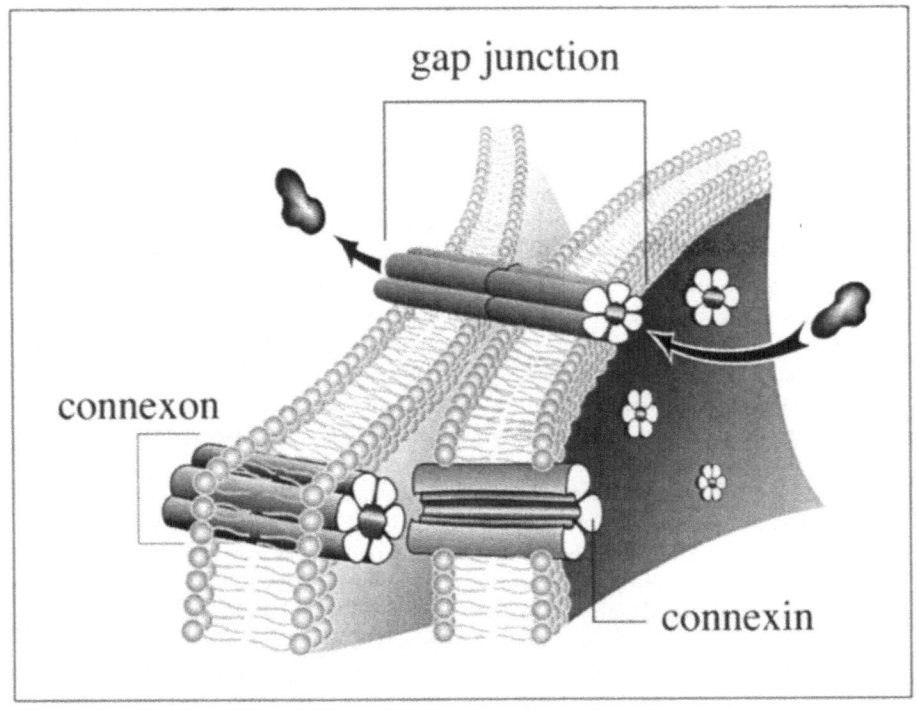

Figure 1. Basic structure of the gap junction. Each intercellular gap junction channel is formed by the pairing of connexon hemichannels, located in the lateral plasma membrane of adjacent hepatocytes. Each connexon in turn is made up of six connexins.

iodide can cross Cx26 but not Cx32. Recent studies in which chimeras of Cx26/Cx30 were transfected into HeLa cells demonstrated that there are differences in diffusion properties and voltage gating of the channels when compared to the properties of cells transfected with the homotypic Cx26 or Cx30 channels.[16] These studies suggest that gap junctions may even show selective permeability to second messengers, and that this specificity depends on the molecular structure of the messenger molecule. For instance, homomeric Cx32 channels in the liver are permeable to both cAMP and cGMP, while heteromeric Cx32/Cx26 channels are permeable to cGMP but not cAMP.[17]

Gap junction plaques comprise approximately 3% of the total hepatocyte surface area.[18,19] Electron microscopic studies have shown that these plaques have a thickness of 15 to 18 nm, and that the extracellular gap between adjacent cell membranes is 2 to 4 nm.[9,20] Molecules of less than 1kD in size typically can be transferred from hepatocyte to hepatocyte by moving through gap junctions.[21-23] Immunostaining of rat liver sections demonstrates that Cx32 is expressed among hepatocytes throughout the hepatic lobule. However, the expression of Cx26 is selectively increased in the periportal region.[24,25] Cholangiocytes appear to express Cx43 predominantly.[26,27]

The formation of a fully functional gap junction is achieved through a series of steps. First, connexins assemble to form a hemichannel, or connexon, while being trafficked from the endoplasmic reticulum to the Golgi.[28] The connexon is then translocated to the plasma membrane, where it joins with that of an adjacent cell, forming a channel, or gap junctional plaque.[29,30] Cx26 is the only connexin that for the most part bypasses the Golgi apparatus for this translocation process.[28] Gap junctions have a very rapid turnover.[31] In fact, Cx32 channels are degraded 1.5 hours after their formation.[24] However, this rate of turnover can be prolonged by increased intracellular concentrations of cAMP[32], and it has been postulated that this second messenger increases the lifetime of the mRNA that encodes for Cx32. cAMP may also act by decreasing the rate of degradation of the gap junction plaque.[32]

The connexin protein is made up of four transmembrane, two extracellular, and three cytoplasmic domains. The cytoplasmic domains are in the amino terminal region of the protein. There also is a loop between two of the membrane spanning domains and the carboxy terminal tail region, which is located in the cytoplasm.[3,33] Most connexins have homologous extracellular domains, which could explain the formation of heterotypic channels. However, most heterogeneity among connexins occurs in the carboxy terminal tail region. Most connexins have phosphorylation sequences in this region, but this is not the case for Cx26, presumably due to its short C-terminal tail region.[24] Expression studies using connexin mutants with truncated C-terminal regions have shown that these phosphorylation sites determine the permeability properties of the channels and modulate the level of activity of the gap junctions.[34]

Abnormal gap junction formation has been linked to disease states. Charcot-Marie-Tooth Disease is a group of disorders characterized by progressive demyelinating peripheral neuropathy. There are several subclasses of this disease, with varying inheritance patterns and phenotypic presentations. The X-linked subclass (CMTX) is associated with point mutations in the gene GJB1, which encodes the Cx32 protein.[35] Over 160 GJB1 mutations have been identified in CMTX patients.[36,37] Electrophysiologic studies in Cx32-deficient mice show that these mice develop a progressive peripheral neuropathy similar to that seen in CMTX patients.[38,39] Studies in HeLa cells transfected with various Cx32 mutations showed that the genetic defect associated with the most severe phenotype of CMTX results in incorrect trafficking of Cx32 to the plasma membrane.[40] Even though some of these mutations prevent gap junction formation and trafficking,[40-44] expression of other mutant forms of Cx32 proteins in Xenopus oocytes do not interfere with the formation of functional channels.[41,45-47] Instead, it has been found that such mutations cause a reduction in the channel's open probability,[44,45] diameter,[37,45] or sensitivity to changes in pH,[46] and therefore might alter its permeability to certain messenger molecules. These findings provide important insights to the molecular pathophysiology of diseases that result from altered gap junction function.

The Role of Gap Junctions in Cell-to-Cell Signaling

An important role of gap junctions is to permit cell-to-cell signaling by direct exchange of cytosolic second messengers. This has been studied in detail in the liver, mostly by examining the role of hepatic gap junctions in regulating intercellular Ca^{2+} signaling. Microinjection studies in freshly isolated rat hepatocyte couplets and triplets showed that liver gap junctions are permeable to the second messengers Ca^{2+} and inositol 1,4,5-trisphosphate (InsP3).[48] Moreover, there is an absolute requirement for gap junctions to be expressed in order for an intercellular spread of Ca^{2+} waves to occur. For example, expression of Cx43 in C6 glioma cells, which lack functional gap junctions, permits Ca^{2+} waves to spread among adjacent cells.[49] Similarly, in HEK293 cells transfected with ryanodine receptors, caffeine-induced Ca^{2+} signals only spread from cell to cell if Cx43 also is expressed. This Ca^{2+} signal then can be abolished by the gap junction blockers octanol or doxyl stearic acid (DSA).[50] The degree of permeability to second messengers varies among connexins. For example, the relative permeability of Cx32, Cx26, and Cx43 to InsP3 and Ca^{2+} was examined in HeLa cells in which InsP3 was microinjected and cell-to-cell Ca^{2+} waves were analyzed. These experiments showed that InsP3-induced Ca^{2+} waves spread a distance that was 3- to 4-fold greater in Cx32 transfectants than in Cx26 cells, and 2.5-fold greater than in Cx43 cells.[51] These findings are consistent with previous work in HeLa cells and Xenopus oocytes transfected with Cx32 or Cx26, in which channels formed by Cx32 were 6 to 9 times more permeable to lucifer yellow than those formed by Cx26.[52] This diversity in permeability among connexins could explain why most tissues express more than one form of these proteins.

The role of Cx32 in regulating intercellular signaling in the liver has also been examined using Cx32-deficient mice.[53] Microinjection studies in hepatocyte couplets isolated from these mice suggest that Cx32-deficient livers require a 25-fold higher InsP3 concentration than couplets from wild type livers to mediate a coordinated intercellular Ca^{2+} wave.[54] Interestingly, these mice show decreased protein expression of the only other known gap junction in hepatocytes, Cx26.[53] This may suggest that Cx32 serves a stabilizing role for the Cx26 protein.

Cell-to-cell signaling via gap junctions provides a mechanism to integrate and coordinate the response of hepatocytes to hormones. For example, studies in isolated rat hepatocyte couplets showed that stimulation with the alpha-adrenergic agonist phenylephrine induces oscillations that are synchronized from cell to cell (Figure 2), and stimulation with vasopressin induces a Ca^{2+} wave that starts in one hepatocyte of a couplet and then spreads without delay to the other cell.[56] Treatment with octanol causes a delay in the spread of Ca^{2+} waves from cell to cell, so that hepatocytes become functionally uncoupled.[56] Moreover, microperfusion of an individual hepatocyte with vasopressin induces a Ca^{2+} signal that spreads to its neighbor without exposing the neighboring cell to vasopressin. Thus, the spread of hormone–induced Ca^{2+} waves is highly organized and coordinated by gap junctions.[56] Increasing the affinity of the InsP3 receptor to InsP3 with thimerosal or 8-Bromo-cAMP causes further synchronization of Ca^{2+} oscillations among communicating hepatocytes. Furthermore, microinjecting the middle cell of a triplet with InsP3 5-phosphatase, which degrades InsP3, abolishes this synchronization.[55] Thus, coordination of Ca^{2+} waves among hepatocytes appears to be mediated by the passage of InsP3 rather than Ca^{2+} through gap junctions.

The synchronized spread of hormone-induced Ca^{2+} waves has been documented not only in isolated hepatocyte couplets, but also in intact perfused rat livers.[57-59] In the liver the vasopressin V1a receptor is concentrated in the pericentral region, with minimal expression in the periportal area.[57] Consequently, vasopressin-induced Ca^{2+} signals spread from pericentral to periportal regions of the liver,[57,59] in the same direction as bile flow.[60] Vasopressin can induce Ca^{2+} oscillations in isolated hepatocytes, and in the perfused liver these oscillations also occur and progress from cell to cell along cords of hepatocytes.[58,59] The lobular pattern of Ca^{2+} waves depends upon the agonist, however. For example, unlike vasopressin, ATP-induced signals begin randomly throughout the hepatic lobule.[59] Studies in populations of isolated pericentral and periportal hepatocytes suggest that this reflects differences in sensitivity among hepatocytes to each agonist.[61] The molecular basis for the organization of cell-to-cell Ca^{2+} waves has been examined in further detail in isolated hepatocyte couplets and triplets. When microinjected caged InsP3 is photoreleased in one hepatocyte of a couplet or triplet, the resulting Ca^{2+} signal spreads simultaneously among connected hepatocytes.[62] Stimulation with vasopressin instead induces Ca^{2+} waves that spread from cell to cell in an organized fashion. Taken together, these data suggest that three factors are necessary to organize Ca^{2+} waves in the liver.[63] First, functional gap junctions must be present, since disrupting such communication causes Ca^{2+} oscillations to become asynchronous. Second, all connected hepatocytes must undergo hormonal stimulation, so that each cell has an elevated concentration of InsP3. Finally, one of the hepatocytes must have increased sensitivity to the agonist relative to its neighbor, and thus serve as a "pacemaker".[62,63]

An alternative non-junctional pathway of intercellular Ca^{2+} signaling has been demonstrated in liver[64-66] and other tissues.[67,68] This pathway first was demonstrated in rat basophilic leukemia cells, where it was shown that individual cells can secrete ATP, which then increases Ca^{2+} in neighboring cells by activating extracellular nucleotide receptors.[67] Mechanical stimulation of isolated rat hepatocytes similarly increases Ca^{2+} not only in coupled hepatocytes, but also in non-adjacent hepatocytes and bile duct cells. This signaling to distant hepatocytes or bile duct cells is abolished by blocking nucleotide receptors or by hydrolyzing extracellular ATP, suggesting that hepatocytes also communicate with each other and with bile duct cells via the release of nucleotides.[64] Furthermore, bile duct cells possess apical purinergic receptors,[65] and ATP is found in bile.[69] Biliary ATP not only induces Ca^{2+} signals in bile duct cells, but also activates

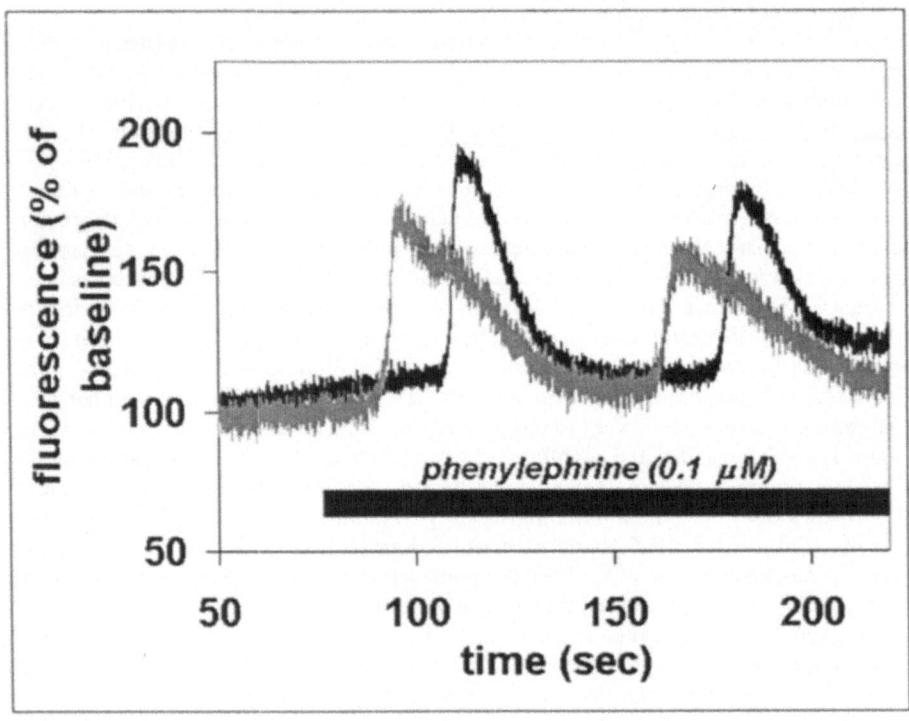

Figure 2. Ca^{2+} oscillations are synchronized in isolated rat hepatocyte couplets. These tracings were obtained by time lapse confocal microscopic examination of a couplet loaded with the Ca^{2+}-sensitive fluorescent dye fluo-4 and stimulated with the alpha-1B-adrenergic agonist phenylephrine.

ductular bicarbonate secretion.[65] Thus, secretion of ATP into bile may provide a novel non-gap junction route by which hepatocytes regulate Ca^{2+} signaling and bicarbonate secretion in bile duct cells that are downstream. Interestingly, the therapeutic bile acid ursodeoxycholic acid (UDCA) and its conjugated form, tauroursodeoxycholic acid (TUDCA) both stimulate hepatocytes to secrete ATP into bile. Direct stimulation of biliary epithelia with TUDCA does not cause an increase in cytosolic Ca^{2+}, whereas stimulation with ATP does.[66] These findings suggest that UDCA might mediate its therapeutic effects on cholangiocytes in part by inducing ATP release from hepatocytes, thus stimulating ductular bile flow and bicarbonate secretion in a paracrine fashion.[66]

Gap junctional and paracrine ATP-mediated intercellular communication may in fact be related. HeLa, C6 glioma, and U373-MG cells transfected with Cx43, 32, or 26 release 10 times more ATP than wild type cells.[70] Thus, it has been proposed that connexins enhance ATP release. However, further studies are needed to explain how and when connexins regulate the release of ATP.

Regulation of Liver Function by Gap Junctions

Glucose Metabolism

There is both indirect and direct evidence that gap junctions play a role in regulating glucose release from the liver. Studies with perfused rat livers showed that 24 to 48 hours after partial hepatectomy, there is a 25% decrease in Cx32 protein levels, and almost complete inhibition of glucose release induced by hepatic nerve stimulation. Once Cx32 protein levels return to baseline, the liver resumes normal glucose output in response to neural stimuli.[71] Furthermore, when gap junctions are blocked with heptanol, carbenoxolone, or 4-beta-PMA, there is complete or partial inhibition of glucose output in response to nerve stimulation.[72] Similarly, perfused livers of Cx32-deficient mice release 78% less glucose than wild type livers after electrical stimulation.[53] Taken together, these studies suggest that the impaired glucose release that is seen after liver damage may in part be due to a decrease in the number of gap junctions, and therefore a reduction in signal propagation across hepatocytes in response to sympathetic nerve stimulation.

Gap junctions also regulate hormone-induced glucose release from the liver. In perfused mouse liver studies, no difference in the rate of glucose release was seen between wild type and Cx32-deficient livers treated with saturating concentrations of glucagon or norepinephrine. However, glucose output was significantly reduced in Cx32-deficient livers when compared to normal livers perfused with physiologic concentrations of these hormones.[73] Furthermore, in isolated rat hepatocyte studies, vasopressin-induced glycogenolysis decreased by 70% in cells in which gap junction intercellular communication was impaired with octanol or 18α glycyrrhetinic acid (αGA). Likewise, vasopressin-induced glycogen breakdown was reduced by 50% if hepatocyte reaggregation was impaired or if gap junction formation was blocked during reaggregation.[74] Similarly, in isolated perfused rat livers αGA caused a significant decrease in glucose release in response to glucagon or vasopressin.[75] Together, these studies suggest that intercellular communication via gap junctions plays a crucial role in facilitating hormone- and neural-induced glycogenolysis and glucose release.

Bile Secretion

Several recent studies have defined the manner in which gap junctions contribute to the regulation of bile secretion. Gap junctions do not appear to affect the basal rate of bile secretion, since bile flow in the isolated perfused rat liver is not altered by αGA.[75] Similarly, basal bile flow is not altered in the Cx32 knockout mouse.[76] However, the cAMP agonist glucagon increases bile flow, and this effect is abolished by αGA. In addition, the Ca^{2+} agonist vasopressin decreases bile flow, and this inhibition is exacerbated by αGA.[75] Similar to the vasopressin receptor, the glucagon receptor is concentrated in the pericentral region,[57,77] and this may be important for αGA to exert its inhibitory effect. For example, dibutyryl cAMP activates protein kinase A in a receptor-independent fashion in liver, and its choleretic effect is not altered by αGA. Similarly, αGA does not alter the cholestatic effect of tBuBHQ, which increases Ca^{2+} throughout the hepatic lobule in a receptor-independent fashion.[75] Although the principal effect of vasopressin is to decrease bile flow, it initially induces a highly transient increase in bile flow,[78] which has been attributed to a wave of canalicular contraction that is driven by a pericentral-to-periportal Ca^{2+} wave.[57] Pretreatment of rats with vasopressin can abolish the vasopressin receptor gradient across the hepatic lobule, and attenuates this transient choleretic effect.[79] Moreover, this choleretic effect is not induced by ATP, which as mentioned above, increases Ca^{2+} uniformly across the hepatic lobule, rather than in a pericentral-to-periportal fashion.[55,59] These findings suggest that gap junctions modulate hormone-mediated changes in bile flow, but only when gradients in hormone receptor expression would lead to gradients in second messengers across the hepatic lobule. Gap junctions also appear to be important for neurally mediated changes in bile flow, since the cholestatic effect of sympathetic nerve stimulation is dramatically attenuated in the Cx32 knockout mouse.[76] Thus, gap junctions are necessary for integration of second messenger signals across the hepatic lobule, which in turn is necessary for physiological regulation of bile secretion.

Tumor Suppression and Hepatocellular Growth

Gap junction intercellular communication plays a role in the regulation of hepatocellular growth, and in the suppression of tumor formation. Most malignant cells lack gap junctions, and also lack the ability to control their spatial spread and growth.[80,81] Several lines of evidence have documented that gap junctions play a role in tumor suppression.[82-88] Studies in which Cx43 was overexpressed in mouse 10T1/2 cells, which lack gap junctions, showed that the growth rate of these cells was reduced.[89] These findings are in agreement with studies in which the human hepatoma cell line SKHep1 was stably transfected with Cx32. Cells were injected into athymic nude mice, and the growth rate of the cells in which gap junctions were expressed was markedly inhibited compared to wild type cells.[82] There also is a significant increase in the number of diethylnitrosamine (DEN)-induced liver tumors in Cx32-deficient mice when compared to wild type mice.[83] Moreover, Cx32-deficient male and female mice have a 25-fold and 8-fold, respectively, higher incidence of developing spontaneous liver tumors than their wild type counterparts.[83] These findings suggest that loss of gap junction intercellular communication increases the chance the likelihood of hepatocellular carcinoma. These results are in agreement with the observation that rat hepatocellular carcinomas have a 71% decrease in gap junctions, as determined by immunofluorescence and immunoblot studies.[84] Similarly, immunohistochemical analysis of resected human hepatocellular carcinomas showed that the expression of Cx32 is decreased significantly relative to the surrounding normal tissue.[85] This decrease in gap junction expression also is seen in early chemically–induced preneoplastic nodules,[86] suggesting that the loss of gap junction intercellular communication occurs during the early stages of hepatic carcinogenesis. Furthermore, the liver tumor promoting agent phenobarbital reduces gap junction expression without affecting the expression of two genes that are often used as markers of tumor formation, glutathione S-transferase (placental form), and gamma-glutamyl transpeptidase.[87] Treatment with the carcinogens DEN, 2-acetylaminofluorene or N-ethyl-N-hydroxyethylnitrosamine also reduces Cx32 mRNA levels.[86] These findings suggest that these tumor-promoting agents may act in part by impairment of gap junction intercellular communication, and that connexins serve an anticarcinogenic function in the liver. This anti-tumorigenic role is not limited to Cx32 and 26. In fact, the forced expression of Cx43, the connexin found in cholangiocyte gap junctions, into tumorigenic mouse MCA-10 cells suppresses tumorigenicity of these cells when injected into nude mice.[88] The fact that restoring gap junction expression re-establishes normal cell growth further supports the idea that connexins act as tumor suppressors. Moreover, various oncogenes down-regulate gap junction intercellular communication. For instance, co-transfection of the myc and ras oncogenes into the rat liver epithelial cell line WB-F344 induced a loss of functional channels and of normal growth regulation. Furthermore, gap junction intercellular communication is significantly impaired, and tumorigenicity is induced in rats into which transfected cells were injected.[89] Conversely, certain chemotherapy agents can up-regulate gap junction intercellular communication, such as the ras oncogene inhibitor lovastatin. In studies in which tumorigenic WB-ras cells were treated with this agent, there was a reversal of the neoplastic phenotype and a restoration of gap junction intercellular communication.[90] These studies indicate the critical role of gap junctions in controlling cell growth.

The marked regenerative capacity of the liver has been widely studied. Following an insult such as partial hepatectomy or exposure to hepatotoxic agents, the remaining intact hepatocytes undergo an accelerated growth process.[91] Within the first 24 hours after such an insult, the remaining healthy hepatocytes lose their gap junctions,[92] perhaps as a means to be isolated from neighboring damaged cells and to protect the liver from further injury. Thus gap junctions may be instrumental in regulating growth of the liver in a range of normal and abnormal conditions.

Altered Gap Junction Expression in Cholestasis

Gap junction function and expression levels are altered in cholestasis and other types of liver damage. In the rat, gap junction expression is decreased after partial hepatectomy as discussed above.[71,93] Decreased expression also occurs during acute inflammation and ischemia.[94,95] Further studies have examined the mechanisms by which inflammation alters gap junction expression and intercellular communication in the liver. Immunoblot, Northern blot, and immunofluorescence analyses of mouse embryonic hepatocytes incubated for 24 hours with different inflammatory cytokines (mIL-1, mIL-6, and TNFα) revealed no significant decrease in Cx32 mRNA levels. However, these cytokines induced a five-fold reduction in Cx32 protein levels, whereas there was a two-fold increase in Cx26 mRNA and protein levels. The decrease in Cx32 levels translated into a significant (70%) reduction in gap junction intercellular communication, determined by intercellular transfer of microinjected lucifer yellow.[96] Thus, the partial up-regulation of Cx26 levels does not compensate for the decreased communication among hepatocytes that results from loss of Cx32. Recent studies in which endotoxin was injected into mice in order to induce an acute phase response also showed a significant decrease in Cx32 expression. Interestingly, injection of the pro-inflammatory cytokines mIL-1, mIL-6, and TNFα into wild type mice did not decrease Cx32 mRNA in liver, but nevertheless reduced Cx32 protein levels by 40-70%.[97] One possible explanation for these findings is that post-transcriptional modification of Cx32 leads to loss of gap junctions during liver damage, and that this is the first response of the liver to injury, even before initiation of an acute phase response.

The effects of obstructive cholestasis on gap junctions have been examined as well. After inducing cholestasis by common bile duct ligation (CBDL), Cx32 mRNA and protein levels in the rat liver decrease and remain low for 14 days.[98] Although Cx26 mRNA levels rise during this period, the expression of Cx26 is initially reduced at the protein level as well. Furthermore, studies with isolated rat hepatocytes after CBDL showed that even in the early stages of liver damage, hormone-induced Ca^{2+} signals became asynchronous when compared to controls.[98] Cx32 expression is reduced during chronic liver disease in humans.[99] This includes patients with chronic viral hepatitis, autoimmune hepatitis, and alcoholic cirrhosis. No changes in Cx26 expression occur in any of these patients, though. Curiously, gap junction expression is not decreased in estrogen cholestasis.[100] In fact, expression of Cx26 is increased in an animal model of this disorder.[100] On the other hand, liver dysfunction during cholestasis and other forms of liver disease may be due in part to impaired communication among hepatocytes caused by a decrease in gap junctions. It is possible that this impairment in gap junction intercellular communication during liver damage serves as a way to isolate, and in this way protect, normal hepatocytes from affected cells.

Conclusions

Liver homeostasis requires a coordinated effort among hepatocytes. One of the ways this is accomplished is via the gap junction. Intercellular communication via gap junctions plays a role in the regulation of intercellular signaling, and facilitates liver functions such as glucose metabolism and bile secretion. Gap junctions also are responsible for maintaining appropriate hepatocellular growth and thus suppressing carcinogenesis. Gap junction expression and function are impaired during cholestasis and other types of liver disease, which may contribute to the pathophysiology of these disorders. It is likely that the use of genetically altered mice and other investigational tools will continue to increase our understanding of the ways in which gap junctions regulate liver function in health and disease.

Acknowledgements

This work was supported by an AASLD-Schering Advanced Hepatology Fellowship Award and a GIDH Basic Science Research Award (to WE) and NIH grants DK45710 and DK57751 (to MHN).

References

1. Revel JP, Karnovsky MJ. Hexagonal array of subunits in intercellular junctions of the mouse heart and liver. J Cell Biol 1967; 33:C7-C12.
2. Goodenough DA, Stoeckenius W. The isolation of mouse hepatocyte gap junctions. Preliminary chemical characterization and x-ray diffraction. J Cell Biol 1972; 54:646-56.
3. Yeager M, Unger VM, Falk MM. Synthesis, assembly and structure of gap junction intercellular channels. Curr Opin Struct Biol 1998; 8:517-24.
4. Swenson KI, Jordan JR, Beyer EC et al. Formation of gap junctions by expression of connexins in Xenopus oocyte pairs. Cell 1989; 7(57):145-55.
5. Werner R, Levine E, Rabadan-Diehl C et al. Formation of hybrid cell-cell channel. Proc Natl Acad Sci USA 1989; 86:5380-4.
6. Stauffer KA. The gap junction proteins beta 1-connexin (connexin-32) and beta 2-connexin (connexin-26) can form heteromeric hemichannels. J Biol Chem 1995; 270:6768-72.
7. Rubin JB, Verselis VK, Bennett MV et al. Molecular analysis of voltage dependence of heterotypic gap junctions formed by connexins 26 and 32. Biophys J 1992; 62:183-93
8. Valiunas V, Niessen H, Willecke K et al. Electrophysiological properties of gap junction channels in hepatocytes isolated from connexin32-deficient and wild type mice. Pflugers Arch 1999; 437:846-56
9. Kojima T, Sawada N, Duffy HS et al. Gap and tight junctions in liver: composition, regulation, and function. In: Arias I, Boyer JL et al, eds. The Liver: Biology and Pathobiology. Fourth edition. New York: Raven Press, 2001:29-46.
10. Beyer EC, Paul DL, Goodenough DA. Connexin 43: a protein from rat heart homologous to a gap junction protein from liver. J Cell Biol 1987; 105:2621-9
11. Kumar NM, Gilula NB. Cloning and characterization of human and rat liver cDNAs coding for a gap junction protein. J Cell Biol 1986; 103:767-76
12. Paul DL. Molecular cloning of cDNA for rat liver gap junction protein. J Cell Biol 1986; 103:123-34
13. Zhang JT, Nicholson BJ. Sequence and tissue distribution of a second protein of hepatic gap junctions, Cx26, as deduced from its cDNA. J Cell Biol 1989; 109:3391-401.
14. Sosinsky G. Mixing of connexins in gap junction membrane channels. Proc Natl Acad Sci USA 1995; 92:9210-4.
15. Elfgang C, Eckert R, Lichtenberg-Frate H et al. Specific permeability and selective formation of gap junction channels in connexin-transfected HeLa cells. J Cell Biol 1995; 129:805-17.
16. Manthey D, Banach K, Desplantez T et al. Intracellular domains of mouse connexin26 and -30 affect diffusional and electrical properties of gap junction channels. J Membr Biol 2001; 181:137-48.
17. Bevans CG, Kordel M, Rhee SK et al. Isoform composition of connexin channels determines selectivity among second messengers and uncharged molecules. J Biol Chem 1998; 273:2808-16.
18. Yee AG, Revel JP. Loss and reappearance of gap junctions in regenerating liver. J Cell Biol 1978; 78:554-64.
19. Yancey SB, Easter D, Revel JP. Cytological changes in gap junctions during liver regeneration. J Ultrastruct Res 1979; 67:229-42.
20. Spray DC, Ginzberg RD, Morales EA et al. Electrophysiological properties of gap junctions between dissociated pairs of rat hepatocytes. J Cell Biol 1986; 103:135-44.
21. Pitts JD, Simms JW. Permeability of junctions between animal cells. Intercellular transfer of nucleotides but not of macromolecules. Exp Cell Res 1977; 104:153-63.
22. Stewart WW. Functional connections between cells as revealed by dye- coupling with a highly fluorescent naphthalimide tracer. Cell 1978; 14:741-59.
23. Simpson I, Rose B, Loewenstein WR. Size limit of molecules permeating the junctional membrane channel. Science 1977; 195:294-6.
24. Traub O, Look J, Dermietzel R et al. Comparative characterization of the 21-kD and 26-kD gap junction proteins in murine liver and cultured hepatocytes. J Cell Biol 1989; 108:1039-51.
25. Rosenberg E, Spray DC, Reid LM. Transcriptional and posttranscriptional control of connexin mRNAs in periportal and pericentral rat hepatocytes. Eur J Cell Biol 1992; 59:21-6.
26. Bode HP, Wang LF, Cassio D et al. Expression and regulation of gap junctions in cholangiocytes. Hepatology 2002; 36:631-40.
27. Zhang M, Thorgeirsson SS. Modulation of connexins during differentiation of oval cells into hepatocytes. Exp Cell Res 1994; 213:37-42.
28. George CH, Kendall JM, Evans WH. Intracellular trafficking pathways in the assembly of connexins into gap junctions. J Biol Chem 1999; 274:8678-85.
29. Evans WH. Assembly of gap junction intercellular communication channels. Biochem Soc Trans 1994; 22:788-92.

30. Zhang JT, Chen M, Foote CI et al. Membrane integration of in vitro- translated gap junctional proteins: co- and post-translational mechanisms. Mol Biol Cell 1996; 7:471-82.
31. Fallon RF, Goodenough DA. Five-hour half-life of mouse liver gap-junction protein. J Cell Biol 1981; 90:521-6.
32. Sáez JC, Gregory WA, Watanabe T et al. cAMP delays disappearance of gap junctions between pairs of rat hepatocytes in primary culture. Am J Physiol 1989; 257:C1-11.
33. Goodenough DA, Goliger JA, Paul DL. Connexins, connexons, and intercellular communication. Annu Rev Biochem 1996; 65:475-502.
34. Fishman GI, Moreno AP, Spray DC et al. Functional analysis of human cardiac gap junction channel mutants. Proc Natl Acad Sci USA 1991; 88:3525-9.
35. Bergoffen J, Scherer SS, Wang S et al. Connexin mutations in X-linked Charcot-Marie-Tooth disease. Science 1993; 262:2039-42.
36. Nelis E, Haites N, Van Broeckhoven C. Mutations in the peripheral myelin genes and associated genes in inherited peripheral neuropathies. Hum Mutat 1999; 13:11-28.
37. Abrams CK, Oh S, Ri Y et al. Mutations in connexin 32: the molecular and biophysical bases for the X-linked form of Charcot-Marie-Tooth disease. Brain Res Brain Res Rev 2000; 32:203-14.
38. Anzini P, Neuberg DH, Schachner M et al. Structural abnormalities and deficient maintenance of peripheral nerve myelin in mice lacking the gap junction protein connexin 32. J Neurosci 1997; 17:4545-51.
39. Scherer SS, Xu YT, Nelles E et al. Connexin32-null mice develop demyelinating peripheral neuropathy. Glia 1998; 24:8-20.
40. Martin PE, Mambetisaeva ET, Archer DA et al. Analysis of gap junction assembly using mutated connexins detected in Charcot-Marie-Tooth X-linked disease. J Neurochem 2000; 74:711-20.
41. Rabadan-Diehl C, Dahl G, Werner R. A connexin-32 mutation associated with Charcot-Marie-Tooth disease does not affect channel formation in oocytes. FEBS Lett 1994; 351:90-4.
42. Bruzzone R, White TW, Scherer SS et al. Null mutations of connexin32 in patients with X-linked Charcot-Marie-Tooth disease. Neuron 1994; 13:1253-60.
43. Omori Y, Mesnil M, Yamasaki H. Connexin 32 mutations from X-linked Charcot-Marie-Tooth disease patients: functional defects and dominant negative effects. Mol Biol Cell 1996; 7:907-16.
44. Abrams CK, Freidin MM, Verselis VK et al. Functional alterations in gap junction channels formed by mutant forms of connexin 32: evidence for loss of function as a pathogenic mechanism in the X-linked form of Charcot-Marie-Tooth disease. Brain Res 2001; 900:9-25.
45. Oh S, Ri Y, Bennett MV et al. Changes in permeability caused by connexin 32 mutations underlie X-linked Charcot-Marie-Tooth disease. Neuron 1997; 19:927-38.
46. Ressot C, Gomes D, Dautigny A et al. Connexin32 mutations associated with X-linked Charcot-Marie-Tooth disease show two distinct behaviors: loss of function and altered gating properties. J Neurosci 1998; 18:4063-75.
47. Castro C, Gomez-Hernández JM, Silander K et al. Altered formation of hemichannels and gap junction channels caused by C-terminal connexin-32 mutations J Neurosci 1999; 19:3752-60.
48. Sáez JC, Connor JA, Spray DC et al. Hepatocyte gap junctions are permeable to the second messenger, inositol 1,4,5-trisphosphate, and to calcium ions. Proc Natl Acad Sci USA 1989; 86:2708-12.
49. Charles AC, Naus CC, Zhu D et al. Intercellular calcium signaling via gap junctions in glioma cells. J Cell Biol 1992; 118:195-201.
50. Toyofuku T, Yabuki M, Otsu K et al. Intercellular calcium signaling via gap junction in connexin-43-transfected cells. J Biol Chem 1998; 273:1519-28.
51. Niessen H, Harz H, Bedner P et al. Selective permeability of different connexin channels to the second messenger inositol 1,4,5-trisphosphate. J Cell Sci 2000; 113:1365-72.
52. Cao F, Eckert R, Elfgang C et al. A quantitative analysis of connexin-specific permeability differences of gap junctions expressed in HeLa transfectants and Xenopus oocytes. J Cell Sci 1998; 111:31-43.
53. Nelles E, Butzler C, Jung D et al. Defective propagation of signals generated by sympathetic nerve stimulation in the liver of connexin32-deficient mice. Proc Natl Acad Sci USA 1996; 93:9565-70.
54. Niessen H, Willecke K. Strongly decreased gap junctional permeability to inositol 1,4,5-trisphosphate in connexin32 deficient hepatocytes. FEBS Lett 2000; 466:112-4.
55. Clair C, Chalumeau C, Tordjmann T et al. Investigation of the roles of Ca(2+) and InsP(3) diffusion in the coordination of Ca(2+) signals between connected hepatocytes. J Cell Sci 2001; 114:1999-2007.
56. Nathanson MH, Burgstahler AD. Coordination of hormone-induced calcium signals in isolated rat hepatocyte couplets: demonstration with confocal microscopy. Mol Biol Cell 1992; 3:113-21.
57. Nathanson MH, Burgstahler AD, Mennone A et al. Ca^{2+} waves are organized among hepatocytes in the intact organ. Am J Physiol 1995; 269:G167-71.

58. Robb-Gaspers LD, Thomas AP. Coordination of Ca^{2+} signaling by intercellular propagation of Ca^{2+} waves in the intact liver. J Biol Chem 1995; 270:8102-7.
59. Motoyama K, Karl IE, Flye MW et al. Effect of Ca^{2+} agonists in the perfused liver: determination via laser scanning confocal microscopy. Am J Physiol 1999; 276:R575-85.
60. Watanabe N, Tsukada N, Smith CR et al. Motility of bile canaliculi in the living animal: implications for bile flow. J Cell Biol 1991; 113:1069-80.
61. Tordjmann T, Berthon B, Combettes L et al.The location of hepatocytes in the rat liver acinus determines their sensitivity to calcium-mobilizing hormones. Gastroenterology 1996; 111:1343-2.
62. Tordjmann T, Berthon B, Jacquemin E et al. Receptor-oriented intercellular calcium waves evoked by vasopressin in rat hepatocytes. EMBO J 1998; 17:4695-703.
63. Tordjmann T, Berthon B, Claret M et al. Coordinated intercellular calcium waves induced by noradrenaline in rat hepatocytes: dual control by gap junction permeability and agonist. EMBO J 1997; 16:5398-407.
64. Schlosser SF, Burgstahler AD, Nathanson MH. Isolated rat hepatocytes can signal to other hepatocytes and bile duct cells by release of nucleotides. Proc Natl Acad Sci USA 1996; 93:9948-53.
65. Dranoff JA, Masyuk AI, Kruglov EA et al. Polarized expression and function of P2Y ATP receptors in rat bile duct epithelia. Am J Physiol Gastrointest Liver Physiol 2001; 281:G1059-67.
66. Nathanson MH, Burgstahler AD, Masyuk A et al. Stimulation of ATP secretion in the liver by therapeutic bile acids. Biochem J 2001; 358:1-5.
67. Osipchuk Y, Cahalan M. Cell-to-cell spread of calcium signals mediated by ATP receptors in mast cells. Nature 1992; 359:241-4.
68. Dubyak GR, el-Moatassim C. Signal transduction via P2-purinergic receptors for extracellular ATP and other nucleotides. Am J Physiol 1993; 265:C577-606.
69. McGill JM, Basavappa S, Mangel AW et al. Adenosine triphosphate activates ion permeabilities in biliary epithelial cells. Gastroenterology 1994; 107:236-43.
70. Cotrina ML, Lin JH, Alves-Rodrigues A et al. Connexins regulate calcium signaling by controlling ATP release. Proc Natl Acad Sci USA 1998; 95:15735-40.
71. Iwai M, Miyashita T, Shimazu T. Inhibition of glucose production during hepatic nerve stimulation in regenerating rat liver perfused in situ. Possible involvement of gap junctions in the action of sympathetic nerves. Eur J Biochem 1991; 200:69-74.
72. Seseke FG, Gardemann A, Jungermann K. Signal propagation via gap junctions, a key step in the regulation of liver metabolism by the sympathetic hepatic nerves. FEBS Lett 1992; 301:265-70.
73. Stumpel F, Ott T, Willecke K et al. Connexin 32 gap junctions enhance stimulation of glucose output by glucagon and noradrenaline in mouse liver. Hepatology 1998; 28:1616-20.
74. Eugenín EA, González H, Sáez CG et al. Gap junctional communication coordinates vasopressin-induced glycogenolysis in rat hepatocytes. Am J Physiol 1998; 274:G1109-16.
75. Nathanson MH, Rios-Velez L, Burgstahler AD et al. Communication via gap junctions modulates bile secretion in the isolated perfused rat liver. Gastroenterology 1999; 116:1176-83.
76. Temme A, Stumpel F, Sohl G et al. Dilated bile canaliculi and attenuated decrease of nerve-dependent bile secretion in connexin32-deficient mouse liver. Pflugers Arch 2001; 442:961-6.
77. Berthoud VM, Iwanij V, Garcia AM et al. Connexins and glucagon receptors during development of rat hepatic acinus. Am J Physiol 1992; 263:G650-8.
78. Nathanson MH, Gautam A, Ng OC et al. Hormonal regulation of paracellular permeability in isolated rat hepatocyte couplets. Am J Physiol 1992; 262:G1079-86.
77. Nathanson MH, Burgstahler AD, Mennone A et al. Characterization of cytosolic Ca^{2+} signaling in rat bile duct epithelia. Am J Physiol 1996; 271:G86-96.
79. Serriere V, Berthon B, Boucherie S et al. Vasopressin receptor distribution in the liver controls calcium wave propagation and bile flow. FASEB J 2001; 15:1484-6.
80. Holder JW, Elmore E, Barrett JC. Gap junction function and cancer. Cancer Res 1993; 53:3475-85.
81. Yamasaki H, Krutovskikh V, Mesnil M et al. Gap junctional intercellular communication and cell proliferation during rat liver carcinogenesis. Environ Health Perspect. 1993; 101:191-7.
82. Eghbali B, Kessler JA, Reid LM et al. Involvement of gap junctions in tumorigenesis: transfection of tumor cells with connexin 32 cDNA retards growth in vivo. Proc Natl Acad Sci USA1991; 88:10701-5.
83. Temme A, Buchmann A, Gabriel HD et al. High incidence of spontaneous and chemically induced liver tumors in mice deficient for connexin32. Curr Biol 1997; 7:713-6.
84. Janssen-Timmen U, Traub O, Dermietzel R et al. Reduced number of gap junctions in rat hepatocarcinomas detected by monoclonal antibody. Carcinogenesis 1986; 7:1475-82.
85. Yamaoka K, Nouchi T, Tazawa J et al. Expression of gap junction protein connexin 32 and E-cadherin in human hepatocellular carcinoma. J Hepatol 1995; 22: 536–9.

86. Fitzgerald DJ, Mesnil M, Oyamada M et al. Changes in gap junction protein (connexin 32) gene expression during rat liver carcinogenesis. J Cell Biochem 1989; 41:97-102.
87. Mesnil M, Fitzgerald DJ, Yamasaki H. Phenobarbital specifically reduces gap junction protein mRNA level in rat liver. Mol Carcinog 1988; 1:79-81.
88. Rose B, Mehta PP, Loewenstein WR. Gap-junction protein gene suppresses tumorigenicity. Carcinogenesis 1993; 14:1073-5.
89. Hayashi T, Nomata K, Chang CC et al. Cooperative effects of v-myc and c-Ha-ras oncogenes on gap junctional intercellular communication and tumorigenicity in rat liver epithelial cells. Cancer Lett 1998; 128:145-54.
90. Ruch RJ, Madhukar BV, Trosko JE et al. Reversal of ras-induced inhibition of gap-junctional intercellular communication, transformation, and tumorigenesis by lovastatin. Mol Carcinog 1993; 7:50-9.
91. Leffert HL, Koch KS, Rubalcava B et al. Hepatocyte growth control: in vitro approach to problems of liver regeneration and function. Natl Cancer Inst Monogr 1978; 48:87-101.
92. Yee AG, Revel JP. Loss and reappearance of gap junctions in regenerating liver. J Cell Biol 1978; 78:554-64.
93. Traub O, Druge PM, Willecke K. Degradation and resynthesis of gap junction protein in plasma membranes of regenerating liver after partial hepatectomy or cholestasis. Proc Natl Acad Sci USA 1983; 80:755-9.
94. Gingalewski C, Theodorakis NG, Yang J et al. Distinct expression of heat shock and acute phase genes during regional hepatic ischemia-reperfusion. Am J Physiol 1996; 271:R634-40.
95. Gingalewski C, Wang K, Clemens MG et al. Posttranscriptional regulation of connexin 32 expression in liver during acute inflammation. J Cell Physiol 1996; 166:461-7.
96. Temme A, Traub O, Willecke K. Downregulation of connexin32 protein and gap-junctional intercellular communication by cytokine-mediated acute-phase response in immortalized mouse hepatocytes. Cell Tissue Res 1998; 294:345-50.
97. Temme A, Ott T, Haberberger T et al. Acute-phase response and circadian expression of connexin26 are not altered in connexin32-deficient mouse liver. Cell Tissue Res 300:111-7.
98. Fallon MB, Nathanson MH, Mennone A et al. Altered expression and function of hepatocyte gap junctions after common bile duct ligation in the rat. Am J Physiol 1995; 268:C1186-94.
99. Yamaoka K, Nouchi T, Kohashi T et al. Expression of gap junction protein connexin 32 in chronic liver diseases. Liver 2000; 20:104-7.
100. Kojima T, Sawada N, Oyamada M et al. Rapid appearance of connexin 26-positive gap junctions in centrilobular hepatocytes without induction of mRNA and protein synthesis in isolated perfused liver of female rat. J Cell Sci 1994; 107:3579-90.

CHAPTER 5

Cholestasis: An Intracellular "Traffic Jam"

Helmut Kipp and Irwin M. Arias

Summary

Mutations in the coding region of BSEP which result in its absence from the bile canalicular membrane are manifested by progressive cholestasis and liver damage. We have proposed that defects in intracellular trafficking and/or posttranslational regulation of BSEP may produce a similar phenotype. To test this hypothesis, it was necessary to determine the pathway utilized by BSEP and other canalicular ABC transporters and its regulation. Our studies reveal that BSEP traffics from Golgi to a subapical recycling endosomal compartment from which the transporter cycles to and from the canalicular membrane; the efflux compartment is cAMP-dependent. A second intracellular pool which responds to taurocholate was identified. Both intracellular sites contain at least six times more BSEP as is present in the canalicular membrane. Using pulse-chase and subcellular fractionation studies in rats and on-line image analysis of the movement of chimeric fluorescent BSEP in WIFb cells, we measured the trafficking rates and demonstrated the critical role of 3′ phosphoinositide products of phosphatidyl (PI) 3-kinase, microtubules and other intracellular components. In addition, the activity of BSEP in the canalicular membrane is regulated by PI 3-kinase products and by cAMP. All of these effects occur in the absence of new protein synthesis. Our studies indicate that the hepatocyte is protected from the detergent effects of retained bile acids by posttranslationally mediated enhanced BSEP transcellular transport and activation within the canalicular membrane. Experimentally induced alterations in trafficking components impaired BSEP delivery to and/or activation in the canalicular membrane. The results suggest that intrahepatic cholestasis may result from an intracellular traffic jam and create exciting opportunities to investigate the interaction between inheritable defects in ABC transporters and acquired conditions in the pathogenesis of cholestasis associated with drugs, viruses, metals and other factors.

Introduction

As discussed elsewhere in this volume, patients having point mutations in the coding regions of SPGP (BSEP, ABC11) and MDR 3, and mice in which these genes have been removed by homologous recombination manifest a progressive cholestatic phenotype. These observations confirm the physiologic importance of canalicular ATP-dependent bile acid and phospholipid translocase activity in bile formation and secretion.

Because the canalicular levels of SPGP, MDR3, other ABC transporters and FIC1, an ATP-dependent P-type aminophospholipid translocase are dynamically controlled by transporter synthesis, trafficking, activation and degradation, inheritable and acquired defects in these processes may produce a phenotype indistinguishable from that resulting from mutated ABC transporter genes and their protein products. Therefore, it is necessary to understand these processes at the molecular and cellular levels before their specific role in intrahepatic cholestasis is characterized and can become a target for more effective therapies.

Molecular Pathogenesis of Cholestasis, edited by Michael Trauner and Peter L.M. Jansen.
©2004 Eurekah.com and Kluwer Academic / Plenum Publishers.

This Chapter presents selected aspects of the complex mechanisms involved in intracellular trafficking of ABC transporters and their activation in the canalicular plasma membrane. These processes require ATP, calcium, microtubules, cytoplasmic motors, PI 3-kinase lipid products, cAMP, guanosine triphosphatases of the rab family and, undoubtedy, many as yet unidentified components.

Trafficking of Newly Synthesized ABC Transporters in Hepatocytes

Membrane targeting of the newly synthesized canalicular ectoenzymes dipeptidylpeptidase IV, aminopeptidase N and 5'-nucleotidase, and the canalicular cell adhesion molecule cCAM105 (also known as HA4) has been studied in rat liver by in vivo metabolic pulse chase labeling. After biosynthesis, these canalicular proteins are transferred from Golgi to the basolateral membrane and subsequently reach the bile canaliculus only by transcytosis.[1,2] Based on these results, it was proposed that all newly synthesized canalicular proteins, including canalicular ABC transporters, are targeted via this indirect route.[3,4]

An important observation from previous studies was that the canalicular cell adhesion molecule cCAM105, but not canalicular ABC transporters, were readily detected in Western blots of highly purified sinusoidal/basolateral membrane vesicles (SMV) from rat liver using Western blots.[5,6] The presence of cCAM105 in SMV can be explained by the fact that cCAM105 is initially transferred to the basolateral membrane after biosynthesis and subsequently reaches the apical pole by transcytosis. This scenario is in accord with detectable steady state levels of cCAM105 in SMV. These observations suggested that canalicular ABC transporters may not undergo transcytosis after biosynthesis. The hypothesis of direct apical targeting of canalicular ABC transporters in rat hepatocytes was subsequently tested using metabolic pulse chase labelling.[6]

Rats were metabolically labeled with ^{35}S-methionine for 15 min and the content of newly synthesized cCAM105, MDR1, MDR2 and SPGP was determined after 15,30,60,120 and 180 min in purified canalicular membrane vesicles (CMV), sinusoidal/ basolateral membrane vesicles (SMV) and Golgi membranes from rat liver using immunoprecipitation with specific antibodies. These studies[6] confirmed the transcytotic pathway for apical targeting of newly synthesized cCAM105 (HA4).[1] In contrast, at no time between passage through Golgi and arrival at the bile canaliculus were apical ABC transporters MDR1, MDR2 and SPGP detected in SMVs, indicating a direct Golgi-to-bile canaliculus pathway for their membrane targeting. Also newly synthesized MDR1, MDR2 and SPGP were not initially transferred to the basolateral membraneand revealed different post-Golgi trafficking patterns. After passage through Golgi, MDR1, MDR2 were rapidly delivered directly to the bile canaliculus, whereas Golgi-to-bile canaliculus trafficking of SPGP involved additional intermediate steps. At 1h after metabolic labeling, only the mature form of SPGP was detected in the homogenate, indicating that processing and passage through the Golgi were complete at this point. This is also supported by decreased radioactivity to background levels in SPGP immunoprecipitates from Golgi membranes after 1h. At this time point, SPGP was not detected in SMV and CMV and, therefore, had not reached the cell surface which occurred only 2h after metabolic labeling. The most likely explanation is that SPGP is sequestered in an intracellular pool prior to delivery to the canalicular membrane. Intrahepatic sequestering of newly synthesized SPGP was demonstrated in a later study, which included a rab 11-enriched endosomal fraction in metabolic labeling experiments.[7]

The membrane targeting pathways of newly synthesized canalicular proteins discoverd by in vivo labeling studies are depicted in Fig. 1. These studies provide direct biochemical evidence for intrahepatic pools of ABC transporters. The characteristics of these intrahepatic ABC transporter pools are described subsequently. Furthermore, the results promoted investigation of direct Golgi-to-bile canaliculus trafficking of MDR1-green fluorescent protein (GFP) in WIF-B cells, a polarized hepatocyte cell culture model.

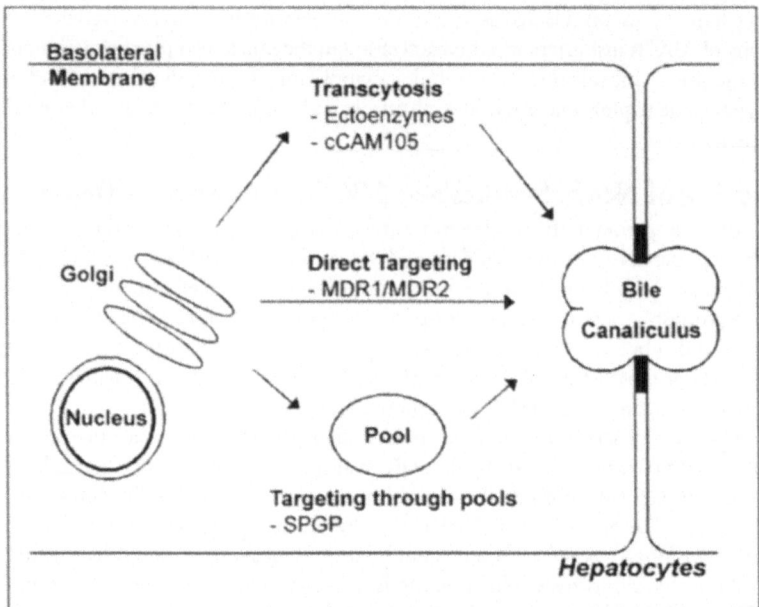

Figure 1. Membrane targeting of newly synthesized canalicular proteins in rat hepatocytes. (From Ref. 6.)

WIF-B cells are a hybrid of rat hepatoma cells and human fibroblasts, have functional bile canaliculi and serve as a useful model for hepatocytes.[8,9] Functional features of hepatocytes are also observed in WIF-B cells: i.e., basolateral to apical membrane transcytosis of canalicular ectoenzymes;[10] secretion of fluorescent bile acids and substrates for MDR1[8,11] and MRP2;[12] inhibition of the secretion of fluorescent substrates by the PI 3-K inhibitor, Wortmannin;[11] and enhancement of the secretion of fluorescent substrates by taurocholate and PI 3-K activating synthetic peptide.[11]

Intracellular distribution and trafficking of MDR1-green fluorescent protein (GFP) was recently studied in stably transfected WIF-B cells.[11] Fluorescence of the MDR1-GFP chimera was exclusively detected in Golgi and bile canalicular membranes, no labeling of basolateral plasma membranes was observed. To visualize movement of MDR1-GFP chimeric protein between Golgi and bile canaliculi, fluorescent images of stably transfected WIF-B cells were serially examined by confocal microscopy. Digital fluorescent images were collected for 20 minutes and converted into a QuickTime movie (see complete movie at http:// www.healthsci.tufts.edu/ LABS/IMArias/Sai_F9.htm). Selected sequences from the movie are depicted in Fig. 2. The upper six panels of Fig. 2A show a time sequence over 2-6 minutes during which MDR1-GFP moved rapidly from the Golgi (G), directly along straight or curvilinear paths and merged with the bile canaliculus (BC). In Fig. 2A, middle right 9 panels, tubulovesicular movement of MDR1-GFP was also observed between the bile canaliculus and pericanalicular region (arrows).

Single long tubules shrank, formed vesicles and subsequently fused with the canalicular membrane. Other tubular structures extended from the canalicular membrane into the subapical region and retracted to the canalicular membrane (arrowhead in Fig. 2A, left middle panel). Fig. 2B shows a series of confocal images with tubular structures reaching directly from Golgi to the bile canaliculus (arrows). Movement of MDR1-GFP was not synchronous. Individual tubulovesicular structures frequently changed shape during translocation. Frequently there was brief delay following which tubular vesicles fused with the canalicular membrane. This event appeared distinct from movements of tubules in other directions (ie., those presumably not

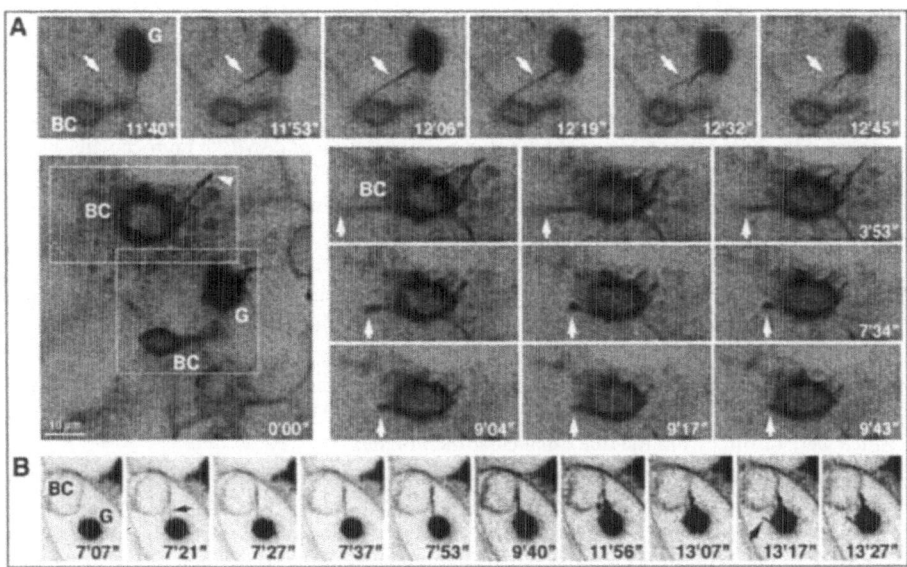

Figure 2. Visualization of intracellular movement of mdr1-GFP protein between Golgi and canalicular membrane in stably transfected WIF-B9 cells. WIF-B9 cells stably transfected with mdr1-GFP were grown on glass cover slips, mounted on the microstage and maintained at 37°C. A and B represent independent experiments using independent cell cultures. The cells which express mdr1-GFP both in canaliculi and perinuclear Golgi region were identified under phase and epifluorescence microscope. Digital images were collected using confocal microscope at 3.24 second intervals for about 20 minutes. Every 4 images were averaged to reduce background noise. See QuickTime movie at http://www.healthsci.tufts.edu/LABS/IMArias/Sai_F9.htm. Smaller panels on top and right were time sequences of clipped images (actual time in observation was indicated) from white rectangles in the large middle-left panel. G, Golgi compartment; BC, canalicular membranes. See text for explanation. (From Ref. 11.)

fusing). Multiple confocal examinations of many cells indicate that the tubule is not moving in and out of focus but fuses with the canalicular membrane.

In addition, incubation of MDR1-GFP stably transfected WIFB9 cells at 15°C for 20 hours revealed only co-localization of MDR1-GFP with Golgi markers. Following increase in incubation temperature to 37°C, MDR1-GFP progressively moved to the canalicular plasma membrane within 30-60min; this process was accelerated on incubation of cells with taurocholate and the entire process including release from Golgi was prevented by preincubation with Wortmannin. At no time was basalateral membrane localization of MDR1-GFP observed. This model provides further opportunity to examine the role of specific candidate participants in intracellular trafficking and membrane localization.

The observed direct Golgi to bile canalicular trafficking of MDR1-GFP in WIFB 9 cells is consistent with the membrane targeting detected using C219 antibody (MDR1, MDR2) in rat metabolic labeling studies in vivo.[6] Furthermore, the movement of MDR1-GFP from Golgi to the canalicular membrane was tubulovesicular in appearance and intermittent (occurring every 5-20min). A speed of 0.02-0.6 μm/ second was slightly less than values obtained from other studies, which ranged from 0.03-1 μm/second.[13,14,15] The process closely resembles the movement of VSV-G protein from ER to Golgi in nonpolarized cells[13,16] and previously described Golgi to plasma membrane trafficking involving large tubular-vesicular structures[14,17,18,19] rather than discrete vesicles which has been the conventional postulate.[20]

Characteristics of Intrahepatic ABC Transporter Pools

Gatmaitan et al[21] and other authors[5] observed that bile secretion was significantly enhanced in isolated perfused rat liver after treatment of rats with the second messenger, cAMP, or the bile salt, taurocholate. Increased bile secretion resulted from increased amounts of ABC transporters in the bile canalicular membrane after administration of cAMP or taurocholate. There was an increase in the specific amounts of canalicular ABC transporters indicating that the responses to cAMP and taurocholate did not result from an increase in total amount of canalicular membrane. Since the increase in canalicular ABC transporter amount was dependent on an intact microtubule system and occurred within minutes, we postulated the existence of intracellular ABC transporter pools from which additional transporters can be rapidly recruited to the canalicular membrane. Previous morphological studies in rats rendered cholestatic by bile duct ligation,[22] phalloidin[23] or lipopolysaccharide[24] suggested that MRP2 and SPGP may traffic from the bile canaliculus to intracellular sites. In addition, MRP2[24] and SPGP[25] were observed by immunogold staining and electron microscopy in undefined vesicular structures which were distinct from the bile canalicular membrane.

ABC transporter trafficking from intracellular sites to the hepatocyte apical domain was stimulated by cAMP and taurocholate and recently used to confirm the presence and define the properties of potential intrahepatic ABC-transporter pools in rats in vivo.[7] Administration of cAMP or taurocholate to rats increased amounts of MDR1, MDR2 and SPGP in the bile canalicular membrane by ~3-fold. These effects abated after 6 h and the bile canalicular content of MDR1, MDR2 and SPGP returned to basal levels. Pretreatment of rats with cycloheximide inhibited protein biosynthesis did not prevent the increase in the amounts of each canalicular ABC transporter following administration of cAMP or taurocholate. These data demonstrate that additional ABC transporters in the bile canalicular membrane do not result from enhanced transcription or translation, but indicate recruitment from existing intracellular pools. Using ^{35}S-methionine metabolic labeling, the overall half life of MDR1, MDR2 and SPGP was 5 days in rat liver, suggesting that ABC transporters cycle between intracellular pools and the bile canalicular membrane prior to degradation.

The kinetics of the intrahepatic distribution of SPGP were further investigated (Fig. 3) after metabolic labeling of rats with ^{35}S-methionine and immunoprecipitation of SPGP from Golgi membranes, a rab 11-enriched combined endosomal fraction (CEF) and canalicular membrane vesicles (CMV). It was observed in a previous study that newly synthesized SPGP was never detected in the sinusoidal/basolateral plasma membrane of the rat hepatocyte[6] indicating non-transcytotic apical targeting of newly synthesized SPGP. Radiolabeled SPGP peaked in Golgi membranes after a chase time of 30 min and thereafter was virtually absent from the Golgi indicating that processing and passage of SPGP through Golgi is complete after 30-60 min (Fig. 3). SPGP peaked in CEF at 1 h and, after a chase time of 2 h, first appeared in CMVs. These experiments demonstrated that newly synthesized SPGP is targeted through an endosomal compartment before reaching the bile canalicular membrane. Furthermore, SPGP was not completely transferred from CEF and a significant amount of SPGP remained in the endosomal fraction. These results suggested distribution of SPGP between canalicular membrane and intracellular pools (2 and 3 h chase) which was also observed after a chase time of 20 h, and presumably represent steady state distribution of SPGP between the bile canaliculus and intracellular pools under basal conditions (Fig. 3).

Previous studies[21] indicate that the effects of cAMP and taurocholate on bile canalicular ABC transporter amount are additive rather than alternative, which suggests the presence of at least two distinct intrahepatic pools of ABC-transporters: one of which is mobilized to the canalicular membrane by cAMP ("cAMP-pool"), and the other by taurocholate ("TC-pool"). The hypothesis of two distinct intrahepatic pools of ABC transporters was further supported by the observation that targeting of newly synthesized SPGP through intrahepatic sites to the bile canalicular membrane is accelerated by cAMP, but not by taurocholate.[7] A tentative model

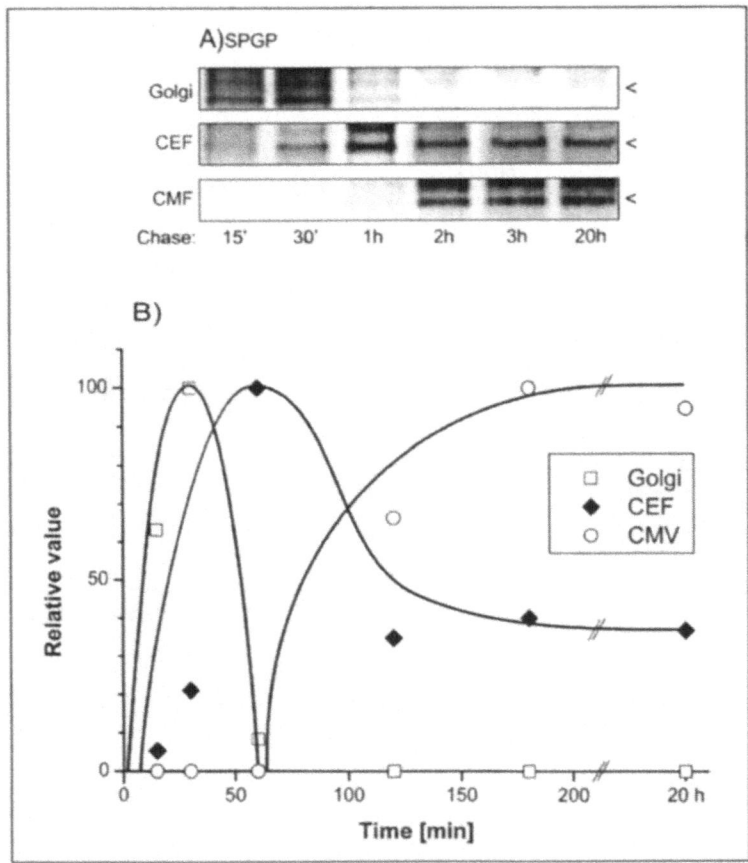

Figure 3. Newly synthesized SPGP is targeted through an endosomal compartment. Rats were pulse-labeled for 15 min with [35S]methionine (5 mCi) and then chased with unlabeled methionine for 15 min, 30 min, 1 h, 2 h, 3 h, and 20 h, respectively. SPGP was then immunoprecipitated from Golgi membranes, CEF and CMVs. Immunoprecipitates were separated by SDS-PAGE and [35S]SPGP was detected in a PhosphorImager. Panel A shows representative results observed in three independent sets of rats; arrowheads indicate the position of mature antigens. To establish the kinetics of newly synthesized SPGP trafficking through cellular compartments (Golgi, squares; CEF, diamonds; CMV, circles), intensities of [35S]SPGP-bands were quantified with a PhosphorImager. The relative intensity (highest reading in each fraction equals 100) was plotted versus labeling time (B); mean values ± S.D., n=3. (From Ref. 7.)

for the intrahepatic pathways of ABC transporters is shown in Figure 4. After passage through Golgi, SPGP accumulates in an intrahepatic cAMP-pool and later equilibrates with the TC-pool. Whether equilibration of newly synthesized SPGP with the taurocholate-pool occurs from the bile canalicular membrane or the cAMP-pool remains unclear. Newly synthesized MDR1,MDR2 bypass the intracellular pools on their journey to the bile canaliculus.[6,11] However, at steady state levels, MDR1,MDR2 are also mobilized to the bile canalicular membrane by cAMP and taurocholate suggesting that these ABC transporters also equilibrate with intrahepatic pools after reaching the bile canalicular membrane.

Upon stimulation with cAMP, trafficking of membrane transporters from intracellular sites to the plasma membrane has been described in several systems, i.e.,: (i) cystic fibrosis transmembrane regulator (CFTR) channel into the apical surface of rat duodenal villous epithelia;[26]

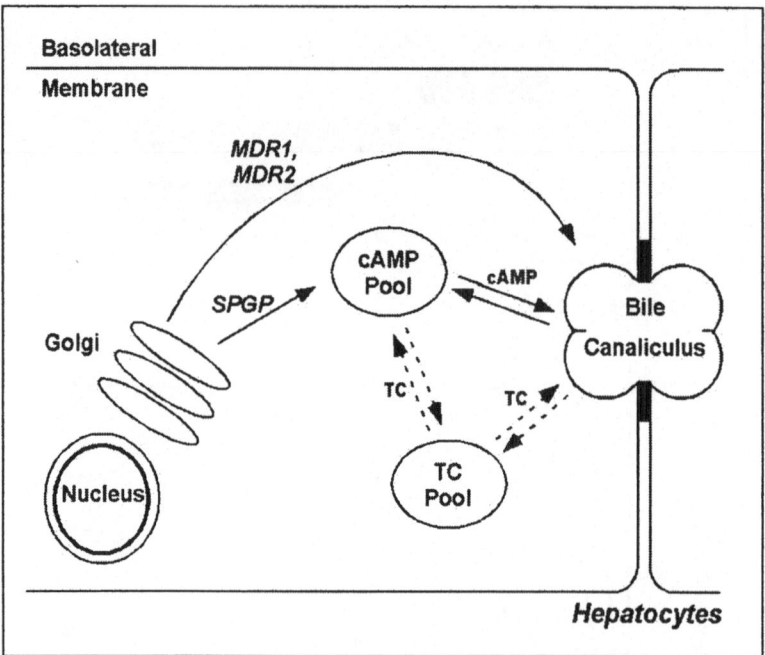

Figure 4. Tentative model for ABC-transporter trafficking in rat hepatocytes. Dashed arrows indicate possible pathways of ABC transporters from the TC pool. See text for explanation. (From Ref. 7.)

(ii) Na^+-taurocholate cotransport protein (ntcp) in the basolateral membrane of rat hepatocytes;[27] (iii) aquaporin-2 water channel into the apical membrane of LLC-PK1 cells, a polarized renal cell line;[28] (iv) H^+/K^+-ATPase into the apical membrane of gastric parietal cells;[29] (v) insulin responsive glucose transporter 4 (GLUT4) into the plasma membrane of rat adipocytes.[30] In each of these examples, recruitment of transporters to the plasma membrane from a recycling endosome has been suggested. In particular, trafficking of GLUT4 in rat adipocytes parallels that of ABC transporter trafficking in rat hepatocytes. GLUT4 traffics from distinct intracellular sites to the plasma membrane in response to cAMP and insulin.[30] In analogy to these other systems, we propose that cAMP recruits ABC transporters to the bile canalicular membrane from a rab 11-enriched recycling endosome, whereas the effect of taurocholate appears to be hepatocyte-specific and involves a yet to be defined different mechanism.

An important question concerns the distribution-ratio of ABC transporters between the bile canalicular membrane and intrahepatic sites. For MRP2, the canalicular to intracellular ratio was calculated to be 1:1 by quantitation of immunogold stained MRP2 in electron microscopy[24] in rat liver under basal conditions. Investigation of ABC transporters by immunoblots at steady state levels revealed that, after stimulation with cAMP or taurocholate, the amounts of MDR1, MDR2 and SPGP significantly increased in the bile canalicular membrane, whereas a change in ABC transporter amount in a combined endosomal fraction prepared from the same rat liver remained below detection limit.[7] This observation is in accordance with the presence of "large" intrahepatic pools. Under basal conditions, most MDR1, MDR2 and SPGP appear to reside in intrahepatic pools rather than in the bile canalicular membrane. However, the increase in the amount of ABC transporter in the canalicular membrane after stimulation with cAMP or taurocholate is approximately 3-fold for each effector. Taking into account that cAMP and taurocholate recruit transporters from different intracellular sources, the intrahepatic pool of ABC transporters is at least 6-times more than the amount present in the bile canalicular membrane. Since this calculation presumes that all intracellular ABC transporters are translocated to the bile canalicular membrane upon stimulation, this number represents a lower limit. Thus, the intrahepatic/canalicular ratio of MDR1, MDR2 and SPGP probably exceeds 6:1.

Phosphatidyl 3-Kinase Regulates Intrahepatic Trafficking and Activity of Bile Canalicular ABC Transporters

PI 3-K was discovered in Cantley's laboratory over ten years ago.[31] Major interest initially focused on its role in growth factor and oncogene tyrosine kinase-mediated signal transduction.[32] Subsequent studies revealed that PI 3-K are ubiquitous lipid kinases that function as signal transducers downstream of the cell surface receptors and are essential for cell proliferation, adhesion, survival, and cytoskeletal rearrangement.[33,34] The products of PI3K-catalyzed reactions are PtdIns(3,4,5)P$_3$, PtdIns(3,4)P$_2$, and PtdIns(3)P, which serve as second messengers in many signal transduction pathways. Serendipitous observation after characterization of yeast proteins that were involved in intracellular trafficking resulted in demonstration that PI 3-K is required for vesicle trafficking in plant, yeast and animal cells.[32] Studies of the function of PI 3-K were greatly facilitated by discovery that, at low nmolar concentration, wortmannin, a fungal component, specifically inhibits PI 3-K.[35] Other PI 3-K inhibitors of the LY series were subsequently synthesized. When used at the proper concentrations, these inhibitors have high specificity for PI 3-K in isolated systems and cells.

Folli et al[36] were among the first to study the effect of wortmannin administration in perfused rat liver. Using physiologic and morphologic techniques, they demonstrated intracellular vacuolation, intracellular and canalicular membrane disruption, and impaired transcytosis. Misra et al investigated the effect of wortmannin on taurocholate-stimulated bile secretion in isolated perfused rat liver[37] (Fig. 5). After the perfusion experiments were performed, liver was subfractionated into CMV and SMV and the content of various membrane proteins was quantified in the fractions by Western blot (Fig. 6).

Different schedules of taurocholate and wortmannin administration in perfused liver (Fig. 5) and comparision with ABC transporter protein levels in CMV (Fig. 6) revealed two unexpected effects of wortmannin on taurocholate-induced bile secretion. When compared with a control experiment, taurocholate perfusion increased bile secretion, which resulted from an increased amount of canalicular ABC transporters.[21] Administration of wortmannin before taurocholate prevented increased bile secretion and simultaneously prevented increase in the amount of canalicular ABC transporters. These observations indicate that active PI 3-K is required for recruitment and vesicular trafficking of additional ABC transporters from intracellular pools to the bile canalicular membrane after stimulation by taurocholate. When wortmannin was administered after taurocholate administration, the levels of canalicular ABC transporters remained elevated; however, bile secretion rapidly decreased by 50% after addition of wortmannin to the perfusion. In these experiments, the amount of canalicular ABC transporters remained elevated; however, transport activity and bile secretion were impaired, which suggests that active PI 3-K is not only required for taurocholate-induced vesicular trafficking but also may play a role in regulation of ABC transporter activity in the canalicular membrane.

The Effects of PI 3-K Modulators on CMV Function In Vitro[38]

In studies in which the transport of [^{3}H]-taurocholate into CMV was measured, preincubation of CMV with wortmannin (50 nM) for 5 min inhibited ATP-dependent [^{3}H]-taurocholate transport by more than 50%, whereas ATP-independent uptake remained unaltered (Fig. 7). The IC$_{50}$ for wortmannin was 25 nM. The same inhibitory effect was observed with a different PI 3-K inhibitor, LY294002 (IC$_{50}$ 20μM). CMV and SMV contain substantial amounts of active PI 3-K.[5] Preincubation for 5 min with wortmannin and LY294002 was necessary to inhibit [^{3}H]-taurocholate transport into CMV. These observations suggest that PI 3-K products, which facilitate ABC transporter activity, undergo rapid turnover in the bile canalicular membrane.

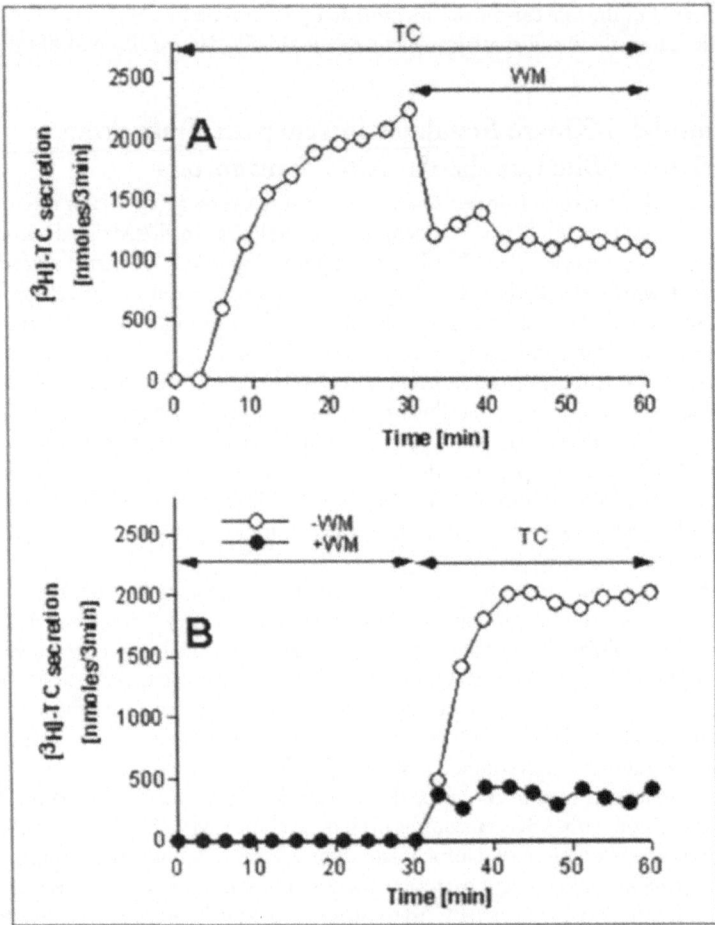

Figure 5. Effect of wortmannin on taurocholate-induced bile acid secretion in isolated perfused rat liver. Bile was collected at 3-min intervals in rats (A) receiving taurocholate (TC) (100 μM at 30 ml/min) that contained a tracer amount of [³H]-taurocholate for 60 min. Wortmannin (WM) (100 nM) was added to the perfusate for the last 30 min. (B) Wortmannin (WM) was administered before taurocholate (TC) (closed circles). In a control experiment, secretion was measured without wortmannin administration (open circles). (From Ref. 5.)

That PI 3-K lipid products are sufficient to enhance ATP-dependent transport activity of canalicular ABC transporters was proven by adding the lipid products to CMV from rats which received various pretreatments, and ATP-dependent [³H]-taurocholate and [³H]-GS-DNP transport were measured (Fig. 8). ATP-dependent transport of [³H]-taurocholate into CMV was enhanced by prior treatment of rats with taurocholate and inhibited by addition of wortmannin. Addition of PI 3-K lipid product PtdIns(3,4)P$_2$ to CMV not only rescued wortmannin inhibition, but restored ATP-dependent [³H]-taurocholate transport above the level induced after taurocholate administration. A similar effect was observed on addition of PI 3-K lipid products PtdIns(3,4,5)P$_3$ and PtdIns(3)P but not for the structurally related PtdIns(4,5)P$_2$, which lacks 3-hydroxyl phosphorylation. These studies reveal the specificity of PI 3-K lipid products on ABC transporter activation. Further evidence for involvement of PI 3-K lipid products in maximal ABC transporter function was gained by investigation of the effect of a synthetic

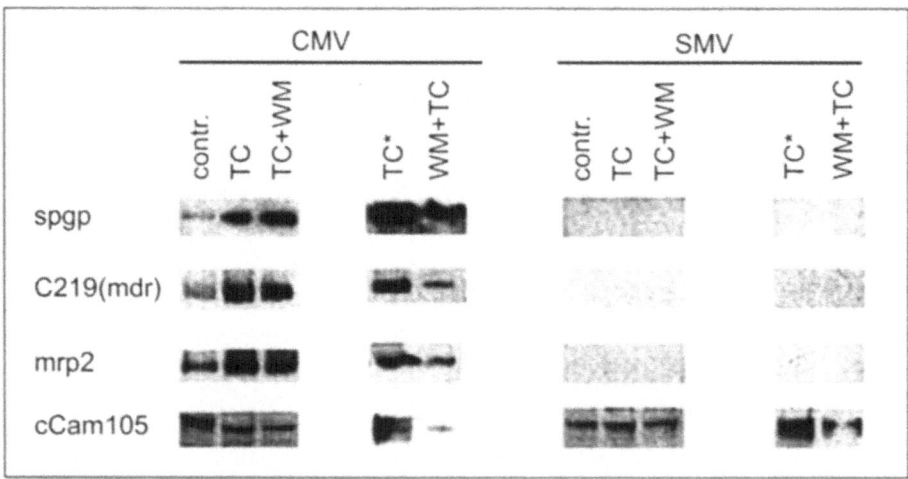

Figure 6. Protein levels (Western Blots) in CMV and SMV prepared from livers after perfusion experiments (comp. Fig. 5). The membrane fractions (equal amount of protein) were separated by SDS-PAGE, transferred onto nitrocellulose, and probed with antibodies against SPGP, MDR proteins (C219), MRP2, and cCAM105. Pretreatment of rats in perfusion experiments: control buffer for 60 min; TC, taurocholate for 60 min; TC + WM, taurocholate for 60 min, wortmannin for the last 30 min; TC*, buffer for 30 min and then taurocholate for 30 min; WM + TC, wortmannin for 30 min and then taurocholate for 30 min. (From Ref. 5.)

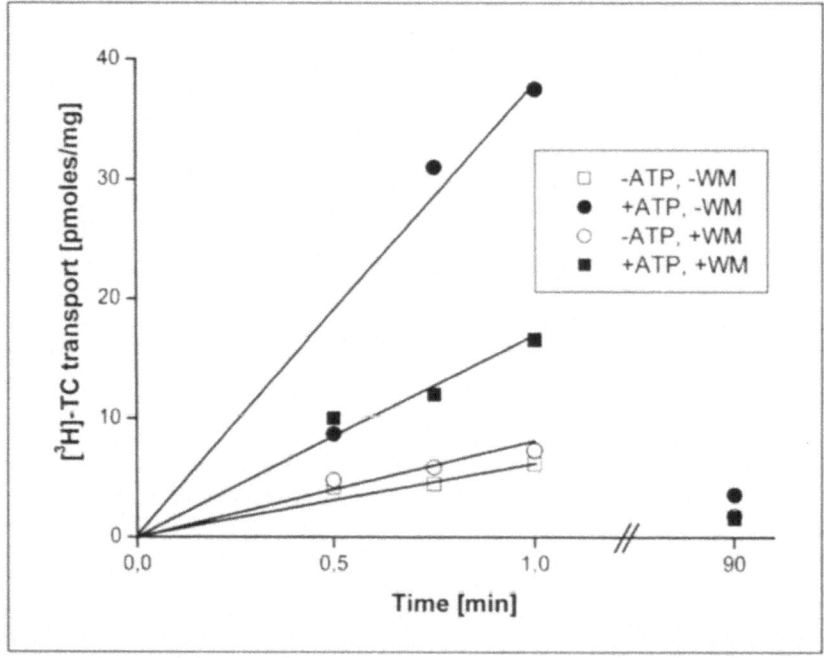

Figure 7. Effect of wortmannin (WM) (50nM) on ATP-dependent transport of [³H]-taurocholate into CMV. (From Ref. 38.)

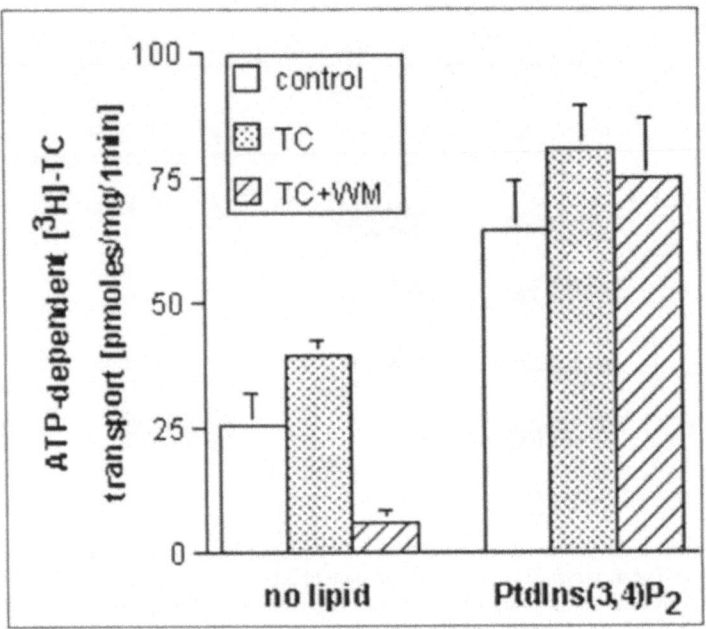

Figure 8. Effect of PtdIns(3,4)P2 on ATP-dependent [^{3}H]-taurocholate transport into CMV prepared from livers receiving pretreatments with taurocholate (TC) and/or wortmannin (WM). CMV were prepared from livers perfused with control buffer (open bars), taurocholate (TC, shaded bars), and taurocholate and wortmannin (TC + WM, hatched bars). ATP-dependent [^{3}H]-taurocholate transport into these CMV was measured with and without preincubation of CMV with PtdIns(3,4)P$_2$. (From Ref. 38.)

peptide, which specifically activates PI 3-K. The rhodamine-linked decapeptide (Janmey PA, Cunningham CC, Stossel TP, et al. U.S. Patent 5,846,743) selectively increases PtdIns(3,4)P$_2$ and PtdINS(3,4,5)P$_3$.[39,40] Addition of the peptide to CMV doubled ATP-dependent [^{3}H]-taurocholate transport into CMV in a dose-dependent and saturable manner (Fig. 9). Studies of the kinetics of ATP-dependent transport of taurocholate by SPGP and GS-DNP by MRP2 in CMV were previously performed at saturating levels of all substrates and kinetic constants, K$_m$ and V$_{max}$, were determined. Addition of the PI3K-stimulating peptide increased ATP-dependent transport of taurocholate and GS-DNP 1.5- to three-fold.[38] The mechanism responsible for this unexpected result is uncertain; however, we speculate that an active PI 3-K signal transduction system within the canalicular membrane result in lipid kinase products which alter and may physiologically regulate transporter activity and bile secretion.

These studies indicate that active PI 3-K is not only required for intracellular vesicular trafficking of ABC transporters in rat liver, but that the lipid products of PI 3-K are necessary for maximal ATP-dependent transport of canalicular ABC transporters in the canalicular membrane. The mechanism by which PI 3-K regulates ATP-dependent transporters is not known; however, a direct interaction with phospholipids has been proposed for the multidrug resistance protein (MDR1).[41,42] Recent studies on the regulation of the K$_{ATP}$ channel by PtdIns(4)P and PtdIns(4,5)P$_2$ suggest that these negatively charged lipids may bind to positive charges at the protein, thereby opening the channel.[43,44] However, because the effect of the PI 3-K lipid products was observed on ATP- dependent transport of [^{3}H]-taurocholate (SPGP), ATP-dependent transport of [^{3}H]-dinitrophenyglutathione (MRP2), and ATP-dependent NBD-PC translocation (MDR2) but not on ATP-dependent [^{3}H]-daunomycin transport (MDR1) (Misra

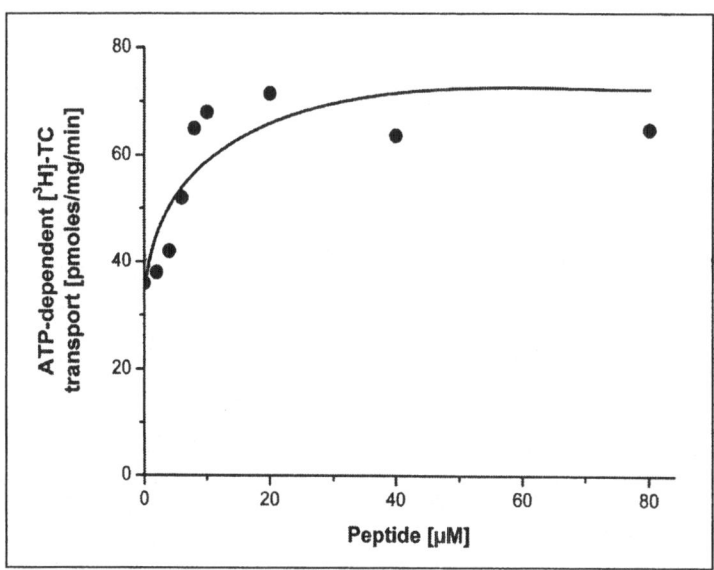

Figure 9. Effect of PI3K activating peptide on ATP-dependent [3H]-taurocholate transport into CMV. (From Ref. 38.)

S, Arias IM, unpublished data, 1999), some but not all canalicular ABC transporters are regulated by PI 3-K lipid products.

Results of Several Recent Studies Support the Thesis Presented in this Review

(i) The function of SPGP and other ABC transporters within the canalicular membrane may be directly activated by cAMP which increases ATP-dependent taurocholate transport in isolated CMV in proportion to activation of PI 3-K; the process is inhibited by wortmannin.[45] These studies support the hypothesis that ABC transporters in the canalicular membrane may exist at different activity levels resulting from posttranscriptional modifications. At least two "crosstalking" signal transduction pathways, cAMP and PI 3-K, participate in this regulation.[38,46,47]

(ii) A point mutation resulting in altered trafficking of the LDL receptor was demonstrated to account for inheritable hypercholesterolemia in Finland.[47] This is the first definitive evidence that mutations affecting intracellular trafficking can produce phenotypes which mimic those resulting from point mutations in a plasma membrane receptor or transporter.

(iii) A molecular defect in a single ABC transporter may profoundly influence the function of other ABC transporters and bile acid homeostasis. For example, mdr 2 -/- mice have reduced cholesterol secretion but manifest liver damage when fed bile acids.[48] Reduced phospholipid secretion in MDR 2/3 mutants may alter bile acid absorption as well as canalicular ABC transporter synthesis, trafficking and regulation thereby producing secondary effects which result in cholestasis and cholesterol lithiasis. Heterozygosity for mutated MDR 3 may predispose individuals to cholestasis and/or cholesterol lithiasis[49,50] when other pathogenic mechanisms alter ABC transporter transcriptional regulation, intracellular trafficking or function within the canalicular membrane.

(iv) "Like Pandora's box, this expanding pathophysiologic panorama poses questions about normal physiology that we are just beginning to answer. By unravelling these interacting skeins, we will better appreciate how acquired conditions interact with genetic susceptibility and produce disease, including cholestasis."[51]

Acknowledgments

This work was supported by Deutsche Forschungsgemeinschaft research grant Ki640 (to H. K.) and by National Institutes of Health grants DK35652 (NIDDK) and 30DK34928 (Digestive Disease Center, NIDDK) (to I. M. A.).

References

1. Bartles JR, Feracci HM, Stieger B et al. Biogenesis of the rat hepatocyte plasma membrane in vivo: comparison of the pathways taken by apical and basolateral proteins using subcellular fractionation. J Cell Biol 1987;105:1241-51.
2. Schell MJ, Maurice M, Stieger B et al. 5'nucleotidase is sorted to the apical domain of hepatocytes via an indirect route [published erratum appears in J Cell Biol 1993;123(3):following 767]. J Cell Biol 1992;119:1173-82.
3. Bartles JR, Hubbard AL. Plasma membrane protein sorting in epithelial cells: do secretory pathways hold the key? Trends Biochem Sci 1988;13:181-4.
4. Roelofsen H, Soroka CJ, Keppler D et al. Cyclic AMP stimulates sorting of the canalicular organic anion transporter (Mrp2/cMoat) to the apical domain in hepatocyte couplets. J Cell Sci 1998;111 (Pt 8):1137-45.
5. Misra S, Ujhazy P, Gatmaitan Z et al. The role of phosphoinositide 3-kinase in taurocholate-induced trafficking of ATP-dependent canalicular transporters in rat liver. J Biol Chem 1998;273:26638-44.
6. Kipp H, Arias IM. Newly synthesized canalicular ABC transporters are directly targeted from the Golgi to the hepatocyte apical domain in rat liver. J Biol Chem 2000;275:15917-25.
7. Kipp H, Pichetshote N, Arias IM. Transporters on demand. intrahepatic pools of canalicular atp binding cassette transporters in rat liver. J Biol Chem 2001;276:7218-24.
8. Ihrke G, Neufeld EB, Meads T et al. WIF-B cells: an in vitro model for studies of hepatocyte polarity. J Cell Biol 1993;123:1761-75.
9. Shanks MR, Cassio D, Lecoq O et al. An improved polarized rat hepatoma hybrid cell line. Generation and comparison with its hepatoma relatives and hepatocytes in vivo. J Cell Sci 1994;107 (Pt 4):813-25.
10. Ihrke G, Martin GV, Shanks MR et al. Apical plasma membrane proteins and endolyn-78 travel through a subapical compartment in polarized WIF-B hepatocytes. J Cell Biol 1998;141:115-33.
11. Sai Y, Nies AT, Arias IM. Bile acid secretion and direct targeting of mdr1-green fluorescent protein from Golgi to the canalicular membrane in polarized WIF-B cells. J Cell Sci 1999; 112 (Pt 24):4535-4545.
12. Nies AT, Cantz T, Brom M et al. Expression of the apical conjugate export pump, Mrp2, in the polarized hepatoma cell line, WIF-B. Hepatology 1998;28:1332-40.
13. Presley JF, Cole NB, Schroer TA et al. ER-to-Golgi transport visualized in living cells. Nature 1997;389:81-5.
14. Toomre D, Keller P, White J et al. Dual-color visualization of trans- Golgi network to plasma membrane traffic along microtubules in living cells. J Cell Sci 1999;112 (Pt 1):21-33.
15. Nakata T, Terada S, Hirokawa N. Visualization of the dynamics of synaptic vesicle and plasma membrane proteins in living axons. J Cell Biol 1998;140:659-74.
16. Hirschberg K, Miller CM, Ellenberg J et al. Kinetic analysis of secretory protein traffic and characterization of golgi to plasma membrane transport intermediates in living cells. J Cell Biol 1998;143:1485-503.
17. McNiven MA. Dynamin: a molecular motor with pinchase action. Cell 1998;94:151-4.
18. Keller P, Simons K. Post-Golgi biosynthetic trafficking. J Cell Sci 1997;110 (Pt 24):3001-9.
19. Lippincott-Schwartz J. Cytoskeletal proteins and Golgi dynamics. Curr Opin Cell Biol 1998; 10:52-59.
20. Traub LM, Kornfeld S. The trans-Golgi network: a late secretory sorting station. Curr Opin Cell Biol 1997;9:527-33.
21. Gatmaitan ZC, Nies AT, Arias IM. Regulation and translocation of ATP-dependent apical membrane proteins in rat liver. Am J Physiol 1997;272:G1041-G1049.
22. Paulusma CC, Kothe MJ, Bakker CT et al. Zonal down-regulation and redistribution of the multidrug resistance protein 2 during bile duct ligation in rat liver. Hepatology 2000;31:684-93.
23. Rost D, Kartenbeck J, Keppler D. Changes in the localization of the rat canalicular conjugate export pump Mrp2 in phalloidin-induced cholestasis. Hepatology 1999;29:814-21.
24. Dombrowski F, Kubitz R, Chittattu A et al. Electron-microscopic demonstration of multidrug resistance protein 2 (Mrp2) retrieval from the canalicular membrane in response to hyperosmolarity and lipopolysaccharide. Biochem J 2000;348 Pt 1:183-8.

25. Gerloff T, Stieger B, Hagenbuch B et al. The sister of P-glycoprotein represents the canalicular bile salt export pump of mammalian liver. J Biol Chem 1998;273:10046-50.
26. Ameen NA, Martensson B, Bourguinon L et al. CFTR channel insertion to the apical surface in rat duodenal villus epithelial cells is upregulated by VIP in vivo. J Cell Sci 1999; 112 (Pt 6):887-894.
27. Mukhopadhayay S, Ananthanarayanan M, Stieger B et al. cAMP increases liver Na+-taurocholate cotransport by translocating transporter to plasma membranes. Am J Physiol 1997;273:G842-G848.
28. Fushimi K, Sasaki S, Marumo F. Phosphorylation of serine 256 is required for cAMP-dependent regulatory exocytosis of the aquaporin-2 water channel. J Biol Chem 1997;272:14800-4.
29. Yao X, Karam SM, Ramilo M et al. Stimulation of gastric acid secretion by cAMP in a novel alpha-toxin- permeabilized gland model. Am J Physiol 1996;271:C61-C73.
30. Pessin JE, Thurmond DC, Elmendorf JS et al. Molecular basis of insulin-stimulated GLUT4 vesicle trafficking. Location! Location! Location! J Biol Chem 1999;274:2593-6.
31. Whitman M, Kaplan DR, Schaffhausen B et al. Association of phosphatidylinositol kinase activity with polyoma middle- T competent for transformation. Nature 1985;315:239-42.
32. Fruman DA, Meyers RE, Cantley LC. Phosphoinositide kinases. Annu Rev Biochem 1998 67:481-507.
33. Toker A, Cantley LC. Signalling through the lipid products of phosphoinositide-3-OH kinase. Nature 1997;387:673-6.
45. Christensen,R.A., Scherpenseel, I. deA., Varticovski, L. Syn thesis of and signaling through D-3 phosphoinositides. In: The Liver: Biology and Pathobiology, 4th edition. Lippincott Williams Wilkins, Phila. PA 495-510.
35. Arcaro A, Wymann MP. Wortmannin is a potent phosphatidylinositol 3-kinase inhibitor: The role of phosphatidylinositol 3,4,5-trisphosphate in neutrophil responses. Biochem J 1993; 296 (Pt 2):297-301.
36. Folli F, Alvaro D, Gigliozzi A et al. Regulation of endocytic- transcytotic pathways and bile secretion by phosphatidylinositol 3-kinase in rats. Gastroenterology 1997;113:954-65.
37. Kipp, H and Arias, IM Regulation of hepatic ABC transporters. Physiol. Reviews 2001 (in press).
38. Misra S, Ujhazy P, Varticovski L et al. Phosphoinositide 3-kinase lipid products regulate ATP-dependent transport by sister of P-glycoprotein and multidrug resistance associated protein 2 in bile canalicular membrane vesicles. Proc Natl Acad Sci U S A 1999;96:5814-9.
39. Hartwig JH, Bokoch GM, Carpenter CL et al. Thrombin receptor ligation and activated Rac uncap actin filament barbed ends through phosphoinositide synthesis in permeabilized human platelets. Cell 1995;82:643-53.
40. Lu PJ, Shieh WR, Rhee SG et al. Lipid products of phosphoinositide 3-kinase bind human profilin with high affinity. Biochemistry (Mosc) 1996;35:14027-34.
41. Doige CA, Yu X, Sharom FJ. The effects of lipids and detergents on ATPase-active P-glycoprotein. Biochim Biophys Acta 1993;1146:65-72.
42. Sharom FJ. The P-glycoprotein multidrug transporter: interactions with membrane lipids, and their modulation of activity. Biochem Soc Trans 1997;25:1088-96.
43. Shyng SL, Nichols CG. Membrane phospholipid control of nucleotide sensitivity of KATP channels. Science 1998;282:1138-41.
44. Baukrowitz T, Schulte U, Oliver D et al. PIP2 and PIP as determinants for ATP inhibition of KATP channels. Science 1998;282:1141-4.
45. Misra S, Varticovski,L, Arias,IM Regulation of taurocholate secretion by cAMP in rat liver (Amer.J. Physiol, in review).
46. Misra,S, Bard,S,Varticovski,L, Arias, IM PI 3-kinase regulates the intracellular trafficking and canalicular membrane activity of BSEP in rat liver (Science, in review).
47. Koivisto,UM, Hubbard, AL, Mellman, I. A novel cellular phenotype for familial hypercholesterolemia due to a defect in polarixing targeting of LDL receptor. Cell 2001;105:575-585.
48. Smit,JJM, Shinkel,AH, Oude Elferink, RPJ et al. Homozygous disruption of the murine mdr2 pglycoprotein gene leads to a complete absence of phospholipid from bile and to liver disease. Cell 1993;75,451-462.
49. Rosmorduc,O, Hermelin,B, Poupon,R. MDR 3 gene defecet in adults with symptomatic intrahepatic and gallbladder cholelithiasis. Gastro. 2001:120:1459-1467.
50. Jacquemin,E, de Vree,JML, Cresteil, D et al. The wide spectrum of multidrug fresistance 3 deficiency: from neonatal cholestasis to cirrhosis of adulthood. Gastro. 2001;120, 1448-1457.
51. Ortiz,D, Arias, IM. MDR3 mutations: a glimpse into Pandora's box and the future of canalicular pathophysiology. Gastro. 2001:120. 1549-1552.

CHAPTER 6

The Pathobiology of Cholangiocytes

Gene LeSage, Shannon Glaser, Heather Francis, Marco Marzioni and Gianfranco Alpini

Summary

In this chapter, we first review the bile duct structure, then the intracellular mechanisms involved in ductal secretion and absorption. The modulation of ductal secretion by hormones, enzymes and neuropeptides is described. Next we describe the specific transporters (e.g., bile acids, chloride channels and Cl^-/HCO_3^- exchanger) that are involved in ductal secretion. We review how the second messenger systems [e.g., cyclic adenosine monophosphate (cAMP), protein kinase C (PKC), and intracellular Ca^{2+}] positively or negatively cooperate to regulate membrane transporter activity. The mechanisms involved in cholangiocyte proliferation, cholangiocyte loss (ductopenia) and cholangiocyte carcinogenesis are reviewed. Finally, we review the disease states associated with cholangiocyte dysfunction and the cholangiocyte response to disease.

Abbreviations

Ach = acetylcholine; ANIT = α-naphthylisothiocyanate; AP = alkaline phosphatase; ATP = adenosine 5'-triphosphate; BDL = bile duct ligation; $[Ca^{2+}]_i$ = intracellular Ca^{2+}; cAMP = adenosine 3', 5'-monophosphate; CCl_4 = carbon tetrachloride; CFTR = cystic fibrosis transmembrane regulator; CF = cystic fibrosis; ET-1 = endothelin-1; IBDU = intrahepatic bile duct units; IP_3 = D-myo-inositol 1,4,5-triphosphate; NHE = Na^+/H^+ exchanger; PBC = primary biliary cirrhosis; PKLD = polycystic kidney liver disease; PKA = protein kinase A; PKC = protein kinase C; PSC = primary sclerosing cholangitis; TC = taurocholic acid; TLC = taurolithocholic; TUDCA = tauroursodeoxycholic acid; VIP = vasoactive intestinal peptide.

Overview of Cholangiocyte Functions

Knowledge of the molecular biology of cholangiocytes is rapidly growing due to increased investigational activity, the development of new experimental models,[1-10] and the characterization of transport systems and second messenger systems. Cholangiopathies include liver diseases that target cholangiocytes such as primary biliary cirrhosis (PBC), primary sclerosing cholangitis (PSC), idiopathic ductopenia syndromes and cystic fibrosis (CF).[3] In other forms of liver diseases, bile ducts proliferate in response to chronic inflammation.[3] Specifically how cholangiocyte function is altered and how they become targets in hepatobiliary diseases remain unknown.

Bile secretion is initiated by the active transport of bile acids from hepatocytes into the bile canaliculus.[11] The osmotic water flow induced by bile acid-dependent canalicular bile flow accounts for approximately 40% of total bile flow.[11] Independently, bile acid-independent secretion (due to transport into bile of organic solutes and electrolytes) contributes to approximately 30% total bile flow.[11] Bile formation is also caused by reabsorption and secretion of fluid and inorganic electrolytes by bile ducts.[3,12,13] Ductal bile flow is primarily regulated by

Molecular Pathogenesis of Cholestasis, edited by Michael Trauner and Peter L.M. Jansen.

the hormone secretin (Fig. 1)[1-5,7-10,12,14-19] and represents 30% of total bile flow.[11] The cholangiocyte transport systems [cAMP-dependent chloride channel cystic fibrosis transmembrane regulator (CFTR),[3,12,13] Cl^-/HCO_3^- exchanger,[3,12,13,20] a Na^+/H^+ exchanger (NHE),[12,21] and a $Na^+:HCO_3^-$ cotransporter],[12,21] cooperate together to produce a bicarbonate-rich bile secretion[3,12,13] (Fig. 1). Secretin stimulates ductal bile flow by increasing intracellular cAMP[7,12] which promotes biliary HCO_3^- secretion,[15,19,20] by stimulating apical Cl^- channels[22] and Cl^-/HCO_3^- exchanger[3,12,19-21] activities (Fig. 1). Cholangiocyte secretion is also increased by adenosine 5'-triphosphate (ATP), which is released into the bile by hepatocytes and cholangiocytes, and activates cholangiocyte apical purinergic receptors, Cl^- channel and NHE activities.[23,24]

Cholangiocytes display an array of membrane transport systems for bile acids,[25-28] glucose,[29] and amino acids[30] on their apical membrane. In addition, water channels are expressed by cholangiocytes,[31-35] which regulate passive water movement in a bi-directional manner producing both water secretion and absorption by bile ducts.[33-35] Transporters may acutely increase their activity by translocation to the apical membrane.[33,35]

Bile Duct Structure

Intrahepatic bile ducts are lined by columnar cells resting on a basement membrane.[36-38] There is a conspicuous ER and Golgi as well as a vacuolar compartment,[36-38] but they are less well developed as compared to hepatocytes. Functional tight junctions have been identified between cholangiocytes.[39,40] The apical membrane has microvilli increasing effective surface area.[39,40]

Intrahepatic bile ducts are classified by diameter: hepatic ducts (> 800 μm), segmental ducts (400-800 μm), area ducts (300-400 μm), septal bile ducts (100-200 μm), interlobular ducts (15-100 μm), and bile ductules (<15 μm).[36-38] Small bile ducts are lined by 4 to 5 cuboidal cholangiocytes and larger bile ducts consist of 10-100 cholangiocytes.[36,37] The lining cholangiocytes are progressively larger and more columnar in shape as ducts become larger.[36,37]

Secretin-, and somatostatin-regulated cholangiocyte secretion occurs exclusively in larger (> 15 μm in diameter) bile ducts in rats (Fig. 2).[1] Large (but not small) cholangiocytes[2,4,5] and intrahepatic bile duct units (IBDU)[1] isolated from rats express the messages for secretin,[1,2,4,5] somatostatin (SSTR_2) receptors,[2] Cl^- channels[4,5] and Cl^-/HCO_3^- exchanger[4,5] and respond to secretin and somatostatin with changes in cAMP levels[1,2,4,5,8,10] (Fig. 2). As a direct demonstration of regionalization of secretory activity in the intrahepatic biliary tree, large (> 15 μm in diameter) but not small (< 15 μm in diameter) isolated IBDU respond to secretin with an increase in duct lumen size.[1] Similar to rats,[1,2,4,5,8,10,25] human bile duct secretion is also regionalized, since only large-size bile ducts express the Cl^-/HCO_3^- exchanger.[41] The exclusivity of secretion to only large ducts may be due to the presence of peribiliary plexus,[42] the vascular element for cholangiocytes, being limited to ducts greater than 15 μm in diameter. This is supported by recent studies showing that following bile duct ligation (BDL), microvascular proliferation occurs only adjacent to large proliferating ducts.[42]

Ductal Secretion

Overview

Studies in purified cholangiocytes[2-5,8,10,12,13,19,25,27,33-35] and IBDU[1,20,21,43] have disclosed the transport mechanisms involved in secretin-stimulated ductal secretion.[1-5,8,10,12,13,19-21,25,27,33-35,43] Cholangiocytes modify canalicular bile by secretion of Cl^-[5,22] and HCO_3^-.[1,12,15,19-21] On the basolateral membrane, NHE[21,44] and the $Na^+:HCO_3^-$ symporter[21] mediate HCO_3^- uptake,[21,44] and on the apical membrane, cAMP-activated Cl^- channel[5,22] and Cl^-/HCO_3^- exchanger[4,12,19-21,43] secrete bicarbonate into the lumen.[1,3,4,8,10,15,19-21] Both CFTR[22,45] and Ca^{2+}-activated Cl^- channels[23,24] are present on the apical membrane. Cholangiocyte secretory functions are increased by the hormones secretin (Fig. 1),[1-5,7,8,10,12,13,15,19,21,45] bombesin[46] and vasoactive intestinal peptide (VIP),[47] whereas the hormones somatostatin, gastrin (Fig. 1),

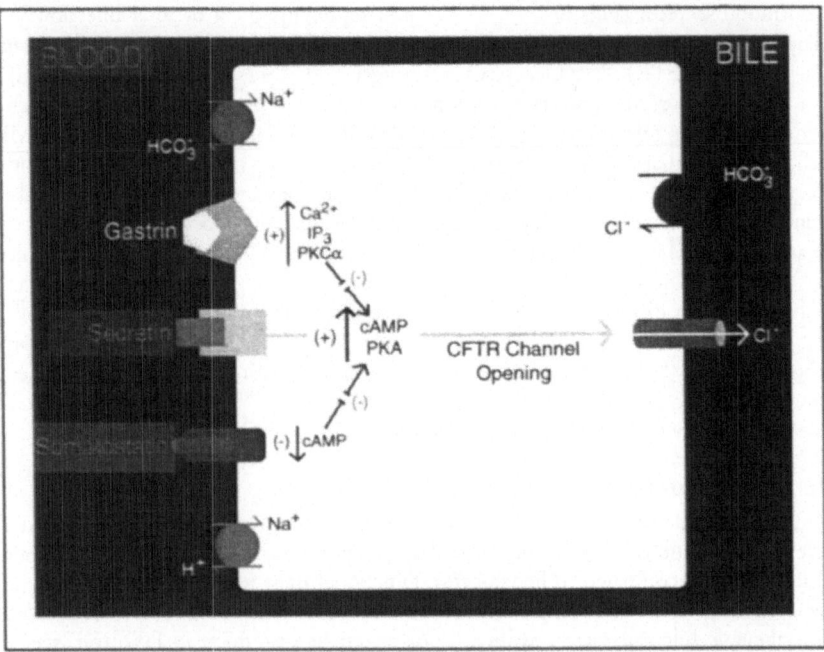

Figure 1. Schematic representation of the intracellular mechanisms by which secretin, somatostatin, and gastrin coordinately regulate bicarbonate secretion in ductal bile. Secretin interaction with its own receptor induces an increase in cAMP-dependent PKA activity. Increased PKA activity stimulates (by phosphorylation) CFTR, leading to activation of the Cl^-/HCO_3^- exchanger with subsequent secretion of bicarbonate in water. Somatostatin (through interaction with $SSTR_2$ receptors) inhibits secretin-stimulated cAMP levels, Cl^- channel activation, and Cl^-/HCO_3^- exchanger activity, leading to decreased ductal bicarbonate secretion. Gastrin inhibitory effect on secretin-stimulated bicarbonate secretion is mediated by activation and membrane translocation of the Ca^{2+}-dependent PKC pathway, leading (by cross talk with adenylate cyclase) to decreased secretin-induced cAMP synthesis and CFTR and Cl^-/HCO_3^- exchanger activity. Reproduced with permission from ref.[12]

endothelin-1 (ET-1) and insulin decrease secretion due to down regulation of cAMP.[2,6,18,48] ATP[23,24] and bile acids[25,27,49] regulate ductal secretion by signaling events at the apical membrane of cholangiocytes. ATP signals through purinergic receptors present on the apical membrane[23,24] and bile acids alter cholangiocyte secretion after uptake by the Na^+-dependent apical membrane bile acid transporter (ABAT).[25]

Regulatory Hormones, Enzymes and Neurotransmitters

Secretin

Secretin, the major regulator of ductal secretion,[1-5,7,8,10,12,15,19,21,45] binds to secretin receptor on the basolateral membrane of cholangiocytes[16] resulting in increased cAMP levels[1,2,4-8,10,18,19] (Fig. 1). In vivo, secretin induces a much larger bicarbonate-rich choleresis in rats with enhanced ductal hyperplasia induced by BDL,[3,8,13-15] partial hepatectomy,[19] cirrhosis,[50,51] or α-naphthylisothiocyanate (ANIT) feeding[9,15] compared to normal rats where secretin-induced choleresis is minimal.[9,14,15,18,19] Studies [52] have shown that secretin administration to normal rats via the hepatic artery induces a much greater choleresis compared to when given in a peripheral vein or portal vein showing that hormone-induced ductal secretion is a significant contributor to bile secretion in normal rats as well.

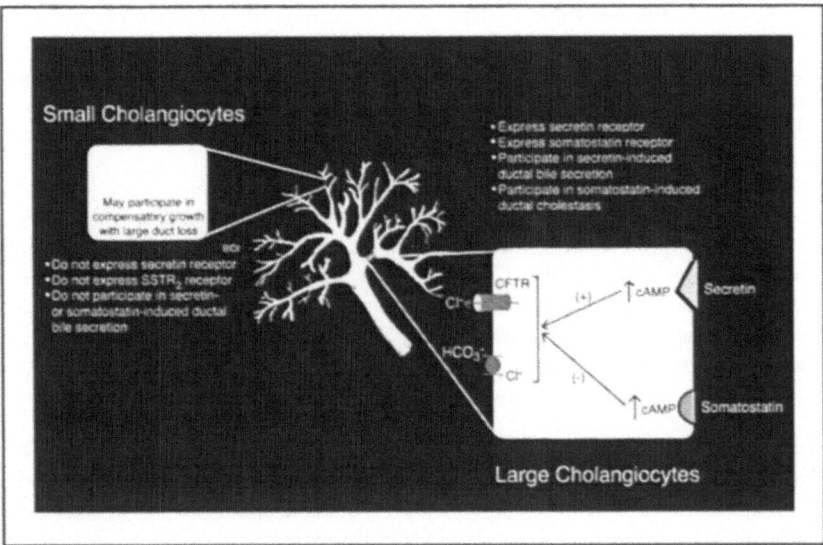

Figure 2. Schematic representation of the secretory heterogeneity of the intrahepatic biliary tree. The cartoon shows that the major sites of secretin- and somatostatin-regulated transport of water and electrolytes involve cholangiocytes in large bile ducts (which contain secretin and somatostatin receptors), whereas small cholangiocytes in small ducts do not express secretin and somatostatin receptors and do not participate in hormone-regulated ductal secretion. Reproduced with permission from ref.[12]

Bombesin and VIP

Bombesin and VIP are neuropeptides that stimulate bile secretion.[46,47] Recent studies have shown that these neuropeptides increase bile flow by augmenting cholangiocyte secretion.[46,47] In the studies, bombesin and VIP were shown to directly increase ductal secretin in IBDU but not hepatocyte couplets from rat liver.[46,47] Similar to secretin[1-10,12-15,18,19,21,50] (Fig. 1), in BDL rats, bombesin and VIP increase bile flow and bicarbonate to a much greater degree as compared to controls,[46,47] consistent with an increased number of cholangiocytes in BDL rats producing more ductal bile flow.[14,15] Stimulation of the Cl^-/HCO_3^- exchanger activity by bombesin was independent of increases in the second messengers cAMP, cGMP or intracellular Ca^{2+} $[Ca^{2+}]_i$.[46] Similarly, VIP increases ductal secretion by a cAMP-independent mechanism.[47] Further studies are necessary to evaluate the intracellular mechanisms by which bombesin and VIP stimulate ductal secretion.

Gastrin, Insulin, ET-1 and Somatostatin

Gastrin (Fig. 1), insulin, ET-1 and somatostatin (Fig. 1) inhibit secretin-stimulated ductal secretion by inhibiting the expression of secretin receptors and secretin-stimulated cAMP synthesis, thus preventing secretin's choleretic effects.[2,6,18,48] Both gastrin and insulin increase $[Ca^{2+}]_i$ and PKC alpha activity in cholangiocytes, which are required for inhibition of secretin stimulated cAMP synthesis and ductal secretion.[17,48] Gastrin receptors and somatostatin receptors are localized on the basolateral membrane,[2,17,18] whereas insulin receptors are present on the apical membrane of cholangiocytes (LeSage, Glaser, and Alpini, unpublished observations). Since insulin significantly inhibits secretin-stimulated ductal secretion in vivo, it is likely that insulin in bile is biologically active.

Alkaline Phosphatase

Cholangiocytes are exposed to high concentrations of alkaline phosphatase (AP) in bile.[53] Recent studies in rats have shown that in vivo, acute and chronic administration of AP decreases basal and secretin-stimulated ductal secretion.[54] In vitro, basal and secretin-stimulated $Cl^-/$ HCO_3^- exchanger activity of IBDU was immediately inhibited by AP intralumenal microinjection (apical exposure) but only after a prolonged exposure to the basolateral domain of cholangiocytes.[54] The studies show that the inhibitory effects of AP on secretin-stimulated ductal secretion may be due to its capacity to block CFTR activity or to hydrolyze ATP bonds.[54]

Acetylcholine

Acetylcholine (ACh) enhances secretin-stimulated ductal secretion.[20] M_3 subtype but not M_1 or M_2 ACh receptors are present in rat cholangiocytes.[20] ACh increases secretin-stimulated (but not basal) activity of the Cl^-/HCO_3^- exchanger in IBDU and secretin-stimulated cAMP synthesis in isolated cholangiocytes.[20] The potentiation of secretin-stimulated ductal bile secretion was dependent on Ca^{2+} but not PKC.[20] FK-506 and cyclosporine inhibit Ach potentiation of secretin stimulated Cl^-/HCO_3^- exchanger, demonstrating that calcineurin most likely mediates the cross talk between the calcium and adenylyl cyclase pathways.[20]

Paracrine and Autocrine

Extracellular ATP and adenosine have established roles as stimulants of electrolyte and fluid secretion in bile ducts.[23,24] In cholangiocytes, ATP binds to apical $P2Y_2$ (P_{2U}) receptors.[23,24] ATP, released from hepatocytes, is capable of stimulating receptors on both adjacent hepatocytes and after traversing to the biliary tree, stimulating receptors on cholangiocytes.[23,24] ATP in bile is a potent stimulus for cholangiocyte Cl^- and fluid secretion and activates basolateral NHE.[24] ATP also increases Cl^-/HCO_3^- exchanger activity in cholangiocytes pretreated with cAMP analogs.[24] Thus, ATP may play a role in coordinating the hepatocyte and ductal components of bile formation, a process that has been termed "hepatobiliary coupling". Recent studies[23,24] have demonstrated constitutive and volume-sensitive ATP release from primary hepatocytes and from apical membranes of polarized normal rat cholangiocytes. Consequently, ATP release into the canalicular space (by hepatocytes) or the release of ATP from cholangiocytes may regulate secretion of cholangiocytes in the downstream biliary tree (paracrine signaling).[23,24] It is likely that ATP released from individual cholangiocytes may bind to their own P_{2U} receptors, thus self-regulating their secretion (autocrine signaling).[23,24] The concentration of biliary ATP will depend on various parameters including the amount of ATP delivered by hepatocytes and the activity of lumenal ATPases.

Cholangiocyte Transporters and Channels

Chloride Transporters

CFTR is a cAMP-regulated chloride channel that resides in the apical membrane of many epithelial cells.[56] The primary mode of regulation of CFTR is by phosphorylation of the R domain by cAMP-dependent protein kinase.[56] Once phosphorylated, CFTR channels require hydrolyzable nucleotides to be active.[56] The second mode of regulation of CFTR is by cAMP-stimulated CFTR apical membrane insertion.[56] Although much is known about the structural properties and regulation of CFTR, little is known of its relationship to cellular functions other than the cAMP-dependent Cl^- secretion.[56] In the liver, cholangiocytes are the sole site of expression of the CFTR.[57] The disease CF is caused by mutations in CFTR, which can result in defective protein production, defective processing and degradation in the endoplasmic reticulum, or defective channel pore properties or gating properties.[56] The primary defect in CF is the altered activity of a cAMP-activated Cl^- channel, the CFTR channel.[56,58] It has been shown that $\Delta F508$, the most common CF mutation, impedes CFTR trafficking to the apical surface of epithelial cells.[58] CF is a disease characterized by abnormalities in the epithelia of the lungs [59], intestine [60], salivary and sweat glands [61], liver [62], pancreas [62], and reproductive systems [63], often as a result of inadequate hydration of their secretions. However, it is not clear how a defect in the CFTR Cl^- channel function leads to the observed pathological changes.

In cholangiocytes, other Cl⁻ channels,[23,24] besides CFTR[5,12,22] are present. A Ca^{2+}-activated Cl⁻ channel is regulated by a Ca^{2+}/calmodulin-dependent protein kinase.[64,65] Cl⁻ channel activity is required for the production of ductal bile secretion, since inhibitors of Cl⁻ channels ablate secretin-stimulated ductal secretion.[45] In CF, Ca^{2+}-activated Cl⁻ channel and voltage-gated Cl⁻ channels may substitute, thus partially maintaining ductal secretion.

Other Ion Transporters

The primary secretion from cholangiocytes is a bicarbonate rich fluid.[3,14,15,18,20,43] This secretion is the result of cooperation between ion transporters on the basolateral and apical membrane of cholangiocytes. The Na^+:HCO_3^- symporter mediates HCO_3^- uptake,[21] and on the apical membrane, the Cl⁻/HCO_3^- exchanger secretes bicarbonate into the lumen.[3,4,12,19-21] These ion transporters appear to be primarily under the control of cAMP.[3,4,12,19-21] Secretin stimulates basolateral secretin receptors[1-5,7-10,12,13,16,19] which subsequently induces elevation of cytosolic cAMP levels[1,2,4,5,7-10,17-19] and activation of intracellular protein kinase A (PKA)[66] (Fig. 1). PKA induces opening of the CFTR by phosphorylation, leading to extrusion of Cl⁻ ions, decreased intracellular Cl⁻ concentration, and depolarization of the cell membrane[22,67] (Fig. 1). Opening of Cl⁻ channels[22,45,67] leads to activation of the apically located Cl⁻/HCO_3^- exchanger activity[1,4,12,14,15,19,21,66] that results in secretin-stimulated bicarbonate-rich choleresis[8-10,14,15,18,19] (Fig. 1). Secretin-induced choleresis is also regulated, by a balance between the activities of activating kinases and inactivating phosphatases, which activate and inactivate CFTR respectively.[66]

However, recent findings[68] show that secretin acts in a DIDS-insensitive fashion, which suggests that secretin may not act through the Cl⁻/HCO_3^- exchanger in cholangiocytes. CFTR may mediate bicarbonate efflux directly, rather than via Cl⁻ efflux or activation of the Cl⁻/HCO_3^- exchanger.[68] These studies are consistent with the other recent findings showing that the CFTR blockers DPC and NPPB both inhibit secretin-induced bicarbonate excretion in cholangiocytes.[68] Cholangiocytes express Ca^{2+}-dependent Cl⁻ channels and secretion via CFTR can be bypassed by Ca^{2+}-mediated activation of these channels in some systems.[23,24,65] Since secretion may occur by Ca^{2+}-dependent Cl⁻ channels, increased ductal secretion by agonists that increase Ca^{2+} may be employing Ca^{2+}-dependent Cl⁻ channels to enhance the activity of the Cl⁻/HCO_3^- exchanger. Recent studies[68] showing that Ach stimulates ductal bile secretion in a DIDS-sensitive fashion are consistent with this explanation.

With bicarbonate extrusion into bile, other ion transport mechanisms come into play to maintain cholangiocyte intracellular pH. Mechanisms that restore cholangiocyte intracellular pH after activation of secretion are important for maintenance of bicarbonate secretion since a higher intracellular pH activates the apical Cl⁻/HCO_3^- exchanger due to the presence of a pH-dependent domain on the exchanger.[21,69] Some studies[21,70] show that the electrogenic Na^+/HCO_3^- cotransporter on the basolateral membrane of rat cholangiocytes causes the entry of HCO_3^- ions into cells and increases the intracellular pH. Other studies,[44] showing that HCO_3^- influx is minimal in human cholangiocytes, suggested that the Na^+/HCO_3^- cotransporter is not needed for bicarbonate secretion. Furthermore, the Na^+/HCO_3^- cotransporter is not active at physiological pH in cholangiocytes.[44] As an alternative to the Na^+/HCO_3^- cotransporter, a basolateral NHE counteracts bicarbonate excretion by extruding acid into the circulation, thus maintaining the intracellular pH.[21] Secretin does not directly regulate the Na^+/HCO_3^- symporter or the NHE in rat cholangiocytes.[21] Other studies in guinea pigs show secretin activates the NHE and causes intracellular alkalinization in cholangiocytes.[71]

Bile Acid Transporters

Up until recently, it was assumed that bile acids, after hepatocyte secretion, were simply conducted to the intestine by bile ducts.[72] Studies with BDL rats suggested that bile acids may move across biliary epithelium and that unique bile acids may passively be absorbed by bile ducts as a protonated species.[73] The absorbed bile acid returns via the peribiliary plexus to the hepatocytes for resecretion into bile.[74,75] This shunting of bile acids back and forth between hepatocytes and cholangiocytes has been termed cholehepatic shunting.[74,75] Most studies have suggested cholehepatic shunting did not significantly contribute to bile formation in the normal state.

More recently, the interest in bile acid transport in bile ducts has markedly increased due to the identification of bile acid transport in cholangiocytes.[25,27,28,49] Recent studies[25] showed genetic and protein expression for ABAT and the 14-kilodalton ileal cytosolic binding protein in cholangiocytes. ABAT is structurally identical to the ileal bile acid transporter, which is the sole transporter involved in the reclamation of bile acids from the ileum.[76] The ABAT has a Km and Vmax very similar to IBAT in the ileum and the Km (33 μmol) is (as in the ileum) much lower than the duct lumen bile acid concentration.[76,77] Presumably, the low Km matches the low perimembrane bile acid concentration due to unstirred effects as it has been proposed for the ileal bile acid uptake mechanism.[76,77] ABAT is expressed in large but not small (less than 15 μm) bile ducts.[25] It has been suggested that the presence of ABAT in small ducts would counterpoise that secretion of bile acids at the level of the canalicular membrane, thus reducing overall bile acid-induced bile flow.[25]

Bile acids interact with cholangiocytes, both in vitro[49] and in vivo in bile acid [i.e., tauro-cholic (TC) and taurolithocholic (TLC) acids] fed animals,[78] resulting in cholangiocyte pro-liferation (Fig. 3) and increases in ductal bile secretion.[49,78] Corresponding to the presence of ABAT only in large cholangiocytes, bile acids stimulate proliferation and secretion only in large cholangiocytes.[49] This observation is probably due to the requirement of bile acid uptake by ABAT[25] since only intracellular bile acids can signal cholangiocyte proliferation and secretion.

Several bile efflux mechanisms have been identified in cholangiocytes.[25,26,28] Recent studies[26] functionally identified an anion exchange mechanism for the efflux of bile acids across the basolateral membrane of cholangiocytes. Recently, an exon-2 skipped, alterna-tively spliced form of ABAT, designated t-ASBT, was found in rat cholangiocytes, ileum and kidney.[28] The t-ASBT was specifically localized to the basolateral domain of cholangiocytes.[28] Transport studies in Xenopus oocytes revealed that t-ASBT functions as a bile acid efflux protein.[28]

Water Channels

Aquaporin-1 (AQP1) water channels are present in the apical membranes of cholangiocytes, and mediate the transport of water in these cells.[31,33-35] Secretin increases cholangiocyte os-motic water permeability and stimulates the redistribution of AQP1 from an intracellular ve-sicular pool to the cholangiocyte apical membrane.[33,35] Membrane insertion of AQP1 is mi-crotubule-, and cAMP-dependent.[35] It has been proposed that insertion of AQP1 into the apical cholangiocyte membrane is at least partially responsible for the ability of secretin to stimulate ductal bile secretion.[35] AQP4, another water channel is constitutively expressed on the basolateral cholangiocyte membrane and is secretin unresponsive.[34] AQP4 may facilitate the basolateral transport of water in cholangiocytes.[34]

Ductal Absorption

Cholangiocytes likely absorb fluid and solutes from primary bile as it passes from the canali-cular membrane to the intestine. Sugars are absorbed by the biliary ductal system by Na^+-dependent and Na^+-independent transport systems.[29] The Na^+-dependent glucose trans-porter (i.e., SGLT1) is present on the apical domain of cholangiocytes, and, together with GLUT1 (a facilitative glucose transporter) on the basolateral domain, regulates glucose absorp-tion from bile.[29] In the isolated perfused rat liver, there is a decrease in bile flow, but an increase in biliary bile acid concentration following increasing glucose concentration in the perfusate, findings which suggest that the reabsorbed glucose induces osmotic induced absorption of water.[79] Biliary glucose concentration is generally very low, which has been attributed the pro-cess of glucose absorption by bile ducts and may help prevent bacterial growth in the biliary system.[80] Studies employing isolated perfused IBDU also demonstrate the capacity of bile ducts to absorb water.[81]

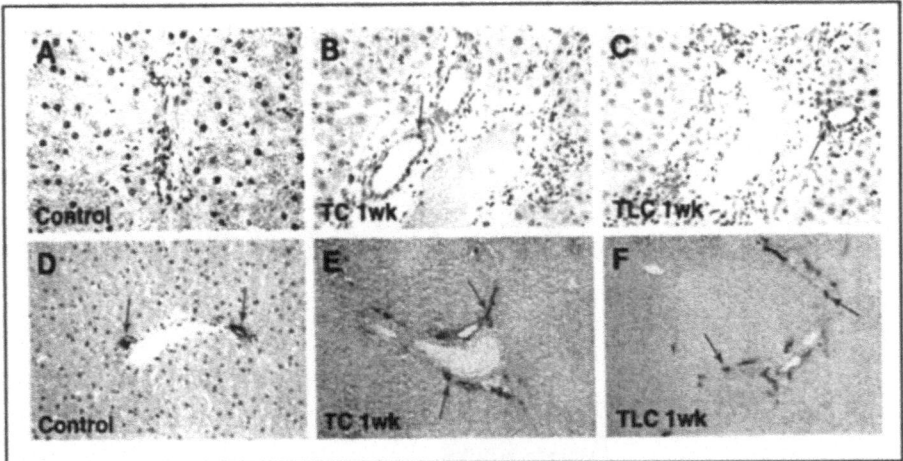

Figure 3. Staining for (A-C) PCNA and (D-F) γ-GT in sections from (A and D) control, (B and E) TC-fed, or (C and F) TLC-fed rats. After TC and TLC feeding (at 1 week, a representative experiment), there was a marked increase in the number of PCNA-positive cholangiocytes and γ-GT-positive bile ducts (original magnifications 125X and 160X). Arrows indicate cholangiocytes and bile ducts positive for PCNA or γ-GT, respectively. Reproduced with permission from ref.[78]

Vesicle Pathways

Cholangiocytes likely have an active plasma membrane turnover due to endocytosis/exocytosis events at both the apical and basolateral membrane domains.[82] Receptor-mediated endocytosis occurs in cholangiocytes.[82] Cholangiocytes also participate in fluid-phase endocytosis.[83,84] Studies have shown that cholangiocytes internalized the fluid phase marker, HRP, exclusively at the apical domain, and consistent with fluid phase endocytosis, internalization was completely blocked at 4°C.[83] After internalization, HRP was seen in transcytotic vesicles and lysosomes.[83] Later, HRP is then discharged by exocytosis at the basolateral cell surface.[83] By endocytosis, cholangiocytes may modify the composition of bile by internalizing both biliary proteins and fluid.[85]

Cholangiocyte Proliferative Responses

Cholangiocytes proliferate in most human liver diseases,[3] with experimental bile duct injury due to BDL[3,14,15,17] or ANIT feeding.[9,15] Cholangiocytes also proliferate following partial hepatectomy, restoring bile duct mass to normal within one week.[19] With all models of cholangiocyte proliferation studied to date, there is associated increased ductal secretion.[2,3,8-10,14-19,50,51,78]

The intracellular signals that control cholangiocyte proliferation are multiple and complex and cross talk between these signals exists. The best-studied second messenger is cAMP, which is increased in proliferating cholangiocytes isolated from all models of ductal hyperplasia.[2,8-10,17-19,78] In animal models where cAMP synthesis is impaired, there is blunted ductal proliferation in response to ductal injury.[86] Increased PKC and Ca^{2+} (induced by the administration of gastrin) inhibit hyperplastic (Fig. 4A) and neoplastic cholangiocyte proliferation[17,87] and secretin-stimulated cAMP levels, a marker of cholangiocyte proliferation (Fig. 4B).[2,3,8-10,17,19] Bile acids increase cholangiocyte proliferation[49,78] (Fig. 3). After uptake by ABAT, TC and TLC increase proliferation[49,78] whereas tauroursodeoxycholate (via a PKC and Ca^{+2}-dependent mechanism) inhibits cholangiocyte proliferation (Alpini, Glaser and LeSage, unpublished observations, 2001). Ductal proliferation, induced by duct ligation, is associated with increase expression of c-myc in cholangiocytes.[88]

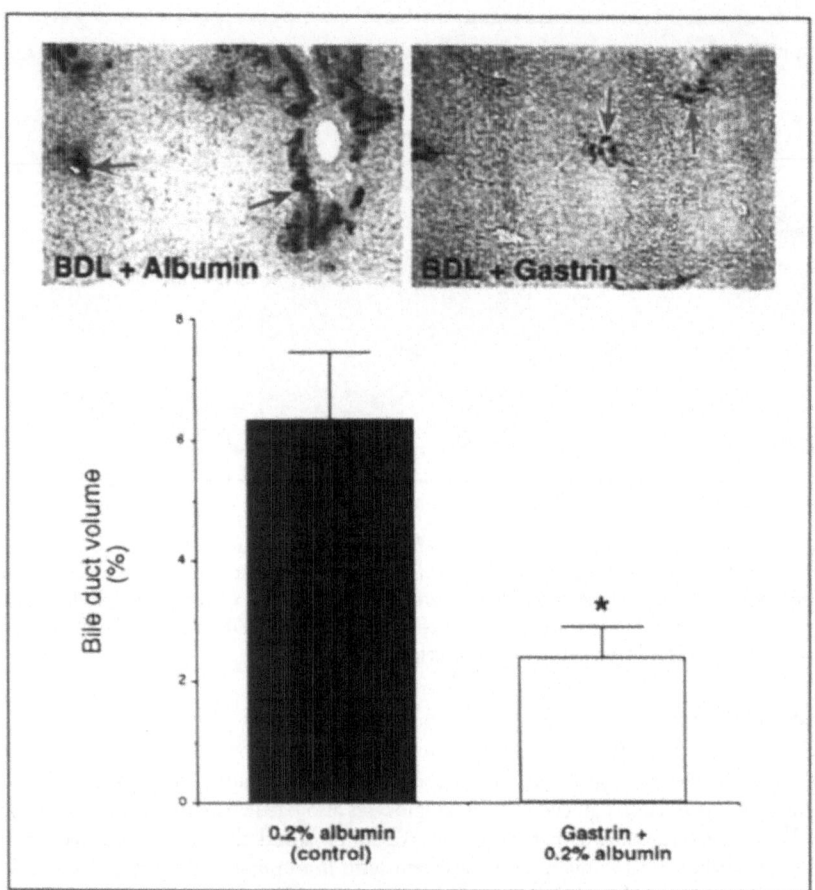

Figure 4A. Histochemistry for γ-GT in liver sections from BDL rats infused for 7 days with gastrin (left) or 0.2% BSA (control [right]). After gastrin infusion, the number of intrahepatic bile ducts markedly decreased compared with BSA-treated BDL rats. Original magnification X125. *P < .05 versus ductal mass of BSA-treated BDL control rats.[17]

Cholangiocytes differentially proliferate in response to liver injury/toxins.[2,8-10] Following BDL, only large cholangiocytes in large ducts undergo mitosis with increases in DNA synthesis and ductal secretion.[2] A single dose of CCl₄ induces damage of large hormone-responsive ducts with loss of proliferative and secretory capacity whereas small cholangiocytes [which are constitutively quiescent and unresponsive to secretin][1,2,4,5,8-10] de novo proliferate and secrete to compensate for loss of large duct function.[8,10] Chronic ANIT feeding stimulates proliferative capacity of both small and large cholangiocytes.[9] The mechanisms by which different sized compartments of the intrahepatic biliary epithelium differentially respond to specific injury/ toxins are undefined.

Overview of Cholangiopathies

Damage to the bile ducts is the cause of several chronic cholestatic disorders (cholangiopathies).[3] In all cholangiopathies, there is the coexistence of cholangiocyte death and proliferation, ductal remodeling, inflammation and fibrosis.[3] Cholestasis is almost always present and may be the cause or promote the progression of the disease.[3] CF is an example of

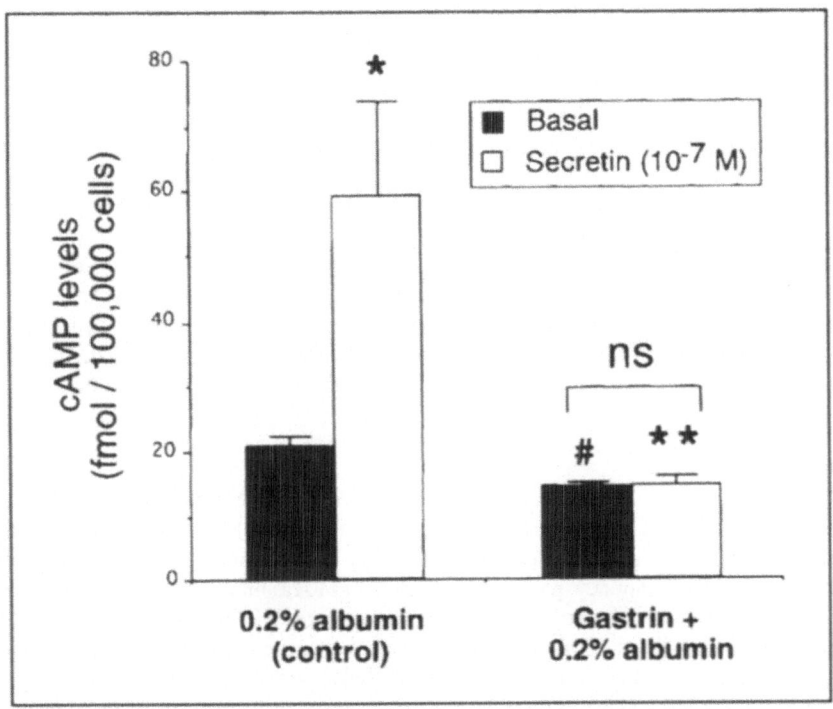

Figure 4B. Intracellular cAMP levels in cholangiocytes from BDL rats treated for 7 days with gastrin or 0.2% BSA (control) by osmotic minipumps. Cholangiocytes were stimulated at 22°C for 5 minutes with secretin and cAMP levels determined by RIA. *P < .05 versus corresponding basal values. #P < .05 versus basal values of BDL control rats. **P < .05 versus secretin-induced cAMP levels of BDL control rats. Data are mean ± SEM of 6 to 9 experiments. Reproduced with permission from ref.[17]

biliary cirrhosis secondary to a dysfunction of cholangiocyte ion transport.[3,89] Thus, dysfunctional biliary electrolyte transport may also promote the cholestasis in other cholangiopathies.[3] Most primary cholangiopathies appear to be due to an autoimmune induced process.[3] Cytokines and proinflammatory mediators likely induce apoptotic and proliferative responses in cholangiocytes, activate fibrogenesis, and alter the transport functions of cholangiocytes.[3,90]

Primary Biliary Cirrhosis

PBC is the prototypic bile duct damage disease in humans.[3] The diagnosis of PBC is most often made in the asymptomatic phase.[3] The antimitochondrial antibody is the predominant serologic marker for the disease, although not all patients test positive.[3,91,92] The etiology of PBC remains elusive; studies suggest that the interlobular bile duct destruction is immune based, and associated autoimmune diseases are common.[92] Patients with PBC have autoantibodies that react with components of mitochondrial multi-enzyme complexes.[92] In addition to binding to mitochondria, patients' autoantibodies to the assumed major auto antigen pyruvate dehydrogenase complex (PDC) dihydrolipoamide acetyltransferase (E2) bind to the plasma membrane of cholangiocytes specifically in PBC.[3] Symptoms common in this disease are fatigue, pruritus and xanthelasma, as well as complications of portal hypertension and osteoporosis.[3] Treatment includes symptomatic and preventive measures, as well as specific therapeutic measures.[3] Immunosuppressive therapy has yielded disappointing results in the long-term management of PBC, and the only therapy shown to improve survival is the hydrophobic dihydroxy bile acid ursodeoxycholic acid.[3]

Primary Sclerosing Cholangitis

PSC is an unusual disease of unknown etiology.[3] The primary pathologic event is inflammation and prominent fibrosis of intrahepatic and extrahepatic bile ducts.[3] The sclerosis of the bile ducts may be the result of multiple factors, including autoimmune, bacterial, congenital, drug, or viral agents.[3] The etiology of PSC remains poorly understood, despite a large number of studies looking at differing hypotheses. The most commonly associated diseases are ulcerative colitis with an incidence that varies from 2.5% to 7.5%.[3,93] Conversely, 50% to 75% of patients with PSC have ulcerative colitis.[93] Most patients present with jaundice, pain, and pruritus, although an increasing number of asymptomatic patients with inflammatory bowel disease and abnormal liver function are being diagnosed.[94] Cholangiography is the key to the diagnosis and is usually pathognomonic except in the rare case where PSC is confused with cholangiocarcinoma.[94] Multiple forms of medical therapy have been tried, including ursodeoxycholic acid.[94] To date, however, none of these medications has altered the course of this disease.

Cholangiocarcinoma

Cholangiocarcinoma occurs frequently in patients with PSC.[3] Most of the other risk factors for cholangiocarcinoma (e.g., liver flukes and intrahepatic lithiasis) have long-standing inflammation and injury of cholangiocytes.[95] Approximately 10% of patients with ulcerative colitis develop cholangiocarcinoma.[95] p53 over expression and K-ras mutations occur commonly in patients with PSC and biliary tract cancer and are associated with a shortened survival.[96] Patients with longstanding PSC are less likely to have these genetic alterations and may have a better prognosis.[96] The mechanisms for control of cholangiocarcinoma growth are poorly understood, however studies have shown gastrin (by increasing PKC alpha),[87] alpha2 adrenergic stimulation (by a raf-dependent, and ras-independent mechanism)[97] and ursodeoxycholate (by a PKC-dependent mechanism) (Kanno, Alpini, and LeSage, unpublished observations, 2001) inhibit the growth of cholangiocarcinoma.

Cystic Fibrosis

Although CF is considered primarily to be a pulmonary disease, liver disease has been increasingly diagnosed during recent years probably due to the increase awareness, more careful hepatic assessment and reduced death from extra-hepatic causes.[3,89] Although liver disease with CF generally runs a mild course, it is a major complication that may limit survival and quality of life of affected patients.[3,89] CF is the first inherited liver disorder in which the primary defect affects cholangiocyte transport.[3,89] Although data assessing the effects of defective CFTR on cholangiocyte pathobiology have not been defined, the impaired secretory function of the cholangiocytes is likely responsible for reduced bile flow and alkalinity.[3,89] No clear association between specific CFTR mutations and the presence of liver disease has been observed. Treatment with ursodeoxycholic acid, aimed at improving biliary secretion in terms of bile viscosity and bile acid composition, is currently the most useful therapeutic approach in CF-associated liver disease.[3,89,98]

Polycystic Kidney Liver Disease

In autosomal dominant polycystic kidney liver disease (PKLD), the genetic defect results in the slow growth of a multiple epithelial cysts within the renal and liver parenchyma.[99] Cysts appear in the intrahepatic biliary tree in PKLD.[100] The cystic ductal cell also secretes Cl^- and HCO_3^- like normal cholangiocytes, but the secretion is diminished, probably as the result of a reduced of Cl^-/HCO_3^- exchangers in the apical membrane of cystic ductal cells as compared with the normal cholangiocytes.[101]

Biliary Atresia

Biliary atresia is a destructive, inflammatory process of the intrahepatic and extrahepatic bile ducts, which leads to obliteration of the biliary tract and biliary cirrhosis.[102] It is the most common cause of chronic cholestasis in infants.[102] The pathogenesis of biliary atresia is unknown. Most recent studies[103] have focused on dysregulation of ductal morphogenesis and environmental factors (viruses or metabolic insults) in combination with genetic or immunologic susceptibility. Reovirus type 3 and Group C rotavrus have been implicated in biliary atresia.[103] Although additional studies will be required to establish a link between viruses and biliary atresia, this disease is the most likely candidate for a virally induced cholangiopathy.[103] A dysregulation of ductal morphogenesis is supported by the frequent coexistence with other developmental anomalies particularly visceral organ symmetry.[103] This association suggests abnormalities in developmental genes that cause the failure of ductular remodeling process at the hilum.[103] Recently a mouse model with insertional mutation in the proximal region of mouse chromosome 4 has been described with features of biliary atresia and abdominal situs inversus.[104]

Future Perspectives

The cholangiocyte, compared to the remainder of the cell population in the liver, has been relatively neglected in terms of study. The interest in cholangiopathies and the pathophysiology of cholangiocytes has rapidly increased in the recent years and investigation will likely continue to accelerate in the future. Studies are needed to delineate the role of dysfunctional ductal secretion, infectious diseases, and genetic factors in cholangiopathies. The elucidation of the mechanisms of bile duct loss either through apoptosis or failure of duct remodeling may lead to breakthroughs in the understanding and treatment of ductopenic liver disorders. Although the role of hormones in ductal secretion has been partly defined, more studies are needed to determine the role of nerves and blood supply in the normal function and dysfunction of cholangiocytes in disease.

References

1. Alpini G, Glaser SS, Robertson W et al. Large but not small intrahepatic bile ducts (IBDU) from bile duct ligated (BDL) rats proliferate and are involved in secretin-induced ductal bile secretion. Hepatology 1996; 24:A84.
2. Alpini G, Glaser SS, Ueno Y et al. Heterogeneity of the proliferative capacity of rat cholangiocytes after bile duct ligation. Am J Physiol 1998; 274:G767-G775.
3. Alpini G, Prall RT, LaRusso NF. The pathobiology of biliary epithelia. In: Arias IM, Boyer JL, Chisari FV et al, eds. The Liver; Biology & Pathobiology. 4th ed. Philadelphia: Lippincott Williams & Wilkins, 2001:421-435.
4. Alpini G, Roberts SK, Kuntz SM et al. Morphological, molecular and functional heterogeneity of cholangiocytes from normal rat liver. Gastroenterology 1996; 110:1636-1643.
5. Alpini G, Ulrich C, Roberts S et al. Molecular and functional heterogeneity of cholangiocytes from rat liver after bile duct ligation. Am J Physiol 1997; 272:G289-G297.
6. Caligiuri A, Glaser S, Rodgers R et al. Endothelin 1 inhibits secretin-stimulated ductal secretion by interacting with ETA receptors on large cholangiocytes. Am J Physiol 1998; 275:G835-G846.
7. Kato A, Gores GJ, LaRusso NF. Secretin stimulates exocytosis in isolated bile duct epithelial cells by a Cyclic AMP-mediated mechanism. J Biol Chem 1992; 267:15523-15529.
8. LeSage G, Glaser S, Marucci L et al. Acute carbon tetrachloride feeding induces damage of large but not small cholangiocytes from bile duct ligated rat liver. Am J Physiol 1999; 276:G1289-G1301.
9. LeSage G, Glaser S, Ueno Y et al. Regression of cholangiocyte proliferation after cessation of ANIT feeding is associated with increased apoptosis. Am J Physiol 2001; 281:G182-G190.
10. LeSage GD, Benedetti A, Glaser S et al. Acute carbon tetrachloride feeding selectively damages large, but not small, cholangiocytes from normal rat liver. Hepatology 1999; 29:307-319.
11. Nathanson MH, Boyer JL. Mechanisms and regulation of bile secretion. Hepatology 1991; 14:551-566.
12. Kanno N, LeSage G, Glaser S et al. Secretin regulation of cholangiocyte bicarbonate secretion. Am J Physiol 2001; 281:G612-G625.
13. Baiocchi L, LeSage G, Glaser S et al. Regulation of cholangiocyte bile secretion. J Hepatology 1999; 31:179-191.
14. Alpini G, Lenzi R, Sarkozi L et al. Biliary physiology in rats with bile ductular cell hyperplasia. Evidence for a secretory function of proliferated bile ductules. J Clin Invest 1988; 81:569-578.
15. Alpini G, Lenzi R, Zhai W-R et al. Bile secretory function of intrahepatic biliary epithelium in the rat. Am J Physiol 1989; 257:G124-G133.
16. Alpini G, Ulrich II C, Phillips J et al. Upregulation of secretin receptor gene expression in rat cholangiocytes after bile duct ligation. Am J Physiol 1994; 266:G922-G928.

17. Glaser S, Benedetti A, Marucci L et al. Gastrin inhibits cholangiocyte growth in bile duct ligated rats by interaction with CCK-B/gastrin receptors via IP_3-, Ca^{2+}-, and PKCa-dependent mechanisms. Hepatology 2000; 32:17-25.
18. Glaser SS, Rodgers R, Phinizy JL et al. Gastrin inhibits secretin-induced ductal secretion by interaction with specific receptors on rat cholangiocytes. Am J Physiol 1997; 273:G1061-1070.
19. LeSage G, Glaser S, Gubba S et al. Regrowth of the rat biliary tree after 70% partial hepatectomy is coupled to increased secretin-induced ductal bile secretion. Gastroenterology 1996; 111:1633-1644.
20. Alvaro D, Alpini G, Jezequel AM et al. Role and mechanisms of acetylcholine in the regulation of cholangiocyte secretory functions. J Clin Invest 1997; 100:1349-1362.
21. Alvaro D, Cho WKC, Mennone A et al. Effect of secretin on intracellular pH regulation in isolated rat bile duct epithelial cells. J Clin Invest 1993; 92:1314-1325.
22. Fitz JG, Basavappa S, McGill J et al. Regulation of membrane chloride currents in rat bile duct epithelial cells. J Clin Invest 1993; 91:319-328.
23. Roman RM, Feranchak AP, Salter KD et al. Endogenous ATP release regulates Cl⁻ secretion in cultured human and rat biliary epithelial cells. Am J Physiol 1999; 276:G1391-400.
24. Zsembery A, Spirli C, Granato A et al. Purinergic regulation of acid/base transport in human and rat biliary epithelial cell lines. Hepatology 1998; 28:914-920.
25. Alpini G, Glaser S, Rodgers R et al. Functional expression of the apical Na^+-dependent bile acid transporter in large but not small rat cholangiocytes. Gastroenterology 1997; 113:1734-1740.
26. Benedetti A, Di Sario A, Marucci L et al. Carrier-mediated transport of conjugated bile acids across the basolateral membrane of biliary epithelial cells. Am J Physiol 1997; 272:G1416-G1424.
27. Lazaridis KN, Pham L, Tietz P et al. Rat cholangiocytes absorb bile acids at their apical domain via the ileal sodium-dependent bile acid transporter. J Clin Invest 1997; 100:2714-2721.
28. Lazaridis KN, Tietz P, Wu T et al. Alternative splicing of the rat sodium/bile acid transporter changes its cellular localization and transport properties. Proc Natl Acad Sci USA 2000; 97:11092-11097.
29. Lazaridis KN, Pham L, Vroman B et al. Kinetic and molecular identification of sodium-dependent glucose transporter in normal rat cholangiocytes. Am J Physiol 1997; 272:G1168-G1174.
30. Eisenmann-Tappe I, Wizigmann S, Gebhardt R. Glutamate uptake in primary cultures of biliary epithelial cells from normal rat liver. Cell Biol Toxicol 1991; 7:315-325.
31. Roberts SK, Yano M, Ueno Y et al. Cholangiocytes express the aquaporin CHIP and transport water via a channel-mediated mechanism. Proc Natl Acad Sci 1994; 91:13009-13013.
32. Nielsen S, Smith BL, Christensen EI et al. Distribution of the aquaporin CHIP in secretory and resorptive epithelia and capillary endothelia. Proc Natl Acad Sci 1993; 90:7275-7259.
33. Marinelli RA, Pham L, Agre P et al. Secretin promotes osmotic water transport in rat cholangiocytes by increasing aquaporin-1 water channels in plasma membrane. Evidence for a secretin-induced vesicular translocation of aquaporin-1. J Biol Chem 1997; 272:12984-12988.
34. Marinelli RA, Pham LD, Tietz PS et al. Expression of aquaporin-4 water channels in rat cholangiocytes. Hepatology 2000; 31:1313-1317.
35. Marinelli RA, Tietz PS, Pham LD et al. Secretin induces the apical insertion of aquaporin-1 water channels in rat cholangiocytes. Am J Physiol 1999; 276:G280-G286.
36. Schaffner F, Popper H. Electron microscopic studies of normal and proliferated bile ductules. Am J Pathol 1961; 38:393-410.
37. Steiner JW, Carruthers JS. Studies on the fine structure of the terminal branches of the biliary tree. I. The morphology of normal bile canaliculi, bile preductules (ducts of Hering) and bile ductules. Am J Pathol 1961; 38:639-661.
38. Ludwig J. New concepts in biliary cirrhosis. Sem Liv Dis 1987; 7:293-301.
39. Benedetti A, Bassotti C, Rapino K et al. A morphometric study of the epithelium lining the rat intrahepatic biliary tree. J Hepatol 1996; 24:335-342.
40. LaRusso NF, Ishii M, Vroman BT. The ins and outs of membrane movement in biliary epithelia. Trans Am Clin Climatol Ass 1991; 102:245-259.
41. Martinez-Anso E, Castillo JE, Diez J et al. Immunohistochemical detection of chloride/bicarbonate anion exchangers in human liver. Hepatology 1994; 19:1400-1406.
42. Gaudio E, Onori P, Pannarale L et al. Hepatic microcirculation and peribiliary plexus in experimental biliary cirrhosis: a morphological study. Gastroenterology 1996; 111:1118-1124.
43. Mennone A, Alvaro D, Cho W et al. Isolation of small polarized bile duct units. Proc Natl Acad Sci 1995; 92:6527-6531.
44. Strazzabosco M, Joplin R, Zsembery k et al. $Na^{(+)}$-dependent and -independent $Cl⁻/HCO_3⁻$ exchange mediate cellular $HCO_3⁻$ transport in cultured human intrahepatic bile duct cells. Hepatology 1997; 25:976-985.
45. McGill JM, Basavappa S, Gettys TW et al. Secretin activates Cl- channels in bile duct epithelial cells through a cAMP-dependent mechanism. Am J Physiol 1994; 266:G731-G736.

46. Cho WK, Mennone A, Ryderg SA et al. Bombesin stimulates bicarbonate secretion from rat cholangiocytes: implications for neural regulation of bile secretion. Gastroenterology 1995; 113:311-321.
47. Cho WK, Boyer JL. Vasoactive intestinal polypeptide is a potent regulator of bile secretion from rat cholangiocytes. Gastroenterology 1999; 117:420-428.
48. LeSage G, Glaser S, Benedetti A et al. Insulin inhibits secretin-induced ductal secretion in bile duct ligated (BDL) rats through a mechanism involving Ca^{2+}-dependent protein kinase C (PKC) alpha isoform activation. Hepatol 2000; 32:A1090.
49. Alpini G, Glaser S, Robertson W et al. Bile acids stimulate proliferative and secretory events in large but not small cholangiocytes. Am J Physiol 1997; 273:G518-G529.
50. Knuchel J, Krahenbuhl S, Zimmermann A et al. Effect of secretin on bile formation in rats with cirrhosis of the liver; structure-function relationship [see comments]. Gastroenterology 1989; 97:950-957.
51. Alpini G, Elias I, Glaser S et al. Gamma interferon inhibits cholangiocyte proliferation and secretin-induced ductal secretion in a novel murine model of cirrhosis. J Hepatol 1997; 27:371-380.
52. Wheeler HO, Mancusi-Ungaro PL. Role of bile ducts during secretin choleresis in dogs. Am J Physiol 1966; 210:1153-1159.
53. Kaplan MM, Righetti A. Induction of rat liver alkaline phosphatase: the mechanism of the serum elevation in bile duct obstruction. J Clin Invest 1970; 49:508-516.
54. Alvaro D, Benedetti A, Marucci L et al. The function of alkaline phosphatase in the liver: regulation of intrahepatic biliary epithelium secretory activities in the rat. Hepatology 2000; 32:174-184.
55. Strazzabosco M. Transport systems in cholangiocytes: their role in bile formation and cholestasis. Yale J Biol Med 1997; 70:427-434.
56. Kirk KL. New paradigms of CFTR chloride channel regulation. Cell Mol Life Sci 2000; 57:623-634.
57. Cohn JA, Strong TV, Picciotto MR et al. Localization of the cystic fibrosis transmembrane conductance regulator in human bile duct epithelial cells. Gastroenterol 1993;105:1857-1864.
58. Pitt BR. CFTR trafficking and signaling in respiratory epithelium. Am J Physiol Lung Cell Mol Physiol 2001; 281:L13-L15.
59. Verkman AS. Lung disease in cystic fibrosis: is airway surface liquid composition abnormal? Am J Physiol Lung Cell Mol Physiol 2001; 281:L306-L308.
60. Grubb BR, Gabriel SE. Intestinal physiology and pathology in gene-targeted mouse models of cystic fibrosis. Am J Physiol 1997; 273:G258-G266.
61. Quinton PM. Physiological basis of cystic fibrosis: a historical perspective. Physiol Rev 1999; 79:S3-S22.
62. Modolell I, Alvarez A, Guarner L et al. Gastrointestinal, liver, and pancreatic involvement in adult patients with cystic fibrosis. Pancreas 2001; 22:395-399.
63. Valverde MA, O'Brien JA, Sepulveda FV et al. Impaired cell volume regulation in intestinal crypt epithelia of cystic fibrosis mice. Proc Natl Acad Sci USA 1995; 26:9038-9041.
64. Kwiatkowski AP, McGill JM. Human biliary epithelial cell line Mz-ChA-1 expresses new isoforms of calmodulin-dependent protein kinase II. Gastroenterology 1995; 109:1316-1323.
65. Schlenker T, Fitz JG. Ca^{2+}-activated Cl^- channels in a human biliary cell line: regulation by Ca^{2+}/calmodulin-dependent protein kinase. Am J Physiol 1996; 271:G304-G310.
66. Alvaro D, Mennone A, Boyer JL. Role of kinases and phosphatases in the regulation of fluid secretion and Cl^-/HCO_3^- exchange in cholangiocytes. Am J Physiol 1997; 273:G303-G313.
67. McGill JM, Yen MS, Cummings OW et al. Interleukin-5 inhibition of biliary cell chloride currents and bile flow. Am J Physiol 2001; 280:G738-G745.
68. Hirata K, Nathanson MH. Bile duct epithelia regulate biliary bicarbonate excretion in normal rat liver. Gastroenterol 2001; 121:396-406.
69. Strazzabosco M, Mennone A, Boyer JL. Intracellular pH regulation in isolated rat bile duct epithelial cells. J Clin Invest 1991; 87:1503-1512.
70. Spirli C, Granato A, Zsembery k et al. Functional polarity of Na^+/H^+ and Cl^-/HCO_3^- exchangers in a rat cholangiocyte cell line. Am J Physiol 1998; 275:G1236-G1245.
71. Hubner C, Stremmel W, Elsing C. Sodium hydrogen exchange type 1 and bile ductular secretory activity in the guinea pig. Hepatology 2000; 31:562-571.
72. Hofmann AF. Bile acids. In: Arias I, Boyer J, Fausto N, eds. The Liver; Biology & Pathobiology, 3rd ed. New York: Raven Press, 1994:677-718.
73. Buscher HP, Miltenberger C, MacNelly S et al. The histoautoradiographic localization of taurocholate in rat liver after bile duct ligation. Evidence for ongoing secretion and reabsorption processes. J Hepatol 1989; 8:181-191.
74. Gurantz D, Hofmann AF. Influence of bile acid structure on bile flow and biliary lipid secretion in the hamster. Am J Physiol 1984; 247:G736-G748.
75. Palmer KR, Gurantz D, Hofmann AF et al. Hypercholeresis induced by norchenodeoxycholate in biliary fistula rodent. Am J Physiol 1987; 252:G219-G228.

76. Aldini R, Roda A, Lenzi PL et al. Bile acid active and passive ileal transport in the rabbit: effect of luminal stirring. Europ J Clin Invest 1992; 22:744-750.
77. Wilson FA, Dietschy JM. The intestinal unstirred layer: its surface area and effect on active transport kinetics. Biochim Biophys Acta 1974; 353:112-126.
78. Alpini G, Glaser S, Ueno Y et al. Bile acid feeding induces cholangiocyte proliferation and secretion: evidence for bile acid-regulated ductal secretion. Gastroenterology 1999; 116:179-186.
79. Lira M, Schteingart CD, Steinbach JH et al. Sugar absorption by the biliary ductular epithelium of the rat: evidence for two transport systems [see comments]. Gastroenterology 1992; 102:563-571.
80. Guzelian P, Boyer JL. Glucose reabsorption from bile: evidence for biliohepatic circulation. J Clin Invest 1974; 53:526-535.
81. Masyuk AI, Gong AY, Kip S et al. Perfused rat intrahepatic bile ducts secrete and absorb water, solute, and ions. Gastroenterol 2000; 119:1672-1680.
82. Ishii M, Vroman B, LaRusso NF. Morphologic demonstration of receptor-mediated endocytosis of epidermal growth factor by isolated bile duct epithelial cells. Gastroenterology 1990; 98:1284-1291.
83. Ishii M, Vroman B, LaRusso NF. Fluid-phase endocytosis by intrahepatic bile duct epithelial cells isolated from normal rat liver. J Histochem Cytochem 1990; 38:515-524.
84. Thomas P, Zamcheck N. Role of the liver in clearance and excretion of circulating carcinoembryonic antigen (CEA). Dig Dis & Sci 1983; 28:216-224.
85. LaRusso NF. Proteins in bile: how they get there and what they do. Am J Physiol 1984; 247:G199-G205.
86. LeSage G, Alvaro D, Benedetti A et al. Cholinergic system modulates growth, apoptosis and secretion of cholangiocytes from bile duct ligated rats. Gastroenterology 1999; 117:191-199.
87. Kanno N, LeSage G, Glaser S et al. Gastrin inhibits growth and induces apoptosis of cholangiocarcinoma Mz-ChA-1 cells associated with membrane translocation of PKCa. Hepatology 1999; 30:A1084.
88. Tracy TFJ, Goerke ME, Bailey PV et al. Growth-related gene expression in early cholestatic liver injury. Surgery 1993; 114:532-537.
89. Colombo C, Battezzati PM, Strazzabosco M et al. Liver and biliary problems in cystic fibrosis. Sem Liver Dis 1998; 18:227-235.
90. Park J, Gores GJ, Patel T. Lipopolysaccharide induces cholangiocyte proliferation via an interleukin-6-mediated activation of p44/p42 mitogen-activated protein kinase. Hepatology 1999; 29:1037-1043.
91. Klein R, Pointner H, Zilly W et al. Antimitochondrial antibody profiles in primary biliary cirrhosis distinguish at early stages between a benign and a progressive course: a prospective study on 200 patients followed for 10 years. Liver 1997; 17:119-128.
92. Tsuneyama K, Van De Water J, Van Thiel D et al. Abnormal expression of PDC-E2 on the apical surface of biliary epithelial cells in patients with antimitochondrial antibody-negative primary biliary cirrhosis. Hepatology 1995; 22:1440-1446.
93. Portmann B, MacSween R. Diseases of the intrahepatic bile ducts. In: MacSween R, Anthony P, Scheuer P, eds. Pathology of the Liver. Edinburgh: Churchill Livingstone, 1994.
94. Zein CO, Lindor KD. Primary sclerosing cholangitis. Sem Gastrointest Dis 2001; 12:103-112.
95. Torok N, Gores GJ. Cholangiocarcinoma. Sem Gastrointest Dis 2001; 12:125-132.
96. Kubicka S, Kuhnel F, Flemming P et al. K-ras mutations in the bile of patients with primary sclerosing cholangitis. Gut 2001; 48:403-408.
97. Kanno N, Glaser S, Chowdhury U et al. The alpha-2 adrenergic agonist, UK 14,304, inhibits growth and increases apoptosis of the cholangiocarcinoma cell line, Mz-Cha-1, through cAMP but not Ca^{2+}-dependent mechanisms. Gastroenterol 2000; 118:A150.
98. Malet PF, Borum M, Fromm H. Cystic fibrosis: another use for urso? Gastroenterol 1991; 100:841-842.
99. Nguyen HQ, Danilenko DM, Bucay N et al. Expression of keratinocyte growth factor in embryonic liver of transgenic mice causes changes in epithelial growth and differentiation resulting in polycystic kidneys and other organ malformations. Oncogene 1996; 16:2109-2119.
100. Calvet JP, Grantham JJ. The genetics and physiology of polycystic kidney disease. Sem Nephrol 2001; 21:107-123.
101. Perrone RD, Grubman SA, Murray SL et al. Autosomal dominant polycystic kidney disease decreases anion exchanger activity. Am J Physiol 1997; 272:C1748-C1756.
102. Bezerra JA, Balistreri WF. Cholestatic syndromes of infancy and childhood. Sem Gastrointest Dis 2001; 12:54-65.
103. Bates MD, Bucuvalas JC, Alonso MH et al. Biliary atresia: pathogenesis and treatment. Sem Liver Dis 1998; 18:281-293.
104. Mazziotti MV, Willis LK, Heuckeroth RO et al. Anomalous development of the hepatobiliary system in the Inv mouse. Hepatol 1999; 30:372-378.

CHAPTER 7

Genetics, Mutations, and Polymorphisms

Laura Bull

Summary

Genetic approaches complement functional approaches to the study of hereditary disease, and have contributed substantially to our understanding of the biology of enterohepatic circulation in health and disease. The basic steps in genetic mapping of a disease gene are reviewed here. They include identification of the mode of inheritance; genetic mapping of the disease gene; identification and screening of candidate genes; and evaluation of the functional consequences of the mutation(s) identified. Ongoing advances in technology and analytic methods have increased the effectiveness and efficiency of genetic mapping approaches.

Introduction

Genetic and functional approaches are the two main ones used in identification of genes that, when mutated, cause disease. In genetic approaches to the study of hereditary disease, genetic mapping methods are employed to identify the position within the human genome of a genetic factor influencing development of a disorder. Then, the gene or genes within that region are screened for evidence of involvement in the disorder. In contrast, in functional approaches to study of genetic disease, 'candidate genes' that may be mutated in individuals with the disease under study are identified based on other information. This information can include the biology of the disorder and the known or predicted function of the protein encoded by the gene, as well as other knowledge about the biology of the gene or protein, such as the times during development and tissues in which the gene or protein is expressed and the subcellular or extracellular location of the protein.

In the past, genetic approaches to disease gene identification were often considered, with some validity, to be less direct and more laborious than functional approaches. However, with recent improvements in laboratory technology, analytic approaches and available data (the latter including the nearly complete sequence of the human genome), genetic approaches can now be applied with steadily increasing efficiency to study of an ever-expanding set of diseases. In recent years, the application of genetic approaches has contributed greatly to our understanding of hereditary disorders, and this productive use of genetic approaches is likely to continue increasing.

Human genetics has changed dramatically during the past two decades, as the following two examples illustrate. In 1983, the initial localization of the gene mutated in Huntington disease (HD) to human chromosome 4p was one of the earliest successes in use of genetic mapping to genetically localize a previously unknown disease gene, and involved a combination of very hard work and substantial luck. Selected members of two large pedigrees (including one with more than 3000 identified members) were genotyped, along with other unaffected relatives. Luck came into play in that one of the first 12 genetic markers (in this case, restriction fragment length polymorphisms analyzed using Southern hybridization techniques) tested was linked to HD.[1] A decade and the work of many dozens of researchers affiliated with a number of collaborating and competing laboratories was required for the subsequent

Molecular Pathogenesis of Cholestasis, edited by Michael Trauner and Peter L.M. Jansen. ©2004 Eurekah.com and Kluwer Academic / Plenum Publishers.

identification of the gene mutated in HD (i.e., the HD 'disease gene') by a collaborative group of 58 authors from six institutions.[2]

In 1994, a gene mutated in some cases of benign recurrent intrahepatic cholestasis (BRIC) was mapped to human chromosome 18q through study of four patients and six unaffected relatives.[3] The following year, the progressive familial intrahepatic cholestasis type 1 (PFIC1) locus was mapped to the same region through study of two patients and their parents.[4] It then took less than three years, and the work of many fewer individuals than for HD, to identify the gene mutated in these two disorders.[5]

Twenty years after the initial mapping of the HD gene, identification of disease genes for disorders with straightforward inheritance patterns (like HD) borders on the routine. Often, the time lapsing between initiation of a mapping project and identification of a disease gene is on the order of a few years; typically, such projects now require the concentrated efforts of only a few individuals to succeed.

One of the key advantages of genetic approaches over many functional ones is that genetic approaches depend very little upon knowledge about the underlying disease biology. Genetic approaches permit discovery of a disorder's genetic etiology, even if the identity of the disease gene could not have been predicted, given current understanding of cell, molecular and disease biology. Therefore, genetic approaches allow a comprehensive screening of the possible genetic factors leading to development of disease, and facilitate unexpected, unpredictable discoveries. This impressive power of genetic approaches has been one factor driving their increased use.

The identification of the BRIC/PFIC1 gene, *ATP8B1* (also called *FIC1*, for familial intrahepatic cholestasis 1), illustrates this point. Prior to identification of *ATP8B1*, it was hypothesized that the BRIC/PFIC1 gene most likely encoded a canalicular bile acid transporter that was a member of the ATP-binding cassette (ABC) protein family.[6] The reasonableness of this hypothesis was borne out when the gene mutated in a different form of PFIC was identified; the PFIC2 gene is indeed an ABC protein that transports bile acids across the canalicular membrane. In contrast, ATP8B1 is not an ABC transporter, but belongs to a different family of membrane transporters, the P-type ATPases.[7] *ATP8B1* is expressed in a wide variety of tissues, suggesting that it plays a role important in many tissues, rather than being specifically a canalicular bile acid transporter.[5] Within the liver, the protein is found in the canalicular membrane of hepatocytes, and in cholangiocytes.[8,9] ATP8B1 may function in transport of aminophospholipids between membrane leaflets.[8] Initial evaluation of a mouse model suggests that ATP8B1 may participate in regulation of intestinal bile salt absorption.[10] Further studies to determine the function of ATP8B1 are currently underway, and promise to shed new light on the biology of bile acid transport.

Genetic and functional approaches to study of hereditary disease complement each other, and can be applied together. Such an approach facilitated rapid identification of the PFIC2 gene; initially, the PFIC2 locus was genetically mapped to chromosome 2q24.[11] An ABC transporter gene lay in the region, and was indeed the PFIC2 gene, *ABCB11* (also called *BSEP*, for bile salt export protein).[12] If no such excellent candidate gene lay in the region, it likely would have taken longer to identify the disease gene.

In recent years, genetic approaches have been applied to the study of cholestatic disorders, with exciting results. In this chapter, the basic concepts in human genetics that are helpful for understanding of these and future advances will be reviewed, using as illustrations examples from hereditary cholestatic disease.

Our DNA: The Basics

The nucleus of each human somatic cell carries 46 chromosomes: two copies of each of the 22 autosomal chromosomes, and either two copies of the X chromosome (in females), or one copy each of the X and Y chromosomes (in males). Each cell also contains numerous mitochondria, which have their own ~16 kb (kilobase pair) genome. A single copy of the human nuclear genome (one copy each of chromosomes 1-22, X and Y) is ~3.2 Gigabase pairs in length and has been estimated to contain approximately 30,000-35,000 protein-coding genes; approximately 1.5% of our DNA is protein-coding sequence, hidden within a large excess of non-coding DNA.[13,14] This non-coding DNA contains introns, as well as genes that produce noncoding RNAs, including transfer RNA, ribosomal RNA, small nuclear RNA, small nucleolar RNA, microRNAs, small interfering mRNAs, and small temporal RNAs.[15,16] Non-coding DNA also contains numerous other functionally important sequences, including those necessary for maintenance of chromosome ends, proper segregation of chromosomes during cell division, and initiation and regulation of gene expression. Much of our genome has no currently known function; this sequence includes pseudogenes, and the roughly 48% of our DNA that consists of interspersed repetitive elements. This latter group includes many types of transposon-derived repeat, as well as simple sequence repeats.[13]

Review of a few terms may aid understanding of the discussion that follows. The term **locus** refers to the chromosomal location of a gene or DNA marker. A specific DNA sequence is **polymorphic** if it varies between individuals, and the different sequence variants are **alleles**. A **haplotype** is a set of specific co-inherited alleles; typically, these alleles are co-inherited because they are present at neighboring genetic markers. An individual carrying the same allele at both copies of a locus is **homozygous** for that allele, while someone carrying two different alleles is **heterozygous**. **Genetic markers** are specific DNA sequences that are polymorphic, and employed in mapping of disease genes. When a single disorder can be caused by mutation in different genes, **locus heterogeneity** is present; low-gamma-glutamyl-transpeptidase (GGT) PFIC exhibits locus heterogeneity, since mutations in either *ATP8B1*, *ABCB11*, or one or more as-yet-unidentified genes can cause the disease.[17] **Allelic heterogeneity** exists if different mutations in a single disease gene can cause the same disorder; many genetic disorders manifest allelic heterogeneity, complicating mutation screens.

Steps in Genetic Mapping

Genetic mapping studies typically involve a series of steps: identification of the likely mode of inheritance of the disease; application of experimental and analytical methods to initially map the disease gene and then refine its location; identification of candidate genes; screening of candidate genes for mutation; and determination of the functional consequences of the mutation(s).

Mode of Inheritance

Understanding the mode of inheritance of a disease can focus genetic studies, permits selection of the most appropriate form of genetic analysis, and is important when counseling affected families. Additionally, many forms of genetic analysis require specification of the mode of inheritance of a disorder. The modes reported for cholestatic diseases to date are mostly autosomal recessive and autosomal dominant (Table 1), so these are discussed in some detail below. Basic features of several other modes are indicated in Table 2.

For an **autosomal recessive (AR)** disorder, disease only develops if both copies of the 'disease gene' in an individual possess deleterious mutations. Generally, AR disorders are seen in one or more siblings in a family, but not in parents or children of patients. Heterozygous carriers are usually clinically normal, but may exhibit biochemical abnormalities. The ratio of males to females affected is typically ~1:1, and on average, one-fourth of the children of two heterozygous carrier parents will be affected. Consanguinity is seen more often in AR disease than in disorders with other modes of inheritance, and has proven helpful in mapping of PFIC1, AR BRIC, PFIC2, LCS and North American Indian childhood cirrhosis (NAIC).[3,4,11,12,38,39,42] In consanguineous families, affected children are typically homozygous for a disease mutation. Patients with an AR disease who carry two different disease mutations are **compound heterozygotes**.

In a simple **autosomal dominant (AD)** disorder, one defective copy of the disease gene is sufficient to produce disease. In a typical pedigree for an AD disease, individuals in multiple

Table 1. *Genetics of cholestatic disorders and disorders of bilirubin metabolism and transport: genes identified or mapped*

Disorder	Chromosomal Location of Gene	Mutated Gene	Primary Approach(es)	Probable Mode of Inheritance
ATP8B1 disease: PFIC1, BRIC, Greenland Familial Cholestasis (GFC)	18q21	*ATP8B1 (FIC1)*[5,18]	Genetic	AR (BRIC: incomplete penetrance)
ABCB11 disease: PFIC2 (BRIC?)	2q24	*ABCB11 (BSEP)*[12,19,20]	Genetic and functional	AR
ABCB4 disease: PFIC3, cholestasis of pregnancy (ICP)	7q21	*ABCB4 (MDR3)*[21-25]	Functional	PFIC3: AR ICP: AD, sex-limited, incomplete penetrance
Alagille syndrome (AGS)	20p12	*JAG1*[26-28]	Genetic	AD
Dubin-Johnson syndrome	10q24	*ABCC2 (cMOAT)*[29]	Functional	AR, incomplete penetrance
Crigler-Najjar syndrome	2q37	*UGT1A1*[30,31]	Functional	AR
Neonatal giant cell hepatitis	8q21.3	*CYP7B1*[32]	Functional	AR
Wilson disease	13q14	*ATP7B*[33,34]	Genetic and functional	AR
Cystic fibrosis	7q31	*ABCC7 (CFTR)*[35-37]	Genetic	AR
Lymphedema-cholestasis syndrome (LCS)	15q[38]	unknown	Genetic	AR
North American Indian Childhood Cirrhosis (NAIC)	16q22[39]	*CIRHIN*[83]	Genetic	AR
ICP	2p13[40]	unknown	Genetic (association study)	Unknown (sex-limited)
Cystic fibrosis modifier locus for meconium ileus (CFM1)	19q13[41]	unknown	Genetic	?

generations will be affected, no generations will be skipped, males and females are equally likely to be affected, and if affected, to transmit the disorder. In a family in which one parent has an AD disorder, 50% of the children, on average, will also have the disorder. Homozygotes, when they occur, are generally more severely affected than heterozygotes. A form of BRIC demonstrating AD inheritance has recently been described.[43]

As **mitochondrial** DNA is maternally transmitted, affected fathers do not transmit disorders with a mitochondrial mode of inheritance to their children. Of note with regard to

Table 2. Mendelian modes of inheritance: typical features

Mode of Inheritance	Disease Seen in Multiple Generations?	Expected % of Siblings Affected	Father-Son Transmission?	Can Affected Mothers Bear Affected Children?	Males and Females Affected?	Comments
Autosomal recessive (AR)	No (unless extensive consanguinity or very common)	25%	No (unless mother is a carrier)	No (unless father is a carrier)	In equal proportions, with similar severity	Consanguinity seen more often than with other modes of inheritance
Autosomal dominant (AD)	Yes	50%	Yes	Yes	In equal proportions, with similar severity	Fewer than 50% of siblings may be affected if parents are unaffected, and one is germline mosaic for a new mutation
X-linked recessive	Yes, but generations may be 'skipped', due to unaffected status of carrier females	Transmission from unaffected mother: 50% of sons, and no daughters, affected.

Transmission from affected father: no offspring are affected. | No | Yes | Females rarely affected, since they have 2 X chromosomes; heterozygous females may occasionally manifest disease due to skewed X-inactivation | |
| X-linked dominant | Yes | Maternal transmission: 50% of sons and daughters are affected.

Paternal transmission: all daughters, but no sons, are affected. | No | Yes | Females at least twice as likely to be affected as males (some such disorders are lethal prenatally in males, so even fewer affected males are seen) | With small amounts of data, it can be difficult to distinguish X-linked dominant from AD inheritance. |

Continued on next page

Table 2. Mendelian modes of inheritance: typical features (continued)

Mode of Inheritance	Disease Seen in Multiple Generations?	Expected % of Siblings Affected	Father-Son Transmission?	Can Affected Mothers Bear Affected Children?	Males and Females Affected?	Comments
Y-linked	Yes	50%	Yes	Not applicable	Only males are affected	Seen very rarely, as there are few genes on the Y chromosome
Mitochondrial	Yes	Maternal transmission: ~all children affected. Affected fathers do not transmit the disorder to their children	No	Yes	Yes, in equal proportions	As mitochondrial DNA is maternally inherited, affected men do not transmit such disorders.

mitochondrial inheritance is that many genetic diseases in which mitochondrial defects occur do not demonstrate mitochondrial inheritance. This is because many proteins essential for normal function of mitochondria are encoded by the nuclear genome. Navaho neurohepatopathy may be an example of such a disorder; it is AR, but appears to involve depletion of mitochondrial DNA.[44]

Chromosomal abnormalities are large enough to be visible using cytogenetic techniques, and include changes in chromosome number (polyploidy, aneuploidy) or structure (i.e., translocations, deletions, inversions, duplications). Some such abnormalities can be genetically transmitted, although review of their inheritance patterns is beyond the scope of this chapter. Before embarking on extensive genetic mapping studies of a disorder, it is sensible to have karyotyping performed on some patients to ensure that they do not have a chromosomal abnormality. For example, early identification of a large deletion in a patient with Alagille syndrome (AGS) narrowed the focus of mapping studies for this disorder to a region on chromosome 20.[45]

Other features can influence the manifestations and apparent inheritance pattern of a genetic disorder. Disorders that present at different rates or with different symptoms in males versus females can be due to mutation in autosomal genes. Such disorders are **sex-influenced** or **sex-limited**; intrahepatic cholestasis of pregnancy (ICP) is a sex-limited disorder. Another important feature of a genetic disease is its **penetrance**—i.e., the proportion of people with a disease-causing genotype who actually develop the disease. Penetrance can be **complete** (i.e., all patients with a disease-causing genotype develop the disease) or **incomplete**, and also may be age-dependent; BRIC due to mutation in *ATP8B1* exhibits incomplete, **age-dependent** penetrance.[46]

In a disorder with **variable expressivity**, patients differ in the severity and/or constellation of disease manifestations they suffer. Sometimes variable expressivity is due to mutations with effects of differing severity on protein function, but in other cases, substantial variability is seen even between patients possessing the same disease mutation. Both BRIC and Alagille syndrome have substantially variable expressivity.[28,46]

Several other features should be kept in mind when considering the pattern of inheritance of a genetic disorder:

1. A **new mutation** may have occurred. For example, a new mutation for a dominant disease may occur in a parent, so that a child is affected, although the parent was not. If the possibility of new mutation is not considered, the disorder in that family might be thought recessive. In Alagille syndrome, as in many AD disorders that reduce reproductive success, a majority of the mutations are 'de novo'.[28] Such a new mutation can occur in a single germ cell in the parent, or the parent can be **mosaic** for it- i.e., a proportion of the parent's cells carry the mutation. In a dominant disorder, parental **germline mosaicism** for a new disease mutation means that multiple children in the family can suffer from an AD disorder, although neither parent is affected.

2. Some disorders show **anticipation**, in which the age of onset decreases, and/or disease severity increases (on average), with each successive generation. Anticipation is seen most typically in disorders caused by expansion of a trinucleotide repeat, such as HD, as the repeat can further expand in successive generations.

3. **Imprinting** is said to occur when inheritance of the same mutation has a different effect on the child, depending upon whether it was inherited from the mother or father.

Sometimes, a disorder's mode of inheritance can be evaluated statistically using a formal segregation analysis. In other cases, especially for rare disorders, and given today's typically smaller families, too few patients and family members are available to permit statistically definitive segregation analysis; nevertheless, it is often possible to identify the most likely mode of inheritance. Knowledge of the mode of inheritance of a disorder can help greatly in identifying the genetic etiology of the disease; for example, if a disorder demonstrates X-linked inheritance, a genetic screen need only be performed for the X-chromosome, rather than for the entire nuclear genome.

Complex disorders are those which do not exhibit simple Mendelian inheritance patterns, but are **multifactorial** (i.e., influenced by multiple genetic and environmental factors). Such disorders may be **oligogenic** (influenced by a small number of genetic loci) or **polygenic** (influenced by many loci). Given an adequate study sample, the extent to which a trait is inherited can be estimated. For example, in one study, sisters and mothers of women with ICP were found to have ~12 times greater risk of developing ICP than were women in the general population;[47] however, both genetic and environmental factors could contribute to this increased relative risk. A number of **susceptibility loci** may exist for a disorder; a susceptibility locus is one at which mutation increases the risk of developing the disease, but does not lead inevitably to disease. **Modifier loci** may influence the phenotype of a disorder; in cystic fibrosis (CF), a modifier locus for meconium ileus has been mapped.[41]

Localizing a Disease Gene: Genetic Markers

A number of experimental and analytic approaches can be used to map a disease gene. Experimentally, these approaches rely upon genotyping of genetic markers, although the number and type of markers vary. The genetic markers most frequently used in mapping studies to date are polymorphic simple sequence repeats (SSRs). These are widely distributed, short, tandemly repeated sequences. The most commonly employed SSRs include dinucleotide and tetranucleotide repeats. The number of copies of the repeat unit varies between alleles, so alleles differ in length. The inheritance pattern can be assessed by amplifying the repeat from genomic DNA using the polymerase chain reaction (PCR), with unique primers flanking the repeat, and then electrophoresing the PCR products to separate the alleles by size. Over 8000 polymorphic SSRs have been positioned on a comprehensive genetic map, for an average density of 1 SSR every ~400 kb;[48] a typical genome screen might involve typing of 200-800 of these markers. A major advantage of SSRs for genetic mapping is that they are highly polymorphic and consequently, very informative in genetic analyses. SSRs have been successfully employed in genetic mapping of PFIC1, AR BRIC, PFIC2, LCS, and NAIC, amongst other disorders.[3,4,11,38,39]

The use of **single nucleotide polymorphisms** (SNPs; genomic sites at which a single base varies between alleles) has been increasing. An advantage of SNPs is that they are extremely common; 2 chromosomes differ from each other at ~1 bp in 1300.[13,14] Also, SNPs have a lower mutation rate than do SSRs. Numerous SNP genotyping technologies are at various stages of development; these include methods employing electrophoresis, oligonucleotide microarrays, mass spectrometry, fluorescent microtiter plate reading, or flow cytometry, among others. Some of these methods are amenable to extremely high-throughput genotyping, compared to what can be achieved with SSRs.[49] A disadvantage of SNPs relative to SSRs is that SNPs are less polymorphic, since most SNPs have only 2 alleles.

Genotyping data can be interpreted using various statistical or empiric forms of analysis; what is most appropriate in a specific situation depends upon characteristics of the disorder, the type and size of sample available, and characteristics of the population(s) from which the sample is derived. Genetic mapping data can be evaluated using linkage analysis and/or population genetic mapping.

Localizing a Disease Gene Part 1: Linkage Analysis

Linkage analysis is a powerful, family-based approach to disease mapping. It makes use of the fact that specific copies of the genomic region containing the disease gene are co-inherited with the disease within a family; this reflects lack of recombination between the disease mutation and neighboring genetic markers, due to their close proximity. Within a family, individuals who share a disease will typically share alleles at markers near the disease gene (Fig. 1). The particular alleles co-inherited with the disease often differ between families, reflecting allelic heterogeneity or ancestral genetic recombination events. Results of linkage analysis are reported as LOD scores representing the relative likelihood that a disease locus and a genetic marker are genetically linked (with a recombination fraction theta), rather than that they are genetically

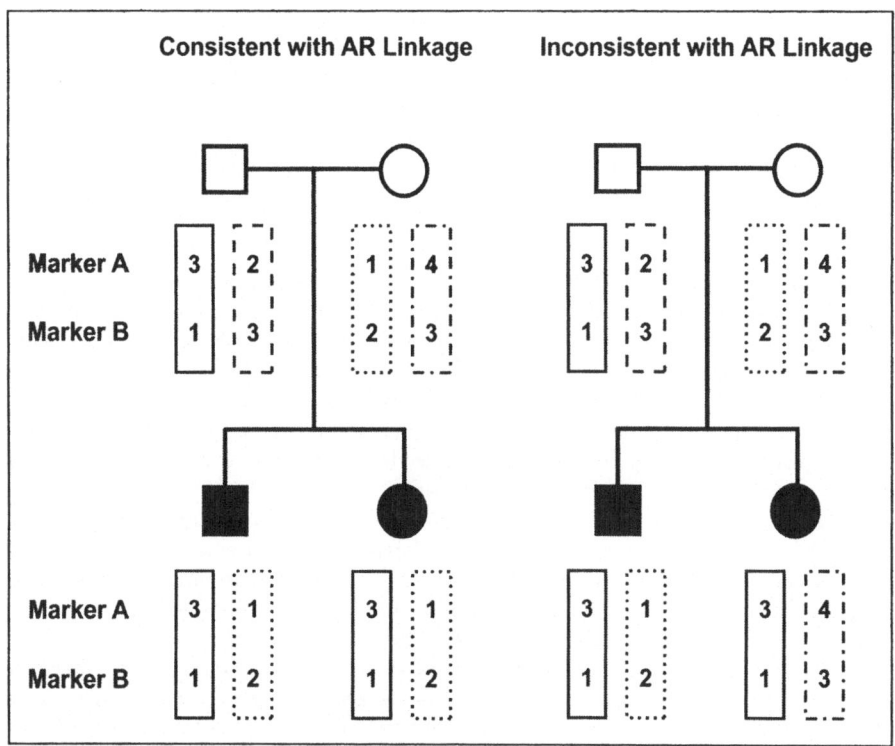

Figure 1. Evidence for or against linkage. This figure shows two pedigrees of families in which an autosomal recessive (AR) disorder occurs. Affected individuals are indicated by black symbols, and unaffected individuals, by unfilled symbols. Data from two genetic markers (A and B) flanking a region of interest are shown. Numbers 1-4 represent different marker alleles. The rectangles of different patterns indicate the different copies of the region present in each family. Data for the pedigree on the left are consistent with linkage to the interval between markers A and B, as both affected children have inherited the same paternal and maternal copies of the region. Data for the pedigree on the right are inconsistent with linkage to this interval; while both children have inherited the same paternal copy of the region, they have inherited different copies of the region from their mother.

unlinked. A LOD score of at least +3.3 is typically considered evidence of linkage from a genome-wide screen. A LOD score of -2 or below excludes disease linkage to a region. Linkage analysis permitted mapping of the CF gene.[50-52]

Standard **parametric** linkage analysis requires specification of a genetic model, including the mode of inheritance, penetrance, and frequency of the disease in the population, and often also necessitates estimation of marker allele frequencies in the population. For disorders with a poorly understood genetic model, other mapping approaches are more appropriate. Additionally, locus heterogeneity within the collection of families studied complicates standard linkage analysis and, if not properly taken into account, can lead to incorrect interpretation of results. Finally, extremely complex pedigrees present computational difficulties for linkage analysis.

Homozygosity mapping is a particularly powerful form of parametric linkage analysis applicable to recessive conditions in consanguineous families. In such a family, patient(s) are likely to be homozygous by descent for a single disease mutation, and for alleles at nearby genetic markers, i.e., both the mutation and marker alleles were inherited from a single ancestor shared by their mother and father. Homozygosity mapping identifies segments of homozygous DNA in patients. Such an approach was employed to map the loci for Wilson disease (WD) and PFIC2.[11,53,54]

When the genetic model of a disorder cannot be determined, modified forms of linkage analysis may be used. When penetrance of a disorder is unknown, an **affecteds-only** parametric linkage analysis can be performed, in which the phenotype of unaffected individuals is considered unknown. This approach results in some loss of statistical power. If the mode of inheritance of a disorder is also unknown, a **nonparametric** linkage approach, such as **affected sib pair** analysis, can be performed. Regions that are shared by affected siblings or other relatives more often than expected by chance are identified; such approaches are useful in mapping susceptibility loci for complex traits.

Localizing a Disease Gene Part 2: Population Genetic Mapping

Population genetic mapping can also be employed to map disease genes. In this approach, the genome is screened in patients from a single population to identify chromosomal segments that the patients share identical by descent (IBD). The concept behind this approach is that members of a population are relatives, even if their exact relationships are unknown. Distantly related patients may share the same disease mutation, and the same version of the chromosomal sequence surrounding the mutation, inherited from a common ancestor, especially if the disease is rare (Fig. 2). Due to genetic recombination, the greater the number of generations that have passed since the introduction of the mutation into the population, the smaller this shared region will be. The presence of a shared haplotype, or set of specific co-inherited alleles, at several consecutive genetic markers indicates IBD sharing of such an ancestral region. Such sharing reflects the presence of **linkage disequilibrium** (LD, i.e., a non-random association) between a disease and particular genetic marker alleles.

Population genetic mapping approaches can be applied in circumstances where successful linkage analysis is difficult or impossible. In population genetic mapping, mode of inheritance and penetrance of the disorder need not be specified and allelic and/or locus heterogeneity may be present without resulting in erroneous interpretation of results (although it does reduce statistical power). Also, precise family relationships need not be known, and it is not necessary to identify large pedigrees with multiple patients; DNA need only be obtained from single patients (and if possible, their parents). Additionally, population genetic approaches often permit further refinement of a disease gene's localization than is possible with linkage analysis.

A disadvantage of population genetic studies is that, if the patients are very distantly related, evidence of LD may only be detectable over a very small genomic region, necessitating screening of a large sample with densely spaced genetic markers. This potential problem can be minimized through careful choice of study populations. Recently founded genetically isolated populations are ideal for population genetic mapping, as the size of the genomic region(s) in LD with the disease is likely to be relatively large, and more easily detectable. If such a population was founded by a relatively small number of individuals, the risk of extensive allelic and/or locus heterogeneity, which reduce power to detect a disease locus, is decreased.

Population genetic mapping was successfully employed in the mapping of BRIC, PFIC1, and LCS.[3,4,38] A modification of this approach was used to map the NAIC locus; a genome screen was performed on three pools of DNA- one each from patients, unaffected siblings, and parents. Markers with enrichment of an allele in the patient pool, as compared to the other two pools, were identified. Those regions were then further characterized to identify the NAIC region.[39]

Typically, **association** studies are used to evaluate candidate genes (or regions) in less consanguineous populations. DNA from a set of patients is collected and data from a polymorphism of interest is generated and statistically evaluated to determine whether one of the alleles is over-represented in patients as compared to controls. Large patient samples are usually needed for such studies, and patients and controls must be carefully matched. False positive results can be caused by factors such as undetected population stratification in the sample used, as well as effects of natural selection. Some association tests, such as the transmission disequilibrium test and the more recently developed genomic control and structured association tests, account for

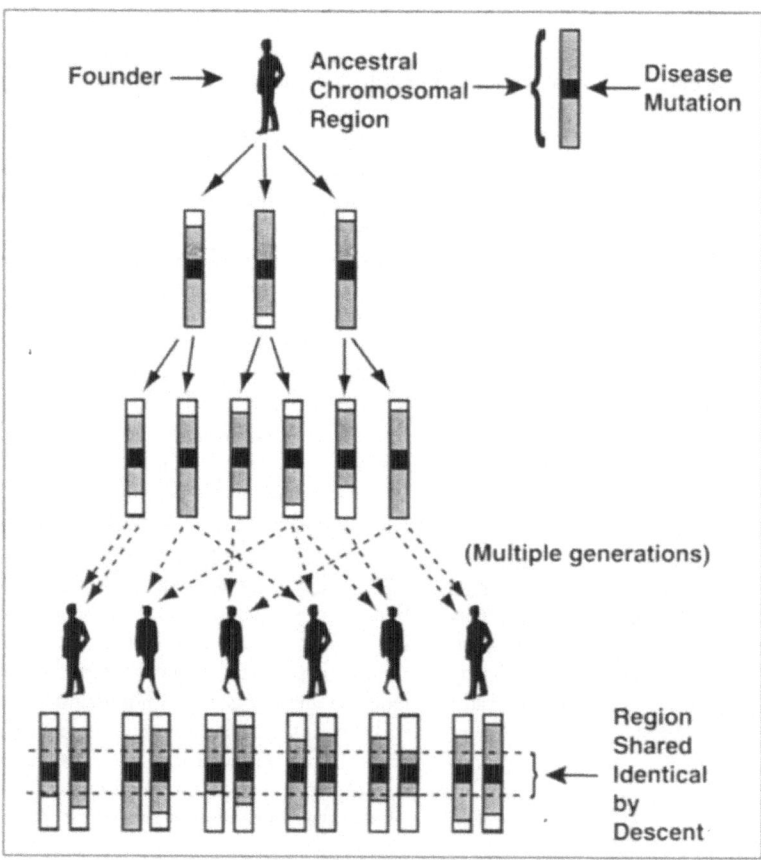

Figure 2. Population genetic mapping. This figure illustrates the principles behind this mapping approach, as applied to an AR disease. A common ancestor, or founder, has introduced a disease mutation into the population. This mutation is represented by the black box on the rectangle representing the chromosomal region in which the mutation occurred. The shaded areas flanking the mutation represent the version of the surrounding chromosomal sequence that was present on the chromosome when it was introduced by the founder- i.e., the ancestral disease-associated haplotype. As the mutation is passed down through the generations, recombination events occur, shrinking the size of this haplotype. The white areas represent those areas that no longer share the ancestral haplotype, due to recombination. Many generations after introduction of the disease mutation into the population, those affected individuals in the population who have inherited the mutation introduced by the common ancestor still share the ancestral haplotype surrounding the disease mutation.

population stratification.[55] In Finnish women, an association study using markers spanning chromosome 2 identified a region that may contain a susceptibility locus for ICP;[40] another study of this same population suggested that a polymorphism in the gene encoding the angiotensin-converting enzyme (not located on chromosome 2) may be associated with ICP.[56] Due to the high rate of false-positive results obtained in association studies, confirmation of these results will be important.

Localizing a Disease Gene Part 3: Fine-Mapping

Often the initial mapping of a disease locus to a region is too imprecise to permit immediate identification of the actual disease gene, particularly if the region contains many genes, and none of them are especially promising from a functional standpoint. The location of a disease gene is genetically further refined through study of additional patients, and/or use of additional genetic markers. Often a combination of analytic approaches and populations is used for study of a single disease, as one approach or population may be most useful for initial mapping of a disease locus, and another for the refinement of localization.

To initially localize *ATP8B1*, for example, haplotypes shared by distantly related patients were identified. [3,4] The position of the locus was further refined using linkage analysis in additional families and finally, detailed haplotype analysis of a larger set of PFIC1 and BRIC patients from multiple populations. [42,57] This analysis permitted identification of shared disease haplotypes, and one disease-associated deletion, and localized the disease gene to a ~1 cM (centimorgan) interval, greatly facilitating identification of *ATP8B1*.

Candidate Gene Identification

Once a disease gene has been mapped, genes within the candidate region must be identified. In the past, this required much laborious and clever laboratory work, but it has become vastly easier. With the availability of the human genome sequence, a substantially complete inventory of the genes in a given region can be obtained without performing any experiments. At a public or private website devoted to the human genome (e.g., http://www.ensembl.org/ Homo_sapiens/, http://www.genome.ucsc.edu/goldenPAth/help/hgTracksHelp.html, http:// www.ncbi.nlm.nih.gov/, or http://www.ncbi.nlm.nih.gov/genome/guide/human/), the genomic region of interest can be selected, and a list of known or predicted genes in the region displayed. Sequence from the region can be used to scan databases of transcribed sequences. To find genes that have not yet been identified through laboratory experiment, sequence from the region can be evaluated using computer programs designed to identify genes within genomic sequence. Nevertheless, difficulties remain. There are still gaps and errors in the genome sequence. Also, the process of computationally predicting the presence and structure of genes, given DNA sequence, is imperfect. Particularly in the case of genes that do not demonstrate substantial sequence homology to known genes, it is possible that a gene might be present, but not predicted to exist.

Usually, one or more promising candidate genes are identified in the region. When multiple genes are present, those screened first for mutation are usually those that data suggest are the most promising candidates, given the known biology of the disease. A gene that encodes a functionally uncharacterized protein can be evaluated as a candidate using information such as whether it has homology to proteins that have known functions consistent with the disease phenotype, and/or whether it is expressed in the tissue(s) most affected by the disease. Genetic and functional approaches to disease gene identification often meet at this point.

Genetic Polymorphisms and Mutations: Types and Identification

The categories of mutation that can occur in our DNA are summarized in Table 3. For most disorders, the most common category of disease mutation identified is the single base pair substitution, in which a single base pair of DNA is replaced with another base pair. Deletions, inversions and insertions (including duplications and repeat expansions) also occur.

Deleterious mutations are changes in the DNA sequence that lead to development of, or increased susceptibility to, disease. **Neutral polymorphisms** or **normal variants** are DNA sequence changes that have no apparent functional significance. When it is unclear whether a sequence change has any functional consequence, it can conservatively be referred to as a **variant**.

A number of methods for identifying disease mutations are available. The types and proportion of total mutations detected depends on the method used. Most commonly used methods involve as an initial step PCR amplification of the coding sequence of the candidate gene; to screen for regulatory mutations, the gene's promoter should also be identified and screened. Genes can be screened for mutation using RNA or genomic DNA. Screening of genomic DNA usually requires more effort, as each exon of the gene is amplified by PCR, then analyzed.

Table 3. Categories of mutation

Category of Mutation	Type of Mutation
Single base-pair change	Missense (altered amino acid)
	Nonsense (stop codon)
	Splicing (prevention of normal splicing, induction of abnormal splicing)
	Regulatory (change transcription levels or pattern, or mRNA stability)
Deletion	Frameshift (change of reading frame)
	In-frame (removal of amino acids)
	Splicing
	Regulatory
	Gene deletion
	Microdeletion/contiguous gene syndrome (loss or disruption of multiple genes)
Insertion (including duplication and repeat expansion)	Frameshift
	In-frame (addition of amino acids)
	Regulatory
	Splicing
Inversion	Disruption of normal gene structure
Chromosomal abnormalities	Changes in chromosome number (polyploidy, trisomy, etc.) or structure (translocations, deletions, inversions, duplications)

Where RNA from tissue in which the candidate gene is expressed is available, the transcript can be amplified from it. Unfortunately, patient RNA from relevant tissues is often not readily available. For example, *ATP8B1* has 27 coding exons, and screening of the genomic DNA encoding it requires generation and evaluation of 24 separate PCR products (in three cases, small introns enable two exons to be included in a single PCR product). The coding portion of the *ATP8B1* transcript is 3.8 kb in length, so when patient RNA from a tissue in which *ATP8B1* is expressed is available, the coding portion of the gene can be amplified in substantially fewer PCR reactions.

Once PCR products have been generated, they can be analyzed using one of several methods; the 'gold-standard' method is DNA sequencing. However, generation and analysis of DNA sequence can be comparatively expensive and time-consuming, so other methods are sometimes used for screening, especially when a large number of samples is involved. Also, heterozygous mutations may occasionally go unnoticed, due to base-calling errors; sequencing both strands of DNA can minimize this problem.

Numerous other approaches to identification of mutations through analysis of PCR products have been developed. One commonly used approach is single strand conformation polymorphism analysis (SSCP); in this approach PCR products are denatured and then electrophoresed through a non-denaturing gel. Mutations may alter the conformation, and thus the mobility, of the single-stranded PCR product; however, some mutations are typically not detected. Three other approaches, denaturing gradient gel electrophoresis (DGGE), mismatch cleavage, and denaturing high performance liquid chromatography (DHPLC), use various techniques to detect differences between homoduplex and heteroduplex DNA molecules. Heteroduplexes are double-stranded DNA molecules in which the sequence differs between the two strands. They are formed by denaturation and reannealing of a PCR product; as a patient may be homozygous for a mutation, the mutation detection rate for these approaches is highest if the patient PCR product is mixed with a PCR product from a control sample prior to denaturation, to ensure that heteroduplexes as well as homoduplexes will form. DHPLC in particular lends itself to automation, and reportedly detects over 95% of mutations; however, the necessary equipment is expensive.[58] In general, these techniques identify samples that possess sequence changes, and then the region containing the sequence change is sequenced from that sample to precisely characterize the mutation. The presence of a frequent neutral polymorphism may necessitate a lot of extra sequencing. In the future, mutation screening may increasingly be performed using oligonucleotide microarrays; these techniques are still being perfected.

Although PCR-based methods of mutation detection are efficient and require only small quantities of patient DNA or RNA, some types of mutation are difficult or impossible to detect using them. Therefore, other methods still play a role. Use of karyotyping and forms of fluorescence in situ hybridization (FISH) allow detection of changes in chromosome number, as well as large deletions or other rearrangements. Preparation of standard genomic Southern blots, followed by hybridization with probes from the candidate region, and/or pulsed field gel electrophoresis (PFGE) of genomic DNA digested with restriction enzymes that yield large fragments, followed by Southern blotting and hybridization, can permit detection of deletions and rearrangements of intermediate size. Although comparatively large amounts of DNA and labor are required for such studies, they can be worthwhile, particularly in patients in whom mutations remain unidentified after use of other screening methods.

Regardless of the method used for mutation detection, findings in patients should be compared with those in a control sample (ideally ethnically matched, although this is not always possible), to help distinguish disease-causing mutations from neutral polymorphisms. If a particular disease mutation is found to occur frequently in patients, a specific assay for its efficient detection can be developed.

Functional Consequences of Mutation

There are a variety of ways in which a disease mutation in a gene can ultimately affect the function of the encoded protein; for example a mutation may affect transcription, mRNA stability, translation, or protein stability, localization, or function. The functional consequences of a mutation can often be predicted based upon the sequence change induced; nevertheless, functional studies are extremely valuable in confirming, refining, or changing these predictions.

1. **Transcription:** a mutation may occur in a promoter or enhancer element of a gene, and prevent transcription of the gene, or alter the levels, timing, and/or tissue distribution of expression. An example of such a mutation was recently described in the Wilson's disease (WD) gene, *ATP7B*; a 15 bp deletion in the promoter region of the gene was present on 60.5% of Sardinian WD chromosomes, and reduced transcriptional activity by 75% in expression assays.[59] Mutations can also change splicing patterns of a gene.

2. **mRNA stability and translation:** Another way in which a mutation may effect mRNA levels is by decreasing the stability of the transcript. Mutation in a polyadenylation site may prevent polyadenylation of the transcript, and lead to decreased stability of the mRNA, and/or inhibition of its translation. The presence of a sequence change that leads to premature termination of translation (such as a nonsense mutation, or an insertion, deletion, or splicing mutation that leads to a frameshift) can also lead to mRNA degradation through nonsense-mediated mRNA decay.[60] Rarely, the initiation codon may be mutated.

3. **Protein stability or localization:** a mutation may have no effect on transcript levels, but alter the stability or localization of the protein. A mutation causing abnormal folding of a protein may result in its ubiquitination and subsequent degradation. Coding sequence mutations may alter localization signals, or prevent post-translational modification, necessary

for delivery of a protein to its correct location. Several examples of mutations leading to liver disease through such mechanisms are known. For instance, the common WD mutation, H1069Q, appears to cause a protein-folding defect resulting in degradation of the protein.[61] Similarly, the common *ABCC7* mutation in CF patients, deltaF508, causes ABCC7 to be incompletely glycosylated, mislocalized, and degraded.[62,63] Two *JAG1* mutations found in AGS lead to abnormal glycosylation and mislocalization of JAG1.[64]

4. **Protein function:** normal levels of a protein may be produced, and the protein may be delivered to its proper location, but a change in the amino acid sequence of the protein may prevent it from functioning properly; its function may be partially or completely destroyed. For example, two *ABCC7* mutations found in CF patients have been shown not to prevent normal protein localization, but to disrupt protein function.[62,65] Occasionally, mutations may instead cause a 'gain of function', in that the protein has higher activity than normal, or functions in ways or circumstances different from those of the normal protein. **Dominant negative** mutations are loss-of-function mutations in which, in heterozygous individuals, the mutated protein interferes with the function of the normal versions of the protein- generally, this effect occurs with multimeric proteins.

Genetics of Cholestatic Disorders

Application of genetic approaches has greatly increased our understanding of the molecular biology of cholestasis. Through use of genetic and functional approaches, the genes mutated in a number of disorders have been identified; others have been genetically mapped (Table1; a helpful source of more information and references on these disorders is the Online Mendelian Inheritance in Man database at http://www.ncbi.nlm.nih.gov/entrez/query.fcgi?db=OMIM). There remains much knowledge to be gained through further genetic study of cholestatic disorders. For example, a recent report indicates that at least one additional low-gamma-GT PFIC gene remains to be identified; in some families segregating low-GGT PFIC, the disease does not map to either *ATP8B1* or to *ABCB11*.[17] Mutation of *ABCB11* in a BRIC patient has recently been reported, and a third BRIC locus waits to be identified, as a family in which BRIC demonstrates an autosomal dominant mode of inheritance unlinked to either *ATP8B1* or to *ABCB11* has been described.[20,43] It is clear that many cholestatic disorders exhibit extensive allelic and/or locus heterogeneity, and this is both a challenge with respect to identifying the genetic etiologies of these disorders, as well as a rich source of information about the liver and the enterohepatic circulation in health and disease. Given the rapid pace and high efficiency with which genetic research can now move, we will soon be learning much that is new about the molecular basis of cholestasis.

This new knowledge will eventually extend beyond increased understanding of rare cholestatic diseases; common genetic polymorphisms associated with increased or decreased susceptibility to relatively common, less obviously 'genetic' disorders will likely be identified, as will factors influencing disease severity, rate of progression, age of onset, and other similar variables. Work in this area is already beginning, with the mapping of a modifier locus for meconium ileus in CF and of potential susceptibility loci for ICP in the Finnish population.[40,41,56] It is likely that some of the sequence variants eventually found to influence more common hepatic disorders will occur in genes that encode proteins involved in hepatic transport and metabolism; these genes may include those initially functionally linked to cholestasis through their identification as disease genes for rare disorders.

Association studies of candidate genes have already revealed genetic factors that potentially influence development of a number of hepatic disorders. In primary biliary cirrhosis (PBC), alleles have been identified that appear associated with increased or decreased disease susceptibility or rate of disease progression; these polymorphisms include ones in the major histocompatibility locus (MHC), as well as one in the cytotoxic T lymphocyte-associated antigen-4 (CTLA-4) gene, one in the interleukin 1 gene, another in the vitamin D receptor, and a 4-SNP haplotype in the mannose-binding lectin (MBL) gene.[66-71] Variation in susceptibility to primary sclerosing cholangitis (PSC), alcoholic liver disease, and chronic viral hepatitis may also be associated with alleles of polymorphisms in the MHC. [72-77] PSC susceptibility also appears associated with a functional polymorphism in the stromelysin (MMP-3) gene;[78] other genetic associations to development of severe alcoholic liver disease have also been identified, including an association with a common amino acid change in manganese superoxide dismutase, one with a variant in the promoter of the CD14 endotoxin receptor, and a third with variants in the interleukin-1beta gene.[79-81] In chronic hepatitis C, progression of liver disease may be associated with particular variants in the transforming growth factor beta 1 and angiotensin II genes.[82] In evaluating these types of study, it is important to remember that false-positive evidence for association is sometimes obtained, and replication is necessary to confirm results. Also, due to the existence of linkage disequilibrium (LD) between closely linked genetic markers, evidence for association between a disease and a polymorphic allele does not necessarily indicate that that allele itself is the functional cause of the association; another nearby genetic variant in LD with the polymorphism studied may be the functional cause of the association- i.e., it may affect production or function of a protein, and thus influence disease development. Understandably, many genetic association studies of common liver diseases have focused on evaluation of genes encoding proteins involved in function of the immune system. It will be interesting to see whether associations between sequence variants in hepatic transporter genes and particular disease features, such as ability to maintain adequate hepatic function despite liver damage, may also be found.

Conclusions and Outlook

In the field of human genetics, rapid progress has occurred in development of technology and amassing of data including sequencing of the human genome, characterization of expressed sequence tags and identification of SNPs. This progress has set the stage for broader application of genetic techniques. Identification of genes mutated in disorders demonstrating simple Mendelian modes of inheritance is now relatively straightforward. Genetic mapping of genes mutated in disorders with more complex genetic etiology, as well as of modifier genes of 'simple traits', is increasingly feasible. In the future, many already-identified DNA sequence variants will likely be shown to influence disease susceptibility and severity for both common and rare disorders. This knowledge will permit better prediction of disease prognosis in individuals, and also, more targeted preventive interventions.

Another area in which research is rapidly expanding is that of pharmacogenetics, the study of genetic factors influencing drug response. Systematic study of known genetic variations that have no currently identified functional consequences will likely lead to identification of a subset that impact response to medications. Genetic factors leading to a greater risk of deleterious side effects, as well as factors influencing efficacy and optimal dosing, may be identified, dramatically altering medical practice.

Developments in cDNA expression array technology are beginning to yield large quantities of information on gene expression patterns in health and disease. Such studies complement genetic approaches and have parallels to sequencing of the human genome, in that they can enable creation of a vast repository of 'functional' data that can be systematically mined to identify genes encoding proteins likely to play a key role in disease processes.

References

1. Gusella JF, Wexler NS, Conneally PM et al. A polymorphic DNA marker genetically linked to Huntington's disease. Nature 1983; 306:234-238.
2. Huntington's Disease Collaborative Group. A novel gene containing a trinucleotide repeat that is expanded and unstable on Huntington's Disease chromosomes. Cell 1993; 72:971-983.
3. Houwen RH, Baharloo S, Blankenship K et al. Genome screening by searching for shared segments: mapping a gene for benign recurrent intrahepatic cholestasis. Nature Genetics 1994; 8:380-6.
4. Carlton VE, Knisely AS, Freimer NB. Mapping of a locus for progressive familial intrahepatic cholestasis (Byler disease) to 18q21-q22, the benign recurrent intrahepatic cholestasis region. Human Molecular Genetics 1995; 4:1049-53.

5. Bull LN, van Eijk MJ, Pawlikowska L et al. A gene encoding a P-type ATPase mutated in two forms of hereditary cholestasis. Nat Genet 1998; 18:219-24.
6. Sela-Herman S, Bull L, Lomri N et al. In search of a gene for hereditary cholestasis. Biochem Mol Med 1996; 59:98-103.
7. Tang X, Halleck MS, Schlegel RA et al. A subfamily of P-type ATPases with aminophospholipid transporting activity [published erratum appears in Science 1996 Dec 6;274(5293):1597]. Science 1996; 272:1495-7.
8. Ujhazy P, Ortiz D, Misra S et al. Familial intrahepatic cholestasis 1: Studies of localization and function. Hepatology 2001; 34:768-775.
9. Eppens EF, van Mil SWC, de Vree JM et al. FIC1, the protein affected in two forms of hereditary cholestasis, is localized in the cholangiocyte and the canalicular membrane of the hepatocyte. J Hepatol 2001; 35:436-443.
10. Pawlikowska L, Ottenhoff R, Looije N et al. *FIC1* mutant mice have a defect in the regulation of intestinal bile salt absorption. Hepatology 2001; 34:240A, Abstr.
11. Strautnieks SS, Kagalwalla AF, Tanner MS et al. Identification of a locus for progressive familial intrahepatic cholestasis PFIC2 on chromosome 2q24. American Journal of Human Genetics 1997; 61:630-3.
12. Strautnieks SS, Bull LN, Knisely AS et al. A gene encoding a liver-specific ABC transporter is mutated in progressive familial intrahepatic cholestasis. Nature Genetics 1998; 20:233-8.
13. International Human Genome Sequencing Consortium. Initial sequencing and analysis of the human genome. Nature 2001; 409:860-921.
14. Venter JC, Adams MD, Myers EW et al. The sequence of the human genome. Science 2001; 291:1304-1351.
15. Eddy SR. Non-coding RNA genes and the modern RNA world. Nature Reviews/Genetics 2001; 2:919-929.
16. Moss EG. RNA interference: It's a small RNA world. Current Biology 2001; 11:R772-R775.
17. Strautnieks S, Byrne J, Knisely AS et al. There must be a third locus for low GGT PFIC. Hepatology 2001; 34:240A, Abstr.
18. Klomp LW, Bull LN, Knisely AS et al. A missense mutation in FIC1 is associated with greenland familial cholestasis. Hepatology 2000; 32:1337-41.
19. Jansen PLM, Strautnieks SS, Jacquemin E et al. Hepatocanalicular bile salt export pump deficiency in patients with progressive familial intrahepatic cholestasis. Gastroenterology 1999; 117:1370-1379.
20. Kullak-Ublick GA, Kerb R, Mullhaupt B et al. A novel R432T mutation in the bile salt export pump gene (BSEP;ABCB11) is associated with recurrent intrahepatic cholestasis in an adolescent patient. Hepatology 2001; 34:216A, Abstr.
21. Deleuze JF, Jacquemin E, Dubuisson C et al. Defect of multidrug-resistance 3 gene expression in a subtype of progressive familial intrahepatic cholestasis. Hepatology 1996; 23:904-8.
22. de Vree JM, Jacquemin E, Sturm E et al. Mutations in the MDR3 gene cause progressive familial intrahepatic cholestasis. Proceedings of the National Academy of Sciences of the United States of America 1998; 95:282-7.
23. Jacquemin E, de Vree JM, Cresteil D et al. The wide spectrum of multidrug resistance 3 deficiency: from neonatal cholestasis to cirrhosis of adulthood. Gastroenterology 2001; 120:1448-1458.
24. Jacquemin E, Cresteil D, Manouvrier S et al. Heterozygous non-sense mutation of the MDR3 gene in familial intrahepatic cholestasis of pregnancy [letter]. Lancet 1999; 353:210-1.
25. Dixon PH, Weerasekera N, Linton KJ et al. Heterozygous MDR3 missense mutation associated with intrahepatic cholestasis of pregnancy: evidence for a defect in protein trafficking. Hum Mol Genet 2000; 9:1209-1217.
26. Li L, Krantz ID, Deng Y et al. Alagille syndrome is caused by mutations in human Jagged1, which encodes a ligand for Notch1. Nature Genetics 1997; 16:243-51.
27. Oda T, Elkahloun AG, Pike BL et al. Mutations in the human Jagged1 gene are responsible for Alagille syndrome. Nature Genetics 1997; 16:235-42.
28. Spinner NB, Colliton RP, Crosnier C et al. Jagged1 mutations in Alagille syndrome. Human Mutation 2001; 17:18-33.
29. Wada M, Toh S, Taniguchi K et al. Mutations in the canalicular multispecific organic anion transporter (cMOAT) gene, a novel ABC transporter, in patients with hyperbilirubinemia II/ Dubin-Johnson syndrome. Hum Molec Genet 1998; 7:203-207.
30. Ritter JK, Yeatman MT, Kaiser C et al. A phenylalanine codon deletion of the UGT1 gene complex locus of a Crigler-Najjar type I patient generates a pH-sensitive bilirubin UDP-glucuronosyltranferase. J Biol Chem 1993; 268:23573-23579.

31. Kadakol A, Ghosh SS, Sappal BS et al. Genetic lesions of bilirubin uridine-diphosphoglucuronate glucuronosyltransferase (UGT1A1) causing Crigler-Najjar and Gilbert syndromes: correlation of genotype to phenotype. Hum Mutat 2000; 16:297-306.

32. Setchell KD, Schwarz M, O'Connell NC et al. Identification of a new inborn error in bile acid synthesis: mutation of the oxysterol 7-alpha-hydorxylase gene causes severe neonatal liver disease. J Clin Invest 1998; 102:1690-1703.

33. Bull PC, thomas GR, Rommens JM et al. The Wilson disease gene is a putative copper transporting P-type ATPase similar to the Menkes gene. Nature Genet 1993; 5:327-337.

34. Tanzi RE, Petrukhin K, Chernov I et al. The Wilson disease gene is a copper transporting ATPase with homology to the Menkes disease gene. Nature Genet 1993; 5:344-350.

35. Riordan JR, Rommens JM, Kerem B et al. Identification of the cystic fibrosis gene: cloning and characterization of complementary DNA. Science 1989; 245:1066-1073.

36. Rommens JM, Iannuzzi MC, Kerem B et al. Identification of the cystic fibrosis gene: chromosome walking and jumping. Science 1989; 245:1059-1065.

37. Kerem B, Rommens JM, Buchanan JA et al. Identification of the cystic fibrosis gene: genetic analysis. Science 1989; 245:1073-1080.

38. Bull LN, Roche E, Song EJ et al. Mapping of the locus for cholestasis-lymphedema syndrome (Aagenaes syndrome) to a 6.6-cM interval on chromosome 15q. Am J Hum Genet 2000; 67:994-9.

39. Betard C, Rasquin-Weber A, Brewer C et al. Localization of a recessive gene for North American Indian childhood cirrhosis to chromosome region 16q22-and identification of a shared haplotype. Am J Hum Genet 2000; 67:222-228.

40. Heinonen ST, Eloranta ML, Heiskanen JTM et al. Maternal susceptibility locus for obstetric cholestasis maps to chromosome region 2p13 in Finnish patients. Scand J Gastroenterol 2001; 36:766-770.

41. Zielenski J, Corey M, Rozmahel R et al. Detection of a cystic fibrosis modifier locus for meconium ileus on human chromosome 19q13. Nat Genet 1999; 22:128-129.

42. Bull LN, Juijn JA, Liao M et al. Fine-resolution mapping by haplotype evaluation: the examples of PFIC1 and BRIC. Hum Genet 1999; 104:241-8.

43. Floreani A, Molaro M, Mottes M et al. Autosomal dominant benign recurrent intrahepatic cholestasis (BRIC) unlinked to 18q21 and 2q24. Am J Med Genet 2000; 95:450-453.

44. Vu TH, Tanji K, Holve SA et al. Navajo neurohepatopathy: a mitochondrial DNA depletion syndrome? Hepatology 2001; 34:116-120.

45. Byrne JL, Harrod MJ, Friedman JM et al. Del(20p) with manifestations of arteriohepatic dysplasia. Am J Med Genet 1986; 24:673-678.

46. van Mil SWC, Klomp LW, Bull LN et al. FIC1 Disease: A spectrum of intrahepatic cholestatic disorders. Sem Liv Dis 2001; 21:535-544.

47. Eloranta MJ, Heinonen S, Mononen T et al. Risk of obstetric cholestasis in sisters of index patients. Clin Genet 2001; 60:42-45.

48. Broman KW, Murray JC, Sheffield VC et al. Comprehensive human genetic maps: individual and sex-specific variation in recombination. Am J Hum Genet 1998; 63:861-9.

49. Gut IG. Automation in genotyping of single nucleotide polymorphisms. Human Mutation 2001; 17:475-492.

50. Tsui LC, Buchwald M, Barker D et al. Cystic fibrosis locus defined by a genetically linked polymorphic DNA marker. Science 1985; 230:1054-1057.

51. White R, Woodward S, Leppert M et al. A closely linked genetic marker for cystic fibrosis. Nature 1985; 318:382-384.

52. Wainwright BJ, Scambler PJ, Schmidtke J et al. Localization of cystic fibrosis locus to human chromosome 7cen-q22. Nature 1985; 318:384-385.

53. Frydman M, Bonne-Tamir B, Farrer LA et al. Assignment of the gene for Wilson disease to chromosome 13: linkage to the esterase D locus. Proc Natl Acad Sci USA 1985; 82:1819-1821.

54. Bonne-Tamir B, Farrer LA, Frydman M et al. Evidence for linkage between Wilson disease and esterase D in three kindreds: detection of linkage for an autosomal recessive disorder by the family study method. Genetic Epidemiology 1986; 3:201-209.

55. Devlin B, Roeder K, Bacanu SA. Unbiased methods for population-based association studies. Genetic Epidemiology 2001; 21:273-284.

56. Heiskanen JTM, Pirskanen MM, Hiltunen MJ et al. Insertion-deletion polymorphism in the gene for angiotensin-converting enzyme is associated with obstetric cholestasis but not with preeclampsia. Am J Obstet Gynecol 2001; 185:600-603.

57. Sinke RJ, Carlton VE, Juijn JA et al. Benign recurrent intrahepatic cholestasis (BRIC): evidence of genetic heterogeneity and delimitation of the BRIC locus to a 7-cM interval between D18S69 and D18S64. Human Genetics 1997; 100:382-7.

58. Xiao W, Oefner PJ. Denaturing high-performance liquid chromatography: a review. Hum Mutat 2001; 17:439-474.
59. Loudianos G, Dessi V, Lovicu M et al. Molecular characterization of Wilson disease in the Sardinian population—evidence of a founder effect. Hum Mutat 1999; 14:294-303.
60. Byers PH. Killing the messenger: new insights into nonsense-mediated mRNA decay. J Clin Invest 2002; 109:3-6.
61. Payne AS, Kelly EJ, Gitlin JD. Functional expression of the Wilson disease protein reveals mislocalization and impaired copper-dependent trafficking of the common H1069Q mutation. Proc Natl Acad Sci USA 1998; 95:10854-10859.
62. Cheng SH, Gregory RJ, Marshall J et al. Defective intracellular transport and processing of CFTR is the molecular basis of most cystic fibrosis. Cell 1990; 63:827-834.
63. Jensen TJ, Loo MA, Pind S et al. Multiple proteolytic systems, including the proteasome, contribute to CFTR processing. Cell 1995; 83:129-135.
64. Morrissette JJD, Colliton RP, Spinner NB. Defective intracellular transport and processing of Jag1 missense mutations in Alagille syndrome. Hum Mol Genet 2001; 10:405-413.
65. Logan J, Hiestand D, Daram P et al. Cystic fibrosis transmembrane conductance regulator mutations that disrupt nucelotide binding. J Clin Invest 1994; 94:228-236.
66. Donaldson PT. TNF gene polymorphisms in primary biliary cirrhosis: a critical appraisal. J Hepatol 1999; 31:366-368.
67. Wilson AG. Genetics of tumour necrosis factor (TNF) in autoimmune liver diseases: red hot or red herring? J Hepatol 1999; 30:331-333.
68. Agarwal K, Jones DEJ, Daly AK et al. *CTLA-4* gene polymorphism confers susceptibility to primary biliary cirrhosis. J Hepatol 2000; 32:538-541.
69. Donaldson P, Agarwal K, Craggs A et al. HLA and interleukin 1 gene polymorphisms in primary biliary cirrhosis: associations with disease progression and disease susceptibility. Gut 2001; 48:397-402.
70. Vogel A, Strassburg CP, Manns MP. Genetic association of vitamin D receptor polymorphisms with primary biliary cirrhosis and autoimmune hepatitis. Hepatology 2002; 35:126-131.
71. Matsushita M, Miyakawa H, Tanaka A et al. Single nucleotide polymorphisms of the mannose-binding lectin are associated with susceptibility to primary biliary cirrhosis. J Autoimmun 2001; 17:251-257.
72. Wiencke K, Spurkland A, Schrumpf E et al. Primary sclerosing cholangitis is associated to an extended B8-DR3 haplotype including particular MICA and MICB alleles. Hepatology 2001; 34:625-630.
73. Norris S, Kondeatis E, Collins RS et al. Mapping MHC-encoded susceptibility and resistance in primary sclerosing cholangitis: the role of MICA polymorphism. Gastro 2001; 120:1475-1482.
74. Bernal W, Moloney M, Underhill J et al. Association of tumor necrosis factor polymorphism with primary sclerosing cholangitis. J Hepatol 1999; 30:237-241.
75. Spurkland A, Saarinen S, Boberg KM et al. HLA class II haplotypes in primary sclerosing cholangitis patients from five European populations. Tissue Antigens 1999; 53:459-469.
76. Grove J, Daly AK, Bassendine MF et al. Association of a tumor necrosis factor promoter polymorphism with susceptibility to alcoholic steatohepatitis. Hepatology 1997; 26:143-146.
77. Thio CL, Thomas DL, Carrington M. Chronic viral hepatitis and the human genome. Heptology 2000; 31:819-827.
78. Satsangi J, Chapman RW, Haldar N et al. A functional polymorphism of the stromelysin gene (MMP-3) influences susceptibility to primary sclerosing cholangitis. Gastro 2001; 121:124-130.
79. Degoul F, Sutton A, Mansouri A et al. Homozygosity for alanine in the mitochondrial targeting sequence of superoxide dismutase and risk for severe alcoholic liver disease. Gastro 2001; 120:1468-1474.
80. Jarvelainen HA, Orpana A, Perola M et al. Promoter polymorphism of the CD14 endotoxin receptor gene as a risk factor for alcoholic liver disease. Hepatology 2001; 33:1148-1153.
81. Takamatsu M, Yamauchi M, Maezawa Y et al. Genetic polymorphisms of interleukin-1beta in association with the development of alcoholic liver disease in Japanese patients. Am J Gastroenterol 2000; 95:1305-1311.
82. Powell EE, Edwards-Smith CJ, Hay JL et al. Host genetic factors influence disease progression in chronic hepatitis C. Hepatology 2000; 31:828-833.
83. Chagnon P, Michaud J, Mitchell G et al. A missense mutation (R565W) in *Cirhin* (FLJ14728) in North American Indian Childhood Cirrhosis. Am J Hum Genet 2002; 71:1443-1449.

CHAPTER 8

Transcriptional Regulation
of Hepatobiliary Transporters

Saul J. Karpen

Summary

The expression and activities of hepatobiliary transporter genes is a critical component of liver function. Both sinusoidal and canalicular membrane transporters are responsible for the coordinated transport of a wide variety of organic anions, drugs, toxins, endobiotics and bile acids that ultimately are metabolized by hepatocytes and secreted into bile. Over the past decade, a tremendous amount of new information has been uncovered regarding the molecular identification of hepatobiliary transporter genes responsible for the delivery of the principal solutes in bile, and therefore the mechanisms underlying the generation of bile. Along with the cloning and identification of critical hepatobiliary transporter genes has come the capability of exploring, and possibly modifying, the molecular mechanisms driving alterations in transporter gene expression in health and disease. Regulation at the level of mRNA transcription initiation is becoming increasingly recognized as the predominant means governing transporter gene expression. An expanding role for Hepatocyte nuclear factor 1a and several members of the nuclear receptor superfamily of ligand-regulatable transcription factors permits coordination of expression of multiple transporter genes. In particular, the recently-identified nuclear receptor for bile acids, FXR, serves as both sensor and effector to help maintain intracellular bile acid homeostasis. These findings have significant importance in our current understanding, and as a means of directing future therapy, of liver diseases where cholestasis is a prominent feature.

Introduction

Over the past decade, a tremendous amount of new information regarding the identity of gene products responsible for hepatobiliary transport has been reported (see recent series in the *Seminars in Liver Disease*).[1] This new information has led to a greater understanding of the mechanisms underlying the important and central role of the hepatocyte in basic metabolism, cholesterol homeostasis, drug/toxin disposal, bile acid recirculation and the generation of bile, and the role of canalicular transporter gene mutations in the pathogenesis of select rare forms of pediatric cholestasis.[2-4]

Concomitant with the identification of transporter genes and their substrates has been the discovery of tissue and developmental restricted transporter gene expression, as well as a recognition of significant alterations in their expression in a variety of experimental and clinical diseases and conditions. Regulation of transporter gene expression has been documented at all levels—transcriptional, post-transcriptional, translational, post-translational modification and alterations of membrane and intracellular subcellular localization. Not surprisingly, of the several dozen transporter genes well-characterized to date, there are data supporting both common and individualized mechanisms of gene regulation. The data sets are being filled out at a

Molecular Pathogenesis of Cholestasis, edited by Michael Trauner and Peter L.M. Jansen.
©2004 Eurekah.com and Kluwer Academic / Plenum Publishers.

rapid pace, yet are far from complete. However, at this time it is apparent that the centerpiece of transporter gene regulation resides at the level of the initiation of mRNA transcription. And since the initiation of mRNA transcription is primarily dependent upon the presence and activities of nuclear DNA-binding transcription factors, the identification and function of these regulatory proteins will be emphasized. Moreover, recent evidence indicates the central and essential involvement of the nuclear receptor (NR) family of gene regulators.[5]

This chapter will review currently available evidence regarding the regulation of hepatobiliary transporter gene transcription, with attention focused mainly upon the factors governing organic anion and bile acid transporter gene regulation in health and disease. However, given the multitude of hepatobiliary transporters, a comprehensive review of all hepatic transporter genes is beyond the scope of this chapter. Most of the data are derived from animal models or cell culture studies, but wherever possible, human data will be described. The recent application of animal gene deletion studies and micro-array gene analysis has uncovered master regulatory roles for a select group of transcriptional regulators, which may allow for linked expression of transporters in response to environmental and pathological changes in state. The majority of regulatory information comes from animal studies of various forms of cholestasis. Several recent, excellent reviews provide detailed information about transporter gene regulation that expands beyond the transcriptional focus of this chapter.[6-9]

Sinusoidal Transporters

NTCP (SLC10A1)

The rat Na$^+$/taurocholate co-transporting polypeptide gene, Ntcp, was the first sinusoidal bile acid transporter cloned.[10] Since its identification in 1991, a substantial amount of detailed information regarding NTCP expression in mice, rats, humans during development, in the adaptive response to cholestasis and during inflammation (reviewed in ref. 11). The NTCP gene product is responsible for the Na$^+$-mediated high affinity import of conjugated bile acids into the hepatocyte, and is considered to be the major means for the hepatic uptake of bile acids from portal blood.[7,9,12] The NTCP gene is expressed exclusively in hepatocytes, and its protein resides only on the sinusoidal surface. Given its localized expression and functional importance, it should come as no surprise that the gene has significant regulation at the transcriptional level, mainly in a negative feedback manner, whereby the intracellular concentration of bile acids appears to be a major determinant of NTCP RNA expression. In addition to rat, NTCP genes have been cloned from human, hamster, zebrafish, and mouse (GENBANK and refs. 13,14).

The rat and mouse Ntcp genes appear regulated in similar fashion in experimental models of inflammation and cholestasis.[11,15] In virtually all models of inflammation, (cecal ligation and puncture, injection of endotoxin or endotoxin-related cytokines Tumor Necrosis Factor α (TNFα), Interleukin 1β (IL-1β), or cholestasis (bile duct ligation, bile acid feeding)), Ntcp RNA levels are rapidly and profoundly repressed, leading to markedly reduced bile acid import during these states.[16-24] Thus, during periods of significant hepatocyte "vulnerability" to the continued importation of bile acids, the hepatocyte responds by transcriptionally suppressing Ntcp expression.

The transcriptional regulation of the Ntcp gene by inflammatory signals and bile acids can be modeled in vitro by studies of the factors activating the Ntcp promoter. The rat Ntcp promoter is activated by 5 positive-acting transcription factors—Hepatocyte nuclear factor 1α (HNF1α), the NR heterodimer Retinoid X Receptor:Retinoic Acid Receptor (RXR:RAR), CAAT/Enhancer Binding Protein α (C/EBPα), the homeodomain protein Hex, and the Signal Transducer and Transactivator 5 (STAT5).[11,25-27] Sixteen hours after low-dose endotoxin treatment, rat Ntcp RNA levels are reduced by 80-90% of basal levels, coincident with significant reductions in the nuclear DNA-binding activities of two positive-acting transcription factors—HNF1α and RXR:RAR.[19] Treatment of human hepatoblastoma-derived HepG2 cells with IL-1β leads to reduced nuclear RXR:RAR DNA-binding activity and subsequent suppression of rat Ntcp promoter activity.[16] The mitogen activated protein kinase (MAPK) c-jun N-terminal kinase (JNK) is an essential intermediate in IL-1β mediated suppression of nuclear RXR:RAR levels.[18] RXR is phosphorylated by activated JNK, which in turn, leads to reduced nuclear RXR:RAR DNA-binding activity by an as yet unknown mechanism. Thus, the transcriptional suppression of the Ntcp gene in response to inflammation involves cross-talk between MAPK and NR regulatory pathways.[28,29]

Ntcp RNA levels are markedly suppressed in conditions where intracellular bile acid concentrations rise, thereby initiating a negative feedback loop to protect the hepatocyte from bile acid-induced hepatocyte damage and apoptosis.[11,15,21,24,30] Interestingly, Ntcp RNA levels do not rise in response to bile acid depletion, suggesting that the gene is maximally activated in the basal state.[31,32] The recent identification of a NR for bile acids, the farnesoid X receptor (FXR; NR1H4) proved vital in the understanding of how the hepatocyte responds to bile acid excess. FXR is essential for protection of hepatocytes from excess bile acid loads. Feeding of FXR +/+ or FXR +/- mice 1% cholic acid leads to a coordinated response designed to help unload the hepatocyte of excess bile acids, including a marked reduction of Ntcp RNA levels.[24] On the other hand, FXR -/- mice fed the same diet did not initiate these hepatoprotective pathways, maintained high levels of Ntcp RNA expression, which led to continued importation of bile acids, accumulation of intracellular bile acids and rapid bile acid induced hepatocellular damage and apoptosis.[24] Bile acid suppression of the rat Ntcp promoter maps to the RXR:RAR element, the same target for inflammation-reduced Ntcp expression.[17] RXR:FXR heterodimers do not bind to the Ntcp RXR:RAR element—rather the FXR-inducible transcriptional repressor small heterodimer partner (SHP; NR0B2) interferes with RXR:RAR activation of the Ntcp promoter, and reduces Ntcp gene expression in response to bile acids (see below and Fig. 1).[17] Thus, bile acid mediated negative feedback regulation of the Ntcp promoter is indirect, via bile acid activation of an FXR-responsive transcriptional repressor, SHP. Interestingly, SHP-/- mice fed bile acids still suppress Ntcp RNA expression, likely due to bile acid activation of the JNK.[18,33,34] All together, bile acid and inflammation-mediated suppression of the Ntcp promoter converge by concerted downregulation of a common Ntcp activator—the NR heterodimer RXR:RAR.

Two other animal gene deletion models of note lead to reduced Ntcp gene expression. As expected, the HNF1α -/- mouse has substantially reduced Ntcp RNA expression and hypercholanemia.[35] Less directly understandable is the reduced Ntcp expression and hypercholanemia seen in the hepatocyte selective conditional HNF4α -/- mouse, since HNF4α is not a known Ntcp regulator.[36] However, absence of the NR family member HNF4α (NR2A1) could lead to reduced HNF1α RNA expression, which, in turn, would lead to reduced Ntcp expression.[37] Further research will be required to elucidate the mechanism whereby the absence of hepatic HNF4α leads to reduced Ntcp expression.

In addition to downregulation during inflammation and cholestasis, the Ntcp gene is regulated during numerous other physiologic and pathophysiologic states. Ntcp expression is suppressed during late stages of rodent pregnancy, CCl_4-mediated toxicity, and during the early phases of regeneration—conditions where serum levels of bile acids are known to be elevated.[38-41] Only two molecules are known to increase Ntcp gene expression—prolactin (likely via induction of prolactin-activated STAT5) and retinoids (via activation of RXR:RAR).[17,42] Finally, the Ntcp gene is developmentally regulated, with marked upregulation towards the end of rat gestation, reaching adult values by approximately 4 weeks post-natally.[43,44] This reduced Ntcp expression in neonates is the likely reason for perinatal hypercholanemia.

NTCP expression in human liver has been examined in detail in only a few studies. NTCP RNA expression has been quantified from percutaneous liver biopsy specimens from patients with a variety of clinical conditions.[45] In states of significant inflammation (acute alcoholic hepatitis) or during the late stages of the progressive cholestatic disease primary biliary cirrhosis, are NTCP RNA levels substantially downregulated. It is quite possible that NTCP expression

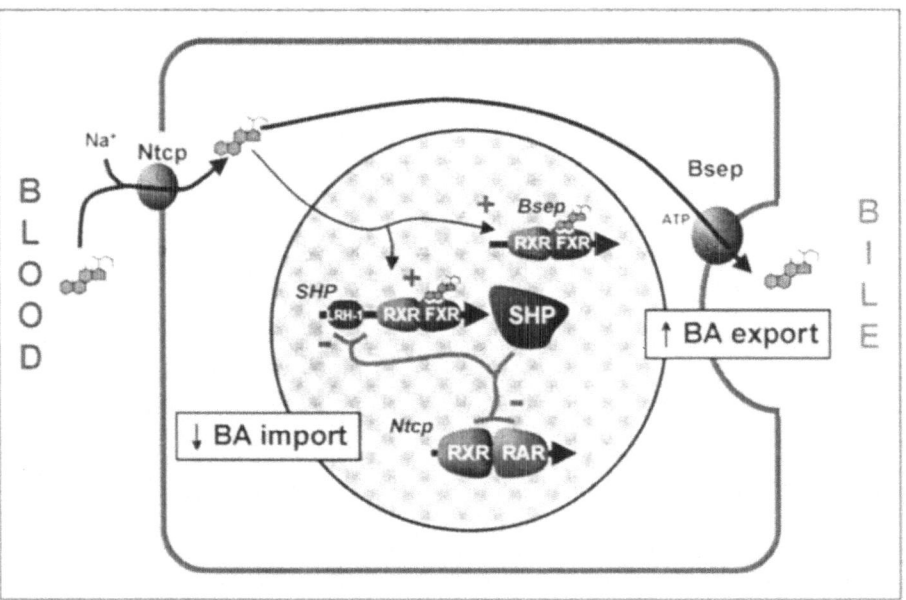

Figure 1. Role of bile acids and FXR in the regulation of bile acid transport. This simplified schematic depicts the regulation of main bile acid transporters by bile acids. Bile acids are taken up from the portal circulation primarily via the Ntcp gene product, rapidly shuttled across the cytosol, and exported via Bsep. A small percentage of bile acids enter the nucleus, moreso during cholestatic periods. Increasing concentrations of nuclear bile acids activate Bsep RNA expression via the FXR response element in its promoter. Expression of the transcriptional repressor, SHP, is also activated by FXR, which, in turn, reduces Ntcp RNA expression by interfering with RXR:RAR activation of the Ntcp promoter. Note that the SHP promoter is self-regulated. All together, the hepatocyte responds to excess accumulation of bile acids by reducing sinusoidal import and increasing canalicular export. It is likely that other transporters are playing roles in this adaptive response to cholestasis, but the molecular mechanisms are currently unknown. Abbreviations: BA, bile acids; Bsep, bile salt export pump; FXR, farnesol X receptor; LRH-1, liver receptor homologue 1; Ntcp, Na⁺/taurocholate co-transporting polypeptide; RAR, retinoic acid receptor; RXR, retinoid X receptor; SHP, small heterodimer partner.

is suppressed in other conditions, but may only be determined by serial biopsies of individual patients, given the marked inter-patient variability of human liver NTCP expression. Finally, the regulation of the human NTCP promoter has not been fully evaluated, but to date is activated by HNF3β and C/EBPα.[46] Any NR-mediated regulation of the human NTCP promoter is currently unknown.

The OATP Family (SLC21A1-14)

The family of organic anion transporting polypeptide (OATP) proteins is comprised of at least a dozen members and are the principal means of transport of a wide variety of substances across the basolateral membrane of the hepatocyte and other, primarily epithelial, tissues including kidney, intestine, choroid plexus, retina (recently reviewed in ref. 7,8). These 12-membrane spanning proteins are multispecific transporters, typically function bidirectionally, and therefore can be generally assigned the possibility of both the import and export of various substances from blood to hepatocyte and vice-versa. Their hepatic expression is therefore critically-linked to the ability of the hepatocyte to adequately metabolize many amphipathic organic molecules including xenobiotics, drugs, hormones, bilirubin and bile acids. A complete understanding of the roles played by the various OATP family members is complicated by our lack of understanding of the full range of transported substrates and these genes' variable expression in differing cells, tissues and species. Moreover, the nomenclature for OATP family members is difficult and complicated by the occasional use of the same designation for distinct family members from different species, (e.g., rat oatp2 and human OATP2 are distinct gene products). Despite these limitations, the transcriptional regulation of some OATP family members are known, mainly OATP-A (SLC21A3), oatp1 (SLC21A1), oatp2 (SLC21A5), OATP-C/oatp4 (SLC21A6), and OATP8 (SLC21A8).[47] Interestingly, some NTCP gene regulators are OATP gene regulators, especially regarding bile acid transport and metabolism, suggesting common nuclear regulation to coordinate physiological processes.

The rat oatp1 and oatp2 genes are downregulated in sepsis, cholestasis, regeneration, and treatment with estrogen, 2,4,6-trinitrobenzenesulfonic acid (TNBS), and CCl_4.[40,41,48-50] The rat oatp2 (slc21a5) gene is expressed only in perivenous hepatocytes and its promoter is upregulated by activation of two nuclear receptors, CAR (constitutive androstane receptor, NR1I3) and PXR (pregnane X receptor; NR1I2), by phenobarbital and pregnenolone-16α-carbonitrile (PCN), respectively.[51-53] Guo et al mapped PXR elements within the rat oatp2 promoter by identifying PCN-responsive sites.[54] The mouse oatp2 gene is also activated by PCN, which has been linked to the coordination of lithocholic acid uptake and detoxification pathways within hepatocytes.[55,56] Of note, lithocholic acid is a PXR ligand, thereby bringing together the regulation and detoxification of a known hepatotoxin through direct NR activation of appropriate target genes. It is important to note that rodent and human PXR (also known as hPXR or SXR) have broad and diverse ligand susceptibilities, leading to significant differences in PXR gene regulation in humans and rodents.[57-59] Rodent oatp1 genes have not been analyzed in detail at the promoter level, however it is likely that HNF1α, HNF1β, and HNF4α are positive regulators of the mouse oatp1 gene, since mouse oatp1 RNA levels are significantly downregulated in HNF1α -/- mice, and in conditional knockouts of HNF1β -/- and HNF4α -/- genes.[35,36,60] Mouse oatp2 gene expression is suppressed in HNF1α -/- mice.[35]

The human OATP-C gene (SLC21A6) has been analyzed at the promoter level. HNF1α overexpression activates the human OATP-C promoter, while deletion of the HNF1α gene markedly suppresses expression of the murine homologue, oatp4/lst-1.[35,61] Jung et al also showed that the mouse oatp4/lst-1 gene, is activated by HNF1α, while sequence analysis of the putative promoter regions of the rat and mouse oatp4/lst-1 genes identified potential binding sites for a variety of transcription factors.[61-63] Interestingly, oatp4/lst-1 RNA levels are rapidly and profoundly suppressed in models of sepsis and cholestasis, suggesting similar means of regulation to the rat Ntcp gene.[64,65] The actual molecular mediators of oatp4/lst-1 regulation in cholestasis and sepsis remain to be determined. In addition to transporting bile acids, the human OATP-C gene product is capable of transporting unconjugated bilirubin.[66] Finally, the related OATP family member OATP8 (SLC21A8) is activated by both HNF1α and FXR.[61,67]

MRP3 (ABCC3) and MRP4 (ABCC4)

Recently, several investigators have identified multidrug resistance protein 3 (MRP3) as an important component of the hepatic response to cholestasis by exporting retained bile acids, and conjugated organic anions, across the basolateral membrane.[68] Both rat and human MRP3 gene products can transport bile acids as well as a variety of conjugated organic anions.[69,70] Normally, Mrp3 expression is low and restricted to the basolateral membrane of cholangiocytes and perivenous hepatocytes.[71-73] However, in cholestatic rat liver, Mrp3 expression is markedly upregulated.[69,71,73-77] Interestingly, MRP3 expression is upregulated in human and rodent models where the canalicular transporter MRP2 (ABCC2, see below) is mutated.[75,78] The significant overlap in substrate specificities between MRP2 and MRP3 transporters thereby permits compensatory basolateral export of typical MRP2 substrates, when canalicular MRP2 expression is impaired. However, the actual role of MRP3 induction in human liver is not clear, since analysis of human liver biopsy samples from cholestatic patients did not show significant activation of MRP3 RNA levels.[45]

Mrp3 RNA levels are upregulated by phenobarbital, several chemotherapeutics and hepatotoxins.[76,79-81] Preliminary investigations of the human and rat MRP3 promoters reveal likely regulation by the nuclear receptors CAR and PXR, although detailed studies have yet to be published.[79,82-86] Sp1/Sp3 response elements have been founds in both human and rat MRP3 promoters.[83,87] It remains to be determined if MRP3 induction in cholestasis, or by various drugs, will be seen as an important mechanism of the hepatic response to cholestasis and toxins. Recent studies suggest that the MRP3 promoter is activated by bile acids in the intestine via activation of the NR liver receptor homologue 1 (LRH-1; NR5A2).[88] The Mrp3 gene appears to be suppressed by sepsis and cytokines, although this has not been evaluated in detail.[89] Finally, it should be recognized that the MRP3 gene is expressed in a variety of epithelial tissues, including high expression in intestine and colon.[85]

Although less is known of the substrate specificities and regulation of the MRP4, it appears to be highly regulated by bile acids, however via FXR-independent pathways.[68,84] Whether or not bile acid-mediated regulation of the MRP4 gene is due to nuclear receptors or signal transduction remains to be determined.

Canalicular Transporters

The transporters responsible for the canalicular efflux of the main solutes in bile—bile acids, cholesterol, phospholipids, conjugated bilirubin, as well as conjugated drugs/toxins/endobiotics—have been identified in recent years.[2,90-93] Although well-characterized functionally, clinically and physiologically, our understanding of the transcriptional regulation of this select group of canalicular transporter genes is still evolving, but does have some links physiological regulators of hepatocyte function, as well as coordination with regulators of sinusoidal transporter genes.

MRP2 (ABCC2)

The multidrug resistance related protein 2, MRP2, is expressed on the canalicular membrane of hepatocytes, as well as the apical membrane of enterocytes and kidney.[94,95] The MRP2 gene product exports glutathione, sulfated and glucuronidated conjugates of drugs, toxins, and bilirubin, and is likely responsible for the majority of bile salt-independent bile flow across the canaliculus.[96] Mutation of MRP2 leads to reduced bile-salt independent bile flow in rodent models, and elevated serum levels of conjugated bilirubin seen in the Dubin-Johnson syndrome.[96,97]

Mrp2 RNA levels are profoundly suppressed in rodent models of sepsis and cholestasis.[23,98-101] Little change in Mrp2 RNA levels are seen in regenerating rat liver.[39,40] MRP2 RNA levels are essentially normal in human liver biopsies from cholestatic patients, but reduced in liver biopsies from patients infected with hepatitis C virus.[45,102] Mice fed diets enriched with cholic acid or ursodeoxycholic acid showed increased hepatic Mrp2 RNA expression.[103] Human and rat MRP2 RNA expression is increased in response to several PXR activators, redox reagents, toxins, and drugs, and decreased in response to inflammatory cytokines.[16,104-110] Interestingly, patients treated with the potent human PXR agonist rifampin had elevated duodenal MRP2 RNA levels, suggesting that the human MRP2 gene is upregulated in vivo by PXR.[104]

The MRP2 promoter from rat and human has been analyzed at the transcriptional level, and is apparently regulated by multiple transcription factors. Two groups have recently shown that PXR is a potent activator of the rat and human MRP2 genes, indicating that PXR is a coordinating regulator of toxin uptake, metabolism and efflux.[109-111] Kast et al report finding a 26 bp element in the rat Mrp2 promoter that is activated by CAR and FXR as well as PXR.[109] This element is adjacent to a previously-identified RAR response element, thereby indicating a highly complex means of Mrp2 gene regulation by multiple NR family members.[16] Pro-inflammatory cytokine-mediated suppression of rat Mrp2 promoter activity mapped to the RXR:RAR element, mediated by repression of RXR and RAR activity, and possibly, expression.[16,107] Whether or not inflammation-based suppression of Mrp2 RNA expression involves alterations in activities of other NR family members FXR, CAR and PXR, remains to be determined.

BSEP(ABCB11)

The bile salt export pump, BSEP was functionally identified by Gerloff et al as the canalicular ATP-dependent bile salt transporter.[112] Virtually concurrent support for BSEP as the bile salt exporter came from the genetic identification of the gene mutated in a rare form of progressive familial intrahepatic cholestasis, PFIC-2 , which has a clinical hallmark of markedly reduced biliary bile salt secretion.[113] Human BSEP transports bile acids with high affinity— Km for taurocholate and taurochenodeoxycholate between 4-8 μM.[114,115] Moreover, Bsep -/- mice have markedly reduced bile salt secretion.[116]

Since its identification in 1998, BSEP, like other transporter genes, has both transcriptional and post-transcriptional regulation. In animal models and cell culture, Bsep RNA levels are increased in response to cholic acid feeding, prolactin, and hypo-osmolarity and suppressed in bile duct ligation, lithocholic acid, endotoxin, biliary depletion, but not significantly altered in response to regeneration, phenobarbital, sirolimus, cyclosporine, rifampin or CCl$_4$ toxicity.[39-41,49,103,117-122]

Several groups have studied the transcriptional regulation of human, rat, and mouse BSEP genes, revealing strong bile acid activation of the BSEP promoter via binding of RXR:FXR heterodimers.[24,84,123-125] Studies by Sinal et al utilizing the FXR -/- mouse clearly demonstrated that FXR is a prime determinant of basal and bile acid activated Bsep expression.[24] Absence of FXR virtually obliterated basal and cholic acid stimulated Bsep RNA expression, while cholic acid feeding of FXR-/- mice led to rapid and significant hepatocellular damage due to an inability to export bile acids. Lithocholic acid is an FXR antagonist, and acts to significantly reduce Bsep expression in cell culture.[121] The regulation of BSEP gene expression in human liver samples from inflammation-based cholestatic diseases is markedly reduced compared to control samples.[45] The FXR target transcriptional repressor SHP is a likely suppressor of Bsep expression, since Bsep RNA levels increase in SHP -/- mice.[34] Taken together, alterations in FXR activity have strong effects on Bsep expression, and the ability of the hepatocyte to export bile acids across the canalicular membrane. Finally, other than FXR, little is known of other Bsep gene regulators, except that conditional deletion of HNF1α or HNF4α genes has little appreciable effect on Bsep expression.[35,36]

MDR3 (ABCB4)

The multidrug resistance protein 3 (MDR3; Mdr2 in rodents) gene product mediates phospholipid transport across the canalicular membrane.[126,127] Deletion or mutation of the MDR3 gene leads to a marked reduction in biliary phospholipid content, significant liver damage, and is the cause of the progressive cholestatic disorder, PFIC-3.[128-130] The rodent Mdr2 gene is activated by statins (pravastatin or squalestatin) and fibrates, while bile acid-mediated regulation remains controversial.[103,131-135]

Detailed studies of the human MDR3, and rat Mdr2 promoters are limited, but suggest Sp1-dependent expression.[136,137] Since it has been known that fibrates and other peroxisomal proliferators activated mouse Mdr2 RNA expression in vivo and in hepatocyte culture, it was assumed that these compounds worked via activation of the NR peroxisomal proliferator and activator α (PPARα; NR1C1).[131,132] Recently, Kok et al show that fibrate-mediated induction of mouse Mdr2 expression is completely dependent upon PPARα, since Mdr2 RNA levels were unchanged by fibrate treatment of PPARα -/- mice.[135]

ABCG5 and ABCG8

These two ATP Binding Cassette family member half-transporters have been recently identified as the gene products mutated in the rare disease of accumulated serum plant sterols, sitosterolemia.[138,139] Abcg5 and Abcg8 are expressed in the canalicular membrane as well as the apical membrane of enterocytes. Together, they apparently function as sterol exporters, and thus when mutated, leads to increased absorption of dietary cholesterol and phytosterols. Hobbs' group has recently shown that dual expression of Abcg5/Abcg8 are necessary for biliary cholesterol secretion, thereby identifying the long-sought biliary cholesterol transporter gene.[92,93,138,140]

The ABCG5 and ABCG8 genes are oriented in a head-to-head configuration on chromosome 2p21, sharing a common short promoter region.[139] In mice, Abcg5 and Abcg8 RNA expression is upregulated by cholesterol feeding and by treatment with an activator of the NR Liver X Receptor, LXR (NR1H3).[139,141] Within this short, shared promoter region is an LXR-response element. This LXR binding element responds to cholesterol loading of hepatocytes by a compensatory transcriptional upregulation of Abcg5 and Abcg8 genes, thereby increasing biliary cholesterol secretion, and maintaining intracellular cholesterol homeostasis. Since these genes, and their regulation by oxysterols, has just been discovered, it is likely that within the next few years more will be found about regulators of ABCG5 and ABCG8 that may have implications for cholesterol homeostasis, biliary tract disease, and gallstone pathogenesis.

Common Regulatory Themes

At least 21 regulators of transcription play significant roles in the expression of hepatobiliary transporter genes (Table 1). To date, it appears that three nuclear factors, HNF1α, RXR, and FXR, are employed in the transcriptional regulation of many critical hepatobiliary transporter genes, whose products reside at both sinusoidal and canalicular membranes. Thus, any alteration in the expression, activities, concentration, cellular location, of any of these 3 key regulators will have significant consequences on the hepatic uptake of bile acids and organic anions, intracellular function, and the canalicular delivery of biliary solutes. Figure 2 schematically depicts the overall known regulation of transporter genes discussed in this Chapter.

HNF1α activates multiple sinusoidal transporter genes (NTCP, oatp1, oatp2, oatp4 in rodents, OATP-C, and OATP8 in humans), and thus is a bona fide master regulator of sinusoidal hepatic transport.[61,142] HNF1α is a well-known regulator of hepatic genes responsible for a wide variety of physiological and nutritional hepatic functions.[143] These recent discoveries place sinusoidal transport along with hepatic glucose, lipid and serum protein expression as main hepatic functions regulated by HNF1α.

Mutations in the human HNF1α gene are associated with a rare form of maturity onset diabetes in the young (MODY).[144] Alterations in bile acid or organic anion transport in patients with MODY have not been reported. Hepatic nuclear HNF1α levels are reduced in response to CCl_4 administration and endotoxin.[19,41,145] Reduced HNF1α levels in the response to endotoxin likely participate in the well-known alterations in bile acid synthesis and import seen during inflammation.

The Central Role of FXR in the Hepatic Response to Cholestasis

FXR is a member of the ligand-regulatable NR family of transcription factors that has recently taken center-stage in the adaptive response to cholestasis (reviewed in refs. 5,146-151). In 1999, three groups identified the primary bile acids cholic and chenodeoxycholic acids as high affinity ligands for FXR, with varying affinities of secondary and other bile acids.[152-154] The potent hepatotoxic bile acid lithocholic acid is an FXR antagonist.[121] Interestingly, the choleretic and therapeutic bile acid ursodeoxycholic acid is not a ligand for FXR, indicating that its functions are FXR-independent. The unique NR family member SHP is a potent transcriptional repressor of several NRs, and is activated by an FXR response element in its promoter.[155,156]

An emerging coordinated role for bile acids, FXR, and SHP strongly suggest these nuclear receptors act as sensors and effectors to maintain "safe" levels of intracellular bile acids (Fig. 1). Once activated, FXR and SHP serve to reduce bile acid import (by reducing Ntcp expression), bile acid synthesis (by suppressing cholesterol 7 alpha hydroxylase, Cyp7a1 expression) and

Table 1. Transcriptional regulators of hepatobiliary transporter genes

Transcription factors	Nuclear receptors
AP-1	**CAR**
C/EBPα	**FXR**
HNF1α	HNF4α
HNF1β	LRH-1
HNF3β	LXR
Hex	PPARα
Sp1/Sp3	PXR/SXR
STAT5	**RAR**
	RXR
Signaling Molecules	**SHP**
JNK	
PKCα	

Note that some transcriptional factors listed are indirect regulators of hepatobiliary transporter genes. Regulators in boldface are have multiple identified target transporter genes. See text and Figure Legends for abbreviations.

increase canalicular bile acid export, by activating Bsep expression. Perhaps the most crucial evidence indicating that FXR is essential to maintaining bile acid homeostasis comes from studies of the FXR -/- mouse.[24] On a normal diet, these mice function without obvious impairments. However, when placed on a 1% cholic acid diet for only 5 days, FXR -/- mice experience increased morbidity due to marked hepatocellular destruction from retained bile acids, while their FXR +/- and +/+ littermates are unaffected. In FXR +/+ mice, a 1% cholic acid diet leads to compensatory reduced hepatic Ntcp and Cyp7a1 RNA levels, and increased Bsep RNA expression. However in FXR -/- mice, Bsep and SHP RNA expression is low, and the hepatic RNA levels of these three core genes, Ntcp, Cyp7a1, and Bsep, are unchanged in response to cholic acid, thereby allowing intracellular accumulation of cholic acid and activation of bile acid-mediated apoptosis and necrosis.

Interestingly, the role for SHP in bile acid homeostasis is complicated by the fact that bile acids suppress still Ntcp and Cyp7a1 RNA levels in SHP -/- mice.[34,157] The degree of suppression of these genes' expression by bile acids in SHP -/- mice is significant, but less than in wild type mice, but clearly suggests that bile acid mediated gene suppression goes beyond FXR induction of SHP.[158] Perhaps, bile acid mediated induction of signal transduction cascades, or cytokine expression in non-parenchymal cells, is a means of SHP-independent Ntcp and Cyp7a1 gene suppression.[18,33,34,159]

Current evidence suggests that therapeutic manipulation of FXR activity may be a novel means of treating cholestasis. There are currently no effective means of treating cholestasis, and perhaps activation of FXR-dependent pathways may provide a means of pharmacologically activating hepatoprotection from excess intracellular bile acids. A potent FXR agonist would increase Bsep RNA levels and reduce both Ntcp and Cyp7a1 RNA expression—all working in concert to reduce intrahepatocytic concentrations of bile acids. FXR agonists GW4064 and 6-α-ethylchenodeoxycholic acid (6αECDCA), both of which act to alter the expected target genes that would reduce intracellular bile acid accumulation.[146,160] Moreover, two FXR antagonists that interference with bile acid activation of FXR target genes, guggulsterone and lithocholic acid, would significantly enhance bile acid accumulation.[121,161] Taken together, it is apparent that modulation of FXR activity by exogenous agents could prove to be a new means of treating cholestasis, or in the case of FXR antagonists, exacerbate it.

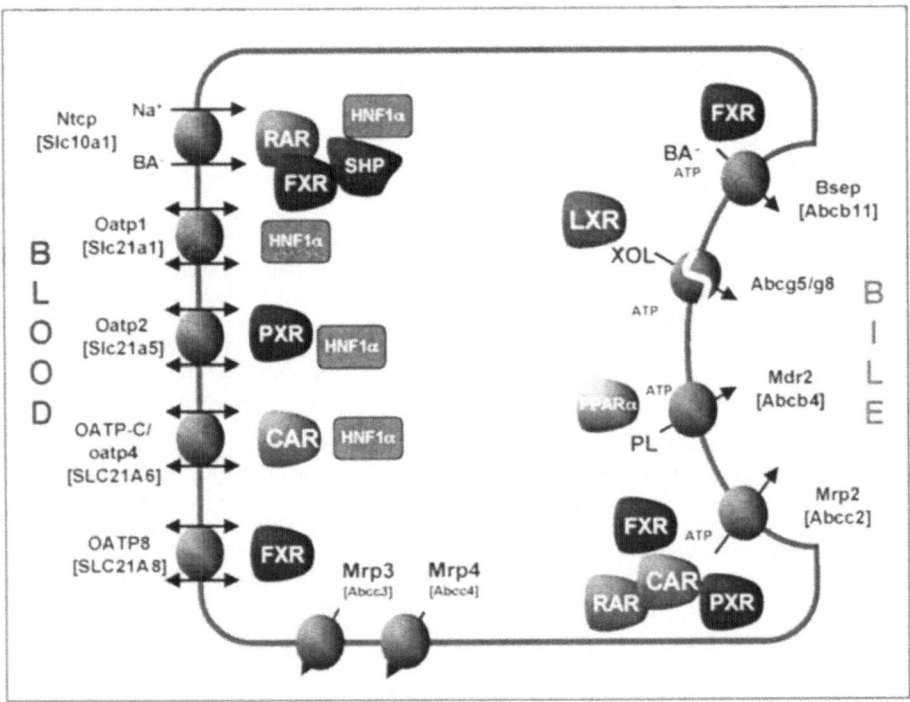

Figure 2. Composite schematic of known transcriptional regulators of hepatobiliary transporter genes. The main regulators of the genes discussed in this Chapter are depicted along their respective membrane of expression. For simplicity, RXR was not included in the Figure, but is a necessary partner for all nuclear receptors except SHP. Thus, it should be remembered that alterations in RXR activity may profoundly affect the expression of virtually all transporter genes. Abbreviations: BA, bile acids; Bsep, bile salt export pump; CAR, constitutive androstane receptor; FXR, farnesol X receptor; HNF1α, hepatocyte nuclear factor 1α; LRH-1, liver receptor homologue 1; LXR, liver X receptor; Mdr2, multidrug resistance protein 2; Mrp, multidrug resistance related protein; Ntcp, Na+/taurocholate co-transporting polypeptide; Oatp, organic anion transporting polypeptide; PL, phospholipid; PPARα; peroxisomal proliferator and activating receptor α; PXR, pregnane X receptor; RAR, retinoic acid receptor; RXR, retinoid X receptor; SHP, small heterodimer partner; XOL, cholesterol.

Conclusions

Hepatobiliary transporter genes are under transcriptional control from a variety of gene regulators. For sinusoidal transporters, HNF1α is a common activator, while members of the NR superfamily regulate virtually every transporter gene studied.[5,162] The nuclear receptor for bile acids, FXR, plays a crucial role in regulating bile acid import, synthesis and export. Further research will be required to understand the detailed means of regulating all hepatobiliary transporter genes, with the potential to explore new means of treating a variety of hepatic diseases and dysfunction—including drug toxicity, metabolic disease, hypercholesterolemia and cholestasis.

References

1. Jansen PL. Foreword: from classic bile physiology to cloned transporters. Semin Liver Dis 2000; 20:245-50.
2. Borst P, Elferink RO. Mammalian ABC transporters in health and disease. Annu Rev Biochem 2002; 71:537-92.
3. Jansen PL, Muller M, Sturm E. Genes and cholestasis. Hepatology 2001; 34:1067-74.

4. Thompson R, Jansen PL. Genetic defects in hepatocanalicular transport. Semin Liver Dis 2000; 20:365-72.
5. Karpen SJ. Nuclear receptor regulation of hepatic function. J Hepatol 2002; 36:832-50.
6. Kullak-Ublick GA, Stieger B, Hagenbuch B et al. Hepatic transport of bile salts. Semin Liver Dis 2000; 20:273-92.
7. Meier PJ, Stieger B. Bile salt transporters. Annu Rev Physiol 2002; 64:635-61.
8. Hagenbuch B, Meier PJ. The superfamily of organic anion transporting polypeptides. Biochim Biophys Acta 2003; 1609:1-18.
9. Wolkoff AW, Cohen DE. Bile acid Regulation of Hepatic Physiology: I. Hepatocyte transport of bile acids. Am J Physiol Gastrointest Liver Physiol 2003; 284:G175-179.
10. Hagenbuch B, Stieger B, Foguet M et al. Functional expression cloning and characterization of the hepatocyte Na+/bile acid cotransport system. Proc Natl Acad Sci USA 1991; 88:10629-33.
11. Karpen SJ. Transcriptional regulation of sinusoidal transporters. In: Matern S, Boyer J, Keppler D, et al, eds. Hepatobiliary transport: From Bench to Bedside. London: Kluwer Academic, 2001:22-31.
12. Hagenbuch B, Scharschmidt BF, Meier PJ. Effect of antisense oligonucleotides on the expression of hepatocellular bile acid and organic anion uptake systems in Xenopus laevis oocytes. Biochem J 1996; 316:901-4.
13. Hagenbuch B, Meier PJ. Molecular cloning, chromosomal localization, and functional characterization of a human liver Na+/bile acid cotransporter. J Clin Invest 1994; 93:1326-31.
14. Cattori V, Eckhardt U, Hagenbuch B. Molecular cloning and functional characterization of two alternatively spliced Ntcp isoforms from mouse liver1. Biochim Biophys Acta 1999; 1445:154-9.
15. Karpen SJ. Bile acid-mediated feedback inhibition of the rat ntcp promoter. In: Henegouwen vB, ed. Falk Symposium: Biology of Bile acids in Health and Disease. Volume 120. London: Kluwer Academic, 2001:95-104.
16. Denson LA, Auld KL, Schiek DS et al. Interleukin-1 beta suppresses retinoid transactivation of two hepatic transporter genes involved in bile formation. Journal of Biological Chemistry 2000; 275:8835-8843.
17. Denson LA, Sturm E, Echevarria W et al. The orphan nuclear receptor, shp, mediates bile acid-induced inhibition of the rat bile acid transporter, ntcp. Gastroenterology 2001; 121:140-7.
18. Li D, Zimmerman TL, Thevananther S et al. Interleukin-1 beta-mediated suppression of RXR:RAR transactivation of the Ntcp promoter is JNK-dependent. J Biol Chem 2002; 277:31416-22.
19. Trauner M, Arrese M, Lee H et al. Endotoxin downregulates rat hepatic ntcp gene expression via decreased activity of critical transcription factors. Journal of Clinical Investigation 1998; 101:2092-2100.
20. Gartung C, Ananthanarayanan M, Rahman MA et al. Down-regulation of expression and function of the rat liver Na+/bile acid cotransporter in extrahepatic cholestasis. Gastroenterology 1996; 110:199-209.
21. Gartung C, Schuele S, Schlosser SF et al. Expression of the rat liver na+/taurocholate cotransporter is regulated in vivo by retention of biliary constituents but not their depletion. Hepatology 1997; 25:284-290.
22. Green RM, Beier D, Gollan JL. Regulation of hepatocyte bile salt transporters by endotoxin and inflammatory cytokines in rodents. Gastroenterology 1996; 111:193-8.
23. Kim PK, Chen J, Andrejko KM et al. Intraabdominal sepsis down-regulates transcription of sodium taurocholate cotransporter and multidrug resistance-associated protein in rats. Shock 2000; 14:176-81.
24. Sinal CJ, Tohkin M, Miyata M et al. Targeted disruption of the nuclear receptor FXR/BAR impairs bile acid and lipid homeostasis. Cell 2000; 102:731-44.
25. Karpen SJ, Sun AQ, Kudish B et al. Multiple factors regulate the rat liver basolateral sodium-dependent bile acid cotransporter gene promoter. J Biol Chem 1996; 271:15211-15221.
26. Ganguly TC, O'Brien ML, Karpen SJ et al. Regulation of the rat liver sodium-dependent bile acid cotransporter gene by prolactin. Mediation of transcriptional activation by Stat5. J Clin Invest 1997; 99:2906-14.
27. Denson LA, Karpen SJ, Bogue CW et al. Divergent homeobox gene Hex regulates promoter of the Na+- dependent bile acid cotransporter. American Journal of Physiology-Gastrointestinal and Liver Physiology 2000; 279:G347-G355.
28. Shao D, Lazar MA. Modulating nuclear receptor function: may the phos be with you. J Clin Invest 1999; 103:1617-8.
29. Kyriakis JM. MAP kinases and the regulation of nuclear receptors. Sci STKE 2000; 2000:PE1.
30. Zollner G, Fickert P, Silbert D et al. Induction of short heterodimer partner 1 precedes downregulation of Ntcp in bile duct-ligated mice. Am J Physiol Gastrointest Liver Physiol 2002; 282:G184-91.

31. Koopen NR, Wolters H, Voshol P et al. Decreased Na+-dependent taurocholate uptake and low expression of the sinusoidal Na+-taurocholate cotransporting protein (Ntcp) in livers of mdr2 P-glycoprotein-deficient mice. J Hepatol 1999; 30:14-21.
32. Wolters H, Elzinga BM, Baller JF et al. Effects of bile salt flux variations on the expression of hepatic bile salt transporters in vivo in mice. J Hepatol 2002; 37:556-63.
33. Gupta S, Stravitz RT, Dent P et al. Down-regulation of cholesterol 7alpha-hydroxylase (CYP7A1) gene expression by bile acids in primary rat hepatocytes is mediated by the c-Jun N-terminal kinase pathway. J Biol Chem 2001; 276:15816-22.
34. Wang L, Lee YK, Bundman D et al. Redundant pathways for negative feedback regulation of bile acid production. Dev Cell 2002; 2:721-31.
35. Shih DQ, Bussen M, Sehayek E et al. Hepatocyte nuclear factor-1alpha is an essential regulator of bile acid and plasma cholesterol metabolism. Nat Genet 2001; 27:375-82.
36. Hayhurst GP, Lee YH, Lambert G et al. Hepatocyte nuclear factor 4alpha (nuclear receptor 2A1) is essential for maintenance of hepatic gene expression and lipid homeostasis. Mol Cell Biol 2001; 21:1393-403.
37. Kuo CJ, Conley PB, Chen L et al. A transcriptional hierarchy involved in mammalian cell-type specification. Nature 1992; 355:457-61.
38. Green RM, Gollan JL, Hagenbuch B et al. Regulation of hepatocyte bile salt transporters during hepatic regeneration. Am J Physiol 1997; 273:G621-7.
39. Vos TA, Ros JE, Havinga R et al. Regulation of hepatic transport systems involved in bile secretion during liver regeneration in rats. Hepatology 1999; 29:1833-9.
40. Gerloff T, Geier A, Stieger B et al. Differential Expression of Basolateral and Canalicular Organic anion transporters During Regeneration of Rat Liver. Gastroenterology 1999; 117:1408-1415.
41. Geier A, Kim SK, Gerloff T et al. Hepatobiliary organic anion transporters are differentially regulated in acute toxic liver injury induced by carbon tetrachloride. J Hepatol 2002; 37:198-205.
42. Ganguly TC, Obrien M, Karpen S et al. Prolactin (PRL) induced transcriptional regulation of Na+/taurocholate cotransporter (Ntcp) gene is mediated by Stat- 5. Hepatology 1996; 24:972-972.
43. Hardikar W, Ananthanarayanan M, Suchy FJ. Differential ontogenic regulation of basolateral and canalicular bile acid transport proteins in rat liver. J Biol Chem 1995; 270:20841-6.
44. Arrese M, Trauner M, Ananthanarayanan M et al. Maternal cholestasis does not affect the ontogenic pattern of expression of the Na+/taurocholate cotransporting polypeptide (ntcp) in the fetal and neonatal rat liver. Hepatology 1998; 28:789-95.
45. Zollner G, Fickert P, Zenz R et al. Hepatobiliary transporter expression in percutaneous liver biopsies of patients with cholestatic liver diseases. Hepatology 2001; 33:633-46.
46. McClure MH, Denson LA, Ananthanaravanan M et al. CCAAT/enhancer binding protein alpha (C/EBP alpha) and hepatocyte nuclear factor 3 beta (HNF3 beta) transactivate the human hepatic Na+/taurocholate cotransporter (Ntcp). Hepatology 1998; 28:1063A.
47. Kullak-Ublick GA, Beuers U, Fahney C et al. Identification and functional characterization of the promoter region of the human organic anion transporting polypeptide gene. Hepatology 1997; 26:991-7.
48. Gartung C, Matern S. Molecular regulation of sinusoidal liver bile acid transporters during cholestasis. Yale J Biol Med 1997; 70:355-63.
49. Geier A, Dietrich CG, Lammert F et al. Regulation of organic anion transporters in a new rat model of acute and chronic cholangitis resembling human primary sclerosing cholangitis. J Hepatol 2002; 36:718-24.
50. Geier A, Dietrich CG, Gerloff T et al. Regulation of basolateral organic anion transporters in ethinylestradiol-induced cholestasis in the rat. Biochim Biophys Acta 2003; 1609:87-94.
51. Reichel C, Gao B, Van Montfoort J et al. Localization and function of the organic anion-transporting polypeptide Oatp2 in rat liver. Gastroenterology 1999; 117:688-95.
52. Guo GL, Johnson DR, Klaassen CD. Postnatal expression and induction by pregnenolone-16alpha-carbonitrile of the organic anion-transporting polypeptide 2 in rat liver. Drug Metab Dispos 2002; 30:283-8.
53. Guo GL, Choudhuri S, Klaassen CD. Induction profile of rat organic anion transporting polypeptide 2 (oatp2) by prototypical drug-metabolizing enzyme inducers that activate gene expression through ligand-activated transcription factor pathways. J Pharmacol Exp Ther 2002; 300:206-12.
54. Guo GL, Staudinger J, Ogura K et al. Induction of Rat Organic anion transporting Polypeptide 2 by Pregnenolone-16alpha-carbonitrile Is via Interaction with Pregnane X Receptor. Mol Pharmacol 2002; 61:832-9.
55. Xie W, Radominska-Pandya A, Shi Y et al. An essential role for nuclear receptors SXR/PXR in detoxification of cholestatic bile acids. Proc Natl Acad Sci USA 2001; 98:3375-80.

56. Staudinger JL, Goodwin B, Jones SA et al. The nuclear receptor PXR is a lithocholic acid sensor that protects against liver toxicity. Proc Natl Acad Sci USA 2001; 98:3369-74.
57. Watkins RE, Wisely GB, Moore LB, Collins JL, Lambert MH, Williams SP, Willson TM, Kliewer SA, Redinbo MR. The human nuclear xenobiotic receptor PXR: structural determinants of directed promiscuity. Science 2001; 292:2329-33.
58. Kliewer SA, Goodwin B, Willson TM. The Nuclear Pregnane X Receptor: A Key Regulator of Xenobiotic Metabolism. Endocr Rev 2002; 23:687-702.
59. Moore JT, Goodwin B, Willson TM et al. Nuclear receptor regulation of genes involved in bile acid metabolism. Crit Rev Eukaryot Gene Expr 2002; 12:119-35.
60. Coffinier C, Gresh L, Fiette L et al. Bile system morphogenesis defects and liver dysfunction upon targeted deletion of HNF1beta. Development 2002; 129:1829-38.
61. Jung D, Hagenbuch B, Gresh L et al. Characterization of the human OATP-C (SLC21A6) gene promoter and regulation of liver-specific OATP genes by hepatocyte nuclear factor 1 alpha. J Biol Chem 2001; 276:37206-14.
62. Choudhuri S, Ogura K, Klaassen CD. Determination of transcription start site and analysis of promoter sequence, splice junction sites, intron sequence and codon usage bias of rat liver-specific organic anion transporter-1 (rlst-1/Oatp-4/Slc21a10) gene. DNA Seq 2002; 13:103-7.
63. Ogura K, Choudhuri S, Klaassen CD. Full-length cDNA cloning and genomic organization of the mouse liver- specific organic anion transporter-1 (lst-1). Biochem Biophys Res Commun 2000; 272:563-70.
64. Kakyo M, Unno M, Tokui T et al. Molecular characterization and functional regulation of a novel rat liver-specific organic anion transporter rlst-1. Gastroenterology 1999; 117:770-5.
65. Abe T, Kakyo M, Tokui T et al. Identification of a novel gene family encoding human liver-specific organic anion transporter LST-1. J Biol Chem 1999; 274:17159-63.
66. Cui Y, Konig J, Leier I et al. Hepatic Uptake of Bilirubin and Its Conjugates by the Human Organic anion transporter SLC21A6. J Biol Chem 2001; 276:9626-9612.
67. Jung D, Podvinec M, Meyer UA et al. Human organic anion transporting polypeptide 8 promoter is transactivated by the farnesoid X receptor/bile acid receptor. Gastroenterology 2002; 122:1954-66.
68. Borst P, Evers R, Kool M et al. A family of drug transporters: the multidrug resistance-associated proteins. J Natl Cancer Inst 2000; 92:1295-302.
69. Hirohashi T, Suzuki H, Takikawa H et al. ATP-dependent Transport of Bile salts by Rat Multidrug Resistance-associated Protein 3 (Mrp3). J Biol Chem 2000; 275:2905-2910.
70. Hirohashi T, Suzuki H, Sugiyama Y. Characterization of the transport properties of cloned rat multidrug resistance-associated protein 3 (MRP3). J Biol Chem 1999; 274:15181-5.
71. Donner MG, Keppler D. Up-regulation of basolateral multidrug resistance protein 3 (Mrp3) in cholestatic rat liver. Hepatology 2001; 34:351-9.
72. Nies AT, Konig J, Pfannschmidt M et al. Expression of the multidrug resistance proteins MRP2 and MRP3 in human hepatocellular carcinoma. Int J Cancer 2001; 94:492-9.
73. Scheffer GL, Kool M, de Haas M et al. Tissue distribution and induction of human multidrug resistant protein 3. Lab Invest 2002; 82:193-201.
74. Kool M, van der Linden M, de Haas M et al. MRP3, an organic anion transporter able to transport anti-cancer drugs. Proc Natl Acad Sci USA 1999; 96:6914-9.
75. Konig J, Rost D, Cui Y et al. Characterization of the human multidrug resistance protein isoform MRP3 localized to the basolateral hepatocyte membrane. Hepatology 1999; 29:1156-63.
76. Ogawa K, Suzuki H, Hirohashi T et al. Characterization of inducible nature of MRP3 in rat liver. Am J Physiol Gastrointest Liver Physiol 2000; 278:G438-46.
77. Soroka CJ, Lee JM, Azzaroli F et al. Cellular localization and up-regulation of multidrug resistance- associated protein 3 in hepatocytes and cholangiocytes during obstructive cholestasis in rat liver. Hepatology 2001; 33:783-91.
78. Akita H, Suzuki H, Sugiyama Y. Sinusoidal efflux of taurocholate correlates with the hepatic expression level of Mrp3. Biochem Biophys Res Commun 2002; 299:681-7.
79. Xiong H, Yoshinari K, Brouwer KL et al. Role of constitutive androstane receptor in the in vivo induction of Mrp3 and CYP2B1/2 by phenobarbital. Drug Metab Dispos 2002; 30:918-23.
80. Schrenk D, Baus PR, Ermel N et al. Up-regulation of transporters of the MRP family by drugs and toxins. Toxicol Lett 2001; 120:51-7.
81. Hinoshita E, Uchiumi T, Taguchi K et al. Increased expression of an ATP-binding cassette superfamily transporter, multidrug resistance protein 2, in human colorectal carcinomas. Clin Cancer Res 2000; 6:2401-7.
82. Fromm MF, Leake B, Roden DM et al. Human MRP3 transporter: identification of the 5'-flanking region, genomic organization and alternative splice variants. Biochim Biophys Acta 1999; 1415:369-74.

83. Takada T, Suzuki H, Sugiyama Y. Characterization of 5'-flanking region of human MRP3. Biochem Biophys Res Commun 2000; 270:728-32.

84. Schuetz EG, Strom S, Yasuda K et al. Disrupted Bile acid Homeostasis Reveals an Unexpected Interaction among Nuclear Hormone Receptors, Transporters, and Cytochrome P450. J Biol Chem 2001; 276:39411-39418.

85. Cherrington NJ, Hartley DP, Li N et al. Organ distribution of multidrug resistance proteins 1, 2, and 3 (Mrp1, 2, and 3) mRNA and hepatic induction of Mrp3 by constitutive androstane receptor activators in rats. J Pharmacol Exp Ther 2002; 300:97-104.

86. Xiong H, Suzuki H, Sugiyama Y et al. Mechanisms of Impaired Biliary Excretion of Acetaminophen Glucuronide after Acute Phenobarbital Treatment or Phenobarbital Pretreatment. Drug Metab Dispos 2002; 30:962-969.

87. Tzeng SJ, Huang JD. Transcriptional regulation of the rat Mrp3 promoter in intestine cells. Biochem Biophys Res Commun 2002; 291:270-7.

88. Inokuchi A, Hinoshita E, Iwamoto Y et al. Enhanced expression of the human multidrug resistance protein 3 by bile salt in human enterocytes. A transcriptional control of a plausible bile acid transporter. J Biol Chem 2001; 276:46822-9.

89. Hartmann G, Cheung AKY, Piquette-Miller M. Inflammatory Cytokines, but Not Bile acids, Regulate Expression of Murine Hepatic Anion Transporters in Endotoxemia. J Pharmacol Exp Ther 2002; 303:273-281.

90. Trauner M, Meier PJ, Boyer JL. Molecular regulation of hepatocellular transport systems in cholestasis. J Hepatol 1999; 31:165-78.

91. Elferink RO, Groen AK. Genetic defects in hepatobiliary transport. Biochim Biophys Acta 2002; 1586:129-45.

92. Yu L, Li-Hawkins J, Hammer RE et al. Overexpression of ABCG5 and ABCG8 promotes biliary cholesterol secretion and reduces fractional absorption of dietary cholesterol. J Clin Invest 2002; 110:671-80.

93. Yu L, Hammer RE, Li-Hawkins J et al. Disruption of Abcg5 and Abcg8 in mice reveals their crucial role in biliary cholesterol secretion. Proc Natl Acad Sci USA 2002; 99:16237-16242.

94. Keppler D, Konig J. Hepatic canalicular membrane 5: Expression and localization of the conjugate export pump encoded by the MRP2 (cMRP/cMOAT) gene in liver. Faseb J 1997; 11:509-16.

95. Kullak-Ublick GA, Beuers U, Paumgartner G. Hepatobiliary transport. J Hepatol 2000; 32:3-18.

96. Keppler D, Konig J, Buchler M. The canalicular multidrug resistance protein, cMRP/MRP2, a novel conjugate export pump expressed in the apical membrane of hepatocytes. Adv Enzyme Regul 1997; 37:321-33.

97. Tsujii H, Konig J, Rost D et al. Exon-intron organization of the human multidrug-resistance protein 2 (MRP2) gene mutated in Dubin-Johnson syndrome. Gastroenterology 1999; 117:653-60.

98. Trauner M, Arrese M, Soroka CJ et al. The rat canalicular conjugate export pump (Mrp2) is down-regulated in intrahepatic and obstructive cholestasis. Gastroenterology 1997; 113:255-64.

99. Vos TA, Hooiveld GJ, Koning H et al. Up-regulation of the multidrug resistance genes, Mrp1 and Mdr1b, and down-regulation of the organic anion transporter, Mrp2, and the bile salt transporter, Spgp, in endotoxemic rat liver. Hepatology 1998; 28:1637-44.

100. Kubitz R, Wettstein M, Warskulat U et al. Regulation of the multidrug resistance protein 2 in the rat liver by lipopolysaccharide and dexamethasone. Gastroenterology 1999; 116:401-10.

101. Wielandt AM, Vollrath V, Manzano M et al. Induction of the multispecific organic anion transporter (cMoat/mrp2) gene and biliary glutathione secretion by the herbicide 2,4,5-trichlorophenoxyacetic acid in the mouse liver [In Process Citation]. Biochem J 1999; 341:105-11.

102. Hinoshita E, Taguchi K, Inokuchi A et al. Decreased expression of an ATP-binding cassette transporter, MRP2, in human livers with hepatitis C virus infection. J Hepatol 2001; 35:765-73.

103. Fickert P, Zollner G, Fuchsbichler A et al. Effects of ursodeoxycholic and cholic acid feeding on hepatocellular transporter expression in mouse liver. Gastroenterology 2001; 121:170-83.

104. Fromm MF, Kauffmann HM, Fritz P et al. The effect of rifampin treatment on intestinal expression of human MRP transporters. Am J Pathol 2000; 157:1575-80.

105. Stockel B, Konig J, Nies AT et al. Characterization of the 5'-flanking region of the human multidrug resistance protein 2 (MRP2) gene and its regulation in comparison with the multidrug resistance protein 3 (MRP3) gene. Eur J Biochem 2000; 267:1347-1358.

106. Hagenbuch N, Reichel C, Stieger B et al. Effect of phenobarbital on the expression of bile salt and organic anion transporters of rat liver. J Hepatol 2001; 34:881-7.

107. Denson LA, Bohan A, Held MA et al. Organ-specific alterations in RAR alpha:RXR alpha abundance regulate rat Mrp2 (Abcc2) expression in obstructive cholestasis. Gastroenterology 2002; 123:599-607.

108. Johnson DR, Klaassen CD. Regulation of rat multidrug resistance protein 2 by classes of proto-typical microsomal enzyme inducers that activate distinct transcription pathways. Toxicol Sci 2002; 67:182-9.
109. Kast HR, Goodwin B, Tarr PT et al. Regulation of multidrug resistance-associated protein 2 (ABCC2) by the nuclear receptors pregnane X receptor, farnesoid X-activated receptor, and constitutive androstane receptor. J Biol Chem 2002; 277:2908-15.
110. Dussault I, Lin M, Hollister K et al. Peptide mimetic HIV protease inhibitors are ligands for the orphan receptor SXR. J Biol Chem 2001; 276:33309-12.
111. Synold TW, Dussault I, Forman BM. The orphan nuclear receptor SXR coordinately regulates drug metabolism and efflux. Nat Med 2001; 7:584-90.
112. Gerloff T, Stieger B, Hagenbuch B et al. The sister of P-glycoprotein represents the canalicular bile salt export pump of mammalian liver. J Biol Chem 1998; 273:10046-50.
113. Strautnieks SS, Bull LN, Knisely AS et al. A gene encoding a liver-specific ABC transporter is mutated in progressive familial intrahepatic cholestasis. Nat Genet 1998; 20:233-8.
114. Noe J, Stieger B, Meier PJ. Functional expression of the canalicular bile salt export pump of human liver. Gastroenterology 2002; 123:1659-66.
115. Byrne JA, Strautnieks SS, Mieli-Vergani G et al. The human bile salt export pump: characterization of substrate specificity and identification of inhibitors. Gastroenterology 2002; 123:1649-58.
116. Wang R, Salem M, Yousef IM et al. Targeted inactivation of sister of P-glycoprotein gene (spgp) in mice results in nonprogressive but persistent intrahepatic cholestasis. Proc Natl Acad Sci USA 2001; 98:2011-6.
117. Lee J, Azzaroli F, Wang L et al. Adaptive regulation of bile salt transporters in kidney and liver in obstructive cholestasis in the rat. Gastroenterology 2001; 121:1473-84.
118. Bramow S, Ott P, Thomsen Nielsen F et al. Cholestasis and regulation of genes related to drug metabolism and biliary transport in rat liver following treatment with cyclosporine a and sirolimus (rapamycin). Pharmacol Toxicol 2001; 89:133-9.
119. Rippin SJ, Hagenbuch B, Meier PJ et al. Cholestatic expression pattern of sinusoidal and canalicular organic anion transport systems in primary cultured rat hepatocytes. Hepatology 2001; 33:776-82.
120. Cao J, Huang L, Liu Y et al. Differential regulation of hepatic bile salt and organic anion transporters in pregnant and postpartum rats and the role of prolactin. Hepatology 2001; 33:140-7.
121. Yu J, Lo J-L, Huang L et al. Lithocholic Acid Decreases Expression of Bile salt export pump through Farnesoid X Receptor Antagonist Activity. J Biol Chem 2002; 277:31441-31447.
122. Warskulat U, Kubitz R, Wettstein M et al. Regulation of bile salt export pump mRNA levels by dexamethasone and osmolarity in cultured rat hepatocytes. Biol Chem 1999; 380:1273-9.
123. Ananthanarayanan M, Balasubramanian N, Makishima M et al. Human Bile salt export pump Promoter Is Transactivated by the Farnesoid X Receptor/Bile acid Receptor. J Biol Chem 2001; 276:28857-28865.
124. Plass JR, Mol O, Heegsma J et al. Farnesoid X receptor and bile salts are involved in transcriptional regulation of the gene encoding the human bile salt export pump. Hepatology 2002; 35:589-96.
125. Gerloff T, Geier A, Roots I et al. Functional analysis of the rat bile salt export pump gene promoter. Eur J Biochem 2002; 269:3495-3503.
126. Smit JJ, Schinkel AH, Oude Elferink RP et al. Homozygous disruption of the murine mdr2 P-glycoprotein gene leads to a complete absence of phospholipid from bile and to liver disease. Cell 1993; 75:451-62.
127. Crawford AR, Smith AJ, Hatch VC et al. Hepatic secretion of phospholipid vesicles in the mouse critically depends on mdr2 or MDR3 P-glycoprotein expression. Visualization by electron microscopy. J Clin Invest 1997; 100:2562-7.
128. de Vree JM, Jacquemin E, Sturm E et al. Mutations in the MDR3 gene cause progressive familial intrahepatic cholestasis. Proc Natl Acad Sci USA 1998; 95:282-7.
129. Jacquemin E, De Vree JM, Cresteil D et al. The wide spectrum of multidrug resistance 3 deficiency: from neonatal cholestasis to cirrhosis of adulthood. Gastroenterology 2001; 120:1448-58.
130. Rosmorduc O, Hermelin B, Poupon R. MDR3 gene defect in adults with symptomatic intrahepatic and gallbladder cholesterol cholelithiasis. Gastroenterology 2001; 120:1459-67.
131. Miranda S, Vollrath V, Wielandt AM et al. Overexpression of mdr2 gene by peroxisome proliferators in the mouse liver. J Hepatol 1997; 26:1331-9.
132. Chianale J, Vollrath V, Wielandt AM et al. Fibrates induce mdr2 gene expression and biliary phospholipid secretion in the mouse. Biochem J 1996; 314:781-6.
133. Gupta S, Todd Stravitz R, Pandak WM et al. Regulation of multidrug resistance 2 P-glycoprotein expression by bile salts in rats and in primary cultures of rat hepatocytes. Hepatology 2000; 32:341-7.

134. Carrella M, Feldman D, Cogoi S et al. Enhancement of mdr2 gene transcription mediates the biliary transfer of phosphatidylcholine supplied by an increased biosynthesis in the pravastatin-treated rat. Hepatology 1999; 29:1825-32.
135. Kok T, Bloks VW, Wolters H et al. Peroxisome proliferator-activated receptor alpha (PPARalpha)-mediated regulation of multidrug resistance 2 (Mdr2) expression and function in mice. Biochem J 2003; 369:539-47.
136. Smit JJ, Mol CA, van Deemter L et al. Characterization of the promoter region of the human MDR3 P- glycoprotein gene. Biochim Biophys Acta 1995; 1261:44-56.
137. Brown PC, Silverman JA. Characterization of the rat mdr2 promoter and its regulation by the transcription factor Sp1. Nucleic Acids Res 1996; 24:3235-41.
138. Hubacek JA, Berge KE, Cohen JC et al. Mutations in ATP-cassette binding proteins G5 (ABCG5) and G8 (ABCG8) causing sitosterolemia. Hum Mutat 2001; 18:359-60.
139. Berge KE, Tian H, Graf GA et al. Accumulation of dietary cholesterol in sitosterolemia caused by mutations in adjacent ABC transporters. Science 2000; 290:1771-5.
140. Graf GA, Li WP, Gerard RD et al. Coexpression of ATP-binding cassette proteins ABCG5 and ABCG8 permits their transport to the apical surface. J Clin Invest 2002; 110:659-69.
141. Repa JJ, Berge KE, Pomajzl C et al. Regulation of ATP-binding Cassette Sterol Transporters ABCG5 and ABCG8 by the Liver X Receptors alpha and beta J. Biol. Chem. 2002; 277:18793-18800.
142. Arrese M, Karpen SJ. HNF-1 alpha: have bile acid transport genes found their "master"? J Hepatol 2002; 36:142-5.
143. Tronche F, Bach I, Chouard T et al. Hepatocyte nuclear factor 1 (HNF1) and liver gene expression. In: Tronche F, Yaniv M, eds. Liver gene expression. Austin: RG Landes Co., 1994:155-181.
144. Yamagata K, Oda N, Kaisaki PJ et al. Mutations in the hepatocyte nuclear factor-1-alpha gene in maturity-onset diabetes of the young (mody3). Nature 1996; 384:455-458.
145. Memon RA, Moser AH, Shigenaga JK et al. In vivo and in vitro regulation of sterol 27-hydroxy-lase in the liver during the acute phase response. potential role of hepatocyte nuclear factor-1. J Biol Chem 2001; 276:30118-26.
146. Willson TM, Jones SA, Moore JT et al. Chemical genomics: Functional analysis of orphan nuclear receptors in the regulation of bile acid metabolism. Med Res Rev 2001; 21:513-22.
147. Arrese M, Karpen SJ. New horizons in the regulation of bile acid and lipid homeostasis: critical role of the nuclear receptor FXR as an intracellular bile acid sensor. Gut 2001; 49:465-6.
148. Chawla A, Repa JJ, Evans RM et al. Nuclear receptors and lipid physiology: opening the X-files. Science 2001; 294:1866-70.
149. Edwards PA, Kast HR, Anisfeld AM. BAREing it all: the adoption of LXR and FXR and their roles in lipid homeostasis. J Lipid Res 2002; 43:2-12.
150. Fitzgerald ML, Moore KJ, Freeman MW. Nuclear hormone receptors and cholesterol trafficking: the orphans find a new home. J Mol Med 2002; 80:271-81.
151. Francis GA, Fayard E, Picard F et al. Nuclear receptors and the Control of Metabolism. Annu Rev Physiol 2002.
152. Makishima M, Okamoto AY, Repa JJ et al. Identification of a nuclear receptor for bile acids. Science 1999; 284:1362-5.
153. Parks DJ, Blanchard SG, Bledsoe RK et al. Bile acids: natural ligands for an orphan nuclear receptor. Science 1999; 284:1365-8.
154. Wang H, Chen J, Hollister K et al. Endogenous bile acids are ligands for the nuclear receptor FXR/BAR. Mol Cell 1999; 3:543-53.
155. Goodwin B, Jones SA, Price RR et al. A regulatory cascade of the nuclear receptors FXR, SHP-1, and LRH-1 represses bile acid biosynthesis. Mol Cell 2000; 6:517-26.
156. Lu TT, Makishima M, Repa J et al. Molecular basis for feedback regulation of bile acid synthesis by nuclear receptors. Mol Cell 2000; 6:507-15.
157. Kerr TA, Saeki S, Schneider M et al. Loss of Nuclear receptor SHP Impairs but Does Not Eliminate Negative Feedback Regulation of Bile acid synthesis. Dev Cell 2002; 2:713-20.
158. Davis RA, Miyake JH, Hui TY et al. Regulation of cholesterol-7alpha-hydroxylase. Barely missing a shp. J Lipid Res 2002; 43:533-43.
159. Miyake JH, Wang SL, Davis RA. Bile acid induction of cytokine expression by macrophages correlates with repression of hepatic cholesterol 7alpha-hydroxylase. J Biol Chem 2000; 275:21805-8.
160. Pellicciari R, Fiorucci S, Camaioni E et al. 6alpha-ethyl-chenodeoxycholic acid (6-ECDCA), a potent and selective FXR agonist endowed with anticholestatic activity. J Med Chem 2002; 45:3569-72.
161. Urizar NL, Liverman AB, Dodds DT et al. A Natural Product That Lowers Cholesterol as an Antagonist Ligand for the FXR. Science 2002; 296:1703-6.
162. Muller M. Transcriptional control of hepatocanalicular transporter gene expression. Semin Liver Dis 2000; 20:323-37.

CHAPTER 9

Signal Transduction in Bile Formation and Cholestasis

M. Sawkat Anwer and Cynthia R.L. Webster

Summary

Bile formation involves vectorial transport of solutes from blood to bile and is dependent on coordinated activities of various solute transporters located at the basolateral and apical membranes of hepatocytes and cholangiocytes. Cholestasis results when the vectorial transport of solutes destined for bile is compromised. Our understanding of various transporters, their substrates and locations has increased steadily as is the cellular mechanism regulating these transporters. It is becoming more evident that choleretic and cholestatic agents modify the function of these transporters through various signal tranduction pathways. Cyclic AMP, acting via PKA and the PI3K signaling pathway, stimulate transhepatic transport of bile acids by translocating Ntcp and Bsep to the sinusoidal and the canalicular membrane, respectively. Cell swelling induced by hypo-osmotic media stimulates hepatic uptake and biliary excretion of bile acids via the PI3K and the ERK signaling pathway, respectively. TUDC and cell swelling also stimulate biliary bile acid excretion via the p38 MAPK signaling pathway. TUDC reverses TLC cholestasis by stimulating PKC-mediated translocation of Mrp2 to the canalicular membrane. Calcium, acting via Ca^{2+}/calmodulin-dependent kinases/phosphatases, augments cAMP-mediated translocation of Ntcp, increases tight-junctional permeability by phosphorylating myosin light-chain and stimulates sinusoidal Na^+/H^+ exchange. PKC stimulates bile acid secretion, most likely by phosphorylating Bsep, and Na^+/H^+ exchange. However, our understanding of the role of specific PKC isoforms involved in bile formation is still lacking. Molecular mechanisms by which PI3K and MAPK signaling pathways stimulate transporter translocation along the cytoskeleton have not been elucidated. Our knowledge of the role of protein phosphatases in bile formation is rather limited. It is, however, anticipated that further understanding in these areas will be forthcoming in the near future.

Introduction

Bile formation is one of the important functions of the liver and the term "cholestasis" is used to describe conditions associated with decreased bile formation. Bile provides a route to excrete a number of endogenous and exogenous substances. Vectorial transport of solutes from the sinusoidal space to the canaliculus provides the osmotic driving force for bile formation and is accomplished by various transporters located at the basolateral and canlanicular membrane of hepatocytes and cholangiocytes.[1-3] It is thus easy to appreciate the paradigm that cholestasis results when the ability of the liver to transport solutes into the canaliculus is compromized.

Our knowledge of various transporters in hepatocytes and cholangiocytes and their functions (Fig. 1) has increased at a steady rate during the last 10 years as we have been able to take advantage of molecular biology techniques to isolate and clone these transporters.[4-7] As we

Molecular Pathogenesis of Cholestasis, edited by Michael Trauner and Peter L.M. Jansen.
©2004 Eurekah.com and Kluwer Academic / Plenum Publishers.

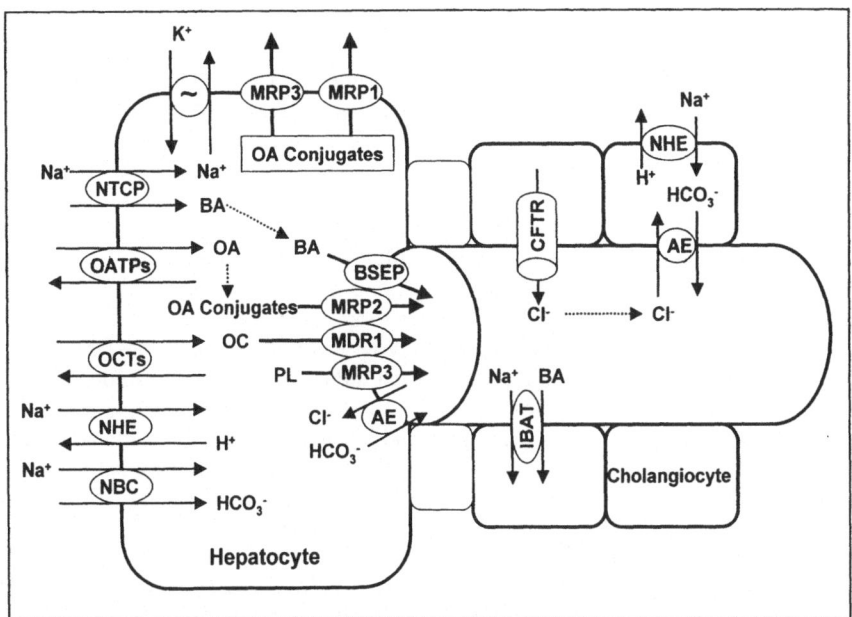

Figure 1. Transporters involved in bile formation: hepatic uptake of bile acid (BA), organic anions (OA) and organic cations (OC) is mediated primarily by Na⁺/taurocholate cotransporting polypeptide (NTCP), the family of organic anion transporting proteins (OATPs) and organic cation transporters (OCTs), respectively. Na⁺/H⁺ exchanger (NHE) and Na⁺/HCO₃⁻ cotransporter (NBC) at the sinusoidal membrane are involved in intracellular pH regulation and HCO₃⁻ uptake. Multi-drug resistance proteins (MRP3 and MRP1) mediate sinusoidal efflux of organic anions, including toxic bile acids, while MRP2 and BSEP (Bile salt export pump) mediate canalicular excretion of conjugated organic anions and bile acids, respectively. MDR1 and MDR2 (multidrug resistance gene products) are involved in biliary excretion of organic cations and phospholipids, respectively. Chloride/bicarbonate exchange is mediated by anion exchanger (AE) at canalicular as well as apical membrane of cholangiocytes. Cystic fibrosis transmembrane conductance regulators (CFTR) act as chloride channels and reabsoprtion of conjugated bile acid from the biliary tree is mediated via Na⁺-dependent bile acid transporter (IBAT).

continue to identify the family members of specific transporters and their substrate specificity, studies are being conducted to elucidate mechanisms of cellular regulation of these transporters in an effort to understand the pathophysiology of cholestasis. Indeed, cholestasis is associated with down-regulation of Ntcp and Mrp2[8,9] and up-regulation of Mrp3[10,11] with a relatively preserved expression of Bsep.[8]

Mutations in a gene encoding for a particular transporter may result in a lack of important transport function leading to cholestasis. For example, mutations of BSEP, MDR3 and MRP2 (cMOAT) are implicated in type 2 progressive familial intrahepatic cholestasis (PFIC2), PFIC3 and Dubin-Johnson syndrome, respectively.[12-14] Apart from genetic defects, regulation of transporters at the level of transcription, translation, post-translational modifications, translocation to the plasma membrane and transporter activity may be altered by chemicals/disease processes leading to decreased bile formation and hence cholestasis.

Recent studies continue to provide insight into the regulation of these transporters at the level of transcription, which is covered elsewhere in this book. Our knowledge of regulation at the level of translation and post-translational modifications is rather limited. Transport proteins, like other proteins, are synthesized in the ER, processed in the golgi complex and are then translocated to their intended site of action, basolateral membrane for Ntcp and canalicular membrane for Bsep, for example. A transporter has to be inserted into the membrane for it to transport its solute across that membrane. This is a complex regulated process, which requires the participation of various signaling molecules along with vesicles, microtubules, microfilaments. A breakdown in this regulated process can lead to a decreased amount or an absence of a transport protein at its intended site resulting in decreased or no transport function and hence cholestasis. In addition, the transporter activity may be decreased directly, leading to cholestasis. This review will focus on recent advances in our understanding of signal transduction pathways involved in translocation and transport activity of various transporters.

Overview of Signal Transduction Pathways

Our understanding of various signaling pathways in hepatocytes has also increased steadily.[2,5,15,16] It is now well established that the interaction of hormones/growth factors with their receptors at the plasma membrane activates a cascade of key enzymes that produce the biological effects. Major signal transduction pathways involved in hepatic bile formation are shown in Figure 2. In hepatocytes, activation of cell surface receptors results in the formation of cAMP or cGMP, increases in cytosolic Ca^{2+} and activation of kinases, such as, PKC, PI3K and MAPK (Fig. 2). These second messengers and kinases are involved in various aspects of bile formation, like organic and inorganic solute transport. It should be noted that these signal transduction pathways are complex and involve a cascade of factors/enzymes, the details of which are still being worked out.

Protein kinase C is a family of at least 12 isozymes.[17-19] These include conventional (cPKCα, β, βI, βII and γ), novel (nPKCδ, ε, η and θ), atypical (aPKCζ and λ) isoforms, and PKCμ. These isoforms differ in their dependency on Ca^{2+} and phospholipids, such that cPKCs are dependent on Ca^{2+} and diacylglycerol (DAG), nPKCs are Ca^{2+}-independent and aPKCs are independent of both Ca^{2+} and DAG. PKCs shown to be present in rat hepatocytes include cPKCζ, nPKCδ, nPKCε and aPKCζ with the presence of cPKCβII being controversial.

PI3K is one of the PI kinases that phosphorylate at specific position(s) of the inositol ring (Fig. 3) and the resulting phosphorylated PIs (PIPs) have been implicated in a wide variety of biological effects produced by hormones/growth factors. PIPs act as second messengers in signal transduction pathways involved in vesicle trafficking, cell survival, cell proliferations, cell migration, and transport of glucose and bile acids.[20,21] PIPs produce these effects by allowing activation of various effector molecules, some of which are shown in Figure 4.

The mitogen activated protein kinases (MAPKs) include a group of protein kinases that are activated by a variety of signals.[22-24] Mammalian cells have 2 major types of MAPK cascade (Fig. 5): 1) ERK1/ERK2 (Extracellular signal-regulated kinase) activated in response to activation of receptor tyrosine kinase by growth factors, 2) SAPKs (stress activated protein kinases) activated by UV radiation, inflammatory cytokines, DNA damaging agents and inhibitor of protein synthesis. SAPKs include two distinct subfamilies: JNKs (c-Jun amino-terminal kinases) and p38 MAPK. These MAPKs mediate signal transduction from the cell surface to the nucleus. They are involved in cell proliferation and differentiation under controlled activation and oncogenesis during uncontrolled activation. Activation of these MAPKs require sequential phosphorylation involving various kinases (Fig. 5).

Role of Cyclic AMP

Glucagon and secretin, acting via cAMP, rapidly stimulate bile formation by hepatocytes and cholangiocytes, respectively. Cyclic AMP stimulates sinusoidal Na^+/Taurocholate (TC) cotransport, transcytotic vesicle trafficking and canalicular secretion of bile acids, organic anions and HCO_3^- in hepatocytes[2,25-27] and HCO_3^- secretion and exocytosis in cholangiocytes.[6,28] Unlike cAMP, cGMP does not stimulate hepatobiliary bile acid transport, but stimulates bile formation by increasing biliary excretion of bicarbonate.[29] The ability of cAMP to stimulate microtubule-dependent vesicle trafficking in hepatocytes[25] led to the suggestion that cAMP stimulates various solute transport by translocating the transporters to the plasma membrane. This is supported by findings that cAMP increases Ntcp[30,31] in sinusoidal membranes and Mrp2,[32] Mdr2 and Mdr3[33] and Bsep[34] in canalicular membranes. However, while the

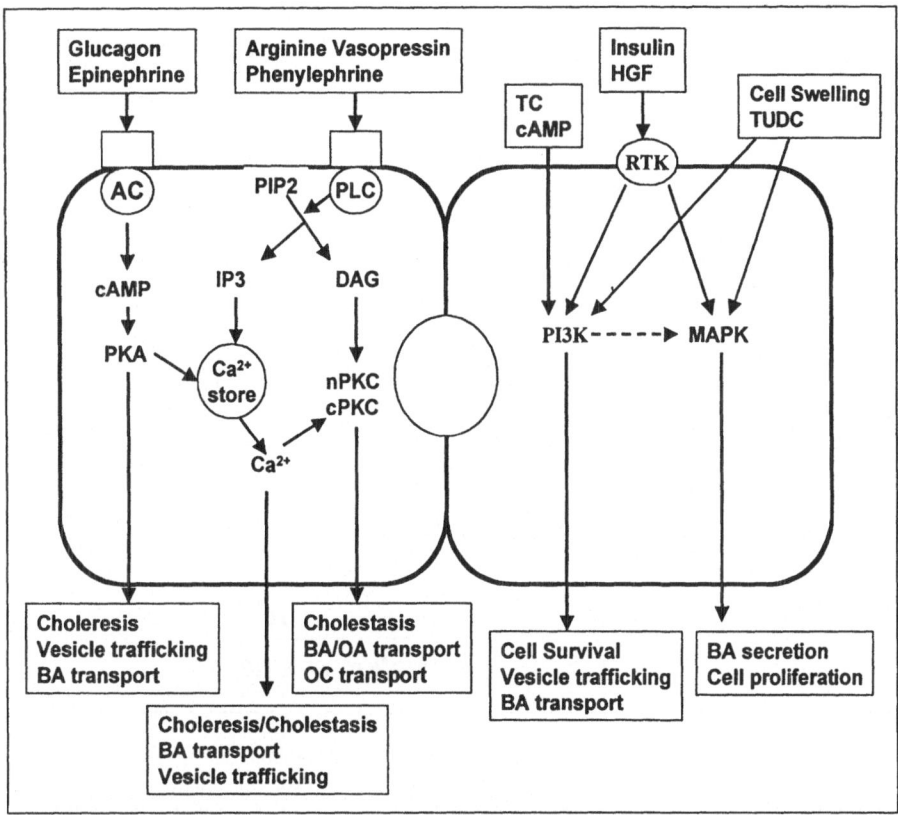

Figure 2. Major signal transduction pathways involved in bile formation: Interaction of glucagons/epineph-rine and Arginine vasopressin/phenylephrine with their receptors leads to the activation of adenylyl cyclase (AC) and phospholipase C (PLC), respectively. AC converts ATP into cAMP and PLC hydrolyses phosphoinositide(4,5)-bisphosphate (PIP2) into inositol(1,4,5)-trisphosphate (IP3) and diacyglycerol (DAG). IP3 and cAMP, acting via PKA, increases $[Ca^{2+}]_i$ by releasing Ca^{2+} from its cytoplasmic store, and DAG activates nPKCs and cPKCs. Insulin and hepatocyte growth factor (HGF), acting via receptor tyrosine kinase (RTK), activate PI3K (phosphoinositide-3-kinase) and MAPK (mitogen activated protein kinase) pathways. TC and cAMP also activate the PI3K pathway and cell swelling and TUDC activate PI3K as well as MAPK pathway. Some of the effects mediated via these signaling pathways are listed in respective boxes. BA=bile acid, OA=organic anion, OC=organic cation.

translocation to canalicular membrane is dependent on microtubules,[33,35] cAMP-mediated translocation to the basolateral membrane is dependent on microfilaments.[31,36] Secretin stimu-lated biliary HCO_3^- secretion is mediated via cAMP, which activates the CFTR Cl^- channel with subsequent activation of Cl^-/HCO_3^- exchanger.[37]

The effect of cAMP is believed to be mediated via cAMP-dependent kinase, also known as protein kinase A or PKA,[2,6] although the role of PKA has been directly evaluated in cAMP mediated stimulation of Na^+/TC cotransport[27] and cholangiocyte Cl^-/HCO_3^- exchange[38] and cAMP-mediated cell survival.[39] In view of the fact that PKA does not account for all intracel-lular targets of cAMP,[40] it is important to experimentally verify whether PKA is the target of a particular effect of cAMP. Signaling pathway(s) downstream of PKA leading to vesicle move-ment and consequent transporter translocation may involve activation of the PI3K pathway and increases in cytosolic Ca^{2+} (Figs. 6 and 7). In addition, PKA may stimulate vesicle move-ment by phosphorylating microtubule-associate proteins.[41]

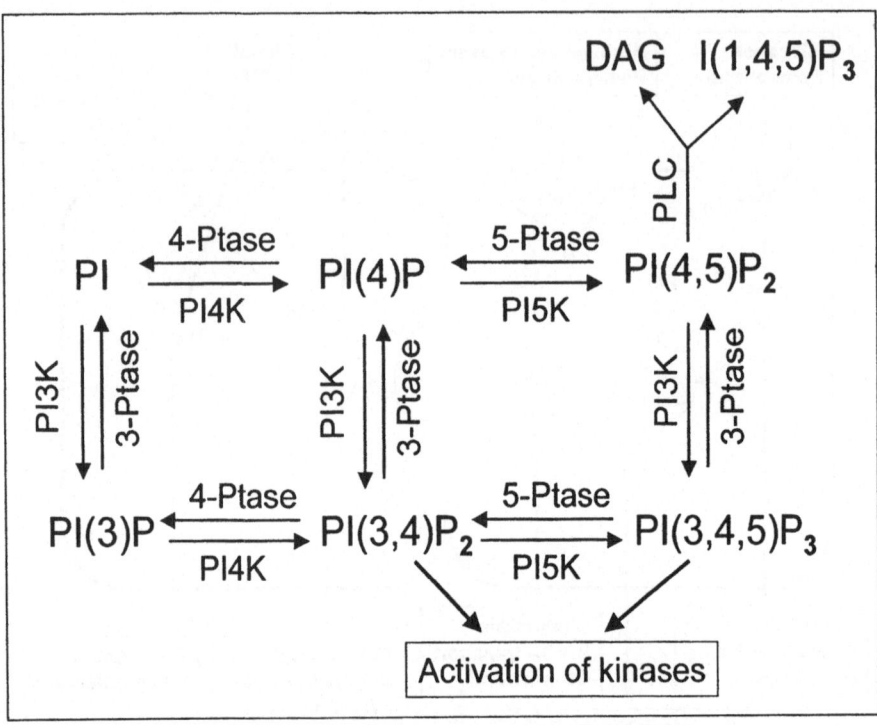

Figure 3. Formation of various phosphoinositide (PI)-phosphates: Phosphoinositide-3-kinase (PI3K) and 3'-PI phosphatase (3-Ptase) can convert PI, PI(4)P and PI(4,5)P$_2$ to the corresponding 3'-phospho-PIs and back, respectively. The lipid products, PI(3,4)P$_2$ and PI(3,4,5)P$_3$, in turn are involved in the activation of various kinases and PI(4,5)P$_2$ is hydrolyzed by phospholipase C (PLC) into I(1,4,5)P$_3$ and DAG.

There is suggestive evidence that the cAMP signaling pathway may be altered in cholestasis. Experimental cholestasis induced by bile duct ligation and endotoxin is associated with decreased expression of Ntcp, Mrp2 and Bsep.[8,9] These decreases are much pronounced for Ntcp and Mrp2 than Bsep. Bile duct ligation also decreases the ability of glucagon to increase cAMP in hamster hepatocytes[42] and this appears to be due to decreased expression of α-subunit(s) of stimulatory as well as inhibitory G proteins[43] coupling the receptor to adenylyl cyclase. Taken together, these studies suggest that cAMP signaling pathway may be down-regulated in cholestasis. It may be noted that a more pronounced decrease in bile acid uptake by Ntcp compared to bile acid secretion by Bsep may represent a hepatocellular defense mechanism against accumulation of intracellular bile acids and hence exacerbation of cholestasis. This is further supported by the finding that the down-regulation of Mrp2 is associated with an up-regulation of Mrp3[11], which exports toxic bile acids across the sinusoidal membrane.[44]

Role of Calcium

Changes in extracellular as well as intracellular Ca^{2+} ([Ca^{2+}]$_i$) have been shown to affect bile formation. A decrease in extracellular Ca^{2+} below 50 μM is associated with a decrease in bile formation. This is due to an increase in tight-junction permeability[45,46] resulting in reflux of secreted solutes and is not due to a direct effect on either hepatic uptake or biliary excretion of TC.[45] Changes in [Ca^{2+}]$_i$ on the other hand, affects hepatic transport of bile acids.

The effect of [Ca^{2+}]$_i$ has been studied using calcium mobilizing agents, like arginine vassopressin (AV), calcium ionophores, like A23187 or ionomycin, and intracellular calcium chelators, like MAPTA or BAPTA. Chelation of [Ca^{2+}]$_i$ by MAPTA, BAPTA or EDTA

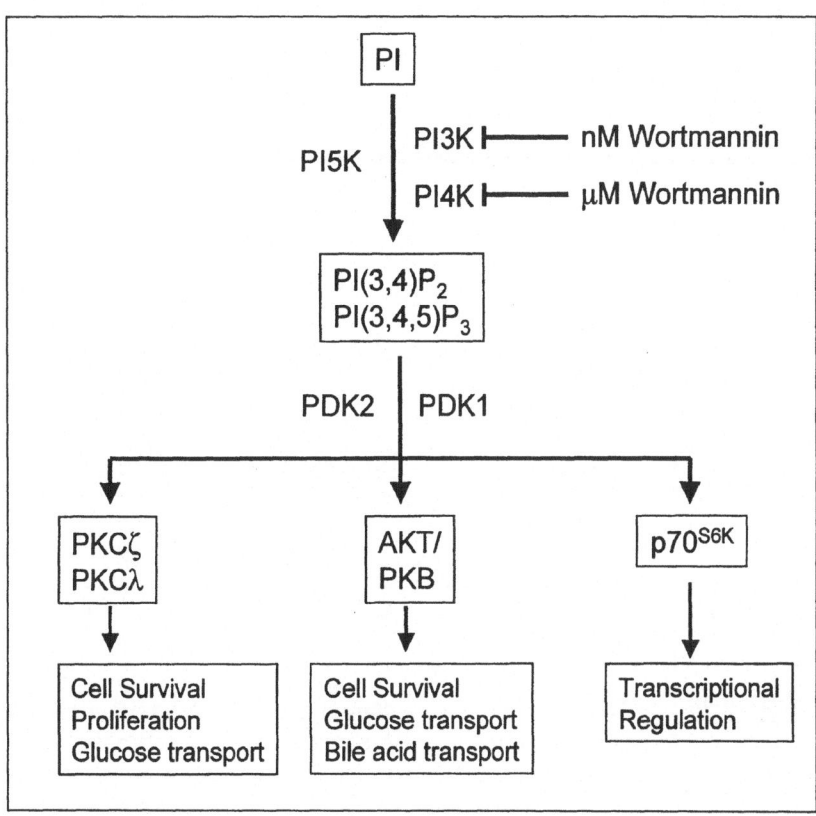

Figure 4. The PI3K signaling pathway: PI is converted to PI(3,4)P$_2$ and PI(3,4,5)P$_3$ by PI3K, PI4K and PI5K. Wortmannin inhibits PI3K and PI4K at nM and μM concentrations, respectively. These PIPs, acting in concert with phosphoinositide depenent kinases (PDK1 & PDK2), are involved in the activation of downstream kinases, such as PKCζ/λ, PKB/Akt and p70S6K, and these kinases have been shown to mediate various cellular processes as indicated.

decreases basal rate of Na$^+$/TC cotransport in hepatocytes.[27,47] Thus, resting [Ca^{2+}]$_i$ plays an important role in maintaining basal Na$^+$/bile acid cotransport. Interestingly, basal Na$^+$/TC cotransport is also inhibited in response to increases in [Ca^{2+}]$_i$ produced by calcium ionophores and AV in hamster hepatocytes,[47] but not in rat hepatocytes.[27,48] This may suggest species differences in the regulation of Na+/bile acid cotransport by [Ca^{2+}]$_i$. Other potential mechanisms have been discussed in a recent review by Bouscarel et al.[15]

In contrast to its effect on uptake, AV increases bile acid excretion transiently in perfused rat livers[49] and bile acid efflux in isolated rat hepatocytes by decreasing substrate affinity for the transporter.[48] Whether these effects are due to increases in [Ca^{2+}]$_i$ or activation of PKC (see Fig. 2) was not investigated. AV and phenylephrine stimulate sinusoidal Na$^+$/H$^+$ exchanger and this effect is mediated via receptor-mediated increases in [Ca^{2+}]$_i$ as well as activation of PKC.[50,51] Intracellular Ca^{2+} also plays a role in cAMP-stimulated bile acid uptake in that this effect of cAMP is dependent on its ability to increase [Ca^{2+}]$_i$ from IP3-sensitive pool.[27]

The calcium signal is transducted via calmodulin to calmodulin-dependent kinases and phosphatases which produce biological effects by phosphorylating and dephosphorylating other proteins, respectively.[52] In hepatocytes, effects of calcium on AV-induced stimulation of Na$^+$/H$^+$ exchanger and cAMP-induced stimulation of Na$^+$/TC cotransporter are mediated via calmodulin (Fig. 6).[27,51] Since cAMP-induced stimulation of Na$^+$/TC cotransporter is associated with dephosphorylation of Ntcp,[53] this may result from the action of a

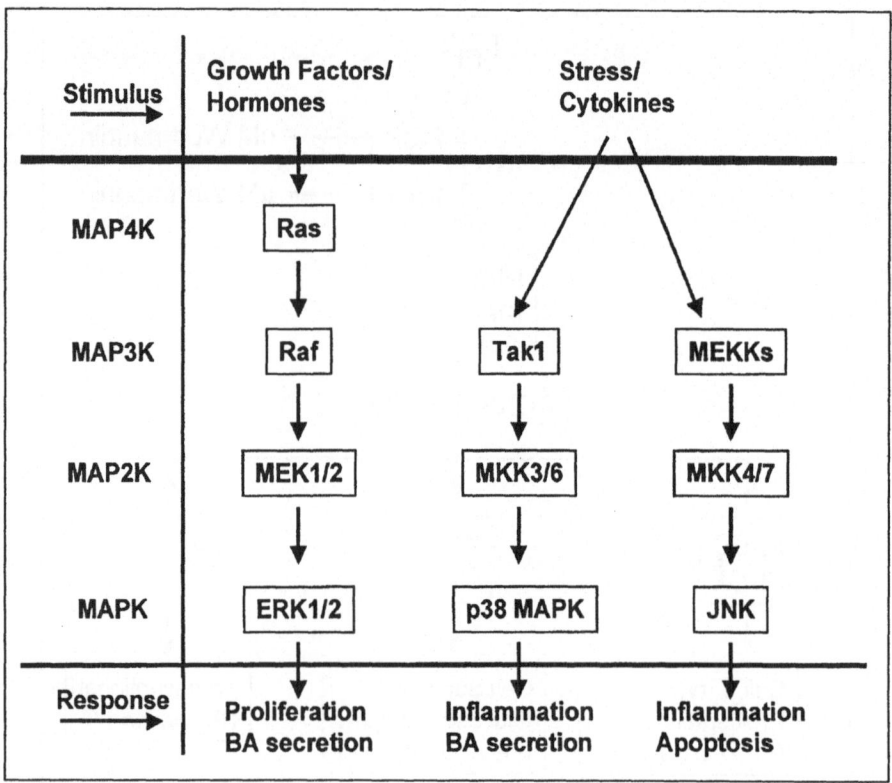

Figure 5. The mitogen activated protein kinase (MAPK) signaling pathway: The activation of MAPK/SAPK cascades, which include ERK1/2 (extracellular signal-regulated kinase 1 and 2), p38 MAPK and JNKs (c-Jun amino-terminal kinases), by growth factors/hormones/stress/cytokines require sequential phosphorylation by two/three-kinase architecture consisting of a MAPK activator (named MAP2K, MEK, MKK or MAPK kinase), a MAPK kinase activator (MAP3K, MEK kinase, MEKK or MAPK kinase kinase) and a MAPK kinase kinase activator (MAP4K or MAPK kinase kinase kinase). Thus, activation of ERK1/2 by growth factors require sequential activation of Ras (rat sarcoma virus, a small G protein), Raf, a serine/threonine kinase, and MEK1/2. Note that an upstream kinase may affect more than one downstream kinase, although Raf does not activate MKKs in the stress pathway.

calmodulin-dependent phosphatase (e.g., protein phosphatase 2B) activated by cAMP-induced increases in $[Ca^{2+}]_i$.

The exact role of calcium-dependent signaling pathway in hepatic bile formation and cholestasis is not clearly understood. Increases in $[Ca^{2+}]_i$ have been associated with choleresis as well as cholestasis. For example, cholestatic (taurolithocholate, TLC) as well as choleretic (tauroursodeoxycholate, TUDC) bile acid increase $[Ca^{2+}]_i$,[54,55] and TUDC can reverse cholestasis produced by TLC.[56] Some studies would suggest that increases in $[Ca^{2+}]_i$ may lead to increased tight-junctional permeability and hence cholestasis[57,58] and this effect may be mediated via myosin light chain phosphorylation (Fig. 6).[59] TUDC may stimulate Ca^{2+}-dependent stimulation of vesicular exocytosis by enhancing Ca^{2+} entry into hepatocytes, which may be compromized in cholestasis.[60] However, increases in $[Ca^{2+}]_i$ do not influence exocytosis in normal hepatocytes.[61] Studies to define the role of $[Ca^{2+}]_i$ have been complicated by the fact that some agents used to increase $[Ca^{2+}]_i$ also activate PKC (Fig. 2). In addition, increases in $[Ca^{2+}]_i$ by different agents may lead to activation of different downstream kinases, including PI3K[62] and PKB,[63] resulting in different effects in normal and cholestatic hepatocytes, and this may be regulated by temporal and spatial changes in $[Ca^{2+}]_i$ induced by different agents.[64,65]

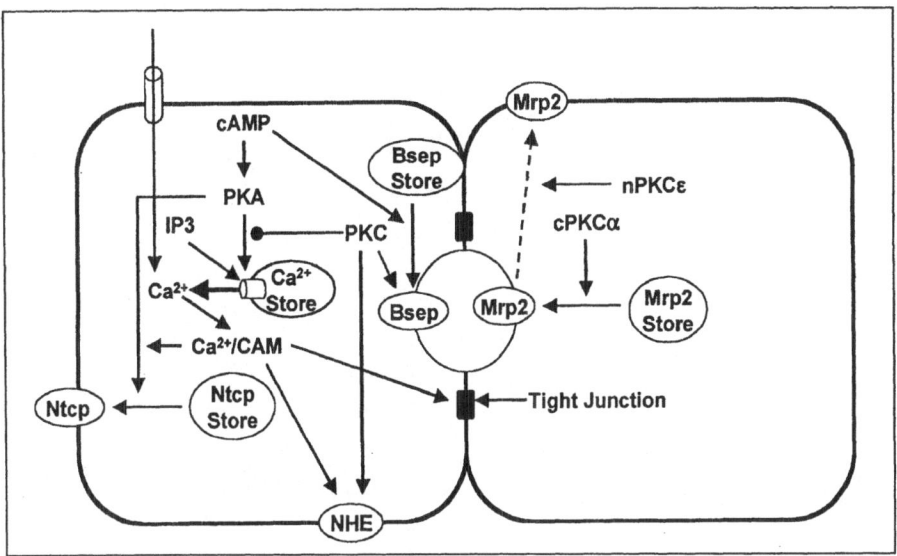

Figure 6. Role of cAMP, Ca^{2+} and PKC in bile formation: Cyclic AMP, acting via PKA, stimulates trans-location of Ntcp to the basolateral membrane and Bsep to the canalicular membrane. Increases in $[Ca^{2+}]_i$, due to influx or release of Ca^{2+} from intracellular store by PKA or IP3, activates Ca^{2+}/calmodulin (Ca^{2+}/CAM)-dependent kinases, which activates Na^+/H^+ exchanger (NHE), augments PKA-induced Ntcp trans-location and increases tight-junction permeability via phosphorylation of myosin light-chain. PKC inhibits cAMP-mediated increases in bile acid uptake by diminishing PKA-mediated increases in $[Ca^{2+}]_i$ and stimu-lates bile acid excretion probably by phosphorylating Bsep. nPKCε may produce cholestasis by retargeting Mrp2 to the basolateral membrane and cPKCα may stimulate MRP2 translocation to the canalicular membrane. TUDC may reverse TLC-induced cholestasis by reversing the effect of TLC on cPKCα and nPKCε.

Role of Protein Kinase C

The role of PKC in bile formation has been studied using known activators (phorbol esters) and inhibitors of PKC. These studies indicate that agents known to activate PKC produce cholestasis,[66] inhibit basal and cAMP-stimulated bile acid uptake,[27,47] stimulate biliary excre-tion of bile acids[49] and organic cations[67] and that PKC is involved in ATP-dependent modula-tion of cation channels.[68] PKC inhibits cAMP-stimulated bile acid uptake, at least in part, by inhibiting the ability of cAMP to increase $[Ca^{2+}]_i$.[27] AV increases bile acid efflux by decreasing Km[48] and cPKCα phosphorylates Bsep.[69] Thus, if the effect of AV is mediated via PKC, cPKCα may stimulate bile acid excretion by phosphorylating Bsep and thereby enhancing substrate affinity (Fig. 6).

Activation of PKCs has been implicated in bile acid-induced cholestasis and apoptosis. GCDC-induced apoptosis requires activation of cPKCα and nPKCδ, but not nPKCε[18] and the non-apoptotic effect of TCDC is due to activation of aPKCζ.[70] On the other hand, a cholestatic bile acid, TLC, activates nPKCε in isolated hepatocytes[17] and inhibits cPKCα in isolated perfused rat livers.[71] TUDC may reverse TLC-induced cholestasis by activating cPKCα and/or inhibiting nPKCε.[71,72] TUDC-induced activation of cPKCα may be involved in the translocation of Mrp2 to the canalicular membrane and consequent restoration of organic anion excretion in cholestatic livers.[71] A recent study shows that PKC (isoforms not studied) retargets canalicular Mrp2 to the basolateral membrane and this may contribute to cholestasis.[73] It is possible that the retargeting may be mediated via nPKCε (Fig. 6). These studies suggest that bile formation and cholestasis can be mediated via PKC isoform-dependent processes, but the role of PKC isoforms as well as the downstream targets of PKC isoforms have not been clearly established.

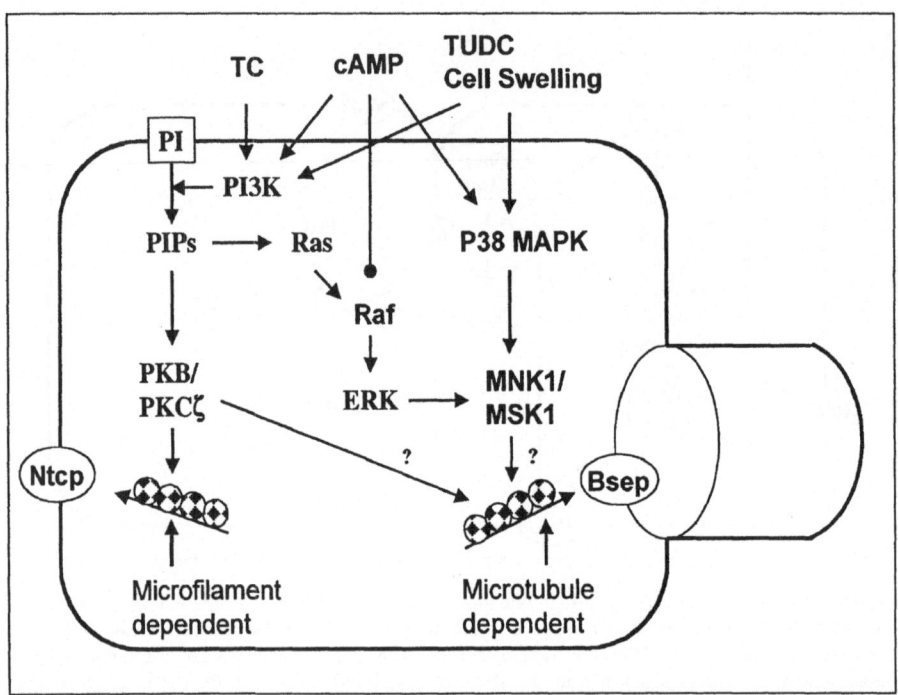

Figure 7. Role of PI3K and MAPK signaling pathway in bile formation: Activation of PI3K by TC, cAMP, TUDC and cell swelling leads to the formation of PIPs which can activate Ras/ERK pathway and PKB/ PKCζ. Cyclic AMP mediated activation of PKB/PKCζ may be involved in microfilament-dependent Ntcp translocation. TUDC and cell swelling induced microtubule-dependent Bsep translocation may involve PI3K/PIPs/Ras/Raf/Erk pathway. The effect of cAMP is not mediated via the Erk pathway, which is inhibited by cAMP. TUDC, cell swelling and cAMP may stimulate Bsep translocation via P38 MAPK. Erk and p38 MAPK pathways may converge on MNK1/MSK1. The role of the PKB/PKCζ pathway in cAMP- and TC-mediated Bsep translocation and the role of the Erk pathway in TC-induced Bsep translocation have not been studied.

Role of PI3K Signaling Pathway

Follio et. al. first demonstrated a role of phosphatidylinositol(PI)-3-kinase (PI3K) in bile formation.[74] In this study, wortmannin, a specific inhibitor of PI3K, inhibited bile formation, bile acid secretion and vesicle trafficking in isolated perfused rat liver. Since then various studies have provided evidence supporting a role for PI3K in the regulation of hepatobiliary transporters. PI3K is involved in 1) Cell swelling and cAMP-mediated translocation of Ntcp,[36,75] 2) ATP-dependent transport of bile acids and other organic anions across the canalicular membrane,[76] 3) TC- and cAMP-induced translocation of Bsep and Mrp2 to the canalicular membrane,[77,78] 4) TUDC -induced increases in bile acid secretion,[79] 5) ATP release and chloride secretion in cholangiocytes[80] and 6) Insulin-mediated membrane recruitment.[81] Cell swelling also stimulates biliary bile acid excretion, translocation of Bsep and Mrp2 to the canalicular membrane[82,83] and activates PI3K.[75,84] Thus, PI3K may also be involved in cell swelling induced translocation of Bsep and Mrp2. However, other mechanisms have been proposed (see below).

Recent studies provide further insight into the PI3K signaling pathway involved in the regulation of transporters involved in bile formation. Cell swelling and cAMP activate wortmannin-sensitive PKB and p70^{S6K}, but p70^{S6K} is not involved in the stimulation of Na$^+$/ TC cotransport and Ntcp translocation.[36,75] The possibility that the PI3K/PKB signaling pathway is involved is supported by results showing that the inhibition of PKB activation decreases cell swelling and cAMP-mediated stimulation of Na$^+$/TC cotransport and Ntcp translocation (unpublished). Translocation of glucose transporter appears to be mediated via PI3K/PKB as well as PI3K/aPKCz signaling pathway.[85,86] Thus, it is possible that the PI3K/aPKCζ signaling pathway may also be involved in Ntcp translocation (Fig. 7). Whether either of these signaling pathways is also involved in PI3K mediated translocation of canalicular transporters has not yet been studied, but is likely.

Mechanism by which translocation of hepatobiliary transporters is mediated via the PI3K signaling pathway is unclear. It is known that translocation of Ntcp to the sinusoidal membrane, and of Bsep and Mrp2 to the canalicular membrane is dependent on microfilaments[31,87] and microtubules,[33,35] respectively, and several canalicular transporters, like Bsep & Mrp2, and the polymeric immunoglubulin A receptor traffic on the same vesicle.[88] PI3K products, like PIP, PIP2 and PIP3, are involved in vesicle trafficking.[20,89] It can thus be speculated that the PI3K/PKB(aPKCζ) signaling pathway stimulates exocytosis by increasing vesicle trafficking along the cytoskeleton. This results in the fusion of intracellular vesicles to the plasma membrane leading to plasma membrane translocation of various transporters stored in the intracellular vesicles (Fig. 7). In addition, PI3K-mediated activation of the MAPK (mitogen activated protein kinase) cascade[21] may also be involved (Fig. 7).

Role of MAPK Signaling Pathway

The MAP signaling pathways involving ERK1/2 and p38 have also been implicated in biliary excretion of bile acids. Hypo-osmotic cell swelling increases the capacity of biliary TC excretion and this effect requires a G-protein- and tyrosine kinase-dependent activation of ERK1/2, but is independent of PKC.[90] Similarly, TUDC-induced increases in bile acid excretion is dependent on ERK1/2 activation, except this activation is not dependent on G-protein, tyrosine kinase or protein kinase C.[91] In addition, cAMP, shown to inhibit ERK1/2 in hepatocytes,[36] inhibited TUDC-induced activation of ERK1/2 and increases in bile acid excretion.[91] The effect of TUDC appears to be mediated via a PI3K-dependent activation of Ras/ERK pathway.[79] Cell swelling- and TUDC-induced increases in bile acid secretion and Bsep translocation to the canalicular membrane also require activation of PI3K-independent activation of p38 MAPK.[92] In contrast to bile acid secretion cell swelling-induced stimulation of hepatic uptake of bile acid is not dependent on ERK1/2.[75] These studies indicate that stimulation of Bsep translocation and the consequent increase in bile acid excretion can be mediated via activation of ERK1/2 as well as p38 MAPK pathway (Fig. 7). Since ERK1/2 and p38 MAPK can activate the downstream kinases,[23] like MAPK signal-integrating kinase 1 (MNK1) and mitogen- and stress-activated protein kinase 1 (MSK1), it is possible that Bsep translocation along the cytoskeleton is mediated via one of these kinases (Fig. 7).

The role of PI3K and MAPK pathways in cholestasis has not been directly evaluated. However, considering their role in bile formation these pathways are likely to be altered in cholestasis. On the other hand, agents that stimulate these pathways should produce anti-cholestatic effects, as is indicated by the anticholestatic effects of ursodeoxycholate.[93]

Role of Protein Phosphatases

Regulation of cellular processes involves protein phosphorylation by kinases and dephosphorylation by phosphatases. While a number of studies investigated the role of various kinases in bile formation, studies to determine the role of protein phosphatases are limited. Studies with inhibitors of protein phosphatase 1 and 2A (PP1/2A), such as okadaic acid and microcystin, suggest that PP2A regulates microtubule-dependent vesicle movement in hepatocytes.[94] Okadaic acid, which inhibits PP2A in hepatocytes, blocks the ability of cAMP to stimulate TC uptake, translocate and dephosphorylate Ntcp, and increase [Ca^{2+}]$_i$ in hepatocytes.[95] Thus, cAMP-induced increases in [Ca^{2+}]$_i$ is dependent on PP2A activity. The ability of cAMP to dephosphorylate Ntcp is inhibited by a calcium chelator.[53] A recent preliminary study showed that cAMP activates Ca^{2+}/calmodulin-dependent PP2B (also known as calcineurin), and an inhibitor of PP2B inhibits cAMP-induced dephosphorylation of Ntcp.[96] Phosphorylation of Oatp1 by extracellular ATP and okadaic acid decreases its activity.[97] Okadaic acid does not affect basal fluid secretion and basal activity of Cl^-/HCO_3^- exchanger, but may be involved in the inactivation of secretin-induced activation of CFTR in cholangiocytes.[38] These studies suggest that protein phosphatases, such as PP2A and PP2B, may regulate vesicular transport and organic anion uptake in hepatocytes and seretin-induced secretion by cholagiocytes by dephosphorylating proteins involved. In view of this, it would of interest to determine whether protein phosphatases are also involved in the regulation of other transporters and are altered in cholestasis.

The existing studies reveal that different signal transduction pathways are involved in bile formation and that a single stimulus can produce multiple effects via different signal transduction pathway. For example, TUDC can stimulate organic anion excretion by a PKC-dependent mechanism to translocate Mrp2,[71] and stimulates Bsep transloctaion by ERK/p38 MAPK-dependent, but PKC-independent pathway.[79,92] Thus, TUDC may affect bile formation via two independent pathways. It is also possible that TUDC reverses cholestasis by acting via PKC[71] while increases bile formation under physiological conditions via the MAPK pathways.[79,92] It may be noted that the effect of PKC was studied using PKC inhibitors, but neither study determined directly whether the inhibitors produced the intended effect under the experimental conditions. Without this information it is difficult to assign a definite role of PKC. On the other hand, different signal transduction pathways can converge on one transporter. For example, TUDC and cell swelling stimulate Bsep translocation via PI3K-dependent ERK pathway as well as via PI3K-independent p38 MAPK pathway. TC- and cAMP-mediated translocation of Bsep is PI3K-dependent,[77,78] and hence can involve the ERK pathway. Cyclic AMP also activates p38 MAPK in hepatocytes and one inhibitor of p38 MAPK (SB203580) inhibits cAMP-mediated activation of PI3K-dependent PKB at higher concentrations (unpublished data). Thus, the effect of cAMP on Bsep translocation may also involve PI3K-indepednent p38 MAPK as well as PI3K-dependent PKB. Since cAMP inhibits ERK pathway in hepatocytes[36,98] at the level of Raf,[99] it is unlikely that the effect of cAMP on Bsep translocation is mediated via PI3K-dependent ERK pathway.

References

1. Anwer MS. Transheptatic solute transport and bile formation. Adv Vet Sci Comp Med 1993; 37:1-29.
2. Anwer MS. Cellular and molecular biology of the liver. Curr Opin Gastroenterol 1998; 14:182-190.
3. Trauner M, Meier PJ, Boyer JL. Molecular regulation of hepatocellular transport systems in cholestasis. J Hepatol 1999; 31:165-178.
4. Meier PJ, Stieger B. Molecular Mechanisms in Bile Formation. News Physiol Sci 2000; 15:89-93.
5. Kullak-Ublick GA, Beuers U, Paumgartner G. Hepatobiliary transport. J Hepatol 2000; 32:3-18.
6. Baiocchi L, Lesage G, Glaser S et al. Regulation of cholangiocyte bile secretion. J Hepatol 1999; 31:179-191.
7. Müller M, Jansen PLM. The secretory function of the liver: new aspects of hepatobiliary transport. J Hepatology 1998; 28:344-354.
8. Lee JM, Stieger B, Soroka CJ et al. Expression of the bile salt export pump is maintained after chronic cholestasis in the rat. Gastroenterology 2000; 118:163-172.
9. Trauner M, Arrese M, Soroka CJ et al. The rat canalicular conjugate export pump (Mrp2) is down-regulated in intrahepatic and obstructive cholestasis. Gastroenterology 1997; 113:255-264.
10. Soroka CJ, Lee JM, Azzaroli F et al. Cellular localization and up-regulation of multidrug resistance-associated protein 3 in hepatocytes and cholangiocytes during obstructive cholestasis in rat liver. Hepatology 2001; 33:783-791.
11. Donner MG, Keppler D. Up-regulation of basolateral multidrug resistance protein 3 (Mrp3) in cholestatic rat liver. Hepatology 2001; 34:351-359.
12. Paulusma CC, Kool M, Bosma PJ et al. A mutation in the human canalicular multispecific organic anion transporter gene causes the Dubin-Johnson syndrome. Hepatology 1997; 25:1539-1542.
13. Strautnieks SS, Bull LN, Knisely AS et al. A gene encoding a liver-specific ABC transporter is mutated in progressive familial intrahepatic cholestasis. Nat Genet 1998; 20:233-238.
14. De Vree JM, Jacquemin E, Sturm E et al. Mutations in the MDR3 gene cause progressive familial intrahepatic cholestasis. Proc Natl Acad Sci U S A 1998; 95:282-287.
15. Bouscarel B, Kroll SD, Fromm H. Signal transduction and hepatocellular bile acid transport: cross talk between bile acids and second messengers. Gastroenterology 1999; 117:433-452.

16. Kipp H, Arias IM. Intracellular trafficking and regulation of canalicular ATP-binding cassette transporters. Semin Liver Dis 2000; 20:339-351.
17. Beuers U, Probst I, Soroka C et al. Modulation of protein kinase C by taurolithocholic acid in isolated rat hepatocytes. Hepatology 1999; 29:477-482.
18. Jones BA, Rao YP, Stravitz RT et al. Bile salt-induced apoptosis of hepatocytes involves activation of protein kinase C. Am J Physiol 1997; 272:G1109-G1115.
19. Stravitz RT, Rao YP, Vlahcevic ZR et al. Hepatocellular protein kinase C activation by bile acids: implications for regulation of cholesterol 7 alpha-hydroxylase. Am J Physiol 1996; 271:G293-G303.
20. Rameh LE, Cantley LC. The role of phosphoinositide 3-kinase lipid products in cell function. J Biol Chem 1999; 274:8347-8350.
21. Toker A. Protein kinases as mediators of phosphoinositide 3-kinase signaling. Mol Pharmacol 2000; 57:652-658.
22. Wilkinson MG, Millar JB. Control of the eukaryotic cell cycle by MAP kinase signaling pathways. FASEB J 2000; 14:2147-2157.
23. Schaeffer HJ, Weber MJ. Mitogen-activated protein kinases: specific messages from ubiquitous messengers. Mol Cell Biol 1999; 19:2435-2444.
24. Seger R, Krebs EG. The MAPK signaling cascade. FASEB J 1995; 9:726-735.
25. Hayakawa T, Bruck R, Ng OC et al. DBcAMP stimulates vesicle transport and HRP excretion in isolated perfused rat liver. Am J Physiol 1990; 259:G727-G735.
26. Benedetti A, Strazzabosco M, Ng OC et al. Regulation of activity and apical targeting of the Cl-/HCO3-exchanger in rat hepatocytes. Proc Natl Acad Sci U S A 1994; 91:792-796.
27. Grüne S, Engelking LR, Anwer MS. Role of intracellular calcium and protein kinases in the activation of hepatic Na⁺/taurocholate cotransport by cyclic AMP. J Biol Chem 1993; 268:17734-17741.
28. Kato A, Gores GJ, LaRusso NF. Secretin stimulates exocytosis in isolated bile duct epithelial cells by a cyclic AMP-mediated mechanism. J Biol Chem 1992; 267:15523-15529.
29. Myers NC, Grüne S, Jameson HL et al. cGMP stimulates bile acid-independent bile formation and biliary bicarbonate excretion. Am J Physiol 1996; 270:G418-G424.
30. Mukhopadhayay S, Ananthanarayanan M, Stieger B et al. cAMP increases liver Na⁺-taurocholate cotransport by translocating transporter to plasma membranes. Am J Physiol 1997; 273:G842-G848.
31. Dranoff JA, McClure M, Burgstahler AD et al. Short-term regulation of bile acid uptake by microfilament-dependent translocation of ntcp to the plasma membrane. Hepatology 1999; 30:223-229.
32. Roelofsen H, Soroka CJ, Keppler D et al. Cyclic AMP stimulates sorting of the canalicular organic anion transporter (Mrp2/cMoat) to the apical domain in hepatocyte couplets. J Cell Sci 1998; 111 (Pt 8):1137-1145.
33. Gatmaitan ZC, Nies AT, Arias IM. Regulation and translocation of ATP-dependent apical membrane proteins in rat liver. Am J Physiol 1997; 272:G1041-G1049.
34. Kipp H, Pichetshote N, Arias IM. Transporters on demand: intrahepatic pools of canalicular ATP binding cassette transporters in rat liver. J Biol Chem 2001; 276:7218-7224.
35. Boyer JL, Soroka CJ. Vesicle targeting to the apical domain regulates bile excretory function in isolated rat hepatocyte couplets. Gastroenterology 1995; 109:1600-1611.
36. Webster CRL, Anwer MS. Role of the PI3K/PKB signaling pathway in cAMP-mediated translocation of rat liver Ntcp. Am J Physiol 1999; 277:G1165-G1172.
37. Kanno N, Lesage G, Glaser S et al. Regulation of cholangiocyte bicarbonate secretion. Am J Physiol Gastrointest Liver Physiol 2001; 281:G612-G625.
38. Alvaro D, Mennone A, Boyer JL. Role of kinases and phosphatases in the regulation of fluid secretion and Cl⁻/HCO3⁻ exchange in cholangiocytes. Am J Physiol 1997; 273:G303-G313.
39. Webster CRL, Anwer MS. Cyclic AMP mediated protection against bile acid induced apoptosis in cultured rat hepatocytes. Hepatology 1998; 27:1324-1331.
40. Richards JS. New signaling pathways for hormones and cyclic adenosine 3',5'- monophosphate action in endocrine cells. Mol Endocrinol 2001; 15:209-218.
41. Davidson HW, McGowan CH, Balch WE. Evidence for the regulation of exocytic transport by protein phosphorylation. J Cell Biol 1992; 116:1343-1355.
42. Matsuzaki Y, Bouscarel B, Le M et al. Effect of cholestasis on regulation of cAMP synthesis by glucagon and bile acids in isolated hepatocytes. Am J Physiol 1997; 273:G164-G174.
43. Bouscarel B, Matsuzaki Y, Le M et al. Changes in G protein expression account for impaired modulation of hepatic cAMP formation after BDL. Am J Physiol 1998; 274:G1151-G1159.
44. Hirohashi T, Suzuki H, Takikawa H et al. ATP-dependent transport of bile salts by rat multidrug resistance- associated protein 3 (Mrp3). J Biol Chem 2000; 275:2905-2910.
45. Anwer MS, Clayton LM. Role of extracellular Ca⁺⁺ in hepatic bile formation and taurocholate transport. Am J Physiol 1985; 249:G711-G718.
46. Reichen J, Berr F, Le M et al. Characterization of calcium deprivation-induced cholestasis in the perfused rat liver. Am J Physiol 1985; 249:G48-G57.

47. Bouscarel B, Reza S, Dougherty LA et al. Regulation of taurocholate and ursodeoxycholate uptake in hamster hepatocytes by Ca^{2+}- mobilizing agents. Am J Physiol 1996; 271:G1084-G1095.
48. Kuhn WF, Gewirtz DA. Stimulation of taurocholate and glycocholate efflux from the rat hepatocyte by arginine vasopressin. Am J Physiol 1988; 254:G732-G740.
49. Kuhn WF, Heuman DM, Vlahcevic ZR et al. Receptor-mediated stimulation of taurocholate efflux from the rat hepatocyte and the ex vivo perfused rat liver. Eur J Pharmacol 1990; 175:117-128.
50. Anwer MS, Atkinson JM. Intracellular calcium mediated activation of Na^+/H^+ exchange by arginine vasopressin and phenylephrine. Hepatology 1992; 15:134-143.
51. Anwer MS. Mechanism of activation of the Na^+/H^+ exchanger by arginine vasopressin in hepatocytes. Hepatology 1994; 20:1309-1317.
52. Lukas TJ, Haiech J, Lau W et al. Calmodulin and calmodulin-regulated protein kinases as transducers of intracellular calcium signals. Cold Spring Harb Symp Quant Biol 1988; 53 Pt 1:185-193.
53. Mukhopadhayay S, Ananthanarayanan M, Stieger B et al. Sodium taurocholate cotransporting polypeptide is a serine, threonine phosphoprotein and is dephosphorylated by cyclic AMP. Hepatology 1998; 28:1629-1636.
54. Anwer MS, Engelking LR, Nolan K et al. Hepatotoxic bile acids increase cytosolic Ca^{++} activity of isolated rat hepatocytes. Hepatology 1988; 8:887-891.
55. Beuers U, Nathanson MH, Boyer JL. Effects of tauroursodeoxycholic acid on cytosolic Ca2+ signals in isolated rat hepatocytes. Gastroenterology 1993; 104:604-612.
56. Scholmerich J, Baumgartner U, Miyai K et al. Tauroursodeoxycholate prevents taurolithocholate-induced cholestasis and toxicity in rat liver. J Hepatol 1990; 10:280-283.
57. Nathanson MH, Gautam A, Ng OC et al. Hormonal regulation of paracellular permeability in isolated rat hepatocyte couplets. Am J Physiol 1992; 262:G1079-G1086.
58. Nathanson MH, Gautam A, Bruck R et al. Effects of Ca2+ agonists on cytosolic Ca2+ in isolated hepatocytes and on bile secretion in the isolated perfused rat liver. Hepatology 1992; 15:107-116.
59. Yamaguchi Y, Dalle-Molle E, Hardison WG. Vasopressin and A23187 stimulate phosphorylation of myosin light chain- 1 in isolated rat hepatocytes. Am J Physiol 1991; 261:G312-G319.
60. Beuers U, Nathanson MH, Isales CM et al. Tauroursodeoxycholic acid stimulates hepatocellular exocytosis and mobilizes extracellular Ca++ mechanisms defective in cholestasis. J Clin Invest 1993; 92:2984-2993.
61. Bruck R, Nathanson MH, Roelofsen H et al. Effects of protein kinase C and cytosolic Ca2+ on exocytosis in the isolated perfused rat liver. Hepatology 1994; 20:1032-1040.
62. Benzeroual K, Pandey SK, Srivastava AK et al. Insulin-induced Ca^{++} entry in hepatocytes is important for PI 3-kinase activation, but not for insulin receptor and IRS-1 tyrosine phosphorylation. Biochim Biophys Acta 2000; 1495:14-23.
63. Yano S, Tokumitsu H, Soderling TR. Calcium promotes cell survival thorugh CaM-K kinase activation of the protein-kinase-B pathway. Nature 1998; 396:584-587.
64. Nathanson MH, Burgstahler AD, Fallon MB. Multistep mechanism of polarized Ca2+ wave patterns in hepatocytes. Am J Physiol 1994; 267:G338-G349.
65. Nathanson MH, Burgstahler AD, Mennone A et al. Ca2+ waves are organized among hepatocytes in the intact organ. Am J Physiol 1995; 269:G167-G171.
66. Corasanti JG, Smith ND, Gordon ER et al. Protein kinase C agonists inhibit bile secretion independently of effects on the microcirculation in the isolated perfused rat liver. Hepatology 1989; 10:8-13.
67. Steen H, Smit H, Nijholt A et al. Modulators of the protein kinase C system influence biliary excretion of cationic drugs. Hepatology 1993; 18:1208-1215.
68. Fitz JG, Sostman AH, Middleton JP. Regulation of cation channels in liver cells by intracellular calcium and protein kinase C. Am J Physiol 1994; 266:G677-G684.
69. Noe J, Hagenbuch B, Meier PJ et al. Characterization of the mouse bile salt export pump overexpressed in the baculovirus system. Hepatology 2001; 33:1223-1231.
70. Rust C, Karnitz LM, Paya CV et al. The bile acid taurochenodeoxycholate activates a phosphatidylinositol 3-kinase-dependent survival signaling cascade. J Biol Chem 2000; 275:20210-20216.
71. Beuers U, Bilzer M, Chittattu A et al. Tauroursodeoxycholic acid inserts the apical conjugate export pump, Mrp2, into canalicular membranes and stimulates organic anion secretion by protein kinase C-dependent mechanisms in cholestatic rat liver. Hepatology 2001; 33:1206-1216.
72. Beuers U, Throckmorton DC, Anderson MS et al. Tauroursodeoxycholic acid activates protein kinase C in isolated rat hepatocytes. Gastroenterology 1996; 110:1553-1563.
73. Kubitz R, Huth C, Schmitt M et al. Protein kinase C-dependent distribution of the multidrug resistance protein 2 from the canalicular to the basolateral membrane in human HepG2 cells. Hepatology 2001; 34:340-350.

74. Folli F, Alvaro D, Gigliozzi A et al. Regulation of endocytic-transcytotic pathways and bile secretion by phosphatidylinositiol 3-kinase in rats. Gastroenterology 1997; 113:954-965.
75. Webster CRL, Blanch CJ, Philips J et al. Cell swelling-induced translocation of rat liver Na$^+$/ taurocholate cotransport polypeptide is mediated via the phosphoinositide 3-kinase signaling pathway. J Biol Chem 2000; 275:29754-29760.
76. Misra S, Ujhazy P, Varticovski L et al. Phosphoinositide 3-kinase lipid products regulate ATP-dependent transport by sister of P-glycoprotein and multidrug resistance associated protein 2 in bile canalicular membrane vesicles. Proc Natl Acad Sci (USA) 1999; 96:5814-5819.
77. Misra S, Ujházy P, Gatmaitan Z et al. The role of phophoinositide 3-kinase in taurocholate-induced trafficking of ATP-dependent canalicular transporters in rate liver. J Biol Chem 1998; 273:26638-26644.
78. Misra,S,Ujhazy,P,Arias,IM et al. CAMP-induced Recruitment of Spgp and Mrp2 is mediated by G-protein associated PI 3-kinase. Hepatology 30, 306A. 1999. Ref Type: Abstract
79. Kurz AK, Block C, Graf D et al. Phosphoinositide 3-kinase-dependent Ras activation by tauroursodesoxycholate in rat liver. Biochem J 2000; 350 Pt 1:207-213.
80. Feranchak AP, Roman RM, Doctor RB et al. The lipid products of phosphoinositide 3-kinase contribute to regulation of cholangiocyte ATP and chloride transport. J Biol Chem 1999; 274:30979-30986.
81. Kilic G, Doctor RB, Fitz JG. Insulin stimulates membrane conductance in a liver cell line: evidence for insertion of ion channels through a phosphoinositide 3-kinase- dependent mechanism. J Biol Chem 2001; 276:26762-26768.
82. Kubitz R, D'Ruso S, Keppler D et al. Osmodependent dynamic localization of the multidrug resistance protein 2 in the rat hepaocyte canalicular membrane. Gastroenterology 1991; 113:1438-1442.
83. Schmitt M, Kubitz R, Lizun S et al. Regulation of the dynamic localization of the rat Bsep gene-encoded bile salt export pump by anisoosmolarity. Hepatology 2001; 33:509-518.
84. Krause U, Rider MH, Hue L. Protein kinase signalling pathway triggered by cell swelling and involved in the activation of glycogen synthase and acetyl-CoA carboxylase in isolated rat hepatocytes. J Biol Chem 1996; 271:16668-16673.
85. Hernandez R, Teruel T, Lorenzo M. Akt mediates insulin induction of glucose uptake and up-regulation of GLUT4 gene expression in brown adipocytes. FEBS Lett 2001; 494:225-231.
86. Standaert ML, Bandyopadhyay G, Perez L et al. Insulin activates protein kinases C-zeta and C-lambda by an autophosphorylation-dependent mechanism and stimulates their translocation to GLUT4 vesicles and other membrane fractions in rat adipocytes. J Biol Chem 1999; 274:25308-25316.
87. Vlahos CJ, Matter WF, Hui KY et al. A specific inhibitor of phosphatidylinositiol 3-kinase, 2-(4-morpholinyl)-8-phenyl-4H-1-benzopyran-4-one (LY294002). J Biol Chem 1994; 269:5241-5248.
88. Soroka CJ, Pate MK, Boyer JL. Canalicular export pumps traffic with polymeric immunoglobulin A receptor on the same microtubule-associated vesicle in rat liver. J Biol Chem 1999; 274:26416-26424.
89. Wurmser AE, Gary JD, Emr SD. Phosphoinositide 3-kinases and their FYVE domain-containing effectors as regulators of vacuolar/lysosomal membrane trafficking pathways. J Biol Chem 1999; 274:9129-9132.
90. Noe B, Schliess F, Wettstein M et al. Regulation of taurocholate excretion by a hypo-osmolarity-activity signal transduction pathway in rat liver. Gastroenterology 1996; 110:858-865.
91. Schliess F, Kurz AK, vom Dahl S et al. Mitogen-activated protein kinase mediate the stimulation of bile acid secretion by tauroursodeoxycholate in rat liver. Gastroenterology 1997; 113:1306-1313.
92. Kurz AK, Graf D, Schmitt M et al. Tauroursodesoxycholate-induced choleresis involves p38(MAPK) activation and translocation of the bile salt export pump in rats. Gastroenterology 2001; 121:407-419.
93. Lazaridis KN, Gores GJ, Lindor KD. Ursodeoxycholic acid 'mechanisms of action and clinical use in hepatobiliary disorders'. J Hepatol 2001; 35:134-146.
94. Hamm-Alvarez SF, Wei X, Bernbdt N et al. Protein phosphatases independently regulate vesicle movement and microtubule subpopulations in hepatocytes. Am J Physiol 1996; 271:C929-C943.
95. Mukhopadhayay S, Webster CRL, Anwer MS. Role of protein phosphatase in cyclic AMP-mediated stimulation of hepatic Na$^+$/taurocholate cotransport. J Biol Chem 1998; 273:30039-30045.
96. Anwer MS, Webster CRL, Blanch CJ. Role of protein phosphatase 2B in cAMP-mediated translocation of rat liver Ntcp. Hepatology 2000; 32:438A.
97. Glavy JS, Wu SM, Wang PJ et al. Down-regulation by extracellular ATP of rat hepatocyte organic anion transport is mediated by serine phosphorylation of Oatp1. J Biol Chem 2000; 275:1479-1484.
98. Schliess F, Kurz AK, Haussinger D. Glucagon-induced expression of the MAP kinase phosphatase MKP-1 in rat hepatocytes. Gastroenterology 2000; 118:929-936.
99. Hafner S, Adler HS, Mischak H et al. Mechanism of inhibition of Raf-1 by protein kinase A. Mol Cell Biol 1994; 14:6696-6703.

CHAPTER 10

Bile Acid-Mediated Apoptosis in Cholestasis

Hajime Higuchi and Gregory J. Gores

Summary

Hepatocyte apoptosis is a common pathologic feature in cholestatic liver diseases. In cholestatic liver injury, elevated tissue bile acid concentrations trigger lethal hepatocyte injury by inducing apoptosis. Toxic bile acids induce apoptosis by activating cell surface membrane death receptors. The activated death receptors stimulate a signaling cascade involving the pro-apoptotic Bcl-2 poteins Bid and Bax. The proapoptotic Bcl-2 proteins initiate mitochondrial injury, the mitochondrial permeability transition (MPT) and oxidative stress. Inhibition of death receptors and their cascades may prove useful in attenuating liver injury during cholestasis.

Bile Acids in the Pathophysiology of Cholestasis

Cholestasis is defined as an impairment in bile formation. Bile is composed of several biliary solutes including bilirubin, cholesterol and bile acids, which are preferentially excreted into bile under physiologic conditions. During cholestasis, these solutes accumulate in the liver and blood because of reduced bile secretion. Indeed, earlier studies have demonstrated that failure of bile acid excretion in cholestasis leads to high concentrations of bile acids within the liver.[1,2] This bile acid accumulation within the liver results in bile acid associated hepatocyte toxicity.[1,3,4] Several clinical examples and animal models highlight the importance of bile acid-mediated hepatotoxicity. First, the pathophysiological importance of bile acid-induced liver injury is demonstrated in children with the progressive familial intrahepatic cholestasis (PFIC) subtype 2 syndrome.[5] These children have mutations in the canalicular transport protein for bile acid secretion into bile and develop progressive liver disease because of the inability to excrete bile acids from the hepatocyte.[6] Second, mice lacking the nuclear receptor FXR are susceptible to hepatotoxicity during bile acid feeding.[7] The FXR nuclear receptor, isolated from a rat liver cDNA library and now recognized as a nuclear bile acid receptor,[8-10] regulates the expression of a number of bile acid transporting proteins including the canalicular bile salt excretory pump (BSEP).[11,12] Mice lacking FXR developed normally but were distinguished from wild-type mice by elevated serum bile acid levels and disorders in bile acid homeostasis after bile acid feeding.[7] When these mice were fed cholic acid, massive liver damage including hepatocyte apoptosis was observed,[7] presumably because of failure to adapt to the bile acids due to their inability to increase BSEP expression. Third, mice lacking the multidrug resistance gene are quite susceptible to bile acid liver injury. The multidrug resistance (MDR) gene knockout mice are an animal model for the PFIC subtype 3 syndrome.[13] In these mice, the absence of phospholipids in bile causes chronic bile acid-induced hepatocyte damage.[13,14] Fourth, Rodrigues et al[15] reported that feeding rats deoxycholate causes hepatocyte apoptosis in vivo in the absence of transport abnormalities. All these data strongly implicate a cytotoxic role for bile acids during cholestatic liver diseases.

Molecular Pathogenesis of Cholestasis, edited by Michael Trauner and Peter L.M. Jansen.
©2004 Eurekah.com and Kluwer Academic / Plenum Publishers.

Bile acids appear to induce liver injury by causing hepatocyte apoptosis. Hepatocyte apoptosis is a common pathologic feature of cholestatic liver diseases.[16] Indeed, it is now recognized that lesions originally termed "acidophillic bodies" are in fact apoptotic bodies.[17] Several data demonstrate that selected bile acids can directly induce hepatocyte apoptosis.[16] Bile acids induce apoptosis in primary cultures of rodent hepatocytes.[18-22] In addition, hepatocyte apoptosis is also observed in bile duct ligated experimental animals, a model of extrahepatic cholestasis (23). In these in vivo studies, inhibition of hepatocyte apoptosis markedly diminishes liver injury as evaluated by serum ALT values.[23] Thus, hepatocyte apoptosis occurs in the cholestatic liver predominantly due to the retention of toxic bile acids.

Mechanisms of Bile Acid Cytotoxicity

Bile acid cytotoxicity has been attributed to their detergent properties.[24-26] Previous studies performed in vivo in rats,[27,28] in perfused rat livers[29] and in vitro with isolated human,[30] or rat hepatocytes or plasma membrane preparations[24,31,32] have shown that hydrophobic bile acids damage hepatocellular membranes via detergent-like action. However, these studies were performed with bile acid concentrations in the milimolar range, considerably higher than the observed bile acid concentrations in vivo, and even nondetergent, hydrophilic bile acids such as hydeoxycholic acid have been demonstrated to induce cellular apoptosis.[33] Bile acid concentrations in the micromoler range are now recognized to induce hepatocyte injury by nondetergent mechanisms.[15,18-22,34-39] These mechanisms include bile acid-induced oxidative stress,[21,34-37] mitochondrial toxicity[15,21,22,38,39] and activation of certain proapoptotic Bcl-2 family proteins (Fig. 1).[21,23] We will now review the current data supporting these seemingly disparate mechanisms of toxicity and develop a unifying hypothesis to explain these apparent divergent observations.

Oxidative Stress

Oxidative stress is thought to occur when a cell, tissue, or organism is exposed to excess oxidant generation, in particular to superoxide anion and its metabolites. It has been postulated that hydrophobic bile acids retained in the hepatocyte during cholestasis initiate the generation of reactive oxygen metabolites from mitochondria, leading to lipid peroxidation and loss of cell viability. Indeed, hydrophobic bile acids stimulate the generation of reactive oxygen species in isolated rat hepatocytes,[38] primary cultured rat hepatocytes,[15] and in purified hepatocyte mitochondria.[21,38] In addition, Sokol et al have reported that oxidative stress occurs in the bile duct ligated rat liver[35,40] and in the rat liver intravenously infused with taurocheonodeoxycholate.[41] Bile acids may impair electron transport in the mitochondrial respiratory chain, thereby, promoting electron leak at the ubiquinone-complex III site, enhancing the formation of superoxide.[42] The bile acid-induced oxidative stress leads to mitochondrial lipid peroxidation, which is abrogated by antioxidants and presumably worsened by cholestatic malabsorption-induced tocopherol deficiency.[34,35,38,40,41] The mitochondrial oxidative stress can lead to the so-called "mitochondrial membrane permeability transition" (MPT).[15] The MPT results in loss of the mitochondrial membrane potential,[21] decreased ATP synthesis, altered ion homeostasis, and protease activation, culminating in necrosis and/or apoptosis. More recently, Yerushalmi et al[36] reported that relatively low concentrations of bile acids induced oxidative stress in hepatocytes prior to the onset of apoptosis, and a significant liner correlation between apoptosis and the magnitude of oxidative stress was observed.[36] However, because oxidative stress may occur as a result of certain cytotoxic events including mitochondrial injury, it is still uncertain if oxidative stress is a critical for the initiation of bile acid-induced cell death, or exaggerate the injury. Indeed, the observation that inhibition of caspase protease activity, essential executioners in apoptosis, attenuates bile acid induced oxidative stress,[36] strongly suggests that oxidative stress is a secondary phenomenon.

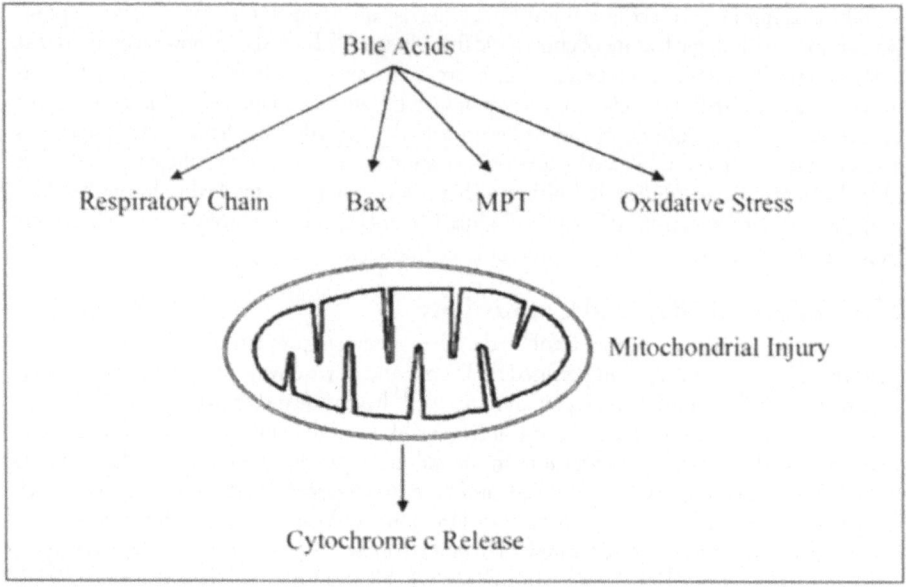

Figure 1. Potential mechanisms for bile acid associated mitochondrial dysfunction and cytotoxicity. Bile acids have been known to cause mitochondrial respiratory chain impairment, Bax transmigration from the cytosol to mitochondria, the mitochondrial permeability transition (MPT), and oxidative stress. All these processes may induce mitochondrial dysfunction resulting in cytochrome c release. Each process may occur independently or more likely in concert in a unified mechanism.

Mitochondrial Injury

Rodrigues et al[15,21,22] have recently reported that the hydrophobic bile acid deoxycholate causes partial loss of the mitochondrial membrane potential, the MPT, mitochondrial swelling, cytochrome c release, leading to caspase activation and apoptosis in rat hepatocytes (Fig. 2). They also reported that both cyclosporin A, a MPT inhibitor, and ursodeoxycholate prevented the MPT and cytochrome c release.[22] The mitochondria-dependent pathway of apoptosis is regulated by the Bcl-2 family proteins. Bcl-2 was discovered to be a protooncogene that inhibited cell death induced by interleukin 3 deprivation,[43] and was shown subsequently to inhibit death induced by many other stimuli. The Bcl-2 family consists of two opposing clans. One subfamily including Bcl-2, Bcl-XL, Ced-9, Bcl-w and Mcl-1 promotes cell survival and the other promotes cell death (Bax, Bok, Bak, Bid, Hrk, Bik, Bim). The pro-apoptotic subfamily can be further divided into those proteins consisting of Bcl-2 homology (BH) 1-3, like Bax, Bak and Bok, and those consisting of only the BH3 region, like Bad, Bid, Bik, Bim and Hrk.[44] Currently, propapoptotic Bcl-2 family proteins are implicated as an upstream mechanism of bile acid-induced mitochondrial cytochrome c release.[21,22,45-47] Rodrigues et al[21,22] demonstrated Bax traslocation to mitochondria during deoxycholate-induced apoptosis in rat hepatocytes. They also reported that increased mitochondrial Bax is accompanied by the MPT and cytochrome c release. Bax is able to bind to constituents of a mitochondrial pore apparatus resulting in pore opening and the associated release of cytochrome c.[48-50] Although Bax activation likely explains the mitochondrial dysfunction occuring during bile acid cytotoxicity, it is unlikely that bile acids themselves directly activate this proapoptotic protein. Bax is likely activated by further upstream mechanisms.

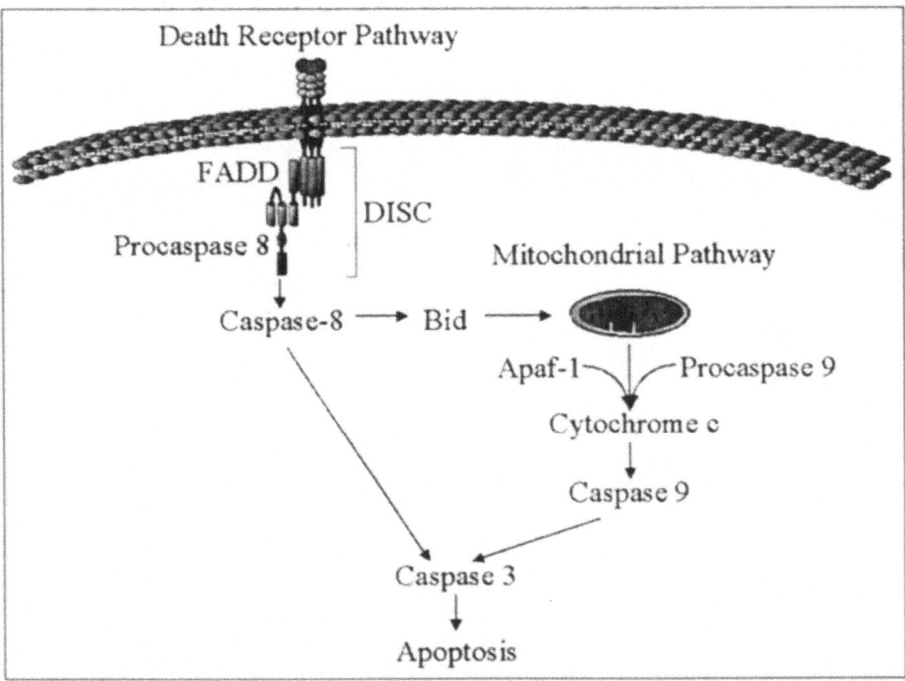

Figure 2. Death receptor and mitochondrial pathways of apoptosis. Apoptosis can be initiated by two fundamental pathways: i) the death receptor pathway; and ii) the mitochondrial pathway. Upon specific ligation of cell surface death receptors and their ligands, apoptosis is initiated by the recruitment of Fas-associated death domain (FADD) to the oligomerized receptor. FADD recruits procaspase 8 to the death inducing signaling complex (DISC). The induced proximity of procaspase 8 molecules is postulated to cause their activation by autocatalytic processes. The mitochondrial pathway of apoptosis is associated with cytochrome c release from mitochondria which then binds to apoptosis activating factor-1 (Apaf-1) resulting in activation of another initiator caspase, caspase 9. Caspase 8 and 9 can activate effector caspases such as caspases 3, 6 and 7 via a caspase cascade which results in execution of the cell death program.

Death Receptor-Mediated Activation as a Unifying Hypothesis for Bile Acid-Induced Apoptosis, Mitochondrial Dysfunction and Oxidative Stress

In addition to mitochondrial dysfunction, apoptosis can also be induced by death receptors (Fig. 2). The cell surface death receptors initiate apoptosis by directly activating the initiator caspase 8.[51] In hepatocytes, the death receptor and mitochondrial pathways of apoptosis are not mutually exclusive, in that death receptor initiated cascades require mitochondrial dysfunction to fully induce apoptosis.[51] In hepatocytes, active caspase 8 cleaves cytosolic Bid, a proapoptotic Bcl-2 family member (Fig.2).[52,53] The carboxyl fragment of truncated Bid (tBid) translocates to mitochondria where it induces cytochrome c release.[52,54,55] Thus, the activation of mitochondrial pathway of apoptosis during bile acid cytotoxicity could easily be explained by stimulation of a cell surface death receptor.

Current concepts suggest that bile acid-associated hepatocyte apoptosis is Fas death receptor initiated rather than due to primary mitochondrial toxicity.[23,56,57] For example, bile acids induce death receptor dependent apoptosis at concentrations below that required for direct mitochondrial dysfunction. For example, micromolar concentrations of glycochenodeoxycholic acid (GCDC) induce hepatocyte apoptosis via Fas death receptor aggregation and subsequent

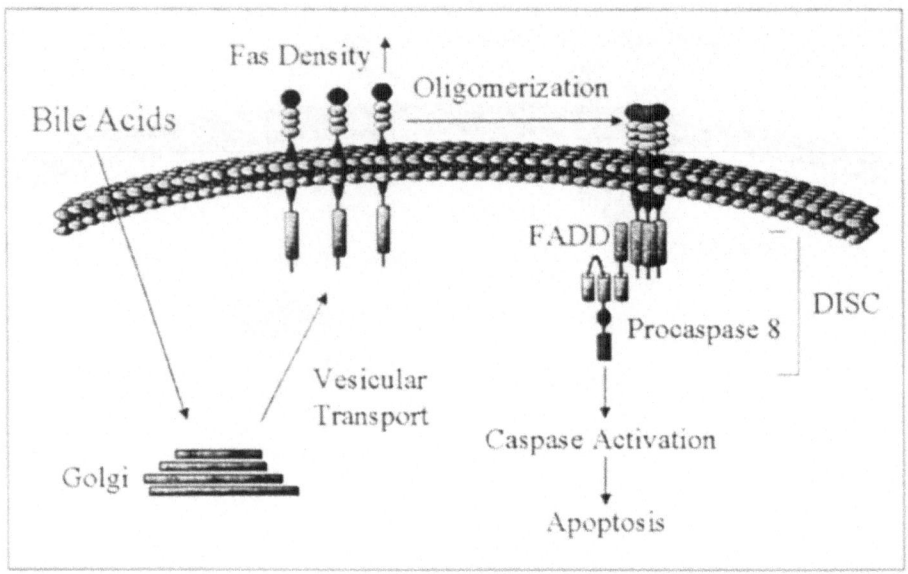

Figure 3. The bile acid-induced Fas death receptor-mediated apoptosis pathway. Death receptor–mediated apoptosis is initiated by oligomerization of cell surface death receptor and formation of DISC. Bile acid stimulates Golgi-associated and microtuble-dependent Fas trafficking to plasma membrane, resulting in an increased density of cell surface Fas. This process likely promotes Fas oligomerization, initiating a caspase-dependent apoptosis pathway.

recruitment of FADD and caspase 8, independent of Fas ligand (Fig.3).[57] Observations in bile duct ligated mice also demonstrate that hepatocellular apoptosis was diminished in bile duct ligated Fas-deficient lpr mice as compared to wild-type animals.[23] Thus, Fas ligand-independent Fas death receptor activation appears to be a common pathway of bile acid-associated hepatocyte apoptosis.

Detailed mechanisms of ligand-independent death receptor activation are currently under investigation by us. A mechanism for the post transcriptional regulation of Fas receptors is to sequester this death receptor within intracellular pools. The receptors can then be shuttled to the plasma membrane, presumably by a vesicular transport pathway, to initiate cell death signals. An increase in the cell surface density of Fas receptors likely promotes their aggregation, causing apoptosis. Indeed, this phenomenon has been observed during bile acid-induced hepatocyte apoptosis.[56] A Golgi-associated and microtuble-dependent pathway appears to be involved in the trafficking of Fas to the cell surface during bile acid induced apoptosis. The increased density of cell surface Fas promotes their oligomerization, initiating a caspase-dependent death signaling pathway (Fig. 3).

Bile acid-induced Fas activation leads to caspase 8-dependent Bid cleavage resulting in transmigration of truncated form of Bid (tBid) to mitochondria.[57,58,62] It has been shown that glycochenodeoxycholate induces Bid cleavage in mouse hepatocytes results in Bid transmigration to mitochondria.[62] Inhibition of Bid by employing selective antisense oligodeoxynucleotides prevents bile acid induced apoptosis.[62] These data further support the importance of death receptor initiated signaling events upstream of mitochondrial pathway in bile acid cytotoxicity. Although Bid had been shown to directly cause release of cytochrome c in cell free systems,[53] Martinou and co-workers have recently demonstrated that Bid and Bax cooperate to induce mitochondrial cytochrome c release.[54,59] Moreover, Fas-mediated apoptosis is inhibited in the Bax/Bak knockout animal despite Bid cleavage and mitochondrial translocation.[60] Bid either helps chaperone Bax to the mitochondria or by directly allosterically modifying Bax promotes

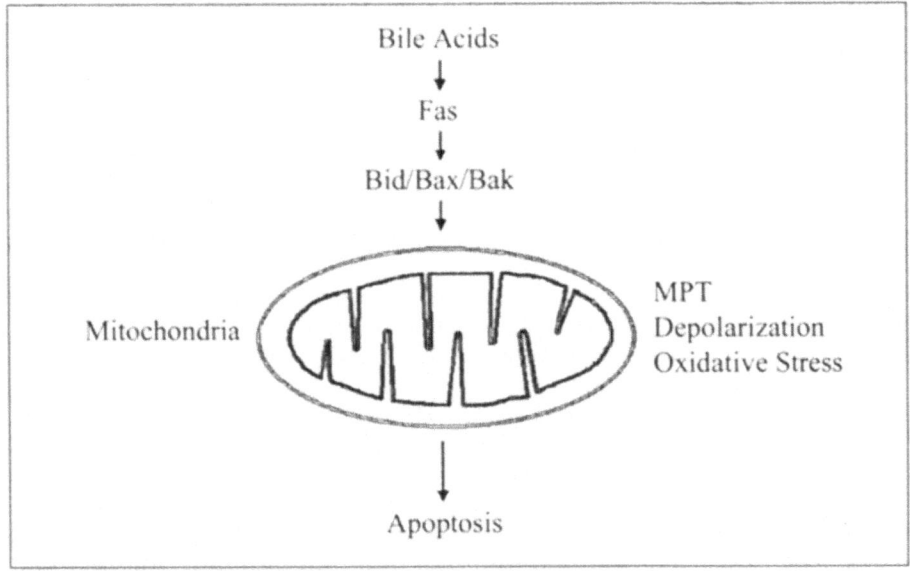

Figure 4. Proposed mechanism unifying the processes resulting in bile acid-induced mitochondrial injury and apoptosis. Bile acid-induced Fas activation initiates a capase 8-dependent apoptosis cascade. In hepatocytes, this signaling leads to the activation and/or transmigration of certain proapoptotic bcl-2 family proteins such as Bid, Bax and Bak. Current evidences have demonstrated that either Bid, Bax or Bak may induce a variety of mitochondrial alterations including the MPT, depolarization and oxidative stress. Thus, although bile acid may induce mitochondrial injury via direct or indirect mechanisms, Fas death receptor initiated Bcl-2 family protein-dependent pathway is a likely upstream mechanism responsible for the mitochondrial injury and oxidative stress induced by bile acids.

its insertion into the outer mitochondrial membrane, a requisite step for Bax-induced cytochrome c release. Thus, Bax translocation to mitochondria by bile acid reported by us and others[21-23] is now a recognized event in death receptor initiated apoptosis (Fig. 4). Cytochrome c release during GCDC-induced apoptosis may be triggered by Bid co-operating with Bax or perhaps Bak to disrupt the integrity of the outer mitochondrial membrane.[60]

The MPT (a phenomenon characterized by permeability of the inner membrane to solutes, collapse of the membrane potential, and uncoupling of mitochondrial oxidative phosphorylation) can occur during apoptosis associated mitochondrial dysfunction. However, we found that mitochondria maintained their membrane potential despite cytochrome c release in hepatocytes and only depolarized late in bile acid-induced apoptosis as a secondary event.[62] The release of cytochrome c in bile acid cytotoxicity may explain the observations by us and others that bile acid cytotoxicity is associated with generation of reactive oxygen species.[37,38] Loss of cytochrome c from the mitochondria results in impairment of electron flow in the respiratory chain enhancing reduction of oxygen by electron.[61] The resultant formation of superoxide anion and generation of other reactive oxygen species would then ensue causing an oxidative stress further contributing to the toxic effects of the bile acids. Although Yerushalmi et al[36] have recently implicated the importance of oxidative stress and the MPT in bile acid induced hepatocyte apoptosis, they have also reported that the selective caspase 8 inhibitor IETD-fmk prevents apoptosis and reduces oxidative stress induced by glycochenodeoxycholate. These data further support the importance of death receptor initiated signaling events upstream of mitochondria during bile acid induced apoptosis. Indeed, bile acid mitochondrial dysfunction and oxidative stress can all be explained as downstream events of Fas activation.

Inhibition of Fas Signaling as a Therapeutic Strategy for Attenuating Cholestatic Liver Injury

Using the bile duct ligated mouse as a model of extrahepatic cholestasis and TUNEL and trypan blue assays to quantitate apoptosis and necrosis, respectively, hepatocyte apoptosis was found to be the predominant mode of cell death.[23] Following bile duct ligation for 3 days, mouse hepatocyte apoptosis increased and was 3-fold more abundant than necrosis. The occurrence of hepatocyte apoptosis was further confirmed biochemically by demonstrating increases in caspase catalytic activity. The hepatocyte apoptosis was Fas death receptor-dependent but Fas ligand-independent consistent with our in vitro observations.[57] Indeed, when Fas-deficient lpr mice were bile duct ligated, hepatocyte apoptosis as well as caspase activation was markedly reduced. In contrast, when Fas ligand-deficient gld mice were subjected to bile duct ligation, hepatocyte apoptosis was not reduced.[23] These data were the first to demonstrate bile acid-associated apoptosis is Fas death receptor mediated in vivo. Moreover, the Fas-dependent apoptosis could be linked to liver injury and animal survival.[23]

Because Bid is essential for Fas-mediated apoptosis in hepatocytes, we examined the role of Bid in bile acid mediated apoptosis in primary mouse hepatocytes and bile duct ligated mice. Translocation of caspase 8 cleaved Bid to mitochondria was observed in primary mouse hepatocytes.[62] These data demonstrate a similarity in ligand activation and bile acid stimulation of the Fas death receptor. Next, we inhibited Bid expression using antisense technology. Bid antisense, but not a scrambled antisense oligonucleotide, inhibited Bid expression and markedly attenuated bile acid induced apoptosis. More importantly, Bid antisense treatment of bile duct ligated mice also reduced liver injury as assessed by serum ALT values. Thus, inhibition of Bid signaling markedly attenuates bile acid-induced hepatocyte apoptosis and liver injury. Bid would appear to be a good pharmacologic target for the treatment of cholestatic liver diseases.

Conclusion

Current studies demonstrate the importance of death receptor-mediated apoptosis in bile salt cytotoxicity. Bile acid associated apoptosis is mediated by FADD and caspase 8 dependent signaling upon aggregation of Fas. This signaling process is mediated by proapoptotic Bcl-2 family proteins (e.g., Bid/Bax interactions) and mitochondrial cytochrome c release, causing down stream caspase activation. Thus, Bid and/or caspase 8 are logical pharmacologic targets for the treatment of cholestatic liver injury. Along these lines, we have demonstrated that Bid antisense therapy reduced Bid protein expression and attenuates bile acid cytotoxicity in vivo and in vitro. Further mechanisms to inhibit bile acid-mediated apoptosis will likely emerge enhancing an ability to treat patients with chronic cholestatic liver diseases.

Acknowledgements

This work was supported by a grant from the National Institutes of Health (DK41876) and the Mayo Foundation.

References

1. Greim H, Trulzsch D, Czygan P et al. Mechanism of cholestasis. 6. Bile acids in human livers with or without biliary obstruction. Gastroenterology 1972;63(5):846-850.
2. Greim H, Trulzsch D, Roboz J et al. Mechanism of cholestasis. 5. Bile acids in normal rat livers and in those after bile duct ligation. Gastroenterology 1972;63(5):837-845.
3. Popper H. Cholestasis. Annu Rev Med 1968;19:39-56.
4. Popper H, Schaffner F. Pathophysiology of cholestasis. Hum Pathol 1970;1(1):1-24.
5. Strautnieks SS, Bull LN, Knisely AS et al. A gene encoding a liver-specific ABC transporter is mutated in progressive familial intrahepatic cholestasis. Nat Genet 1998;20(3):233-238.
6. Kullak-Ublick GA, Meier PJ. Mechanisms of cholestasis. Clin Liver Dis 2000;4(2):357-385.
7. Sinal CJ, Tohkin M, Miyata M et al. Targeted disruption of the nuclear receptor FXR/BAR impairs bile acid and lipid homeostasis. Cell 2000;102(6):731-744.

8. Makishima M, Okamoto AY, Repa JJ et al. Identification of a nuclear receptor for bile acids. Science 1999;284(5418):1362-1365.

9. Parks DJ, Blanchard SG, Bledsoe RK et al. Bile acids: natural ligands for an orphan nuclear receptor. Science 1999;284(5418):1365-1368.

10. Wang H, Chen J, Hollister K et al. Endogenous bile acids are ligands for the nuclear receptor FXR/BAR. Mol Cell 1999;3(5):543-553.

11. Lu TT, Makishima M, Repa JJ et al. Molecular basis for feedback regulation of bile acid synthesis by nuclear receptors. Mol Cell 2000;6(3):507-515.

12. Hylemon PB, Stravitz RT, Vlahcevic ZR. Molecular genetics and regulation of bile acid biosynthesis. Prog Liver Dis 1994;12:99-120.

13. de Vree JM, Jacquemin E, Sturm E et al. Mutations in the MDR3 gene cause progressive familial intrahepatic cholestasis. Proc Natl Acad Sci U S A 1998;95(1):282-287.

14. De Vree JM, Ottenhoff R, Bosma PJ et al. Correction of liver disease by hepatocyte transplantation in a mouse model of progressive familial intrahepatic cholestasis. Gastroenterology 2000; 119(6):1720-1730.

15. Rodrigues CM, Fan G, Ma X et al. A novel role for ursodeoxycholic acid in inhibiting apoptosis by modulating mitochondrial membrane perturbation. J Clin Invest 1998;101(12):2790-2799.

16. Patel T, Gores GJ. Apoptosis and hepatobiliary disease. Hepatology 1995;21(6):1725-1741.

17. Klion FM, Schaffner F. The ultrastructure of acidophilic "Councilman-like" bodies in the liver. Am J Pathol 1966;48(5):755-767.

18. Patel T, Bronk SF, Gores GJ. Increases of intracellular magnesium promote glycodeoxycholate-induced apoptosis in rat hepatocytes. J Clin Invest 1994;94(6):2183-2192.

19. Kwo P, Patel T, Bronk SF et al. Nuclear serine protease activity contributes to bile acid-induced apoptosis in hepatocytes. Am J Physiol 1995;268(4 Pt 1):G613-621.

20. Jones BA, Rao YP, Stravitz RT et al. Bile salt-induced apoptosis of hepatocytes involves activation of protein kinase C. Am J Physiol 1997;272(5 Pt 1):G1109-115.

21. Rodrigues CM, Fan G, Wong PY et al. Ursodeoxycholic acid may inhibit deoxycholic acid-induced apoptosis by modulating mitochondrial transmembrane potential and reactive oxygen species production. Mol Med 1998;4(3):165-178.

22. Rodrigues CM, Ma X, Linehan-Stieers C et al. Ursodeoxycholic acid prevents cytochrome c release in apoptosis by inhibiting mitochondrial membrane depolarization and channel formation. Cell Death Differ 1999;6(9):842-854.

23. Miyoshi H, Rust C, Roberts PJ et al. Hepatocyte apoptosis after bile duct ligation in the mouse involves Fas. Gastroenterology 1999;117(3):669-677.

24. Scholmerich J, Becher MS, Schmidt K et al. Influence of hydroxylation and conjugation of bile salts on their membrane-damaging properties—studies on isolated hepatocytes and lipid membrane vesicles. Hepatology 1984;4(4):661-666.

25. Scharschmidt BF, Keeffe EB, Vessey DA et al. In vitro effect of bile salts on rat liver plasma membrane, lipid fluidity, and ATPase activity. Hepatology 1981;1(2):137-145.

26. Billington D, Evans CE, Godfrey PP et al. Effects of bile salts on the plasma membranes of isolated rat hepatocytes. Biochem J 1980;188(2):321-327.

27. Miyai K, Richardson AL, Mayr W et al. Subcellular pathology of rat liver in cholestasis and choleresis induced by bile salts. 1. Effects of lithocholic, 3beta-hydroxy-5- cholenoic, cholic, and dehydrocholic acids. Lab Invest 1977;36(3):249-258.

28. Drew R, Priestly BG. Failure of hypoactive hypertrophic smooth endoplasmic reticulum to produce cholestasis in rats. Toxicol Appl Pharmacol 1978;45(1):191-199.

29. Schmucker DL, Ohta M, Kanai S et al. Hepatic injury induced by bile salts: correlation between biochemical and morphological events. Hepatology 1990;12(5):1216-1221.

30. Miyazaki K, Nakayama F, Koga A. Effect of chenodeoxycholic and ursodeoxycholic acids on isolated adult human hepatocytes. Dig Dis Sci 1984;29(12):1123-1130.

31. Yousef IM, Fisher MM. In vitro effect of free bile acids on the bile canalicular membrane phospholipids in the rat. Can J Biochem 1976;54(12):1040-1046.

32. Ohta M, Kanai S, Kitani K. The order of hepatic cytotoxicity of bile salts in vitro does not agree with that examined in vivo in rats. Life Sci 1990;46(21):1503-1508.

33. Araki Y, Fujiyama Y, Andoh A et al. Hydrophilic and hydrophobic bile acids exhibit different cytotoxicities through cytolysis, interleukin-8 synthesis and apoptosis in the intestinal epithelial cell lines. IEC-6 and Caco-2 cells. Scand J Gastroenterol 2001;36(5):533-539.

34. Sokol RJ, Devereaux M, Khandwala R et al. Evidence for involvement of oxygen free radicals in bile acid toxicity to isolated rat hepatocytes. Hepatology 1993;17(5):869-881.

35. Sokol RJ, Devereaux MW, Khandwala R. Effect of oxypurinol, a xanthine oxidase inhibitor, on hepatic injury in the bile duct-ligated rat. Pediatr Res 1998;44(3):397-401.

36. Yerushalmi B, Dahl R, Devereaux MW et al. Bile acid-induced rat hepatocyte apoptosis is inhibited by antioxidants and blockers of the mitochondrial permeability transition. Hepatology 2001;33(3):616-626.
37. Patel T, Gores GJ. Inhibition of bile-salt-induced hepatocyte apoptosis by the antioxidant lazaroid U83836E. Toxicol Appl Pharmacol 1997;142(1):116-122.
38. Sokol RJ, Winklhofer-Roob BM, Devereaux MW et al. Generation of hydroperoxides in isolated rat hepatocytes and hepatic mitochondria exposed to hydrophobic bile acids. Gastroenterology 1995;109(4):1249-1256.
39. Sokol RJ, Straka MS, Dahl R et al. Role of oxidant stress in the permeability transition induced in rat hepatic mitochondria by hydrophobic bile acids. Pediatr Res 2001;49(4):519-531.
40. Sokol RJ, Devereaux M, Khandwala RA. Effect of dietary lipid and vitamin E on mitochondrial lipid peroxidation and hepatic injury in the bile duct-ligated rat. J Lipid Res 1991; 32(8):1349-1357.
41. Sokol RJ, McKim JM, Jr., Goff MC et al. Vitamin E reduces oxidant injury to mitochondria and the hepatotoxicity of taurochenodeoxycholic acid in the rat. Gastroenterology 1998; 114(1):164-174.
42. Shivaram KN, Winklhofer-Roob BM, Straka MS et al. The effect of idebenone, a coenzyme Q analogue, on hydrophobic bile acid toxicity to isolated rat hepatocytes and hepatic mitochondria. Free Radic Biol Med 1998;25(4-5):480-492.
43. Vaux DL, Cory S, Adams JM. Bcl-2 gene promotes haemopoietic cell survival and cooperates with c-myc to immortalize pre-B cells. Nature 1988;335(6189):440-442.
44. Adams JM, Cory S. The Bcl-2 protein family: Arbiters of cell survival. Science 1998; 281(5381):1322-1326.
45. Harada K, Iwata M, Kono N et al. Distribution of apoptotic cells and expression of apoptosis-related proteins along the intrahepatic biliary tree in normal and non-biliary diseased liver. Histopathology 2000;37(4):347-354.
46. Stahelin BJ, Marti U, Zimmermann H et al. J. The interaction of Bcl-2 and Bax regulates apoptosis in biliary epithelial cells of rats with obstructive jaundice. Virchows Arch 1999;434(4):333-339.
47. Im EO, Choi YH, Paik KJ et al. Novel bile acid derivatives induce apoptosis via a p53-independent pathway in human breast carcinoma cells. Cancer Lett 2001;163(1):83-93.
48. Marzo I, Brenner C, Zamzami N et al. Bax and adenine nucleotide translocator cooperate in the mitochondrial control of apoptosis. Science 1998;281(5385):2027-2031.
49. Shimizu S, Shinohara Y, Tsujimoto Y. Bax and Bcl-xL independently regulate apoptotic changes of yeast mitochondria that require VDAC but not adenine nucleotide translocator. Oncogene 2000; 19(38):4309-4318.
50. Shimizu S, Narita M, Tsujimoto Y. Bcl-2 family proteins regulate the release of apoptogenic cytochrome c by the mitochondrial channel VDAC. Nature 1999;399(6735):483-487.
51. Scaffidi C, Fulda S, Srinivasan A et al. Two CD95 (APO-1/Fas) signaling pathways. Embo J 1998; 17(6):1675-1687.
52. Li H, Zhu H, Xu CJ et al. Cleavage of BID by caspase 8 mediates the mitochondrial damage in the Fas pathway of apoptosis. Cell 1998;94(4):491-501.
53. Luo X, Budihardjo I, Zou H et al. Bid, a Bcl2 interacting protein, mediates cytochrome c release from mitochondria in response to activation of cell surface death receptors. Cell 1998; 94(4):481-490.
54. Eskes R, Desagher S, Antonsson B et al. Bid induces the oligomerization and insertion of Bax into the outer mitochondrial membrane. Mol Cell Biol 2000;20(3):929-935.
55. Wei MC, Lindsten T, Mootha VK et al. tBID, a membrane-targeted death ligand, oligomerizes BAK to release cytochrome c. Genes Dev 2000;14(16):2060-2071.
56. Sodeman T, Bronk SF, Roberts PJ et al. Bile salts mediate hepatocyte apoptosis by increasing cell surface trafficking of Fas. Am J Physiol Gastrointest Liver Physiol 2000;278(6):G992-999.
57. Faubion WA, Guicciardi ME, Miyoshi H et al. Toxic bile salts induce rodent hepatocyte apoptosis via direct activation of Fas. J Clin Invest 1999;103(1):137-145.
58. Takikawa Y, Miyoshi H, Rust C et al. The bile acid-activated phosphatidylinositol 3-kinase pathway inhibits Fas apoptosis upstream of bid in rodent hepatocytes. Gastroenterology 2001; 120(7):1810-1817.
59. Eskes R, Antonsson B, Osen-Sand A et al. Bax-induced cytochrome C release from mitochondria is independent of the permeability transition pore but highly dependent on Mg2+ ions. J Cell Biol 1998;143(1):217-224.
60. Wei MC, Zong WX, Cheng EH et al. Proapoptotic BAX and BAK: a requisite gateway to mitochondrial dysfunction and death. Science 2001;292(5517):727-730.
61. Fernandez-Checa JC, Kaplowitz N, Garcia-Ruiz C et al. GSH transport in mitochondria: defense against TNF-induced oxidative stress and alcohol-induced defect. Am J Physiol 1997; 273(1 Pt 1):G7-17.
62. Higuchi H, Miyoshi H, Bronk SF et al. Bid antisense attenuates bile acid induced apoptosis and cholestatic liver injury, J Pharmacol Exp Ther 2001; 299:866-873.

CHAPTER 11

Hepatic Drug Metabolism

Hiroshi Suzuki and Yuichi Sugiyama

Summary

The liver plays an important role in the metabolism of xenobiotics and the metabolic reactions are mediated by many different kinds of enzymes. These enzymes mediate the detoxification in most cases, but are sometimes responsible for the bioactivation of xenobiotics. In this chapter, the pharmacological role of clinically important enzymes such as cytochrome P450s, UDP-glucuronosyl transferases, sulfotransferases and glutathione-S-transferases will be summarized. As discussed in detail, drug metabolizing enzymes are characterized by (1) their broad substrate specificity, (2) the presence of many isozymes with overlapping substrate specificity and (3) the presence of interindividual differences in the metabolic activity. Since the plasma concentrations and the extent of pharmacological actions are generally markedly affected by the metabolic activity, the interindividual difference in drug metabolism is one of the most important factors in adapting treatment to suit individual patients. The molecular mechanisms accounting for such interindividual differences are also discussed.

Introduction

Hepatic metabolism is involved in the elimination of many kinds of drugs from the body. Drug metabolism can be classified into phase I and II reactions: the former includes oxidation, reduction and hydrolysis, whereas the latter consists of the conjugation. There have been extensive molecular biological studies of the enzymes involved in these reactions and it is now possible for us to use recombinant human enzymes to quantitatively predict in vivo drug disposition, along with possible drug-drug interactions.[1,2] It has also been found that there are genetic polymorphisms associated with some of these enzymes, which results in inter-individual differences in the action and/or adverse effects of drugs.[3-7] Inter-individual differences in the amounts of the different enzymes may also be related to inter-individual differences in the extent of enzyme induction, the mechanism of which has also been the subject of extensive investigations.[8] In this chapter, the pharmacological role of drug metabolizing enzymes will be summarized.

Phase I Metabolism

The enzymes which are responsible for phase I metabolism include cytochrome P450 (CYP) enzymes, flavin-containing monooxygenases, alcohol and aldehyde dehydrogenases, monoamine oxidases, NADPH-P450 reductase, aldehyde and ketone reductases, esterases, NAD(P)H:quinone oxidoreductase (NQO), epoxide hydrolase, β-glucuronidase, sulfatase and peptidases. Although carboxylesterases are important in the metabolic activation of many ester-type prodrugs (such as angiotensin-converting enzyme inhibitors and irrinotecan),[9] the CYP enzymes have been studied most extensively, since various kinds of clinically used drugs are metabolized by CYP family proteins. Drug metabolizing CYP enzymes are classified into

Molecular Pathogenesis of Cholestasis, edited by Michael Trauner and Peter L.M. Jansen.
©2004 Eurekah.com and Kluwer Academic / Plenum Publishers.

four groups (CYP1, 2, 3 and 4), and each group consists of a subfamily of proteins, whereas CYP7A is responsible for the formation of bile salts from cholesterol. Among the drug metabolizing CYP enzymes, some of the CYP1, 2 and 3 family members are expressed in the liver. CYP enzymes are located in the endoplasmic reticulum membrane and their sites of metabolic reaction face on to the cytosolic side.

CYP1 Family

Among the CYP1 family enzymes, CYP1A2 is the most important hepatic enzyme for drug metabolism, since approximately 4 % of P450 substrate drugs are metabolized by this isozyme, and this isozyme accounts for approximately 12% of hepatic P450.[10] Drugs which are substrates of CYP1A2 include acetaminophen, phenacetin, caffeine and theophylline (Table 1).[10] CYP1A2 polymorphism is related to the toxicity of chlorpromazine (tardive dyskinesia).[11] Among these drugs, caffeine may be used as a probe to examine the in vivo function of CYP1A2.[10]

In addition, the CYP1 family proteins are also involved in the metabolic activation of toxins and/or carcinogens (Table 1).[12,13] For example, CYP1A1 mediates the oxygenation of polycyclic aromatic hydrocarbons (PAHs) contained in cigarette smoke, which results in the formation of reactive electrophiles.[12] In addition, environmental toxins such as 2,3,7,8-tetrachlorodibenzo-p-dioxin (TCDD) are high affinity substrates of CYP1A1.[13] Benzo(a)pyrene is also activated by CYP1A1,[13] while CYP1A2 is involved in the metabolic activation of arylamine carcinogens (such as 4-aminobiphenyl which is a major component of tobacco smoke) and heterocyclic amine food mutagens (such as 2-amino-1-methyl-6-phenylimidazo[4,5-b]pyridine), along with aflatoxin B_1.[12,13]

CYP1A1, 1A2 and 1B1, along with some phase I (such as NQO-1 and aldehyde dehydrogenase 3) and phase II enzymes (such as glutathione-S-transferase Ya subunit and UDP-glucuronosyl transferase 1A6 and 1A9), undergo enzyme induction controlled by ligands of the aryl hydrocarbon receptor (AhR).[14] In particular, the promoter of CYP1A1 is under the control of its enhancer, this enzyme exhibits low constitutive expression and is induced by AhR ligands such as TCDD and 3-methyl cholanthrene.[12,13] In addition, both CYP1A1 and 1A2 are induced by tobacco smoking.[12,13] Under physiological conditions, AhR is located in the cytosol as a protein complex with cytosolic proteins such as hsp90 and AhR-interacting protein.[14] In the presence of inducers, AhR dissociates from these cytosolic proteins and enters the nucleus where it binds to AhR nuclear translocator (ARNT).[14] The AhR/ARNT heterodimer binds to the xenobiotic responsible element (XRE; 5'-dTNGCGTG-3') to promote gene transcription. It has also been shown that AhR repressor is also induced by AhR/ARNT heterodimer, which in turn suppresses induction by sequestrating ARNT.[15] The toxicity of TCDD in AhR (-/-) mice is significantly lower than in wild type mice, due to the lack of induction of CYP1A1.[14]

CYP2 Family

Among the CYP2 family enzymes, the subfamily CYP2C9 and 19 in particular are clinically important, since approximately 11% of P450 substrate drugs are metabolized by these isozymes (Table 1).[10] CYP2C9/19 accounts for approximately 20% of hepatic P450.[10] Clinically significant polymorphisms have been reported for both CYP2C9 and 19; for CYP2C9, the polymorphism is associated with the appearance of adverse effects of warfarin (haemorrhage), tolbutamide (hypoglycemia) and phenytoin (mental confusion etc).[4,11,16,17] In addition, poor CYP2C9 metabolizers exhibit a lower antihypertensive effect of losartan due to reduced metabolic activation of this prodrug.[4,17] (S)-Warfarin 7-hydroxylation or diclofenac 4'-hydroxylation has been used to probe the in vivo function of CYP 2C9.[10] The polymorphism in CYP2C19 is also related to the toxicity of diazepam (prolonged sedation) and mephenytoin (neurotoxicity).[4,11,16,17] It has also been reported that the pharmacological effect of omeprazole also depends on the metabolic activity of CYP2C19.[4,16,17] Metabolic activation of proguanil also depends on the polymorphism of CYP2C19.[17] (S)-Mephenytoin is used as a marker to probe the in vivo function.[10] The role of CYP2C19 polymorphism in drug metabolism/oral clearance has been extensively reviewed by Wedlund.[18]

Table 1. Typical substrates for human cytochrome P450 (CYP) enzymes

CYP1A1
Benzo[a]pyrene, 2,3,7,8-Tetrachlorodibenzo-*p*-dioxin
CYP1A2
Acetaminophen, Caffeine, Chlorpromazine, Phenacetin, Theophylline
Aflatoxin B_1, 4-Aminobiphenyl, 2-Amino-1-methyl-6-phenylimidazo[4,5-b]pyridine
CYP2C9
Antipyrine, Carbamazepine, Diclofenac, Ibuprofen, Indomethacin, Phenytoin, Tolbutamide, (S)-Warfarin
CYP2C19
Diazepam, Imipramine, (R)-Hexobarbital, (S)-Mephenytoin, Omeprazole, Propranolol, (R)-Warfarin
CYP2D6
Bufuralol, Codein, Debrisoquine, Dextromethorphan, Flecainide, Haloperidol, Imipramine, Propranolol, Spartein
CYP2E1
Acetaminophen, Benzene, Toluene
CYP3A4
Alprazolam, Carbamazepin, Codein, Cyclosporin, Dextromethorphan, Diazepem, Diltiazem, Docetaxel, Erythromycin, Etoposide, Haloperidol, Imipramine, Indinavir, Ivermectin, Lidocaine, Nifedipine, Paclitaxel, Quinidine, Ritonavir, Saquenavir, Simvastatin, Tacrolimus, Terfenadine, Theophylline, Triazolam, Verapamil, Vinblastine

Typical substrates for P450 enzymes were taken from ref. 10.

Moreover, these two enzymes are known to undergo induction by ligands such as phenobarbital and rifampicin, although the precise mechanism for this remains unknown. As far as CYP2C9 induction is concerned, it has recently been suggested that orphan receptors may be involved,[19] as discussed for CYP2B and CYP3A4 in detail. In the case of CYP2B, it has been shown that the induction is mediated by nuclear receptors (Fig. 1).[8] Upon exposure to inducers (such as phenobarbital), constitutive active receptor (CAR) translocates to the nucleus, where a heterodimer is formed with retinoid X receptor (RXR).[8] The heterodimer binds to nuclear receptor binding direct repeat (DR)-4 motifs to activate transcription (Fig. 1).[8] Other co-regulators may be required for the activation of transcription.[8] Direct binding of phenobarbital may not required for the nuclear translocation of CAR and it has been suggested that the multiple phosphorylation / dephosphorylation may be involved in regulating the nuclear translocation and/or activation of transcription.[8]

Approximately 30% of P450 substrate drugs are metabolized by CYP2D6 (Table 1), and this enzyme accounts for approximately 4% of the hepatic P450.[10] Since CYP2D6 is not inducible, inter-individual differences in CYP2D6 activity may result from the presence of a genetic polymorphism.[10] Spartein, debrisoquine, and dextromethorphan are used as marker ligands to detect the in vivo function of CYP2D6.[10] As far as CYP2D6 is concerned, human subjects can be classified as poor metabolizers with a defective allelic function, extensive metabolizers with wild type alleles and ultrarapid metabolizers with multiple genes.[10] This polymorphism is also related to the adverse effects of antiarrhythmics (arrhythmias), β-blockers (bradycardia), tricyclic antidepressants (confusion and cardiotoxicity), opioids (dependence), phenformin (lactic acidosis), perhexiline (neuropathy) and perhexilene (hepatotoxicity), along with the interindividual difference in the metabolic activation of prodrugs such as encainide, codeine and ethylmorphine.[11,17]

CYP2E1 is involved in the metabolism of benzene, alcohols and halogenated solvents, along with some drugs such as acetaminophen (Table 1), and is subject to induction by volatile

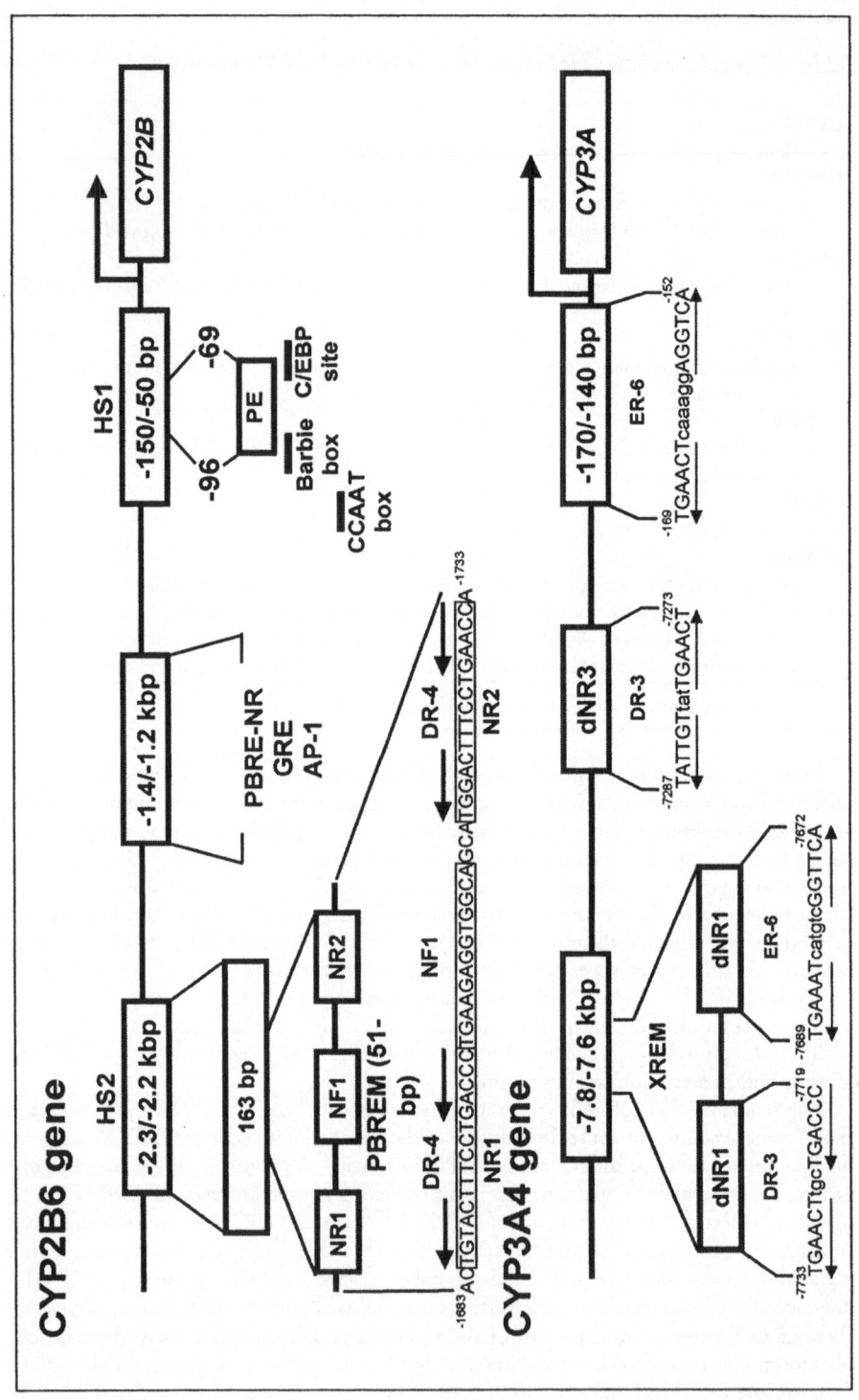

Figure 1. DNA sequence for the upstream regions of human CYP2B6 and 3A4.
For CYP2B6, the phenobarbital-responsive enhancer module (PBREM) is located at 2.2-2.3 kb upstream of the transcription initiation site. This is composed of nuclear receptor (NR)-binding sites (NR1 and 2) and nuclear factor 1 (NF1) binding site. Both NR1 and NR2 are composed of the DR-4 motief, where the imperfect repeated sequences are separated by four nucleotides. In addition to PBREM, glucocorticoid response element (GRE) and a Activator protein 1 (AP-1) site may also be involved in the transcriptional regulation. The proximal promoter contains the Barbie box, CCAAT box and CCAAT-enhancer binding protein (C/EBP) binding site.
For CYP3A4, the xenobiotic-responsive enhancer module (XREM) is located at 7.6-7.8 kb upstream of the transcriptional initiation site. XREM contains two NR binding sites which are separated by 29 bp. NR1 is composed of the DR-3 motief, where the imperfect repeated sequences are separated by three nucleotides, whereas NR2 is composed of the ER-6 motief, where the imperfect everted repeat sequences are separated by six nucleotides. In addition to XREM, DR-3 and ER-6 are located at -7273/-7287 and -152/-169, respectively.
Taken from ref. 8.

organic compounds such as ethanol and acetone due to pre-transcriptional, transcriptional and post-transcriptional regulation.[12,13,20] Chlorzoxazone is used to probe the in vivo function of CYP2E1.[10] CYP2E1 is also known to be responsible for the toxicity of acetaminophen (Fig. 2). At low doses, most of acetaminophen is metabolized by conjugation with sulfate and glucuronic acid (Fig. 2). However, at very high (toxic) concentrations, acetaminophen undergoes metabolism mediated by CYP enzymes to form a reactive electrophilic metabolite (Fig. 2). Although this reactive metabolite can be conjugated with glutathione (GSH), this metabolite can also bind to cellular components to induce toxicity when the cellular GSH concentration is reduced (Fig. 2). Although both CYP1A2 and 2E1 are responsible for the metabolic activation of acetaminophen, the role of CYP2E1 was found to be of great importance by examining the hepatic toxicity in gene knock out mice where acetaminophen toxicity was found to be in the order, wild type mice > CYP1A2 (-/-) mice >> CYP2E1 (-/-) >> CYP1A2/2E1 (-/-) mice.[12,13,21]

CYP2A6 is also polymorphic and poor metabolizers exhibit higher nicotine levels after smoking.[22] Like CYP2D6, duplicated genes have been observed for CYP2A6.[17]

CYP3 Family

CYP3A4 is the most important isozyme, since approximately 52% of P450 substrate drugs are metabolized by it (Table 1), and it accounts for approximately 30% of the enzymes present in the liver.[10] The range of inter-individual differences in CYP3A4 activity and expression level can be as much as 60-fold.[10] In addition, its polymorphism is responsible for the adverse effect of anti-leukaemic agents.[11] Nifedipine, erythromycin, alprazolam, and dextrometorphan are used as probes to study the in vivo function of CYP3A4.[10]

CYP3A4 is susceptible to induction. Typical potent inducers include rifampicin and pregnenolone 16α-carbonitrile (PCN), and in addition, various compounds including phenobarbital, taxol, clotrimazole, hyperforin and ritonavir are also known inducers (Fig. 1).[8,23,24] The induction of human CYP3A4 largely depends on pregnane X receptor (PXR), a nuclear receptor.[8] In contrast to CAR, PXR is constitutively located in the nucleus and the inducers directly bind to PXR.[8] The PXR/RXR heterodimer binds to the enhancer region to stimulate the transcription of the CYP3A4 gene (Fig. 1).[8] In the enhancer region of CYP3A4, the everted repeat (ER)-6 motif is located in the proximal promoter regions (-169 to 152 bp) and its distal element (-7836 to -7607 bp) contains the xenobiotic responsive module (XREM) sequence which includes the DR-3 and ER-6 motifs (Fig. 1).[8] These three sites may be essential for induction by phenobarbital.[8] Since the inducers directly bind to PXR, the in vitro binding studies of PXR may be used to predict the inducibility of compounds under in vivo conditions.[25] It has also been demonstrated that PXR acts as a bile acid sensor.[26,27] The highly hepatotoxic lithocholic

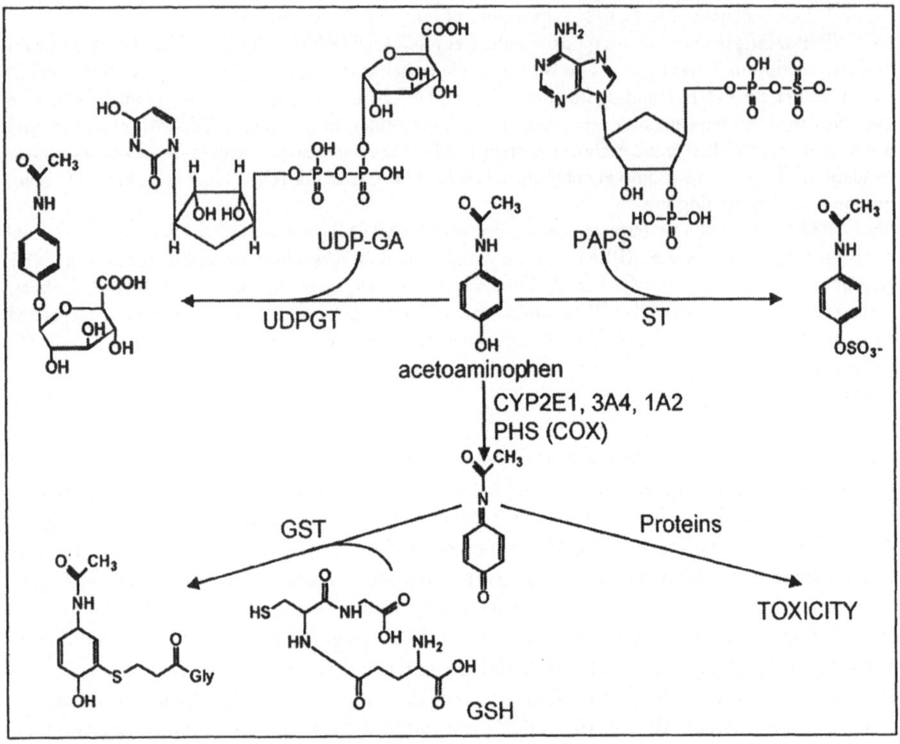

Figure 2. Metabolic pathway of acetaminophen. At therapeutic doses, acetaminophen is metabolized by glucuronide and sulfate conjugation. At higher doses, a reactive electrophilic metabolite is formed with the aid of CYP2E1 and 1A2. Although this reactive metabolite can be conjugated with glutathione (GSH), this metabolite can also bind to cellular components to induce toxicity when the cellular GSH concentration is reduced. Modified from ref. 10.

acid induces CYP3A via activation of PXR and, in turn, lithocholic acid undergoes 6-hydroxy-lation by CYP3A.[26,27] The interspecies differences in the extent of induction of CYP3A en-zymes results from differences in the binding affinity for PXR.[28]

The inducers of CYP2B6 and CYP3A4 overlap each other. Although some controversial reports have appeared, it seems likely that many inducers activate both CAR and PXR.[29,30] After activation, either CAR or PXR, or both, induce the CYP2B6 and/or CYP3A4 genes depending on the experimental conditions.[8,29,30] It has also been demonstrated that dexam-ethasone is involved in the induction of CYP3A4. Some reports suggest that, at submicromolar concentrations, dexamethasone induces the expression of PXR and/or CAR which, in turn, activates the transcription of CYP3A4.[31,32] At higher concentrations (> 10 μM), dexamethazone may induce the expression of CYP3A4 by activating PXR due to direct binding to this receptor.[32]

Both CAR and PXR may also be involved in the constitutive expression of CYP3A4. In human liver, it has been reported that the levels of CYP3A4 mRNA are significantly correlated with the levels of both CAR and PXR mRNA levels.[32] This observation is consistent with the report that the expression level of CYP3A is markedly reduced in RXR (-/-) mouse liver.[33]

Phase II Metabolism

The conjugation reaction is classified as phase II metabolism. Phase II metabolism includes conjugation with glucuronic acid, sulfate, gutathione, glycine, taurine, and the acetyl and me-thyl groups. The SNPs of N-acetyltransferase 2 and thiopurine S-methyltransferase have been

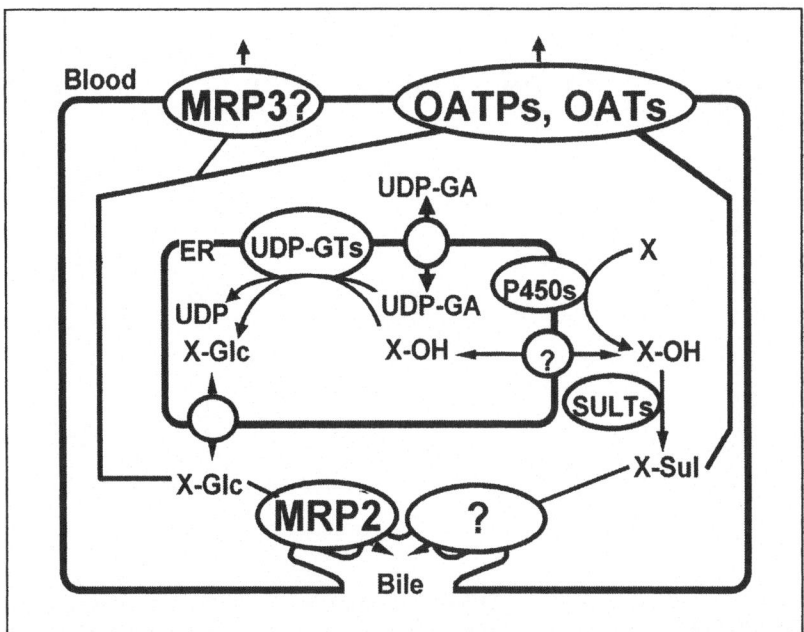

Figure 3. Schematic diagram illustrating the hepatocellular detoxification mechanism. Drug transporters and drug metabolizing enzymes act synergistically to detoxificate the drug molecules. See text for the detailed explanation.

studied extensively as a factor to account for the interindividual variation in the detoxification of isoniazide and 6-mercaptopurine, respectively.[34,35]

In general, many of xenobiotic metabolites, which have had polarized groups introduced by phase I metabolism, undergo phase II metabolism. Phase II metabolites formed in hepatocytes are excreted into the bile and into the blood circulation across the bile canalicular membrane and sinusoidal membranes, respectively (Fig. 3). Although the biliary excretion of glucuronide and glutathione-conjugates and sulfated bile acids involves the action of multidrug resistance associated protein 2 (MRP2),[36,37] the mechanism for the biliary excretion of sulfated conjugates of non-bile acid compounds remains unknown (Fig. 3). The excretion of conjugates across the sinusoidal membrane may be mediated by organic anion transporting polypeptides (OATPs) and/or organic anion transporters (OATs) (Fig. 3), since these transporters may act bidirectionally. Indeed, the cellular uptake by some OATP family proteins is coupled to the efflux of intracellular reduced glutathione and/or bicarbonate ion.[38] In addition, MRP3, which transports MRP2 ligands and monovalent bile acids, may also be involved in the sinusoidal efflux.[38]

Glucuronidation

Glucuronidation is mediated by UDP-glucuronosyl transferases (UGTs). UGTs are located in the endoplasmic reticulum membrane and their sites of metabolism face the inside of the endoplasmic reticulum (ER) (Fig. 3).[39] UGTs require UDP-glucuronic acid as a cofactor, which may be transported into ER from the cytosol via a specific mechanism (Fig. 3).[40] It is not known how the substrates, including the metabolites of phase I reactions, enter the ER and how the glucuronide conjugates can leave the ER,[41] although it has been hypothesized that the efflux of glucuronide conjugates from the ER is coupled with the influx of UDP-glucuronic acid into the ER (Fig. 3).[42] UGTs are classified into two families, UGT1 and 2.[39,43,44]

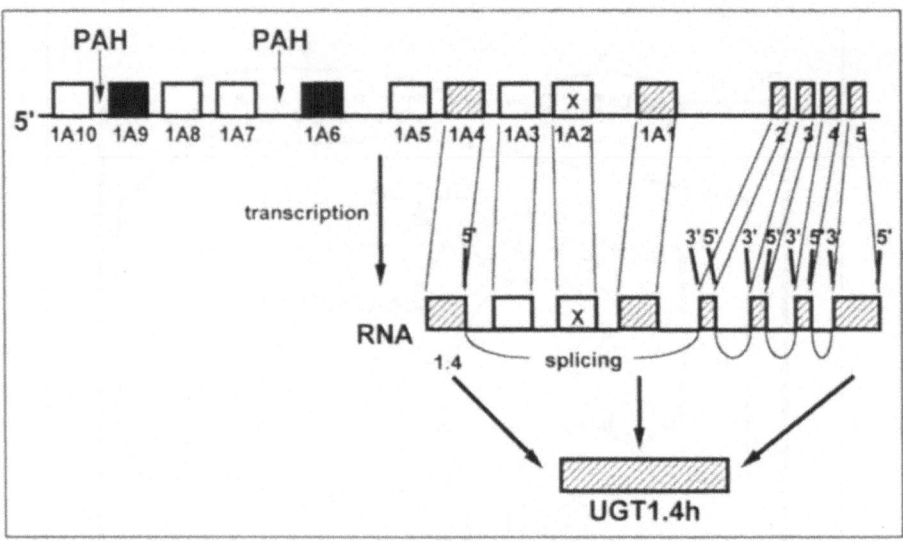

Figure 4. Structure of human UGT1A gene. Four 3' exons are flanked with 12 cassette exons, each encodes one of the UGT1A first exons and, in addition, the first exons are under the control of individual promoters. Eight active gene transcripts (UGT1A1, 1A3, 1A4, 1A6, 1A7, 1A8, 1A9 and 1A10) share the same carboxy 245 amino acid sequence, which is encoded by exons 2-5. UGT1A6 and 1A9 are induced by PAH. Modified from refs. 43 and 44.

Genetic Organization of UGTs and Inherited Diseases

A schematic diagram illustrating the human UGT1A locus is shown in Figure 4. Four 3' exons are flanked with 12 cassette exons, each encodes one of the UGT1A first exons and, in addition, the first exons are under the control of individual promoters.[43,44] Eight active gene transcripts (UGT1A1, 1A3, 1A4, 1A6, 1A7, 1A8, 1A9 and 1A10) have been reported in humans.[39,43,44] They share the same carboxy 245 amino acid sequence, which is encoded by exons 2-5.[39,44] Among them, UGT1A1, 1A3, 1A4, 1A6 and 1A9 are expressed in the human liver.[39,44] UGT1A6 and 1A9 are induced by PAH, due to the presence of XRE in their 5'-flanking regions.[43] The high activity of glucuronidation of paracetamol and propranolol, substrates of UGT1A6 and UGT1A9, respectively, in cigarette smokers is consistent with the fact that these two isoforms are induced by the PAH present in cigarette smoke.[39] These two isoforms are also involved in the glucuronidation of benzo(a)pyrene.[39] In rats, the same organization of UGT1A has been reported.[45] In Gunn rats, which have elevated plasma concentrations of unconjugated bilirubin, a nonsense mutation has been found in the common exon 4.[45] Therefore, the metabolic activity of all UGT1A family enzymes is abolished.[45]

Since UGT1A1 is involved in the glucuronidation of bilirubin, its hereditary defect results in the acquisition of inherited unconjugated hyperbilirubinemia.[46] In Crigler-Najjar syndrome type I, no UGT1A1 activity toward bilirubin is observed and, therefore, patients require immediate orthotopic liver transplantaion.[46] In contrast, in Crigler-Najjar syndrome type II, the UGT1A1 activity toward bilirubin is maintained at least 10% of the normal activity.[44] Patients with type II syndrome are treated by induction therapy or phototherapy.[44] Phenobarbital is often used in the induction therapy, due to the inducible nature of UGT1A1 confirmed by the presence of the CAR/RXR binding sequence in its 5'-flanking region.[47] The UGT1A1 activity in Gilbert syndrome is 60-70% of the normal activity, and therapy is not necessarily required in young adults unless the activity is further reduced under certain conditions (such as infection, fasting, psychological stress or reduced physical activity).[44]

Table 2. Typical substrates for UGTs and SULTs expressed in the human liver

UGT1A1

Bilirubin, Estradiol (3-OH), all-trans-Retinoic acid, SN-38

UGT1A5/1A6

(R)/(S)-Naproxen

UGT1A9

Non-steroidal anti-inflammatory drugs, furosemide

UGT2B7

Non-steroidal anti-inflammatory drugs, Morphine (3-OH/6-OH), Buprenorphine, 17β Estradiol (17β-OH), all-trans-Retinoic acid, Valproic acid

UGT2B15

Phenolphthalein, Dihydrotestosterone

UGT2B17

C19 steroids (Androsterone, Dihydrotestosterone, Testosterone)

SULT1A1

4-Nitrophenol, 1-Napththol, 2-Naphthol, Acetaminophen, many monocyclic phenols, Minoxidil, 17β Estradiol, Estrone

SULT1A2

2-Naphthol > Minoxidil

SULT1B1

Naphthol > 4-Nitrophenol

SULT1E1

17β Estradiol, Estrone, Pregnenolone, 17-Ethinyl-17β estradiol, Dehydroepiandrosterone, 1-Naphthol

SULT2A1

Dehydroepiandrosterone > Epiandrosterone > Androsterone > Testosterone > 17β Estradiol; many bile salts; Pregrenolone > 17-Ethinyl-17β estradiol > Cortisol; Minoxidil

Typical substrates for UGTs and SULTs were cited from refs. 39 and 50.

In contrast to UGT1, UGT2 family enzymes are encoded by independent genes. In humans, the presence of six members (UGT2B4, 2B7, 2B10, 2B11, 2B15 and 2B17) has been reported, all of them being expressed in the liver.[44] Three isozymes are clustered in the order, UGT2B7-2B4-2B15.[44] Although some UGT2B family enzymes are induced by phenobarbital, the precise mechanism for the tissue-specific regulation of UGT2 isozymes has not yet been clarified.[43]

Substrate Specificity of UGTs

The substrate specificity of human UGTs expressed in the liver is summarized in Table 2. The major UGT species in human liver is UGT1A1, which is responsible for the glucuronidation of bilirubin[39,43,44,46] and xenobiotics such as SN-38,[48] an active metabolite of an anticancer agent (irrinotecan). The polymorphisms in the UGT1A1 are summarized by Tukey et al.[44] TA insertion into the wild type A(TA)6TAA promoter has been observed frequently in patients with Gilbert syndrome.[44,46] It has been suggested that this TA insertion results in reduced UGT1A1 activity; indeed, the transcriptional activity of this A(TA)7TAA promoter is only 10-33% of the wild type promoter.[46] It has also been suggested that this mutation in the UGT1A1 promoter may be related to the incidence of the lethal adverse effect of irrinotecan.[48,49]

UGT2B7 is largely responsible for the glucuronidation of morphine.[39] Since the pharmacological activity of morphine-6-glucuronide is more potent than that of morphine, the interindividual differences in the analgesic effect after administration of morphine may be accounted for by interindividual differences in UGT2B7 activity.[39] Although SNPs in UGT2B7 largely affect the metabolic activity toward buprenorphine, no difference was observed for morphine.[39] Since interindividual differences are observed in the expression level of UGT2B7 mRNA in the human cerebellum, it is possible that this difference may affect the pharmacological activity of morphine.[39] UGT2B7 may also make a significant contribution to the formation of bioactive retinoid glucuronides.[39]

Many human UGT isoforms are involved in the D-ring glucuronidation of estrogens (estriol, estrone and 17β-estradiol).[39] Although it has been suggested that UGT1A1 and UGT2B7 may play important roles in 3-O-glucuronidation and 17-O-glucuronidation of these endogenous substrates, the quantitative contribution of each isoform remains unclear.[39] UGTs are also involved in the formation of electrophilic metabolites.[39] For example, N-hydroxy-2-acetylaminofluorene (N-hydroxy-2-AAF) is activated to the N-O-glucuronide of 2-AAF which can bind to guanosine residues, although it exhibits only weak tumorgenicity in rats.[39] In addition, the acyl glucuronides, formed by esterification of carboxylic acids, are also electrophilic.[39] Glucuronidation of carboxylic acid moieties is observed for many drugs including non-steroidal anti-inflammatory drugs (NSAIDs), as well as clofibrate and valproic acid.[39] Although many UGTs catalyze the formation of acyl glucuronides, UGT2B7 may be one of the major contributors to the glucuronidation of carboxylic acid drugs.[39,44]

Sulfation

Sulfation of xenobiotics and small endogenous compounds is mediated by the cytosolic sulfotransferases.[50] Sulfotransferases (SULTs) are responsible for the transfer of the sulfo moiety from 5'-phosphoadenosine-3'-phosphosulfate to the nucleophilic groups of their substrates.[50] Their systemic nomenclature has not established yet. In humans, 10 sulfotransferase genes have been identified: SULT1A1, 1A2, 1A3, 1B1, 1C1, 1C2, 1E1, 2A1, 2B1a/1b and 4A1, where 2B1a and 1b are synthesized by using the alternative first exons.[50] Among them, SULT1A1, 1A2, 1B1, 1E1 and 2A1 have been reported to be expressed in the human liver.[50] The molecular mechanisms for the transcription have not yet been characterized extensively.[51]

The substrate specificity of hepatic SULTs overlap.[50] For example, although 17β estradiol is a high affinity substrate of 1E1, this compound is also efficiently metabolized by 1A1 and 2A1 at higher concentrations.[50] SULT 1E1 also accepts estrone as a high affinity substrate.[50] Iodothyronines are conjugated by 1A1 and 1B1, and efficiently by 1E1.[50] Dehydroepiandrosterone is conjugated by 2A1 and 1E1, and bile salt sulfation is mediated by 2A1.[50] SULT 1A1 has been shown to be responsible for the conjugation of 4-nitrophenol, 1-naphthol, 2-naphthol, acetaminophen, minoxidil and monocyclic phenols with high affinity.[50]

Sulfation is sometimes associated with the formation of reactive products.[50,52] For example, benzylic alcohols, aromatic hydroxylamines, aromatic hydroxamic acids, nitroalkanes and some polycyclic aromatic compounds such as 1-hydroxymethylpyrene and 7-hydroxy-7,8,9,10-tetrahydrobenzo[a]pyrene.[50,52] Polymorphisms have been reported for SULT1A1, 1A2 and 2A1.[50,51] The mutation in SULT 1A1 is associated with that in SULT 1A2.[50,51]

Glutathione Conjugation

Glutathione-S-transferases (GSTs) mediate the conjugation of electrophilic substrates with glutathione. The substrates include arene oxides, unsaturated carbonyls and organic halides, and the precise list of substrate specificity of each isozyme has been given by Strange et al.[53] The conjugation reaction mediates the detoxification and, sometimes, the bioactivation of substrates.[54,55] GSTs can be divided into two superfamilies. The first are the soluble and dimeric enzymes.[56,57] Among them, the Alpha, Mu, Pi, Sigma, Theta, Zeta and Omega families are located in the cytosol, whereas Kappa is located in the mitochondria.[56,57] The second are the trimeric microsomal enzymes primarily involved in the metabolism of arachnoid acids.[56,57] Polymorphisms have been found in many GST genes.[53] Although it has been suggested that

GSTP1 polymorphism may be related to the significantly reduced risk of asthma and GSTT1 null may be associated with basal cell carcinoma, GSTs are not susceptibility factors for tobacco- or diet-associated cancers.[57]

The mechanism for the induction of GSTs has also been studied. In the presence of electrophilic inducers, Nrf2 dissociates from Keap 1, a cytosolic protein, and locates to the nucleus.[54] Nrf2, together with MafK,[58] binds to an antioxidant responsible element (ARE), which is located in the promoter region of some GSTs, including mouse GST A1 and human GST Pi.[54] The consensus sequence of ARE, 5'-TGACNNNGC-3', is shared by the TRE (Tumor Promoting Activity Responsible Element) sequence where AP-1 binds.[58,59] It has been demonstrated that, in Nrf2 (-/-) mice, the induction of GST A1 is markedly reduced, whereas the constitutive expression level of GST A1 is reduced to approximately 50% that of normal mice.[60,61] This may be accounted for by the hypothesis that the constitutive expression of GST A1 may also be mediated by factors such as AP-1 which bind to the ARE/TRE sequence.[60,61]

Nrf2 is also involved in the expression of Ugt1a6 and the heavy chain of γ-glutamylcysteine synthetase, a rate-determining enzyme for the synthesis of GSH.[62] Indeed, in Nrf2 (-/-) mice, the expression level of Ugt1a6 and the intracellular GSH concentration were reduced compared with the values in the wild type.[62] Severe acetaminophen toxicity has been observed in Nrf2(-/-) mice, resulting from the increased amount of N-acetyl-*p*-benzoquinone-imine (NAPQI), a reactive metabolite of acetaminophen, resulting from the reduced formation of acetaminophen glucuronide and the glutathione conjugation of NAPQI.[62]

Drug Metabolism under Cholestatic Conditions

Although Bertolotti et al[63] demonstrated that the bile acid synthesis, but not hepatic CYP7A1 expression, is suppressed in obstructive cholestasis in humans, alterations in the expression level of human hepatic drug metabolizing enzymes have been reported only for severe disease states such as cirrhosis. Using surgically dissected human cirrhotic liver specimens, the protein levels of CYP enzymes, along with the metabolism of marker substrates, have been compared among the control and the cholestatic (CHOL) and non-cholestatic cirrhotic (HC) liver portions. It has been demonstrated that, compared to the control, CYP1A2 was reduced in both CHOL and HC, and CYP2C and 2E1 were reduced only in CHOL, whereas CYP3A was reduced only in HC.[64] In bile duct ligated experimental animals, a reduction in the expression of many drug metabolizing CYP enzymes has been reported.[65-68] In addition, alterations in enzyme expression in experimental animals[69] and in drug disposition in humans[70] after the administration of lipopolysaccharide (LPS) have been demonstrated. The molecular mechanisms for the reduced expression of CYP enzymes have been studied in rats. It has been suggested that, following stimulation by LPS, cytokines (such as interleukin (IL) 1, 2 and 6, tumor necrosis factor α (TNFα) and transforming growth factor β) are released from Kuppfer cells, which subsequently interact with hepatocytes.[69,71-74] Indeed, the administration of these cytokines to rats results in the reduced expression of CYP enzymes.[69,71-74] However, since the response to LPS is also found in TNFα receptor knock out and IL-6 knock out mice,[74-76] it is possible that some complex mechanism(s) is involved in the acute phase response. The mechanism(s) by which cytokines reduce the transcription of CYP enzymes is not fully understood, irrespective of the recent extensive studies.[77-80] Although it is controversial, a contribution by nitric oxide has also been suggested.[71,74]

Concluding Remarks

In this chapter, the characteristics of drug metabolizing enzymes have been summarized. It has been demonstrated that the interindividual differences in drug disposition can be accounted for by considering differences in metabolic activity due to their SNPs, along with differences in the expression level of enzymes. Moreover, although the molecular mechanism(s) are not yet fully understood, the expression levels of enzymes are also altered in disease states. Since the drug concentrations in plasma and, consequently, the extent of the pharmacological action and/or adverse effect of drugs are markedly affected by the activity of drug enzymes, an appropriate dose adjustment is needed for the efficient and safe treatment of patients receiving drugs that have narrow therapeutic ranges.

References

1. Ito K, Iwatsubo T, Kanamitsu S et al. Quantitative prediction of in vivo drug clearance and drug interactions from in vitro data on metabolism, together with binding and transport. Ann Rev Pharmacol Toxicol 1998; 38:461-499.
2. Ito K, Iwatsubo T, Kanamitsu S et al. Prediction of pharmacokinetic alterations caused by drug-drug interactions:metabolic interaction in the liver. Pharmacol Rev 1998; 50:387-412.
3. Evans WE, Johnson JA. Pharmacogenomics: The inherited basis for interindividual differences in drug response. Ann Rev Gen Hum Genet 2001; 2:9-39.
4. Ingelman-Sundeberg M. Pharmacogenetics: An opportunity for a safer and more efficient pharmacotherapy. Int J Med 2001; 250:186-200.
5. Lin JH, Lu AYH. Interindividual variability in inhibition and induction of cytochrome P450 enzymes. Ann Rev Pharmacol Toxicol 2001; 41:535-567.
6. Phillips KA, Veenstra DL, Oren E et al. Potential role of pharmacogenomics in reducing adverse drug reactions:a systemic review. JAMA 2001; 286:2270-2279.
7. Xie HG, Kim RB, Wood AJ et al. Molecular basis of ethnic differences in drug disposition and response. Ann Rev Pharmacol Toxicol 2001; 41:815-850.
8. Sueyoshi T, Negishi M. Phenobarbital response elements of cytochrome P450 genes and nuclear receptors. Ann Rev Pharmacol Toxicol 2001; 41:123-143.
9. Sato T, Hosokawa M. The mammalian carboxylesterases: from molecules to functions. Annu Rev Pharmacol Toxicol 1998; 38:257-88.
10. Anzenbacher P, Anzenbacherova E. Cytochromes P450 and metabolism of xenobiotics. Cell Mol Life Sci 2001; 58:737-747.
11. Pirmohamed M, Park BK. Genetic susceptibility to adverse drug reactions. Trends Pharm Sci 2001; 22:298-305.
12. Kimura S, Gonzalez FJ. Applications of genetically manipulated mice in pharmacogenetics and pharmacogenomics. Pharmacol 2000; 61:147-153.
13. Ghanayem BI, Wang H, Sumner S. Using cytochrome P-450 gene knock-out mice to study chemical metabolism, toxicity, and carcinogenecity. Toxicol Pathol 2000; 28:839-850.
14. Whitlock JP. Induction of cytochrome P4501A1. Ann Rev Pharmacol Toxicol 1999; 39:103-125.
15. Gu Y-Z, Hogenesch JB, Bradfield CA. The PAS superfamily: Sensors of environmental and developmental signals. Ann Rev Pharmacol Toxicol 2000; 40:519-561.
16. Goldstein JA. Clinical relevance of genetic polymorphisms in the human CYP2C subfamily. Br J Clin Pharmacol 2001; 52:349-355.
17. Ingelman-Sundeberg M. Genetic susceptibility to adverse effects of drugs and environmental toxicants. The role of the CYP family of enzymes. Mut Res 2001; 482:11-19.
18. Wedlund PJ. The CYP2C19 enzyme polymorphism. Pharmacology 2000; 61:174-183.
19. Gerbal-Chaloin S, Daujat M, Pascussi J-M et al. Transcriptional regulation of CYP2C9 gene: Role of glucocorticoid receptor and constitutive androstane receptor. J Biol Chem 2002; 277:209-217.
20. Lucas D, Ferrara R, Gonzales et al. Cytochrome CYP2E1 phenotyping and genotyping in the evaluation of health risks from exposure to polluted environments. Toxicol Lett 2001; 124:71-81.
21. Gonzalez FJ. The use of gene knockout mice to unravel the mechanisms of toxicity and chemical carcinogenesis. Toxicol Lett 2001; 120:199-208.
22. Oscarson M. Genetic polymorphisms in the cytochrome P450 2A6 (CYP2A6) gene:Implications for interindividual difference in nicotine metabolism. Drug Metab Dispos 2001; 29:91-95.
23. Dussault I, Lin M, Hollister K et al. Peptide mimetic HIV protease inhibitors are ligands for the orphan receptor SXR. J Biol Chem 2001 276:33309-33312.
24. Ekins S, Erickson JA. A pharmacophore for human pregnane X receptor ligands. Drug Metab Dispos 2002; 30:96-99.
25. Moore JT, Kliewer SA. Use of the nuclear receptor PXR to predict drug interactions. Toxicol 2000; 153:1-10.
26. Staudinger JL, Goodwin B, Jones SA et al. The nuclear receptor PXR is a lithocholic acid sensor that protects against liver toxicity. Proc Natl Acad Sci USA 2001; 98:3369-3374.
27. Xie W, Radominska-Pandya A, Shi Y, et al. An essential role for nuclear receptors SXR/PXR in detoxification of cholestatic bile acids. Proc Natl Acad Sci USA 2001; 98:3375-3380.
28. LeCluyse EL. Pregrane X receptor: Molecular basis for species differences in CYP3A induction by xenobiotics. Chemico-Biol Int 2001; 134:283-289.

29. Moore LB, Parks DJ, Jones SA et al. Orphan nuclear receptors constitutive androstane receptor and prognane X receptor share xenobiotic and steroid ligands. J Biol Chem 2000; 275:15122-15127.
30. Smirlis D, Muangmoonchai R, Edwards M et al. Orphan receptor promiscuity in the induction of cytochrome P450 by xenobiotics. J Biol Chem 2001; 276:12822-12826.
31. Quattrochi LC, Guzelian PS. CYP3A regulation: From pharmacology to nuclear receptors. Drug Metab Dispos 2001; 29:615-622.
32. Pascussi JM, Drocourt L, Gerbal-Chaloin S et al. Dual effect of dexamethasone on CYP3A4 gene expression in human hepatocytes:Sequential role of glucocorticoid receptor and pregnane X receptor. Eur J Biochem 2001; 268:6346-6357.
33. Wan YJ, An D, Cai Y et al. Hepatocyte-specific mutation establishes retinoid X receptor alpha as a heterodimeric integrator of multiple physiological processes in the liver. Mol Cell Biol 2000; 20:4436-4444.
34. Spielberg SP. N-Acetyltransferases:Pharmacogenetics and clinical consequences of polymorphic drug metabolism. J Pharmacokinet Biopharm 1996; 24:509-519.
35. Krynetski EV, Evans WE. Genetic polymorphism of thiopurine S-methyltransferase:Molecular mechanism and clinical importance. Pharmacol 2000; 61:136-146.
36. Suzuki H, Sugiyama Y. Single nucleotide polymorphisms in Multidrug Resistance Associated Protein 2 (MRP2/ABCC2): Its impact on drug disposition. Adv Drug Deliv Rev 2002; 54:1311-1331.
37. Keppler D, Konig J. Hepatic secretion of conjugated drugs and endogenous substances. Semin Liver Dis 2000; 20:265-272.
38. Suzuki H, Sugiyama Y. Transport of drugs across the hepatic sinusoidal membrane:sinusoidal drug influx and efflux in the liver. Semin Liver Dis 2000; 20:251-263.
39. Ritter JK. Roles of glucuronidation and UDP-glucuronosyltranferases in xenobiotic bioactivation reactions. Chem-Biol Int 2000; 129:171-193.
40. Muraoka M, Kawakita M, Ishida N. Molecular characterization of human UDP-glucuronic acid/ UDP-N-acetylgalactosamine transporter, a novel nucleotide sugar transporter with dual substrate specificity. FEBS Lett 2001; 495:87-93.
41. Waddell ID, Robertson K, Burchell A et al. Evidence for glucuronide (small molecule) sorting by human hepatic endoplasmic reticulum. Mol Memb Biol 1995; 12:283-288.
42. Banhegyi G, Braun L, Marcolongo P et al. Evidence for an UDP-glucuronic acid/phenol glucuronide antiport in rat liver microsomal vesicles. Biochem J 1996; 315:171-176.
43. Bock KW, Gschaidmeier H, Heel H et al. Functions and transcriptional regulation of PAH-inducible human UDP-glucuronosyltranferases. Drug Metab Rev 1999; 31:411-422.
44. Tukey R, Strassburg CP. Human UDP-glucuronosyltranferases:Metabolism, expression and disease. Ann Rev Pharmacol Toxicol 2000; 40:581-616.
45. Iyanagi T, Emi Y, Ikushiro S. Biochemical and molecular aspects of genetic disorders of bilirubin metabolism. Biochim Biophys Acta 1998; 1407:173-184.
46. Strassburg CP, Manns MP. Jaundice, genes and promoters. J Hepatol 2000; 33:476-479.
47. Sugatani J, Kojima H, Ueda A et al. The Phenobarbital response enhancer module in the human UDP-glucuronosyltranferase UGT1A1 gene and regulation by the nuclear receptor CAR. Hepatol 2001; 33:1232-1238.
48. Ando Y, Saka H, Ando M et al. Polymorphisms of UDP-glucuronosyltranferase gene and irinotecan toxicity:A pharmacogenetic analysis. Cancer Resw 2000; 60:6921-6926.
49. Mani S. UGT1A1 polymorphism predicts irinotecan toxicity: Evolving proof. AAPS PharmSci 2001; 3:2.
50. Glatt H, Boeing H, Engelke CEH et al. Human cytosolic sulphotransferases: Genetics, characteristics, toxicological aspects. Mut Res 2001; 482:27-40.
51. Nagata K, Yamazoe Y. Pharmacogenetics of sulfotransferase. Ann Rev Pharmacol Toxicol 2000; 40:159-176.
52. Glatt H, Engelke CEH, Pabel U et al. Sulfotransferases: Genetics and role in toxicology. Toxicol Lett 2000; 112/113:341-348.
53. Strange RC, Spiteri MA, Ramachandram S et al. Glutathione S-transferase family of enzymes. Mut Res 2001; 482:21-26.
54. van Bladeren P.J. Glutathione conjugation as a bioactivation reaction. Chem-Biol Int 2000; 129:61-76.
55. Dekant W. Chemical-induced nephrotoxicity mediated by glutathione S-conjugate formation. Toxicol Lett 2001; 124:21-36.
56. Strange RC, Jones PW, Fryer AA. Glutathione S-transferase:genetics and role in toxicology. Toxicol Lett 2000; 112/113:357-363.
57. Hayes JD, Strange RC. Glutathione S-transferase polymorphisms and their biological consequences. Pharmacol 2000; 61:154-166.

58. Nguyen T, Huang HC, Pickett CB. Transcriptional regulation of the antioxidant response element:Activation by Nrf2 and repression by MafK. J Biol Chem 2000; 275:15466-15473.
59. Wasserman WW, Fahl WE. Comprehensive analysis of proteins which interact with the antioxidant responsive element:correlation of ARE-BP-1 with the chemoprotective induction response. Arch Biochem Biophys 1997; 344:387-396.
60. Ito K, Chiba T, Takahashi S et al. An Nrf2/small Maf heterodimer mediates the induction of phase II detoxifying enzyme genes through antioxidant response elements. Biochem Biophys Res Commun 1997; 236:313-322.
61. Ramos-Gomez M, Kwak M-K, Dolan PM et al. Sensitivity to carcinogenesis is increased and chemoprotective efficacy of enzyme inducers is lost in nrf2 transcription factor-deficient mice. Proc Natl Acad Sci USA 2001; 98:3410-3415.
62. Enomoto A, Itoh K, Nagayoshi E et al. High sensitivity of Nrf2 knockout mice to acetaminophen hepatotoxicity associated with decreased expression of ARE-regulated drug metabolizing enzymes and antioxidant genes. Toxicol Sci 2001; 59:169-177.
63. Bertolotti M, Carulli L, Concari M et al. Suppression of bile acid synthesis, but not hepatic cholesterol 7α-hydroxylase expression, by obstructive cholestasis in humans. Hepatol 2001; 34:234-242.
64. George J, Murray M, Byth M et al. Differential alterations of cytochrome P450 proteins in livers from patients with severe chronic liver disease. Hepatol 1995; 21:120-128.
65. Chen J, Murray M, Liddle C et al. Downregulation of male-specific cytochrome P450s 2C11 and 3A2 in bile duct-ligated male rats:Importance to reduced hepatic content of cytochrome P450 in cholestasis. Hepatol 1995; 22:580-587.
66. Chen J, Robertson G, Field J et al. Effects of bile duct ligation on hepatic expression of female-specific CYP2C12 in male and female rats. Hepatol 1998; 28:624-630.
67. Tateishi T, Watanabe M, Nakura H et al. Liver damage induced by bile duct ligation affects CYP isoenzymes differently in rats. Pharmacol Toxicol 1998; 82:89-92.
68. Aguero RM, Favre C, Rodriguez-Garay EA. Inhibitory effect of short-term bile duct ligation on hepaticic cytochrome P450 of bile acid-depleted rats. Pathobiol 2001; 69:30-35.
69. Renton KW, Nicholson TE. Hepatic and central nervous system cytochrome P450 are down-regulated during lipopolysaccharide-evoked localized inflammation in brain. J Pharmacol Exp Ther 2000; 294:524-530.
70. Shedlofsky SI, Israel BC, Tosheva R et al. Endotoxin depresses hepatic cytochrome P450-mediated drug metabolism in women. Br J Clin Pharmacol 1997; 43:627-632.
71. Monshouwer M, Witkamp RF. Cytochromes and cytokines: Changes in drug disposition in animals during an acute phase response: A mini-review. Vet Quart 2000; 22:17-20.
72. Barouki R, Morel Y. Repression of cytochrome P4501A1 gene expression by oxidative stress: Mechanism and biological implications. Biochem Pharmacol 2001; 61:511-516.
73. Morgan ET. Regulation of cytochromes P450 during inflammation and infection. Drug Metab Rev 1997; 29:1129-1188.
74. Morgan ET. Regulation of cytochrome P450 by inflammatory mediators. Why and how? Drug Metab Dispos 2001; 29:207-212.
75. Warren GW, Poloyac SM, Gray DS et al. Hepatic cytochrome P-450 expression in tumor necrosis factor-α receptor (p55/p75) knockout mice after endotoxin administration. J Pharmacol Exp Ther 1999; 288:945-950.
76. Siewert E, Bort R, Kluge R et al. Hepatic cytochrome P450 down-regulation during aseptic inflammation in the mouse is interleukin 6 dependent. Hepatol 2000; 32:49-55.
77. Pascussi JM, Gerbal-Chaloin S, Pichard-Garcia L et al. Interleukin-6 negatively regulates the expression of pregnane X receptor and constitutively activated receptor in primary human hepatocytes. Biochem Biophys Res Commun 2000; 274:707-713.
78. Iber H, Chen Q, Cheng PY et al. Suppression of CYP2C11 gene transcription by interleukin-1 mediated by NF-κB binding at the transcription start site. Arch Biohem Biophys 2000; 377:187-194.
79. Peng HM, Coon MJ. Promoter function and the role of cytokines in the transcriptional regulation of rabbit CYP2E1 and CYP2E2. Arch Biochem Biophys 2000; 382:129-137.
80. Ke S, Rabson AB, Germino JF et al. Mechanism of suppression of cytochrome P450 1A1 expression by tumor necrosis factor-α and lipopolysaccharide. J Biol Chem 2001; 276:39638-39644.

CHAPTER 12

Pathology of Cholestasis

James M. Crawford

Summary

Cholestasis to the pathologist is the visible manifestation of a broad array of pathophysiologic derangements. Non-obstructive impairment of hepatocellular bile formation may give rise to extensive parenchymal changes, without significant alterations in biliary tree morphology. Sepsis or systemic inflammatory conditions also can cause severe hepatocellular cholestasis without obstruction, although there may be attendant inflammatory changes in portal tracts. There is a striking set of bile duct diseases leading eventually to destruction of the intrahepatic and/or extrahepatic biliary tree, with characteristic and oft-times diagnostic histologic features evident on percutaneous liver biopsy. While non-obstructive cholestasis may be relatively mild and reversible depending upon the underlying cause, obstructive cholestasis will lead to substantial hepatic compromise from retained bile. Because of the central role of bile formation in normal hepatic function, and the severe hepatic and systemic toxicity of relentless bile secretory failure, there is a great need to correct the cause of cholestasis if at all possible. Morphological assessment of liver tissue, either through percutaneous liver biopsy or open exploration with biopsy, can play a major role in the clinical assessment and monitoring of these patients. As is evident from discussions elsewhere in this volume, the morphology of cholestasis is also a major clue to the underlying etiologies of disease.

Introduction

Bile formation facilitates the digestion and absorption of lipids from the gut, and provides a mechanism for the excretion of a wide variety of potentially toxic compounds, including cholesterol, bile pigments, and xenobiotics. Although the pathologist, clinical hepatologist, and basic scientist may agree that secretion of bile is a critically important liver function, agreement upon what constitutes cessation of bile flow, or "cholestasis," is more difficult to obtain. The term was first coined five decades ago to describe the histopathological light microscopic findings of bilirubin pigment stagnation in the hepatocyte and bile ducts.[1] The important conceptual advance was the recognition that morphological findings of "cholestasis" might be caused not only by obstruction of large bile ducts, but also by intrinsic hepatocellular dysfunction.

The term "cholestasis" was quickly adopted for use by basic scientists and clinicians as well. Physiologically, cholestasis denotes an impairment of bile flow and failure to secrete the inorganic and organic constituents of bile. In particular and as discussed at length elsewhere in this volume, cholestasis arises from molecular and ultrastructural changes that impair the entry of small organic molecules, inorganic salts, proteins, and ultimately water into the biliary space. Clinically, the physical findings of jaundice and pruritus are accompanied by elevated serum concentrations of bilirubin, bile salts, and alkaline phosphatase.

The clinical features of jaundice and hepatic failure have changed little since ancient times,[2] and pathologists have marveled at the profound hepatic alterations in jaundiced patients for almost two centuries.[3] Nevertheless, recent molecular insights have provided enormous hope

Molecular Pathogenesis of Cholestasis, edited by Michael Trauner and Peter L.M. Jansen.
©2004 Eurekah.com and Kluwer Academic / Plenum Publishers.

for successful treatment of cholestatic patients,[4] and the complex histological events accompanying cholestatic liver disease are slowly yielding their secrets.[5] Excellent discussions of the molecular pathogenesis of cholestasis are given elsewhere in this volume. The particular aim of this chapter is to present the key morphological features of cholestasis.

Anatomic Considerations

Macroscopic Anatomy of the Biliary Tree

It is worth considering first the anatomy of the biliary tree.[6] The extrahepatic biliary tract consists of the common bile duct, cystic duct and gallbladder, and the common hepatic duct. Approximately 60% to 70% of the time, the common hepatic duct bifurcates into the right and left hepatic ducts before entering the liver.[7] The predominant anatomic variation is absence of the right hepatic duct. Instead, posterior and anterior branches of bile ducts supplying the right portion of the liver arise from a hilar confluence with the left hepatic bile duct. This occurs in the form of a three-way branch point with the left hepatic bile duct or variations of two-way confluences of the anterior or posterior branches with the left hepatic duct.[7] Finally, while the common hepatic duct and its branches lie ventral to the portal vein system, the right posterior bile duct may wrap in an inferior/ventral or a superior/dorsal fashion around the right portal vein. This last variation must be kept in mind when performing surgery in the region of the liver hilum, so as to avoid transecting the portal vasculature.

The large intrahepatic bile ducts are defined as follows:[8] right and left hepatic ducts (with origin just outside the liver corpus); segmental ducts (the first major branches of each hepatic duct: left medial and lateral, right anterior and posterior); and area ducts (the first major branches of each segmental duct: superior and inferior). The segmental bile ducts of the caudate lobe of the liver drain directly into the right or left hepatic duct or their major branches.[9] All of these large ducts—right and left hepatic, segmental, and area—are grossly visible and are characterized by association with intrahepatic mucin-secreting peribiliary glands.[8] Within the liver, the biliary tree beyond the area ducts does not branch dichotomously, in that radial trees of bile ducts do not divide symmetrically.[10,11] As a result, there are considerable variations in the branching of the biliary tree within the liver as well as at its hilum.[8]

Microscopic Anatomy of the Biliary Tree

The finer branches of the biliary system are identifiable by light microscopy only, and are not associated with peribiliary glands. By convention, conducting bile ducts are all branches down to the last bifurcation of the biliary system. The most terminal branches are commonly called interlobular bile ducts, based on the concept that it is these branches which supply the lobules of the liver. Saxena et al[12] recommended calling these smallest branches terminal bile ducts, in part to recognize the fact that the microarchitectural units of the hepatic parenchyma go by many names other than "lobule", and to refocus the terminology on the architecture of the biliary tree rather than the hepatic parenchyma. However, within the lexicon of histopathology, 'interlobular' remains the term in common usage. While larger bile ducts may receive their vascular supply from as many as four companion hepatic artery branches,[13] terminal bile ducts maintain a 1:1 pairing with hepatic arteries.[14] Moreover, the diameters of the hepatic artery and terminal bile duct (as delineated by the basement membrane) are essentially the same (Fig. 1). Hence, identification in a portal tract of a hepatic artery unaccompanied by terminal bile duct is a sign of bile duct loss. Identification of bile ducts in less then 90% of portal tracts is evidence of paucity of bile ducts.[14]

The connection between biliary tree and hepatocellular parenchyma has long been the subject of study. In the stereoptic depiction of normal liver anatomy by Elias in 1949,[15] the biliary tree was seen to drain the hepatic parenchyma via tubular structures, cholangioles, emerging from deep within the hepatic lobule. This concept has been propogated in liver textbooks with inconsistent fidelity in the ensuing years (reviewed in ref. 16), such that as recent as 1997 the final connection between the terminal bile duct and hepatocellular parenchyma was deemed

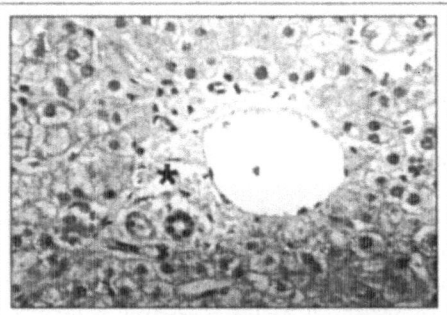

Figure 1. Normal liver. Hematoxylin and eosin stain of a portal tract, showing the three essential elements of a portal "triad": a large diameter portal vein, and the smaller hepatic artery:bile duct pair (*). Bile ductules are not evident in this medium-power image.

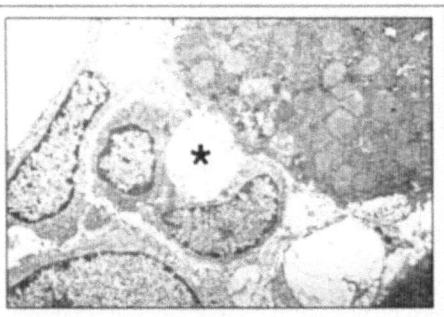

Figure 2. Electron microscopic image of a canal of Hering at the rim of a portal tract, showing two bile duct epithelial cells abutting a hepatocyte to form a central channel (*) for passage of bile. Courtesy of Dr. Donna Beer Stolz, University of Pittsburgh, USA.

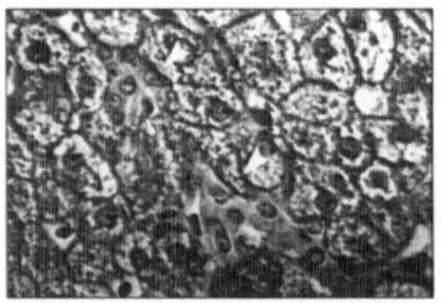

Figure 3. Normal liver. Masson trichrome stain of periportal parenchyma, showing a string of smaller bile duct epithelial cells extending into the parenchyma between hepatocytes and undergoing a bifurcation (arrowheads). This is the canal of Hering, evident by light microscopy using a routine stain.

inapparent by light microscopy.[7] Ultrastructural studies by Steiner and Carruthers in 1961[17] demonstrated that ductular channels could be demonstrated lined partially by hepatocytes and partially by bile ductular epithelial cells. This ultrastructural criterion has become the established definition of a canal of Hering, supporting Hering's original concept of a short straight canal as linking the intralobular network of "bile capillaries" between hepatocytes to the interlobular bile duct in the portal tract.[18] While the initial observation by Steiner and Carruthers[17] pointed out that hemiductular structures extended within the hepatic parenchyma, over the ensuing 40 years these connections between biliary tree and parenchyma were generally viewed as occuring only at the interface between portal tract mesenchyme and parenchyma (e.g., ref. 7). An excellent example of this latter relationship is given in Figure 2.

In a light microscopy study of normal adult liver anatomy, Crawford et al[14] found that isolated ductular structures within the parenchyma were actually readily observed in routine histological sections, at the rate of about 1 intralobular bile ductule per portal tract (Fig. 3). Simultaneously, Theise et al[19] found that staining of normal adult human liver sections with CK-19 revealed about 10 intraparenchymal bile ductule:canal of Hering systems per portal tract in any given tissue section. More importantly, superposition of up to 60 consecutive tissue sections stained for CK-19 demonstrated that the intraparenchymal bile ductule:canal of Hering systems all arose from terminal bile ducts, extending into the parenchyma for up to one third the distance to the terminal hepatic vein. In the normal liver intraparenchymal systems were present but scattered; severe liver damage induced a massive proliferation and ramification of the intralobular ductal system.[19] In all instances, the intraparenchymal system marked by CK-19 immunostaining connected up to terminal bile ducts by bile ductule:canal of Hering units which traversed the portal tract mesenchyme. In no instances were intraparenchymal ductular "arcades" observed, as proposed by Landing and Wells.[11]

With these considerations in mind, the following terminology is recommended,[12] recognizing that this terminology is still under active debate (Roskams T, Desmet V, personal communication). In general, bile ductules are channels branching off the terminal bile ducts that collect bile directly from the hepatocellular parenchyma. In specific, the bile ductule segment that traverses the mesenchyme of the portal tract is the intraportal bile ductule. The segment that penetrates the parenchyma yet remains a tubular structure is the intraparenchymal bile ductule; this segment may not be present. The hemi-tubular segment penetrating the parenchyma further, or remaining at the rim of portal tract and parenchyma, is the canal of Hering. To the best of current knowledge, these three structures—intraportal bile ductule, intraparenchymal bile ductule, and canal of Hering—are a single unit that drains bile from within the parenchyma to the biliary tree, and so for this chapter will be referred to as a bile ductule:canal of Hering unit.

With regards to the vascular supply, Ekataksin et al[20] has demonstrated that the bile ductules emerge from the terminal bile ducts accompanied by portal venous tributaries (termed septal veins). They propose that the local perfusion and drainage of hepatic parenchyma by vein/ductule, respectively, constitute a cholehepaton, the smallest architectural unit of the liver. The entry of hepatic arterial blood into the parenchyma is highly variable between species, with the following general patterns:[21] capillaries from the hepatic arteries in portal tracts supply the conducting and terminal bile ducts, and then dump into the companion portal vein; capillary beds from the hepatic artery feed the capsule, the subcapsular parenchyma, and the vasa vasorum of the hepatic vein system; and rare arterioles may feed sinusoids in the parenchyma directly. Regardless of the arterial component of blood flow, the apparent association of portal vein tributaries with the bile ductule:canal of Hering unit thereby provides a vascular supply for this highly proliferative compartment. In particular, the bile ductule:canal of Hering unit harbors progenitor cells of hepatic lineage, and is the apparent site of entry of extrahepatic bone marrow stem cells into the liver for restitution of the injured liver.[22]

It should be noted that the terms "bile ductule", "cholangiole", and "canal of Hering" are used somewhat interchangeably at the current time. While this chapter presents a specific recommendation for usage, it is clear that the developmental biology and pathobiology of the bile ductule/canal of Hering unit are in need of further elucidation.

The Hepatocyte

As discussed elsewhere in this volume, hepatocytes are polarized cells, possessing a basolateral plasma membrane surface that faces the sinusoidal vascular space and an apical surface that faces the bile canaliculus. The bile canaliculus is a channel by which secreted bile can be drained into the biliary tree, and thence into the gut. Bile canaliculi form between adjacent hepatocytes, with each half of the canalicular channel delineated by the apical, canalicular face of one hepatocyte. The canalicular space is separated from the lateral intercellular space by tight junctions between abutting hepatocytes, which functionally delimit the canalicular channel. The canaliculi run between hepatocytes to drain into the canal of Hering in the periportal parenchyma or at the parenchymal:portal tract interface; this is the anatomic site where bile canaliculi of hepatocytes first encounter the most peripheral conduits of the biliary tree.

Hepatocellular bile secretory function exhibits an zonal distribution, from the parenchyma adjacent to portal tracts (zone 1) to that adjacent to the terminal hepatic veins (zone 3), similar to many other hepatic functions.[23] A zonal gradient for the basolateral multidrug resistance-associated protein (Mrp3) has been observed, and is altered during cholestasis.[24] However, the most striking zonation is physiologic: a decreasing gradient of bile salt uptake into hepatocytes moving from the periportal to the perivenular region, best demonstrated by autoradiographic studies.[25,26] This gradient flattens at higher bile salt loads.[6] The predominant basolateral transporter for bile salts, the sodium-taurocholate cotransporter (Ntcp), is expressed throughout the liver acinus, with similar expression in periportal and perivenous hepatocytes,[27] and hepatocytes throughout the parenchyma are capable of secreting bile salts into bile.[28] Thus, gradients of bile salt concentration within sinusoidal blood must be postulated to result from the high capacity of normal hepatocytes to take up and secrete bile salts.

From the standpoint of canalicular topology, the logic of fluid dynamics would suggest that there is less bile flow in the perivenular canaliculi than in the periportal canaliculi, since the former are more upstream. Documentation that the diameter of bile canaliculi is less in the perivenular zone supports this inference.[29] Whether there are intrinsic zonal differences in canalicular transport capacity remains unclear, although experiments with hepatocyte couplets isolated from different parts of the hepatic acinus reveal differences in the canalicular secretion of fluorescent bile salts.[30]

The Morphology of Cholestasis

The morphologic features of cholestasis depend somewhat on its severity, duration, and underlying cause. The changes in the hepatocellular parenchyma are described first, as they are common to both non-obstructive and obstructive cholestasis. Portal tract changes are then described more briefly, recognizing that specific diseases affecting bile ducts are addressed in subsequent chapters.

Parenchymal Features

Hepatocytes in cholestatic livers exhibit a striking array of histological features, as summarized in Table 1. The dominant themes are accumulation of substances normally secreted in bile and toxic degeneration of hepatocytes.[31] The distinction between the morphologic change induced from retention of bilirubin pigments—brown coloration—and that from retention of bile salts—toxic degeneration, should be kept in mind. Bile salts themselves are colorless, and the correlation between bilirubin and bile salt accumulation is not strict. The most pure example of bilirubin pigment retention without retention of bile salts, Dubin Johnson syndrome, generates a darkly brown liver with numerous brown globules within hepatocytes (Fig. 4). Curiously, these retained pigments are not bilirubin per se but appear to be polymers of epinephrine.[32] Some cholestatic conditions are essentially without histologic changes, such as intrahepatic cholestasis of pregnancy.[33]

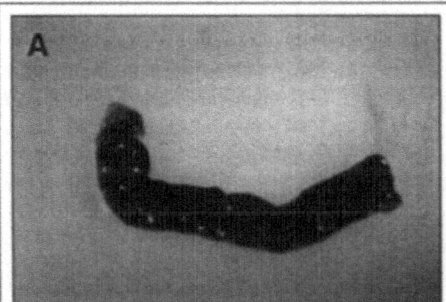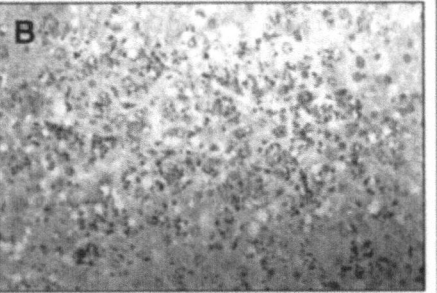

Figure 4. Dubin-Johnson syndrome. A) Photograph of liver tissue core obtained via percutaneous needle biopsy. The liver is darkly pigmented. B) Photomicrograph of the hepatic parenchyma, showing numerous dark brown globular inclusions within hepatocytes.

Table 1. Histological features of cholestasis

Histological Feature	Distinctive Features
Bilirubin pigment accumulation	Hepatocyte cytoplasm pigmentation
	Bile canalicular dilatation with inspissated bile
	Regurgitated pigment: Kupffer cell pigmentation
	Acinar distribution: perivenular > periportal
Hepatocellular degeneration	Foamy degeneration: flocculent cytoplasm of hepatocytes; perivenular > periportal
	Ballooning degeneration: swollen hepatocytes; periportal > perivenular ("cholate stasis")
	Bile infarcts: coalescent periportal necrosis with retained pigmented material
	Mallory body formation
Metal accumulation	Periportal accumulation of copper and copper-binding protein
Giant cell transformation	Coalescence of hepatocytes, with multiple nuclei and free-floating "canaliculi"
Liver cell rosettes	Dilated bile canaliculi surrounded by more than two hepatocytes, in a pseudotubular arrangement

With these caveats aside, the most obvious feature of the cholestatic liver is retained bile within hepatocytes and bile canalicular spaces (Fig. 5).[34] The retention of bile salts within hepatocytes along with bilirubin pigments imparts a fine foamy appearance to the cytoplasm (foamy degeneration), owing to the toxic effects of bile salts. This is more commonly observed in the perivenular region (Fig. 6). Cholestasis may cause hepatocyte dropout scattered through the parenchyma in the form of lysed hepatocytes, ballooning degeneration, or outright apoptosis (Fig. 7).[35] Death of confluent regions of hepatocytes may produce a bile infarct, in which large masses of pigmented material are surrounded by a rim of necrotic hepatocytes or reactive mesenchyme (Fig. 8).

It is not unusual to find Mallory bodies (intracellular strands of clumped intermediate filaments) in periportal areas, which is in contradistinction to the perivenular Mallory body distribution characteristic of alcoholic liver disease. As copper is a heavy metal normally secreted in bile,[36] special stains for copper or copper-binding protein may demonstrate copper accumulation in periportal hepatocytes. Periportal hepatocyte swelling in cholestasis, without or with Mallory body formation, accumulation of copper and copper-binding proteins, and formation of liver cell rosettes, is termed cholate stasis based on the presumption that it is retention of toxic bile salts that is causing this change (Fig. 9).[37]

A peculiar feature of chronic cholestasis of neonates, children, and adults is the presence of so-called cholestatic liver cell rosettes consisting of dilated bile canaliculi surrounded by more than two hepatocytes in a pseudo-tubular arrangement.[38] These rosettes express some cytokeratins characteristic of biliary epithelium, and are thus thought to be an adaptive mechanism to chronic obstructive cholestasis. A further feature of neonatal cholestatic syndromes is the frequent appearance of giant cells; foamy-appearing multinucleated hepatocytes with intracellular pigmented inclusions having the appearance of "floating" bile canaliculi (Fig. 10).[39] On occasion, giant cell transformation also may be seen in adult cholestatic livers. Although the origin of such cells is unclear, it is thought that the detergent action of retained bile salts leads to dissolution of the lateral plasma membranes and coalescence of adjacent hepatocytes. Such giant cell transformation also can occur in biliary atresia (see below).

One disease in which there is a paucity of bile ducts and persistent cholestasis without substantial hepatocellular degeneration is Alagille's syndrome.[40] Hepatocytes may exhibit mild

Figure 5. Cholestasis, non-obstructive, with inspissated bile in bile canaliculi. The bile canalicular spaces between hepatocytes are cut in cross section, and contain darkly pigmented bile (arrowheads). This is a case of hepatitis B viral infection with cholestasis, hence the diffusely granular change to the cytoplasm of some hepatocytes owing to viral particles. Masson trichrome stain.

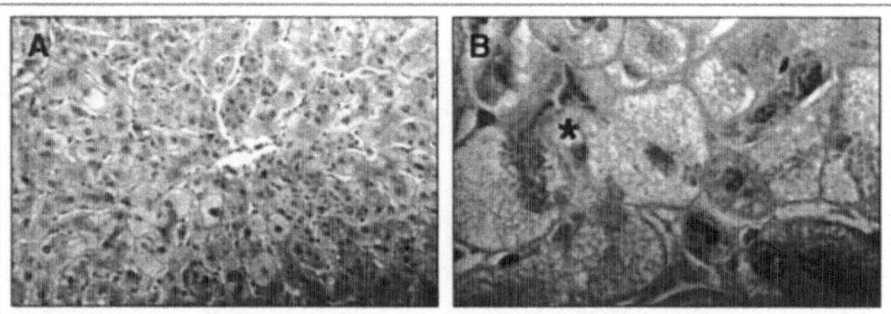

Figure 6. Cholestasis, obstructive. A) Low power image of region of parenchyma around terminal hepatic vein (center of image). Hepatocytes are swollen in this region. B) Higher power image, showing swelling of hepatocytes with a finely granular cytoplasm (foamy change); bile pigment discoloration also is present. A bile canaliculus is cut in longitudinal section, and contains inspissated bile (*). Kupffer cells within the sinusoids also have phagocytosed regurgitated bile.

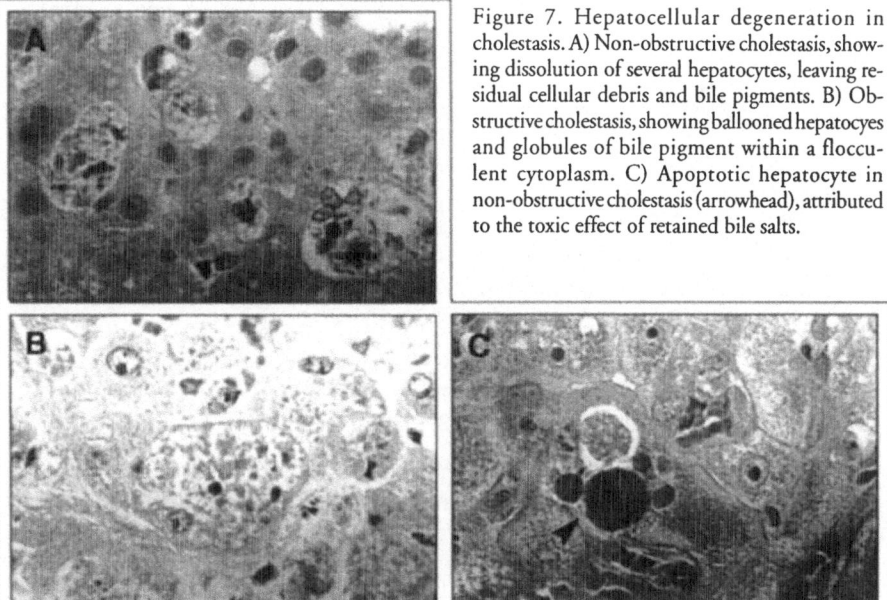

Figure 7. Hepatocellular degeneration in cholestasis. A) Non-obstructive cholestasis, showing dissolution of several hepatocytes, leaving residual cellular debris and bile pigments. B) Obstructive cholestasis, showing ballooned hepatocyes and globules of bile pigment within a flocculent cytoplasm. C) Apoptotic hepatocyte in non-obstructive cholestasis (arrowhead), attributed to the toxic effect of retained bile salts.

Figure 8. Bile infarct (*), present in a needle biopsy of an obstructed liver. A darkly pigmented necrotic center is surrounded by a rim of cellular debris, creating a spherical defect within the hepatic parenchyma. Portal tracts exhibit a significant degree of fibrosis, reflecting long term biliary obstruction.

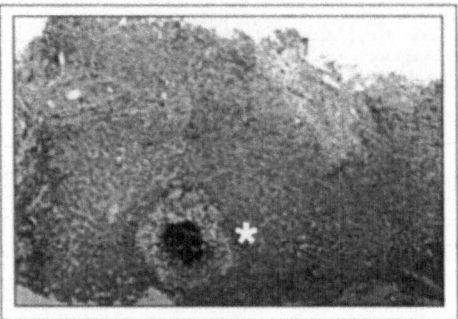

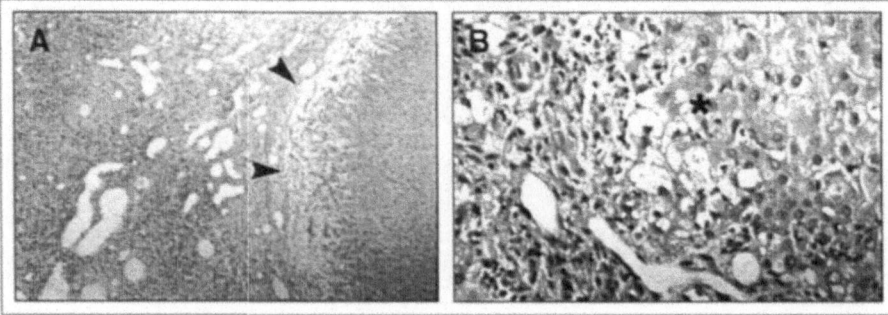

Figure 9. Cholate stasis. A) A low power photomicrograph of a broad fibrous tract in a cirrhotic liver shows a cleared-out space at the septal rim (arrowheads), the result of edema and hepatocyte swelling. B) Medium power image of the portal tract:parenchymal interface of an inflamed cholestatic liver from thorazine toxicity, showing the ballooning of hepatocytes at the interface (*).

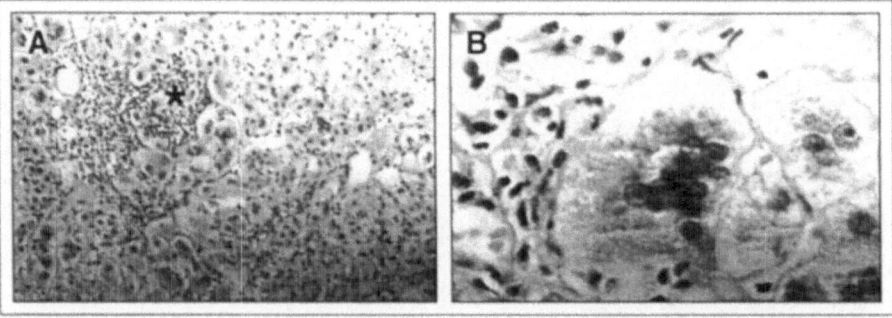

Figure 10. Neonatal hepatitis. A) Lower power image of portal tract and surrounding parenchyma, showing some inflammatory cells in an ill-defined portal tract (*), and marked variability in hepatocyte size. B) High power image of same case, showing giant cell transformation of an hepatocyte. Multiple nuclei are present, and there is faint discoloration by bile pigment.

ballooning and cholestatic changes, and scattered apoptosis, but the severe degenerative changes characteristic of most forms of bile duct obstruction as described above are not observed. A striking ultrastructural observation is the retention of bile pigment in lysosomes and vesicles of the outer convex face of the Golgi apparatus (cis-Golgi); there is no pigment accumulation in bile canaliculi or the pericanalicular region.[41]

Acinar Gradient

Alterations in the acinar distribution of bile secretory function and expression of transport proteins occur under cholestatic conditions, and can be demonstrated by molecular techniques.[42] The elevated bile salt concentrations encountered in portal blood will lead to both greater perivenular uptake of bile salts[26] and dilatation of perivenular bile canaliculi.[29] "Downstream" hepatocytes in the perivenular region of the acinus exhibit increased rates of bile salt uptake, particularly when overall hepatocellular secretory function is decreased, as with ethinylestradiol.[43] Bile drainage also leads to a loss of zonal heterogeneity.[44] Moreover, hepatocellular uptake and secretion of bile salts may actually increase throughout the acinus when bile outflow is obstructed,[45] possibly contributing to canalicular dilatation and suggesting a role for downstream reabsorption of biliary constituents in the biliary tree.

Portal Tract Changes

Obstruction to the biliary tree, either intrahepatic or extrahepatic, induces distention of upstream bile ducts and ductules by bile. The bile stasis and back-pressure induces proliferation of the biliary epithelial cells and looping and reduplication of ducts and ductules (Fig. 11).[46] Unlike the extensive ductular regenerative reaction that occurs in the setting of massive hepatic necrosis (Fig. 12), in obstruction the ductular proliferation is confined to portal tracts. The labyrinthine ductules reabsorb secreted bile salts, serving to protect the downstream obstructed bile ducts from the toxic detergent action of bile salts.[47] Associated histological findings include portal tract edema and periductular infiltrates of neutrophils.[48] Unrelieved obstruction leads to ectasia of larger bile ducts and portal tract fibrosis, which initially extends

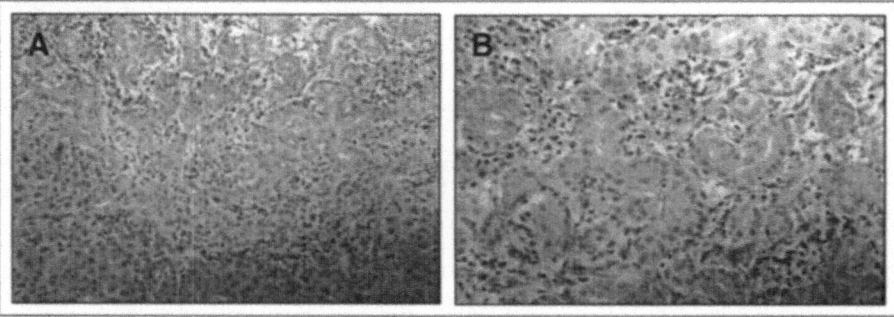

Figure 11. Bile ductular proliferation in obstructive cholestasis. A) Medium power view of a portal tract, which has been markedly expanded by proliferating bile ductules, inflammatory cells, and mild edema. B) Higher power view of the same portal tract, showing racemose proliferating bile ductules.

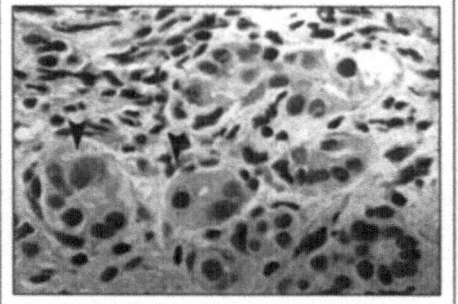

Figure 12. Ductular regenerative reaction in massive hepatic necrosis. Proliferating bile ductules, arising from the bile ductule:canal of Hering unit, are present within the parenchyma amidst inflammatory cells. This is the regenerative response of the parenchyma to severe damage, and is not the portal tract-based bile ductular proliferation characteristic of biliary obstruction. Note that ductules exhibit some partial differentiation into hepatocytes (arrowheads).

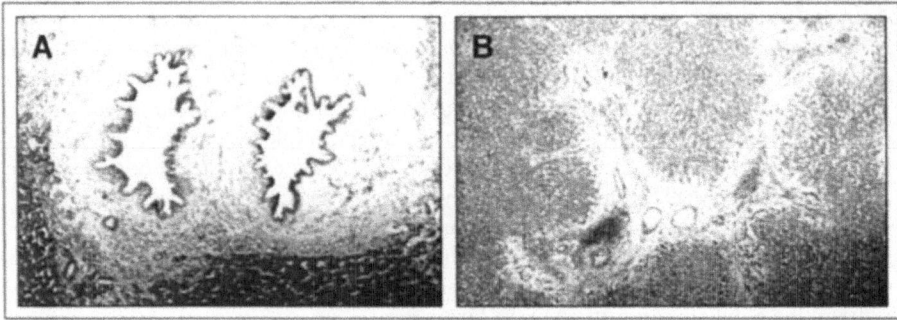

Figure 13. Portal tract changes in longer term biliary obstruction. A) Markedly ectatic large bile ducts, with severe periductal edema. B) Subdivision of the parenchyma by large fibrous septa connecting adjacent portal tract changes, giving rise to a "jigsaw" pattern of fibrosis.

Figure 14. Biliary cirrhosis from long-standing extrahepatic biliary obstruction, coronal section. The liver is coarsely subdivided by dense fibrous bands, and darkly pigmented.

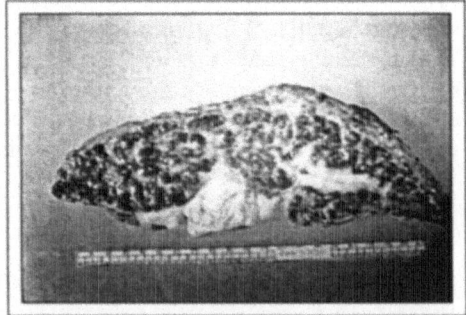

into and subdivides the parenchyma with relative preservation of hepatic architecture (Fig. 13). Ultimately, an end-stage bile-stained, cirrhotic liver is created (biliary cirrhosis, Fig. 14). Partial reversal of severe biliary fibrosis can occur with correction of the underlying obstruction.[49]

It should be noted (although not discussed here) that intrahepatic bile ducts are frequently damaged as part of more general liver disease, as in drug toxicity and viral hepatitis. Bile duct loss leading to paucity does not generally occur,[50] although occasionally there may be progressive bile duct injury.[51]

Secondary Biliary Cirrhosis

Prolonged obstruction to the extrahepatic biliary tree results in profound alteration of the liver itself. The most common cause of obstruction in adults is extrahepatic cholelithiasis (gallstones), followed by malignancies of the biliary tree or head of the pancreas, and strictures resulting from previous surgical procedures. Obstructive conditions in children include biliary atresia, cystic fibrosis, choledochal cysts (a cystic anomaly of the extrahepatic biliary tree), and syndromes in which there are insufficient intrahepatic bile ducts (paucity of bile duct syndromes). The initial morphological features of cholestasis—parenchymal cholestasis as described above, and portal tract bile ductular proliferation, edema, and neutrophilic inflammation—are entirely reversible with correction of the obstruction. However, secondary inflammation resulting from biliary obstruction initiates periportal fibrogenesis, which eventually leads to hepatic scarring and nodule formation, generating secondary biliary cirrhosis. Uncorrected obstruction may promote secondary bacterial infection of the biliary tree (ascending cholangitis), which aggravates the inflammatory injury. Enteric organisms such as coliforms and enterococci are common culprits.

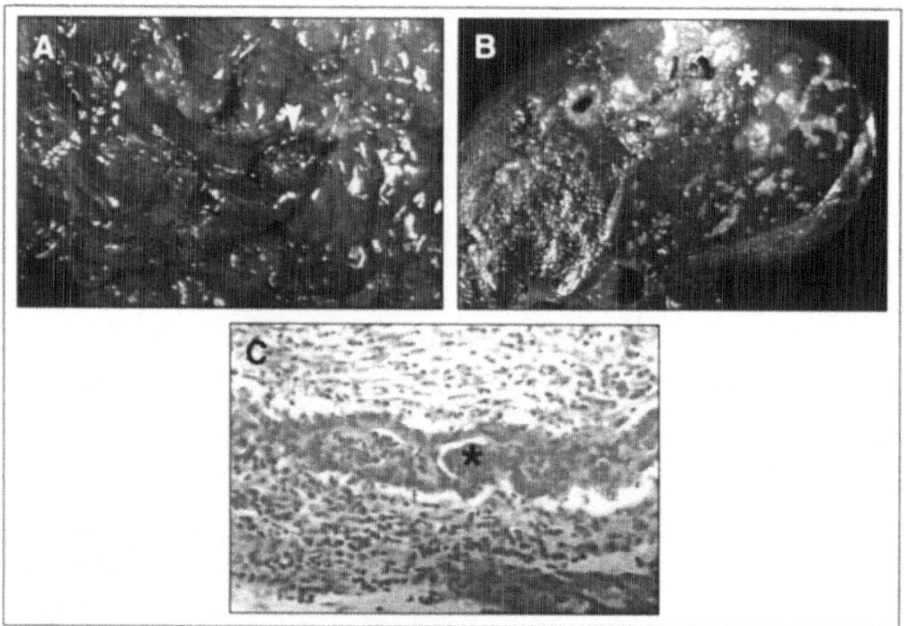

Figure 15. Biliary obstruction with ascending cholangitis. A) Post-mortem image of opened common bile duct, showing a retained stone in the common bile duct (arrowhead) following laparoscopic cholecystectomy. B) Parasagittal section of right lobe of liver, showing extensive hepatic abscess formation from ascending bacterial infection (*). C) Photomicrograph of inflamed interlobular bile duct, containing abunding intraluminal neutrophils (cholangitis, *).

The end-stage obstructed liver exhibits extraordinary yellow-green pigmentation, and is accompanied by marked icteric discoloration of body tissues and fluids. On cut-surface, the liver is hard, with a finely granular appearance. The histology is characterized by coarse fibrous septae that subdivide the liver in a jigsaw-like pattern. Embedded in the septa are distended small and large bile ducts, which frequently contain inspissated pigmented material. There is extensive proliferation of smaller bile ductules and edema, particularly at the interface between septa (formerly portal tracts) and the parenchyma. Cholestatic features in the parenchyma may be severe, with extensive feathery degeneration and formation of bile lakes. However, once regenerative nodules have formed, bile stasis may become less conspicuous. Ascending bacterial infection incites a robust neutrophilic infiltration of bile ducts; severe portal tract inflammation and cholangitic abscesses may develop (Fig. 15).

Primary Biliary Cirrhosis

Primary biliary cirrhosis is a chronic, progressive and often fatal cholestatic liver disease, characterized by the destruction of intrahepatic bile ducts, portal inflammation and scarring, and the eventual development of cirrhosis and liver failure (see related chapter by Bassendine). The primary feature of this disease is a non-suppurative, inflammatory destruction of medium-sized intrahepatic bile ducts. As cirrhosis develops only after many years, the disease name is somewhat misleading for those patients diagnosed early in their course.

Primary biliary cirrhosis is the prototype of conditions leading to small-duct biliary fibrosis and cirrhosis. Primary biliary cirrhosis is a focal and variable disease, exhibiting different degrees of severity in different portions of the liver. During the precirrhotic stage, portal tracts are infiltrated by a dense portal tract infiltrate of lymphocytes, macrophages, plasma cells, and

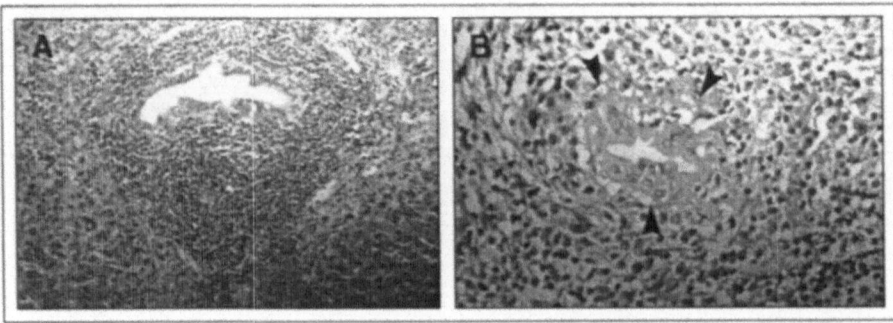

Figure 16. Primary biliary cirrhosis. A) Portal tract expanded by a dense infiltrate of lymphocytes and plasma cells. B) Higher power image of portal tract and interlobular bile duct, showing destructive changes in the bile duct epithelium with disruption of the epithelial barrier, in the setting of dense mononuclear inflammation and periductular histiocytic reaction (arrowheads).

Figure 17. Primary biliary cirrhosis, parasagittal section through right lobe. The liver is finely subdivided by fibrous tissue, and darkly pigmented.

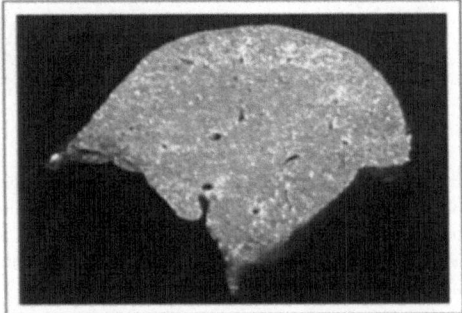

occasional eosinophils. Terminal and conducting bile ducts are infiltrated by lymphocytes and may exhibit non-caseating granulomatous destruction, and undergo progressive destruction (Fig. 16). With time, the obstruction to intrahepatic bile flow leads to progressive secondary hepatic damage. Portal tracts upstream from damaged bile ducts exhibit bile ductular proliferation, inflammation and necrosis of the adjacent periportal hepatic parenchyma. The parenchyma develops generalized cholestasis, and canals of Hering also appear to be destroyed.[52] A ductopenic variant without evident duct destruction also can occur.[53]

Over years to decades, relentless portal tract scarring and bridging fibrosis lead to cirrhosis (Fig. 17). Macroscopically, the liver does not at first appear abnormal, but as the disease progresses, bile stasis stains the liver green. The capsule remains smooth and glistening until a fine granularity appears, culminating in a well-developed, uniform micronodularity. Liver weight is at first normal to increased (owing to inflammation); ultimately liver weight is slightly decreased. In most cases, the end-stage picture is indistinguishable from secondary biliary cirrhosis, or the cirrhosis which follows chronic hepatitis from other causes.

Primary Sclerosing Cholangitis

Primary sclerosing cholangitis is characterized by inflammation and obliterative fibrosis of intrahepatic and extrahepatic bile ducts, with dilatation of preserved segments (see related chapter by Björnsson and Chapman). Characteristic "beading" of a barium column in radiographs of the intra- and extrahepatic biliary tree is attributable to the irregular strictures and dilatations of affected bile ducts.

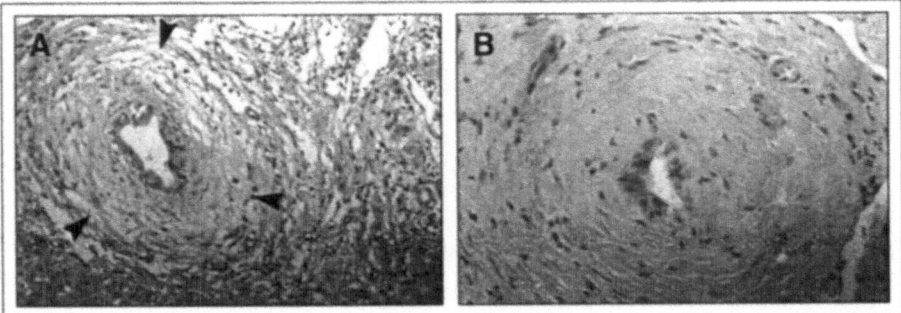

Figure 18. Primary sclerosing cholangitis. A) Early lesion of portal tract, showing periductal edema and "onion-skin fibrosis", with scattered mononuclear inflammatory cells (arrowheads). B) Later lesion, showing dense periductal fibrosis with near-complete obliteration of the bile duct epithelial layer.

Primary sclerosing cholangitis is a fibrosing cholangitis of bile ducts, with a lymphocytic infiltrate, progressive atrophy of the bile duct epithelium and obliteration of the lumen (Fig. 18). The concentric periductal fibrosis around affected ducts ("onion skin fibrosis") is followed by their disappearance, leaving behind a solid, cord-like fibrous scar. In between areas of progressive stricture, bile ducts become ectatic and inflamed, presumably the result of downstream obstruction. As the disease progresses, the liver becomes markedly cholestatic, culminating in biliary cirrhosis much like that seen with primary and secondary biliary cirrhosis. It should be noted that identification of characteristic bile duct lesions in percutaneous liver biopsy specimens is uncommon, since the major bile ducts are the more notable sites of injury. Hence, the more often histologic finding on percutaneous liver biopsy is non-specific features of biliary obstruction.

Transplantation

The liver has the unenviable position of being attacked by graft-versus-host and host-versus-graft mechanisms, in the setting of bone marrow transplantation (BMT) and liver transplantation, respectively. Both conditions—graft-versus-host disease (GVHD) and allograft rejection, respectively—may give rise to severe cholestasis. Liver damage in acute GVHD (10 to 50 days after bone marrow transplantation) is dominated by direct attack of donor lymphocytes on epithelial cells of the liver. This results in a hepatitis with necrosis of hepatocytes and bile duct epithelial cells, and inflammation of the parenchyma and portal tracts. Chronic GVHD is conventionally defined as a syndrome arising after day 100. It may arise as an inexorable extension of acute GVHD, after a disease-free interval, or de novo, without previous episodes of acute GVHD. As the syndrome of chronic GVHD may develop as early as 40 to 50 days after transplantation, the time frames for acute and chronic GVHD can overlap. However, unlike acute GVHD, chronic GVHD is a heterogeneous disease that involves a much wider range of organ systems, including skin, gastrointestinal tract, liver, minor salivary glands, lymph nodes, mouth, eyes, lungs, and musculoskeletal system, thereby constituting a blend of autoimmune syndromes.[54,55]

While acute and chronic GVHD appear to be related, there are sufficient differences to indicate that they represent distinct processes.[55] In general, acute GVHD is dominated by necrosis of the undifferentiated proliferating epithelial cells of the skin, gastrointestinal tract and liver. Chronic GVHD tends to affect more differentiated cells and eventually leads to fibrosis.[56,57] In both cases, it is the disruption of the protective mucosal epithelial barriers which may have the most dire clinical consequences.

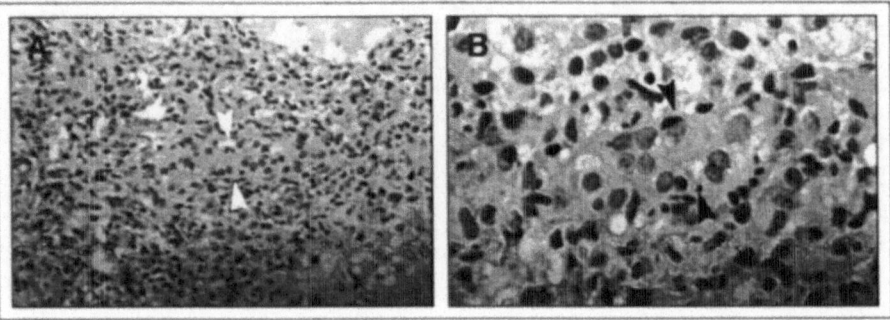

Figure 19. Acute graft-versus-host disease. A) Lower power image of portal tract showing robust mononuclear inflammatory infiltrate; the interlobular bile duct is difficult to identify (arrowheads). B) High power image of the interlobular bile duct (center). The duct is difficult to identify, and is infiltrated by lymphocytes (arrowheads).

Acute Graft-versus-Host Disease

The non-specific features of acute GVHD are a hepatitis-like picture, with cholestatic and ballooning hepatocytes, frequent apoptotic hepatocytes, and lymphocytic infiltration of the lobule, resembling drug- or virally-induced hepatitis. Endothelial attack, manifest as endotheliitis similar to that observed in allograft rejection of the transplanted liver , is an uncommon finding in hepatic GVHD.[58] The sine qua non of acute GVHD of the liver is direct attack of donor lymphocytes on bile duct epithelial cells (Fig. 19).[55,56,59,60] The bile ducts most frequently involved are small caliber.[61] Lymphocytic infiltrates are seen surrounding, invading, and disrupting the walls of interlobular bile ducts. Lymphocytic attachment is accompanied by necrosis of bile duct epithelial cells, evidenced as cytoplasmic vacuolization, nuclear pleomorphism or loss of nuclei, and sloughing of epithelial cells into the bile duct lumens. Residual duct epithelial cells may become attenuated to the point of appearing squamous around a portion of the duct circumference. The withered appearance of the ductal epithelium is to be distinguished from the heaped –up, reactive duct epithelial cells commonly encountered in viral hepatitis, particularly with infection of hepatitis C virus.[50] Because patients with acute GVHD are usually pancytopenic, the degree of bile duct and portal tract inflammation may be quite minimal, despite obvious damage to bile ducts.

It is important to note that the characteristic morphological changes due to GVHD may not be obvious in the early stage of the diseases. For example, the bile duct damage can be identified only rarely in the first 35 days after the bone marrow transplantation in patients.[59] In addition, with the clinical improvements in grafting regimens, widespread use of prophylaxis and improved clinical management have not only reduced the incidence and severity of hepatic GVHD, but at the same time have rendered the histological diagnosis of GVHD more difficult. A more pure hepatitic form without obvious bile duct damage of GVHD also occurs rarely.[62-65]

Chronic Graft-versus-Host Disease

Chronic GVHD is chiefly characterized by portal infiltration by lymphocytes without or with plasma cells or eosinophils and damage to interlobular bile ducts. These bile ducts are generally of small size (<45 um in diameter).[66] Although bile duct epithelial degeneration resembling that of acute GVHD may be observed, more commonly damaged bile duct epithelial cells appear eosinophilic and coagulated when compared to healthy neighboring cells. As with acute GVHD, lymphocytes are seen in close point contact with bile duct epithelial cells. Loss of bile ducts is a relatively late phenomenon in chronic GVHD, although it has been observed as early as 1 month after BMT.[67,68] Regardless of the time frame, bile duct loss may

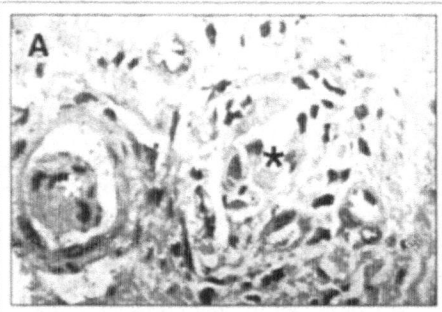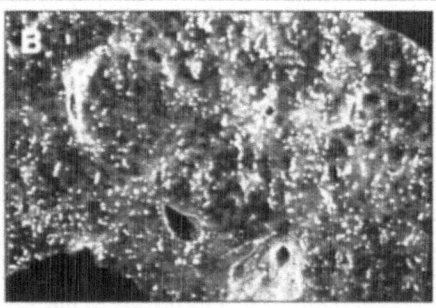

Figure 20. Chronic graft-versus-host disease. A) High power image of portal tract, showing a minimal inflammatory infiltrate. Two bile ducts were present, but are now largely destroyed (*). B) Liver section of end-stage chronic graft-versus-host disease, showing the severely pigmented cirrhotic liver.

lead to a vanishing bile duct syndrome or outright cirrhosis (Fig. 20).[69-72] Unlike acute GVHD, hepatocellular damage is minimal in most cases of chronic GVHD, except for the fact that the loss of interlobular bile ducts gives rise to progressive hepatocellualr cholestasis. In the more advanced stages of chronic GVHD, hepatocellular cholestasis may be so severe as to lead to actual degeneration of hepatocytes, particularly along the portal tract margins. Venous endotheliitis is not a prominent feature of chronic GVHD. Noncaseating granulomata of the portal tract are encountered only rarely.

Allograft Rejection
Acute cellular rejection of implanted livers exhibits features common to all solid organ transplants, including infiltration of a mixed population of inflammatory cells into portal tracts, bile duct and hepatocyte injury, and endotheliitis. With chronic rejection, a severe obliterative arteritis of small and larger arterial vessels will generate ischemic changes in the liver parenchyma. Alternatively, bile ducts are progressively obliterated, either due to direct attack or obliteration of their arterial supply. Both may lead to loss of the graft.

Biliary Atresia
Biliary atresia as a disease entity is not discussed elsewhere in this volume. As a major cause of infantile cholestasis, however, comment is merited.[73] Liver dysfunction in the neonate is commonly associated with the failure of bile secretion and conjugated hyperbilirubinemia. Accordingly, cholestasis is a common presenting feature for many disorders of the neonate. Because of an immaturity of hepatic excretory function, susceptibility to infection during the neonatal period, and early presentation of inborn errors of metabolism, the number of non-obstructive disorders presenting with cholestasis in the neonate is considerable, and the reader is referred to two thorough discussions.[74,75] Differentiation of extrahepatic causes of neonatal cholestasis, principally biliary atresia, from intrahepatic non-obstructive causes is critical for establishing a diagnosis upon which prompt treatment can be instituted.

Extrahepatic biliary atresia (EHBA) accounts for over 30% of all neonates presenting with cholestasis. Its incidence is 1 in 10,000 live births; usually on a sporadic basis without a family history. Despite the fact that infants with biliary atresia are of normal gestational age and birth weight and look well other than cholestatic jaundice at the time of presentation in the first few weeks of life, death usually occurs by 12 - 24 months, owing to complete obliteration of the extrahepatic biliary tree, without or with extension of the destructive process to involve the intrahepatic biliary tree. The median age of death for untreated infants is 8 months.

After non-invasive clinical testing including laboratory evaluation and imaging fails to exclude extrahepatic biliary obstruction in the jaundiced neonate, percutaneous liver biopsy may be needed to determine whether histologic features of large bile duct obstruction are present. Noting that the disease is destruction of the major extrahepatic bile ducts, the intrahepatic features are those of obstruction: bile plugs within bile duct lumina and proliferating bile ductules within portal tracts (see Fig. 8).[76] Edema and neutrophilic inflammation in portal tracts are to be expected. Progressive findings, which will increase with increasing age of the child, include: progressive portal and periportal fibrosis, ballooning degeneration of hepatocytes at the portal tract:parenchymal interface (cholate stasis), and at the extreme copper accumulation and Mallory body formation in periportal hepatocytes, and bile lakes in the parenchyma. Paucity of intrahepatic bile ducts is supposed to occur by 4 to 5 months, with progression to virtual absence of intrahepatic bile ducts by 8 to 9 months. Loss of bile ducts is attributed to persistent downstream obstruction to bile flow, recurrent ascending cholestasis, and intrahepatic progression of the primary inflammatory destructive process. Non-specific findings include parenchymal cholestasis, and giant cell transformation in 15% of cases.

Extensive lobular disease, more characteristic of neonatal hepatitis (NH) without or with giant cells, can be seen in EHBA. In this situation, the only suggestion that the underlying etiology is EHBA is an inability to identify bile ducts within portal tracts (Fig. 21). EHBA and neonatal hepatitis with giant cell transformation have been thought to represent points on a continuum of manifestations of the same disorder.[77] Lobular changes should not be allowed to distract attention from the portal tract features.

The accuracy of percutaneous liver biopsy in the diagnosis of EHBA is reported as 60% to 95%, "depending on the skill of the pathologist."[78] However, major pitfalls are encountered which may lead to the mis-diagnosis of "neonatal hepatitis", seriously delaying definitive treatment of the affected child. First, classic obstructive findings are encountered in only 65% of biopsies;[78] unclear features of neonatal hepatitis constitute the remainder. Second, there are in fact two forms of biliary atresia: the classical form in which the extrahepatic biliary tree is involved primarily with progression to the intrahepatic biliary tree only with time, and "early severe biliary atresia", in which there appears to be intrinsic malformation of the intrahepatic biliary tree.[78,79] In this latter form, the biliary tree does not form normally from the embryonic ductal plate,[6] resulting in portal tracts which do not contain interlobular bile ducts but rather have residual peripheral ductal elements of the embryonic "ductal plate" concentrically placed around the periphery of the portal tract.[80,81] Accompanying this "ductal plate malformation" is hypertrophy of hepatic arterial elements within portal tracts and dense portal tract fibrosis. The percutaneous liver biopsy will therefore exhibit portal tracts with paucity of bile ducts without or with ductal plate remnants, portal tract fibrosis without significant inflammation, and arterial hypertrophy.[79] For these reasons, EHBA can be exceedingly difficult to diagnose, even for the most experienced hepatopathologists.

Other Conditions

Severe cholestasis may develop as a result of inflammatory conditions which are not limited to the liver. The chief mechanism appears to be cytokine-mediated down-regulation of hepatic bile formation, either as the result of endotoxemia stimulating local cytokine release within the liver, or elevated circulating levels of cytokines as in malignant conditions.[48,82] Although the liver may be generally well-preserved, two striking histological findings may be encountered. The first is a neutrophilic pericholangitis, in which neutrophils are seen to surround portal tract bile ducts and ductules (Fig. 22). The second is massive dilatation of bile ductules at the margin of portal tracts, a condition encountered usually only in sepsis and termed cholangitis lenta (Fig. 23). In both instances there appears to be a more general shutdown of hepatocellular bile secretory function as well.[48] Lastly, over and above malignancies of the biliary tree leading to obstruction, the biliary tree may be obliterated in the condition of Langerhans cell histiocytosis.[83] In the very young child, this condition may mimic biliary atresia; in the older individual the clinical features may resemble sclerosing cholangitis.

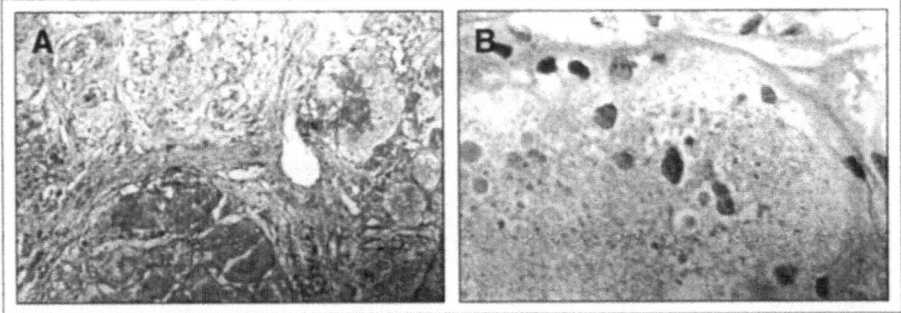

Figure 21. Biliary atresia, hepatic tissue at 12 weeks. A) Medium power image of portal tract region; no bile ducts are identifiable in this fibrotic liver, indicating extension of the destructive process into the liver. B) High power of same case, showing giant cell transformation of hepatocyte and retained globules of bile pigment, similar to that seen in non-obstructive neonatal hepatitis. Masson trichrome stain.

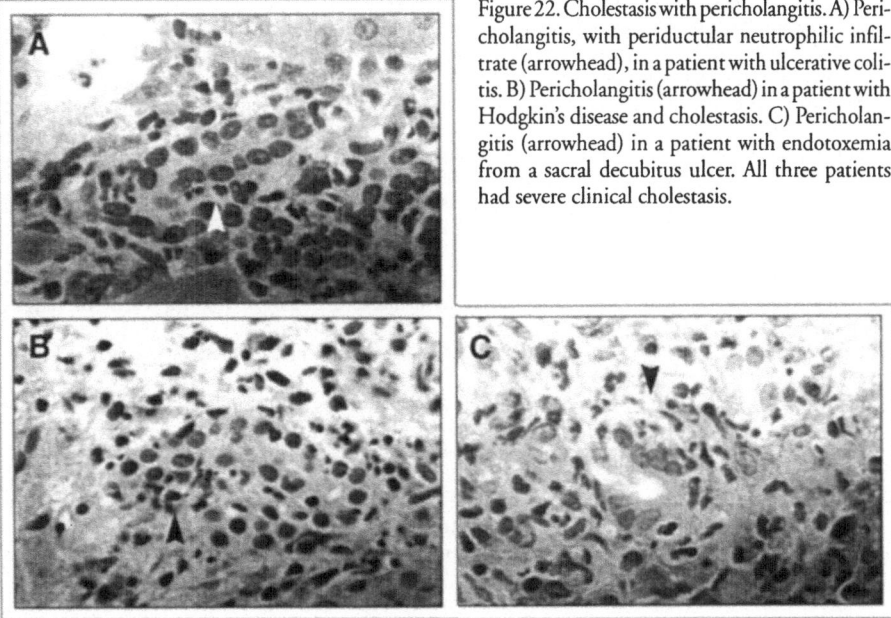

Figure 22. Cholestasis with pericholangitis. A) Pericholangitis, with periductular neutrophilic infiltrate (arrowhead), in a patient with ulcerative colitis. B) Pericholangitis (arrowhead) in a patient with Hodgkin's disease and cholestasis. C) Pericholangitis (arrowhead) in a patient with endotoxemia from a sacral decubitus ulcer. All three patients had severe clinical cholestasis.

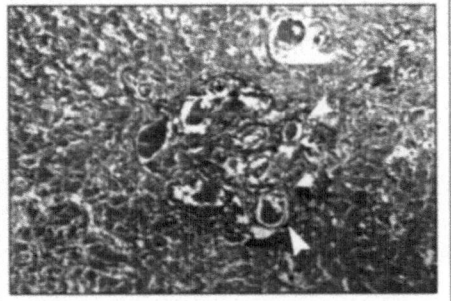

Figure 23. Cholangitis lenta. This low power image of a portal tract shows severe dilatation of bile ductules with inspissated bile (arrowheads).

Conclusion

This chapter provides an overview of the histological features of cholestasis, both as a guide to diagnosis and as a frame of reference for discussions of the pathobiology of cholestasis elsewhere in this volume. Ultrastructural features are not illustrated here, as electron microscopy plays much less of a diagnostic role in the current era of molecular diagnosis than it had previously. From the standpoint of clinical management, morphological features of cholestasis must always be interpreted in the context of clinical and laboratory findings. Findings diagnostic for any specific etiology are far outweighed by the many non-specific features of both non-obstructive and obstructive cholestasis.

References

1. Popper H, Szanto PB. Intrahepatic cholestasis (cholangiolitis). Gastroenterology 1956; 31:683-700.
2. Chen TS, Chen PS. Understanding the liver. A history. Westport: Greenwood Press, 1984:99-108.
3. Anderson J. On hemorrhage from the umbilicus after the separation of the fetus. Boston Medical and Surgical Journal 1850; 41:440-442.
4. Trauner M, Meier PJ, Boyer JL. Molecular pathogenesis of cholestasis. New Eng J Med 1998; 339:1217-1227.
5. Green RM, Crawford JM. Hepatocellular cholestasis: Pathobiology and histological outcome. Semin Liver Dis 1995; 15:372-389.
6. Crawford JM. Development of the intrahepatic biliary tree. Semin Liver Dis 2002; 22:213-226.
7. Nakanuma Y, Hoso M, Sanzen T et al. Microstructure and development of the normal and pathologic biliary tract in humans, including blood supply. Microsc Res Tech 1997; 38:552-570.
8. Healey JE, Schroy PC. Anatomy of the biliary ducts within the human liver. Arch Surg 1953; 66:599-616.
9. Desmet VJ. Congenital diseases of intrahepatic bile ducts: variations on the theme "ductal plate malformation". Hepatology 1992; 16:1069-1083.
10. Terada T, Nakanuma Y, Ohta G. Glandular elements around the intrahepatic bile ducts in man: their morphology and distribution in normal livers. Liver 7:1-8, 1987.
11. Landing BH, Wells TR. Considerations of some architectural properties of the biliary tree and liver in childhood. In: Abramowsky CR, Bernstein J, Rosenberg HS, eds. Transplantation Pathology—Hepatic Morphogenesis. Perspectives in Pediatric Pathology. Vol. 14. Basel: S. Karger, 1991:122-142.
12. Saxena R, Theise ND, Crawford JM. Microanatomy of the human liver: Exploring the hidden interfaces. Hepatology 1999; 30:1339-13469.
13. Washington K, Clavien P-A, Killenberg P. Peribiliary vascular plexus in primary sclerosing cholangitis and primary biliary sclerosis. Human Pathol 1997; 28:791-795.
14. Crawford AR, Lin XZ, Crawford JM. The normal adult human liver biopsy: A quantitative reference standard. Hepatology 1998; 28:323-331.
15. Elias H. A re-examination of the structure of the mammalian liver: II, The hepatic plate and its relation to the vascular and biliary systems. Am J Anat 1949; 85:379-456.
16. Ekataksin W, Wake K. New concepts in biliary and vascular anatomy of the liver. Prog Liver Dis 1997; 15:1-29.
17. Steiner JW, Carruthers JS. Studies on the fine structure of the terminal branches of the biliary tree. 1. The morphology of the normal bile canaliculi, bile pre-ductules (ducts of Hering) and bile ductules. Am J Pathol 1961; 38:639-661.
18. Roskams T, Van Eyken P, Desmet V. Human liver growth and development. In: Strain A, Diehl AM, eds. Liver Growth and Repair. London: Chapman & Hall, 1998:541-557.
19. Theise ND, Saxena R, Portmann BC et al. Canals of Hering and hepatic stem cells in humans. Hepatology 1999; 30:1425-1433.
20. Ekataksin W, Zou ZZ, Wake K et al. The hepatic microcirculatory subunits: an over-three-century-long-search for the missing link between an exocrine unit and an endocrine unit in mammalian liver lobules. In: Motta PM, ed. Recent Advances in Microscopy of Cells, Tissues and Organs. Rome: University of Rome La Sapienza, 1996:375-380.
21. Ekataksin W. The isolated artery: an intrahepatic arterial pathway that can bypass the lobular parenchyma in mammalian livers. Hepatology 2000; 31:269-279.
22. Sell S. Heterogeneity and plasticity of hepatocyte lineage cells. Hepatology 2001; 33: 738-750.
23. Jungermann K, Kietzmann T. Zonation of parenchymal and nonparenchymal metabolism in liver. Annu Rev Nutr 1996; 16:179-203.

24. Donner MG, Keppler D. Up-regulation of basolateral multidrug resistance protein 3 (Mrp3) in cholestatic rat liver. Hepatology 2001; 34:351-359.
25. Suchy FJ, Balistreri WF, Hung J et al. Intracellular bile acid transport in rat liver as visualized by electron microscope autoradiography using a bile acid analogue. Am J Physiol 1983; 245:G681-G689.
26. Buscher H-P, Schramm U, MacNelly S et al. The acinar location of the sodium-independent and the sodium- dependent component of taurocholate uptake. A histoautoradiographic study of rat liver. J Hepatol 1991; 13:169-178.
27. Stieger B, Hagenbuch B, Landmann L et al. In situ localization of the hepatocytic Na+ cotransporting polypeptide in rat liver. Gastroenterology 1994; 107:1781-1787.
28. Aiso M, Takikawa H, Yamanaka M. Biliary excretion of bile acids and organic anions in zone 1- and zone 3-injured rats. Liver 2000; 20:38-44.
29. Layden TJ, Boyer JL. Influence of bile acids on bile canalicular membrane morphology and the lobular gradient in canalicular size. Lab Invest 1978; 39:110-119.
30. Wilton JC, Chipman 3K, Lawson CJ et al. Periportal- and perivenous-enriched hepatocyte couplets: differences in canalicular activity and in response to oxidative stress. Biochem J 1993; 292:773-779.
31. Popper H, Schaffner F. Pathophysiology of cholestasis. Human Pathol 1970; 1:1-24.
32. Kitamura T, Alroy J, Gatmaitan Z et al. Defective biliary excretion of epinephrine metabolites in mutant (TR-) rats: relation to the pathogenesis of black liver in the Dubin-Johnson syndrome and Corriedale sheep with an analogous excretory defect. Hepatology 1992; 15:1154-1159.
33. Germain AM, Carvajal JA, Glasinovic JC et al. Intrahepatic cholestasis of pregnancy: an intriguing pregnancy-specific disorder. J Soc Gynecol Investig 2002; 9:10-14.
34. Stevens A. Pigments and minerals. In: Bancroft 3D, Stevens A, eds. Theory and practice of histological techniques. New York: Churchill Livingstone, 1982:245-249.
35. Searle S, Harmon BY, Bishop CJ et al. The significance of cell death by apoptosis in hepatobiliary disease. J Gastroenterol Hepatol 1987; 2:77-96.
36. Gross JB, Myers BM, Kos U et al. Biliary copper excretion by hepatocyte lysosomes in the rat: major excretory pathway in experimental copper overload. J Clin Invest 1989; 83:30-39.
37. Schaffner F, Bacehin PG, Hutterer F et al. Mechanism of cholestasis. 4. Structural and biochemical changes in the liver and serum in rats after bile duct ligation. Gastroenterology 1971; 60:888-897.
38. Nagore N, Howe 5, Boxer L et al. Liver cell rosettes: structural differences in cholestasis and hepatitis. Liver 1989; 9:43-51.
39. Clayton PT, Casteels M, Mieli-Vergani G et al. Familial giant cell hepatitis with low bile acid concentrations and increased urinary excretion of specific bile alcohols: a new inborn error of bile acid synthesis. Pediatr Res 1995; 37:424-431.
40. Alagille D, Odievre M, Gautier M et al. Hepatic ductular hypoplasia associated with characteristic facies, vertebral malformations, retarded physical, mental and sexual development, and cardiac murmur. J Pediatr 1975; 86:63-71.
41. Valencia-Mayoral P, Weber J, Cutz E et al. Possible defect in the bile secretory apparatus in arteriohepatic dysplasia (Alagille's syndrome): a review with observations on the ultrastructure of the liver. Hepatology 1984; 4:691-698.
42. Zollner G, Fickert P, Zenz R et al.. Hepatobiliary transporter expression in percutaneous liver biopsies of patients with cholestatic liver diseases. Hepatology 2001; 33:633-646.
43. Buscher H-P, Meder I, MacNelly S et al. Zonal changes of hepatobiliary taurocholate transport in intrahepatic cholestasis induced by 17-ethinyl estradiol: a histoautoradiographic study in rats. Hepatology 1993; 17:494-499.
44. Baumgartner U, Seholmerich I, Karseh 1 et al. Loss of zonal heterogeneity and cell polarity in rat liver with respect to bile acid secretion after bile drainage. Gastroenterobogy 1991; 100:1054-1061.
45. Buscher H-P, Miltenberger C, MacNelly S et al. The histoautoradiographic localizatian of taurocholar in rat liver after bile duct ligation. J Hepatal 1989; 8:181-191.
46. Tavoloni N, Schaffner F. The intrahepatic biliary epithelium in the guinea pig: is hepatic artery blood flow essential in maintaining its function and structure? Hepatology 1985; 5:666-672.
47. Benedetti A, Di Sario A, Marucci L et al. Carrier-mediated transport of conjugated bile acids across the basolateral membrane of biliary epithelial cells. Am J Physiol 1997; 272:G1416-G1424.
48. Crawford JM. Cellular and molecular biology of the inflamed liver. Curr Op Gastroenterol 1997; 13:175-185.
49. Hammel P, Couvelard A, O'Toole D et al. Regression of liver fibrosis after biliary drainage in patients with chronic pancreatitis and stenosis of the common bile duct. N Engl J Med 2001; 344:418-423.
50. Bach N, Thung SN, Schaffner F. The histological features of chronic hepatitis C and autoimmune chronic hepatitis: a comparative analysis. Hepatology 1992; 15:572-577.

51. Kumar KS, Saboorian MH, Lee WM. Cholestatic presentation of chronic hepatitis C. A clinical and histological study with a review of the literature. Dig Dis Sci 2001; 10:2066-2073.
52. Saxena R, Hytiroglou P, Thung SN et al. Destruction of canals of Hering in primary biliary cirrhosis. Hum Pathol 2002; 33:983-988.
53. Vleggaar FP, van Buuren HR, Zondervan PE et al. Jaundice in non-cirrhotic primary biliary cirrhosis: the premature ductopenic variant. Gut 2001; 49:276-281.
54. Teshima T, Ferrara JL. Understanding the alloresponse: new approaches to graft-versus-host disease prevention. Semin Hematol 2002; 39:15-22.
55. Snover DC. Acute and chronic graft-versus-host disease: histopathological evidence for two distinct pathogenetic mechanisms. Hum Pathol 1984; 15:202-205.
56. Snover DC. Acute and chronic graft-versus-host disease. In: Burakoff SJ, Geeg HJ, Ferrar J et al, eds. Graft-versus-host disease: Immunology, Pathophysiology, and Treatment. New York: Marcel Dekker, 1990:337-353.
57. Shulman HM. Pathology of chronic graft-vs.-host disease. In: Burakoff SJ, Geeg HJ, Ferrar J et al, eds. Graft-versus-host disease: Immunology, Pathophysiology, and Treatment. New York: Marcel Dekker, 1990:587-614.
58. Snover DC, Wiesdorf SA, Ramsay NK et al. Hepatic graft versus host disease: a study of the predictive value of liver biopsy in diagnosis. Hepatology 1984; 4:123-130.
59. Shulman HM, Sharma P, Amos D et al. A coded histologic study of hepatic graft-versus-host disease after human bone marrow transplantation Hepatology 1988; 8:463-470.
60. Shulman HM, Gooley T, Dudley MD et al. Utility of transvenous liver biopsies and wedged hepatic venous pressure measurements in sixty marrow transplant recipients. Transplantation 1995; 59:1015-1022.
61. Tanaka M, Umihara J, Shimmoto K et al. The pathogenesis of graft-versus-host reaction in the intrahepatic bile duct. An immunohistochemical study. Acta Pathol Jpn 1989; 39:648-655.
62. Strasser SI, Shulman HM, Flowers ME et al. Chronic graft-versus-host disease of the liver: presentation as an acute hepatitis. Hepatology 2000; 32:1265-1271.
63. Fujii N, Takenaka K, Shinagawa K et al. Hepatic graft-versus-host disease presenting as an acute hepatitis after allogeneic peripheral blood stem cell transplantation. Bone Marrow Transplant 2001; 27:1007-1010.
64. Akpek G, Boitnott JK, Lee LA et al. Hepatitic variant of graft-versus-host disease after donor lymphocyte infusion Blood 2002; 100:3903-3907.
65. Malik AH, Collins RH, Saboorian H et al. Chronic graft-versus-host disease after hematopoietic cell transplantation presenting as an acute hepatitis. Am J Gastroenterol 2001; 96:588-590.
66. Vierling JM. Immune disorders of the liver and bile duct. Gastroenterol Clin North Am 1992; 21:427-449.
67. Andersen CB, Horn T, Sehested M et al. Graft-versus-host disease: liver morphology and pheno/genotypes of inflammatory cells and target cells in sex-mismatched allogeneic bone marrow transplant patients. Transplant Proc 1993; 25:1250-1254.
68. Yeh KH, Hsieh HC, Tang JL et al. Severe isolated acute hepatic graft-versus-host disease with vanishing bile duct syndrome. Bone Marrow Transplant 1994; 14:319-321.
69. Yau JC, Zander AR, Srigley JR et al. Chronic graft-versus-host disease complicated by micronodular cirrhosis and esophageal varices. Transplantation 1986; 41:129-130.
70. Knapp AB, Crawford JM, Rappeport JM et al. Cirrhosis as a consequence of graft-versus-host disease. Gastroenterology 1987; 92:513-519.
71. Stechschulte DJ Jr, Fishback JL, Emami A et al. Secondary biliary cirrhosis as a consequence of graft-versus-host disease. Gastroenterology 1990; 98:223-225.
72. Urban CH, Deutschmann A, Kerbl R et al. Organ tolerance following cadaveric liver transplantation for chronic graft-versus-host disease after allogeneic bone marrow transplantation. Bone Marrow Transplant 2002; 30:535-537.
73. Knisely AS, Crawford JM. Inherited, metabolic and development disorders of the liver. In: Odze RD, Goldblum J, Crawford JM, eds. Surgical Pathology of the Alimentary Tract, Liver, Biliary Tree, and Pancreas. Philadelphia: WB Saunders, (in press).
74. Ishak KG, Sharp HL, Schwarzenberg. Metabolic errors and liver disease. In: MacSween RMN, Burt AD, Portmann BC et al, eds. Pathology of The Liver, 4th Ed. London: Churchill Livingstone, 2002:155-255.
75. Jonas MM, Perez-Atayde AR. Liver disease in infancy and childhood. In: Schiff ER, Sorrell MF, Maddrey WC, eds. Diseases of the Liver, 9th Ed. Philadelphia: Lippincott Williams & Wilkins, 2003:1459-1496.
76. Cocjin J, Rosenthal P, Buslon V et al. Bile ductule formation in fetal, neonatal, and infant livers compared with extrahepatic biliary atresia. Hepatology 1996; 24:568-574.

77. Koukoulis G, Mieli-Vergani G, Portmann B. Infantile liver giant cells: Immunohistological study of their proliferative state and possible mechanisms of formation. Pediatr Dev Pathol 1999; 2:353-359.

78. Raweily EA, Gibson AA, Burt AD. Abnormalities of intrahepatic bile ducts in extrahepatic biliary atresia. Histopathology 1990; 17:521-527.

79. Desmet VJ. Ludwig symposium on biliary disorders—part I. Pathogenesis of ductal plate abnormalities. Mayo Clin Proc 1998; 73:80-89.

80. Low Y, Vijayhan V, Tan CEL. The prognostic value of ductal plate malformation and other histologic parameters in biliary atresia: An immunohistochemical study. J Pediatr 2001; 139:320-322.

81. Tan CEL, Moscoso GJ. The developing human biliary system at the porta hepatis level between 29 days and 8 weeks of gestation: A way to understanding biliary atresia. Part 1. Pathol Int 1994; 44:587-599.

82. Crawford JM, Boyer JL. Clinical-Pathologic Conference: Inflammation-induced cholestasis. Hepatology 1998; 28:253-260.

83. Braier J, Ciocca M, Latella A et al. Cholestasis, sclerosing cholangitis, and liver transplantation in Langerhans cell histiocytosis. Med Pediatr Oncol 2002; 38:178-182.

CHAPTER 13

Disorders of Bile Acid Transport

Peter L.M. Jansen, Ekkehard Sturm and Michael Müller

Summary

B ile salts take part in a rather efficient enterohepatic circulation in which most of the secreted bile salts are reclaimed by absorption in the terminal ileum. In the liver the sodium dependent taurocholate transporter (NTCP) at the basolateral (sinusoidal) membrane and the bile salt export pump (BSEP) at the canalicular membrane mediate hepatic uptake and hepatobiliary secretion of bile salts. Transporter genes are transcriptionally regulated by members of the nuclear hormone receptor family. These are ligand activated transcription factors that respond to bile salts. At high bile salt concentrations the expression of NTCP is reduced and that of BSEP increased.

Canalicular secretion is the dominant element in the enterohepatic cycling of bile salts and most genetic diseases are caused by defects of canalicular secretion. Progressive familial intrahepatic cholestasis results from mutations in the *FIC1* gene. This, in a not well understood way, leads to relapsing or permanent cholestasis. The relapsing disease is called benign recurrent intrahepatic cholestasis (BRIC), the permanent cholestasis PFIC type 1. Byler disease is a PFIC type 1 disease. PFIC type 2 results from mutations in the *BSEP* gene. This leads to permanent cholestasis since birth.PFIC type 1 and type 2 are characterized by a low to normal serum gamma-glutamyltransferase activity. Bile diversion procedures, causing a decreased bile salt pool, have a beneficial effect in a number of patients with these diseases. However, liver transplantation is often necessary.

PFIC type 3 is caused by mutations in the *MDR3* gene. MDR3 is a phospholipid translocator in the canalicular membrane. Because of the inability to secrete phospholipids, patients with PFIC type 3 produce bile acid-rich toxic bile that damages the intrahepatic bile ducts. Ursodeoxycholic acid therapy is useful for patients with a partial defect. Liver transplantation is a more definite therapy for these patients. Among patients with intrahepatic cholestasis of pregnancy, heterozygosity for *MDR3* gene mutations is frequently found. In patients with MDR3 defects, the serum gamma-glutamyltransferase activity is elevated.

Introduction

Bile salts are the predominant organic solutes in bile, and their vectorial secretion from blood into bile represents the major driving force for hepatic bile formation. Although bile is isoosmotic in relation to plasma, bile salts are 100 to1000-fold concentrated in bile, necessitating active transport by hepatocytes against a concentration gradient. The total bile salt pool size in adult humans amounts to 50-60 mmol/kg body weight, corresponding to 3-4 g, and is largely stored in the gallbladder during the fasting state. Rats lack this reservoir function because of absence of a gallbladder. The human bile salt pool circulates 6 to 10 times per 24 hours, resulting in a daily bile salt secretion of 20-40 g. Despite a high degree of intestinal bile salt conservation, about 0.5 g of bile salts are lost through fecal excretion. This loss is compensated for by de novo hepatic bile salt synthesis, which contributes less than 3% of bile salts

---

Molecular Pathogenesis of Cholestasis, edited by Michael Trauner and Peter L.M. Jansen. ©2004 Eurekah.com and Kluwer Academic / Plenum Publishers.

secreted with hepatic bile. The intrinsic link between intestinal bile salt absorption and hepatic synthesis has recently been delineated by the discovery that hydrophobic bile salts can upregulate the ileal bile acid binding protein and downregulate hepatic cholesterol 7a-hydroxylase through the action of a nuclear bile salt receptor on gene transcription. By this mechanism, bile salts can regulate their own enterohepatic circulation.

Disturbances of bile salt transport across the basolateral and apical domains of the hepatocyte plasma membrane are important causes of acquired and genetic forms of cholestatic liver disease in humans. The consequences of defective bile salt secretion are not only progressive liver damage, but also metabolic derangements and malnutrition secondary to reduced intestinal absorption of lipids and fat-soluble vitamins. However, what holds for humans not necessarily is true for rats and mice. Humans with a genetic deficiency of the canalicular bile salt pump (BSEP or ABCB11) have severe cholestatic liver disease[1] but disrupting *bsep* expression in mice leads to a much more benign phenotype.[2] To explain this one has to realize that rodents have a considerably higher bile salt independent bile flow as humans. Thus a genetic deficiency of the bile salt export pump will cause a more severe breakdown of bile flow in humans than in rodents. Furthermore, mice produce less toxic bile salts and mice are able to activate alternate metabolic pathways for the detoxification of bile salts. Thus species differences are relevant.

Hepatic Transport Proteins

Hepatic Uptake

The Na$^+$/taurocholate cotransporting polypeptide (NTCP; SLC10A1) represents the major bile salt uptake system of hepatocytes[3] and is localized in the basolateral membrane of hepatocytes.[4-6] Ntcp preferentially mediates Na$^+$-dependent transport of conjugated bile salts such as taurocholate and this transport comprises the predominant fraction in hepatic bile salt uptake.[7] NTCP in the human liver (77 % homology with rat Ntcp) also transports conjugated bile salts, but NTCP has a higher affinity for taurocholate than Ntcp (K_M 6 µM compared with 25 µM).[8]

However, other uptake transporters exist for bile salts. In human liver (Table 1), OATP-C *(SLC21A6)* exhibits the highest (64%) amino acid identity with Oatp4 (Slc21a10) of rat liver. Transport substrates of OATP-C include a number of organic anions and OATP-C likely is the predominant Na$^+$-independent bile salt uptake system of human liver. OATP-A *(SLC21A3)* is a second human Na$^+$-independent bile salt uptake system. OATP-A was originally isolated from human liver but it is mainly expressed in human cerebral endothelial cells. Its substrates include bile salts and other organic anions as well as certain organic cations. OATP-B (SLC21A9) is most abundant in human liver but in contrast to OATP-C and OATP8 (SLC21A8), OATP-B transports no bile salts. OATP-B, OATP-C, and *OATP8* exhibit broad overlapping substrate specificities and account for the major part of sodium-independent organic anion and drug clearance of human liver.[9]

In rats, the organic anion transporting polypeptide 1 (Oatp1; *Slc21a1)* mediates Na$^+$-independent uptake of a broad spectrum of substrates.[3] Oatp2 *(Slc21a5)* exhibits a 77% amino acid identity with Oatp1. Its spectrum of transport substrates is similar, but not identical, to Oatp1. Substrates such as bilirubin monoglucuronide, BSP, or leukotriene C_4 are transported by Oatp1 but not by Oatp2. The cardiac glycosides ouabain and digoxin represent typical high affinity oatp2 substrates.[10] Treatment of rats with phenobarbital resulted in a significant increase in Oatp2 expression. Also pregnenolone-16α-carbonitrile, spironolactone and dexamethasone induce oatp2.[11] Thus, Oatp2 behaves as an inducible transporter. This is in line with the fact that the pregnane X receptor PXR, a member of the superfamily of nuclear hormone receptors responsible for cytochrome P450 3A induction, also induces Oatp2.[12] Induction of oatp2 with the prototypical PXR ligand pregnenolone-16α-carbonitrile was abolished in PXR-null mice.[12] Oatp4 *(Slc21a10) also* mediates Na$^+$-independent uptake of bile salts in rat hepatocytes. Similar to its homologues (43% with Oatp1; 64% with OATP2) Oatp4 has a broad substrate

Table 1. Human hepatic transporter proteins

Name	Gene	Chromosome	Size (TMS)	Localization	Transport Function	Phenotype
NTCP	*SLC10A1*	14q24.1-24.2	349[7]	H-BL	BS	Decreased in cholestasis
ASBT	*SLC10A2*	13q33	348[7]	CH-A	BS	Bile acid diarrhea
OCT1	*SLC22A1*	6q26	554[12]	H-BL	OC	
OAT-2	*SLC22A7*	6p21.2-21.1	548[12]	H-BL	OA	
OATP-A	*SLC21A3*	12p12	670[12]	H-BL	BS / OA / OC	
OATP-C	*SLC21A6*	12p	691[12]	H-BL	BS / OA / OC	
OATP8	*SLC21A8*	12p12	702[12]	H-BL	BS / OA / OC	
OATP-B	*SLC21A9*	11q13	709[12]	H-BL	OA	
ABC1	*ABCA1*	9q22-q31	2261[12]	H	PS, PC?	High density lipoprotein deficiency, Tangier type
MDR1	*ABCB1*	7q21	1279[12]	H-A, CH-A	Multispecific	Multidrug resistance
MDR3	*ABCB4*	7q21	1279[12]	H-A	PC	PFIC3, ICP
BSEP	*ABCB11*	2q24	1321[12]	H-A	BS	PFIC2
MRP1	*ABCC1*	16p13.1	1531[17]	H-BL, CH-BL	OA	Increased expression during cholestasis, Multidrug resistance
MRP2	*ABCC2*	10q24	1545[17]	H-A	OA	Dubin-Johnson syndrome
MRP3	*ABCC3*	17q21.3	1527[17]	H-BL, CH-BL	OA, BS	Increased expression during cholestasis
MRP6	*ABCC6*	16p13.1	1503[17]	H-BL	Peptides ?	Pseudoxanthoma elasticum
CFTR	*ABCC7*	7q31.2	1480[12]	CH-A	Chloride	Cystic fibrosis
ABCG2	*ABCG2*	4q22	655[6]	H-A	Chemotherapeutics	Multidrug resistance
ABCG5	*ABCG5*	2p21	651[6]	H-A	Sterols	Sitosterolemia
ABCG8	*ABCG8*	2p21	673[6]	H-A	sterols	Sitosterolemia
FIC1	*ATP8B1*	18q21-q22	1251[10]	H-A, Ch-A	Aminophospholipid translocation	PFIC1, BRIC
WND	*ATP7B*	13q14.3	1465[10]	H-INT	Copper secretion	Wilson disease

H, Hepatocytes; CH, cholangiocytes; A, apical; BL, basolateral; INT, intracellular. BS, bile salts; OA, organic anions; OC, organic cations; PC, phosphatidylcholine; PS, phosphatidylserine. PFIC, progressive familial intrahepatic cholestasis; ICP, intrahepatic cholestasis of pregnancy; MDR, multidrug resistance; TM, most probable number of transmembrane segments. Size in number of amino acids. The standard name in bold, the gene name assigned by the nomenclature committee (http://www.gene.ucl.ac.uk/nomenclature/genefamily/abc.html) is mentioned.

specificity.[13] The prostaglandin transporter (rPGT) (Slc21a2), OAT-K1/2 (Slc21a4) and Oatp3 (Slc21a7) are additional members of the SLC21A family. Oatp3 is also expressed in small intestine and may play a role in intestinal bile salt transport.[14]

Canalicular Bile Salt Transport

The bile salt excretory pump (BSEP, ABCB11) is critical for ATP-dependent transport of bile acids across the hepatocyte canalicular membrane and for generation of bile acid-dependent bile secretion.[15-18] Plasma membrane vesicles from Sf9 cells transfected with rat *Bsep* cDNA take up taurocholate in an ATP-dependent fashion with a Km of 5.3 μM. These vesicles also transport the taurine-conjugates of chenodeoxycholate and ursodeoxycholate, but not glycocholate and cholate. Murine Bsep was shown to transport taurocholate, taurochenodeoxycholate and glycocholate.[17,18] *Bsep* (-/-) mice are cholestatic in the sense that taurocholate accumulates in their plasma because its secretion into bile is strongly impaired.[2] However, in contrast to patients, the mice excrete substantial amounts of tauromuricholate as well as a hitherto undefined tetrahydroxy bile salt. Apparently, Bsep is not the only bile salt transporting system in the canalicular membrane since another system appears capable of excreting these hydrophilic bile salts. This escape route prevents severe and progressive cholestasis in *Bsep* (-/-) mice and as a consequence the animals have hardly any histopathological signs of liver injury. Because humans are not capable of converting bile salts into muricholate or tetrahydroxy bile salts to any significant extent, this escape route is not present in man(Fig. 1). Interestingly, *Bsep* (-/-) mice were found to have an increased biliary secretion of phospholipid and cholesterol.[2]

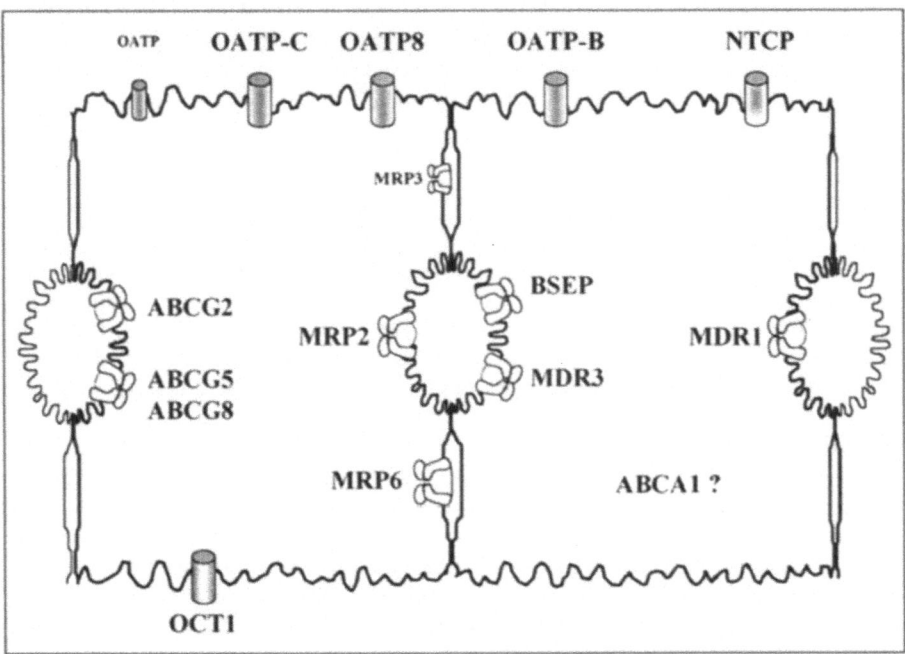

Figure 1. Human hepatic transporter proteins involved in bile formation: Transporter proteins located in the basolateral membrane are responsible for the the uptake of bile salts (NTCP, OATPs), bulky organic anions, uncharged compounds (OATPs) and cations (OATPs, OCT1). Transporter proteins located in the canalicular membrane are responsible for the biliary secretion of bile salts, PC, cholesterol and GSH and the excretion of drugs and toxins. These are: the bile salt transporter BSEP; the PC translocator MDR3; the anionic conjugate transporter MRP2; and the multidrug transporter MDR1. The organic anion transporter MRP3 is present at very low levels in normal human level but this level is strongly increased during cholestasis.

The regulation of rat Bsep has been studied under conditions of endotoxin-treatment,[19-21] bile-duct ligation and ethinylestradiol-induced cholestasis.[20] In the above mentioned cholestatic and stress-models Bsep mRNA and protein expression levels only slightly decreased compared to levels of the basolateral bile salt carriers Ntcp,[22] Oatp1 and Oatp2, or the canalicular transporter Mrp2.[20,23,24] Thus Bsep may continue to secrete bile salts, although at impaired rates. Remarkably, after partial hepatectomy the mRNA level of Bsep is only slightly decreased and the protein level of Bsep was unaffected in contrast to the bile salt uptake transporter Ntcp.[23,25] This may explain the fact that after partial hepatectomy the remnant liver is not cholestatic and not damaged by excess bile salts.

The Basolateral Escape Transporter MRP3

MRP3 (ABCC3) is a transporter protein that supports the basolateral export of organic anions, including bile salts, from hepatocytes.[26,27] Its expression in normal liver is low. Together with MRP2 it is also expressed in human gallbladder epithelia.[28] Mrp3 was found to be up-regulated in the Mrp2 deficient EHBR rat and in bile-duct ligated cholestatic rats.[29-31] Increased amounts of MRP3 are also detected in livers of Dubin-Johnson patients.[32] Considering the cellular localization of Mrp3, its up-regulation during cholestasis, and its substrate specificity, it is postulated that Mrp3 may play a significant role in the basolateral export of organic anions under conditions in which Mrp2 (or Bsep) is down-regulated. The inducible nature of the rat Mrp3 has recently been investigated. An increase in Mrp3 expression was observed in Gunn rats exhibiting hyperbilirubinemia due to a defect UDP-glucuronosyl transferase. Te elevated level of Mrp3 in the Gunn rat was associated with an increased level of unconjugated bilirubin, and after bile duct ligation with increased levels of bilirubin glucuronides.[26] These bilirubin species may induce the hepatic expression of Mrp3. The human MRP3 promoter has been cloned and several putative binding sites for transcription factors, including AP1, AP2, and SP1, have been identified.[33,34] A putative bile-salt responsive element was indentified in the MRP3 5'-UTR that may explain bile salt dependent upregulation under cholestatic conditions.[35]

Intestinal Reabsorption

In the ileum bile salts are reabsorbed. A sodium-dependent apical bile salt transporting protein ASBT (SLC10A2) and an intracellular bile salt binding protein play an important role in reclaiming bile salts from the intestinal lumen but other mechanisms for bile salt reabsorption probably co-exist.[36,37] About 90 percent of the total biliary bile salts are reabsorbed in the ileum. Other bile components and some drugs are also more or less avidly reabsorbed. In drug therapy the participation of drugs in the enterohepatic cycling adds to their biological half-life.

Regulation of Bile Secretion

Hepatocytes are strictly polarized cells. They absorb substrates from the blood and secrete metabolites into the bile. The supply of bile acids is highly variable. Absorption from the portal venous blood is nearly complete and, in the fasting state, mainly occurs in periportal hepatocytes. Thus periportal hepatocytes are continuously exposed to high bile salt concentrations. After a meal the portal venous bile salt concentration increases and hepatocytes, more down-stream into the hepatic acinus, are exposed to bile salts. Bile salts are cytotoxic and cellular homeostasis demands maintenance of intracellular bile salt concentrations within certain limits. Intracellular binding proteins act as buffer but bile salt pumps, in particularly BSEP, needs to pump a bile salt overload out of the cell. Post-translational regulations with recruitement of BSEP from intracellular stores to the canalicular membrane may be operational for this purpose.[38] Obstruction of bile efflux by gallstones or tumors will lead to a more chronic exposure of liver cells to bile salts. Transcriptional regulation may operate to support alterations of bile salt export pump expression in order to support cellular homeostasis and defense against bile salt toxicity. Thus, the cellular bile salt export pumps most probably are under the regulatory influence of short-term and long-term mechanisms.

Regulation of Gene Expression

Nuclear hormone receptors have been identified as important transcription factors in lipid metabolism. Many high-affinity ligands of NHRs are substrates for members of the ABC-transporter superfamily and their activity and functions are strongly interdependent. This relation is important for the physiological regulation of ABC-transporter genes and other NHR-target genes in vivo. The human bile salt export pump, BSEP, is under the transcriptional control of FXR (farnesoid X receptor)(Fig. 2).[39] The multispecific organic anion transporter Mrp2 (abcc2) is under the control of FXR, PXR (pregnane X receptor) and CAR (constitutive androstane receptor), at least in mice.[40] FXR is a ligand-activated transcription factor. Chenodeoxycholic acid, cholic acid, deoxycholic acid and lithocholic acid (and their taurine-conjugates) bind and activate FXR. FXR, PXR and CAR form complexes with the retinoid X-receptor and the heterodimers interact with response elements in the promotor region of the *BSEP* and *mrp2* genes. In addition, the FXR:RXR heterodimer binds to the *SHP-1* (small heterodimer protein-1) gene. SHP-1 suppresses the transcription of CYP7A1 and CYP8B by binding to a transcription factor called liver receptor homolog 1 (LRH-1).[41,42] SHP-1 also suppresses NTCP transcription.[43] Thus, FXR controls several key steps in bile salts metabolism: not only bile salt synthesis but also transporters involved in the enterohepatic cycling of bile salts. Studies in mice with a genetic disruption of FXR showed that the FXR-mediated response is particularly important in dealing with a bile salt load as occurs when feeding mice a high cholesterol- or cholate-containing diet. In FXR null mice the expression of Ntcp, Cyp7a1 and Cyp8b fails to be down-regulated and the expression of Bsep and SHP-1 is not increased, as occurs in wild type mice under these conditions.[44,45]

Infection, inflammation, and trauma induce a wide array of metabolic changes in the liver. The acute phase response is associated with a decrease of an array of NHR proteins such as RXRα, RXRβ, RXRγ, LXRα, PPARα, PPARγ, HNF1 and HNF4. Reduction of RXR levels, interference of cytokines with binding to its response element, as well as down-regulation of other NHRs, could all be mechanisms leading to the down-regulation of a large number of genes in the liver during the acute phase response.[46,47]

Cholestasis in rats is associated with a decreased expression of Ntcp.[48] This probably results from enhanced expression of SHP-1 through activation of FXR by retained bile salts.[43] Also in humans, NTCP expression in cholestatic liver disease is decreased.[49] Down-regulation of NTCP and CYP7A1 in cholestatic liver disease may be cytoprotective, reducing the entry and the synthesis of bile acids when intracellular bile acid levels are already elevated.

Genetic Transport Defects

The spectrum of diseases caused by defects of ABC-transporter proteins is diverse (Table 2, Fig. 3) and includes the liver diseases: progressive familial intrahepatic cholestasis,[50,51] benign recurrent intrahepatic cholestasis,[52-56] intrahepatic cholestasis of pregnancy,[57,58] cystic fibrosis,[59] adrenoleukodystrophy[60] and Dubin-Johnson syndrome.[61,62]

Progressive familial intrahepatic cholestasis, PFIC, constitutes a group of autosomal recessive diseases characterized by cholestasis starting in infancy. For a first differentiation of various PFIC subtypes measurement of the serum gamma-glutamyltransferase (gamma-GT) activity is useful. Diseases associated with low bile salt concentration in bile, have a low serum gamma-GT activity. These diseases have an intrahepatocellular blockade of bile salt secretion in common and could be called intrahepatocellular cholestasis rather than intrahepatic cholestasis. Gamma-GT in human liver is mainly located in the membranes lining the biliary tree. Elevation of serum gamma-GT results from a detergent, membranolytic effect of bile salts on these membranes. Thus either a blockade of bile flow downstream of the location of gamma-GT or bile containing bile salts not antagonized by neutralizing phosphatidylcholine causes gamma-GT to be released in the circulation. Elevated serum gamma-GT therefore is seen in both intrahepatic cholestasis as well as extrahepatic cholestasis.

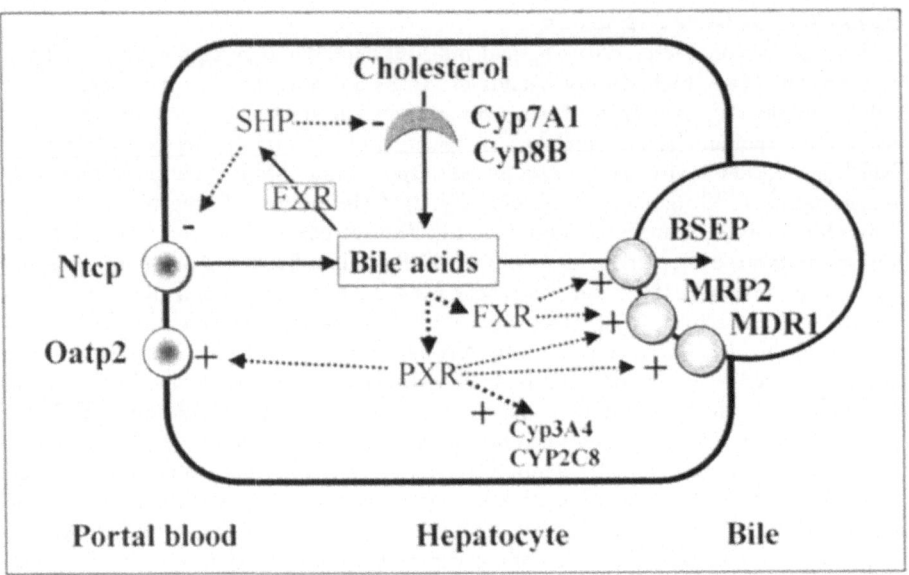

Figure 2. Hepatic gene regulation by bile salts and role of members of the nuclear hormone superfamily. Bile salts are taken up by Ntcp, and members of the Oatp family, and secreted into bile by BSEP. Bile salts serve as ligands for PXR and FXR that induce expression of SHP-1 (SHP). SHP-1 expression leads to a decrease in the expression of bile acid biosynthetic enzymes CYP7A1 and CYP8B as well as of NTCP. FXR activation by bile salts will induce *BSEP* and *MRP2* gene expression. PXR activation will induce *Oatp2, MRP2, MDR1* and *CYP3A4* and *CYP2C8* expression. OATP-C, is a bidirectional transporter and may facilitate efflux of hydroxylated bile salts into blood. MRP2 transports conjugates (sulfates, glucuronides) of CYP-metabolized bile salts. As a consequence of this bile salt-dependent gene regulation intracellular levels of potential toxic bile salts will decrease.

Progressive Familial Intrahepatic Cholestasis Type 1

PFIC type 1 or Byler disease often begins with episodes of cholestasis progressing to permanent cholestasis with fibrosis, cirrhosis and liver failure necessitating liver transplantation in the first two decades of life.[50,51,63-67] Children with PFIC are small for their age and, in addition to cholestasis and pruritus, they often have diarrhea, pancreatitis and occasionally also hearing loss.[68] The larger bile ducts are anatomically normal and liver histology shows bland canalicular cholestasis without much bile duct proliferation, inflammation, fibrosis or cirrhosis.[63,69,70] On electronmicroscopy there is a paucity of canalicular microvilli, a thickened pericanalicular network of microfilaments with in the canaliculi coarse granular bile called "Byler bile". Characteristically the serum gamma-GT activity is not elevated whilst parameters of cholestasis such as alkaline phosphatase and serum primary bile acids (in particular chenodeoxycholic acid) are strongly increased. Serum cholesterol levels are usually normal.

Patients belonging to the Byler kindred are descendants of Jacob and Nancy Byler who emigrated in the late 18th century from Germany to the United States. The PFIC syndrome has also been described in families in The Netherlands, Sweden, Greenland and an Arab population.[50,63-65,67,71,72] Most patients outside the United States are unrelated to the Amish. In the Amish and in some of the non-Amish families the genetic defect could be mapped to the *FIC1* locus on chromosome 18q21-q22. This *FIC1* locus could by narrowed by detailed homozygosity mapping and gene scanning studies to a region encoding a member of a recently defined subfamily of P-type ATPases. A number of *FIC1* mutations have been described. A single mutation is responsible for Byler's disease among the Amish.

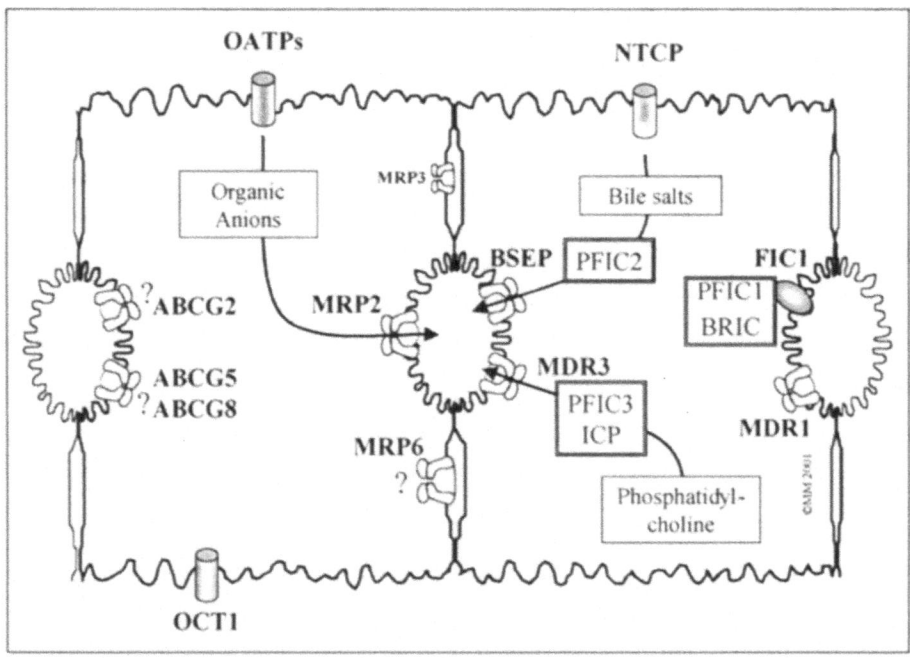

Figure 3. Genetic defects in genes encoding hepatic transporter proteins.
For more information see Table 2.

P-type ATPases belong to a large family of transport pumps such as Na^+/K^+ ATPase, Ca^{2+} ATPase and the copper transporting Wilson protein ATP7B. The function of FIC 1 is uncertain. Its homologue, the bovine P-type ATPase II, mediates the transport of aminophospholipids (i.e., phosphatidylserine) from the outer to the inner leaflet of plasma membranes.[73] In humans, FIC1 is highly expressed in pancreas, small intestine, urinary bladder, stomach and prostate. This may explain the increased frequency of diarrhea and pancreatitis in these patients but the relation with cholestasis, the hallmark of the disease, is not immediately apparent. In the liver the protein is apically located in cholangiocytes and hepatocytes. Fic1(-/-) knock-out mice do not suffer from cholestasis but upon feeding a cholate-containing diet they accumulate much more bile salts in their serum and liver than wild-type mice (Pawlowska et al; AASLD abstract 2001). How this relates to the phenotype observed in humans needs to be delineated.

Benign Recurrent Intrahepatic Cholestasis

Recurrent familial intrahepatic cholestasis is a term recently coined by Tygstrup et al.[55] This disease is also known under the name benign recurrent intrahepatic cholestasis (BRIC) or Summerskil syndrome. Despite recurrent attacks of cholestasis there is no progression to chronic liver disease. During the attacks the patients are severely jaundiced and have pruritus, steatorrhoea and weight loss. As in PFIC 1 the serum gamma-glutamyltransferase is not elevated. Some patients also have renal stones, pancreatitis and diabetes. The gene involved in recurrent familial intrahepatic cholestasis has been mapped to the *FIC1* locus.[53,54,56] This suggests that recurrent familial intrahepatic cholestasis and PFIC type I are genetically related. However not in all BRIC patients could the defect be traced to chromosome 18 mutations.[74] Ursodeoxycholic acid is of no benefit in BRIC.[75] Case reports indicate that rifampicine may reduce the number of cholestatic episodes.[76,77]

Table 2. Genetic cholestasis

Disease	Chromosome	Gene	Phenotype	Therapy
PFIC type 1	18q21	FIC 1 (ATP8B1) P-type ATPase, acts as a aminophospholipid translocator	First recurrent, later permanent cholestasis, bile duct proliferation is late phenomenon. Diarrhea, pancreatitis, pruritus, short stature. Coarse granular bile on EM. Normal GGT.	Ursodeoxycholic acid, bile diversion, liver transplantation
Benign recurrent intrahepatic cholestasis	18q21	FIC1	Recurrent episodes of cholestasis with severe pruritus, steatorrhea and weight loss. Nomal GGT.	Cholestyramine and/or rifampicine as symptomatic antipruritus therapy
PFIC type 2	2q24	BSEP (ABCB11), bile salt Export pump	Neonatal hepatitis, progressive cholestasis, pruritus, short stature, bile duct proliferation is late phenomenon, lobular and portal fibrosis. BSEP protein absent. Amorphous bile on EM. Normal GGT.	Ursodeoxycholic acid bile diversion, liver transplantation
PFIC type 3	7q21	PGY3 (ABCB4, MDR 3), P-glycoprotein 3	Cholestasis, portal hypertension, extensive bile duct proliferation and periportal fibrosis. MDR3 is not expressed. Elevated GGT.	Ursodeoxycholic acid, liver transplantation
Intrahepatic cholestasis of pregnancy	e.g., 7q21	e.g., MDR3	Cholestasis in third trimester of pregnancy. High GGT in case of MDR3 defect; low GGT cases may be caused by genetic defects of other transporter proteins.	Ursodeoxycholic acid causes symptomatic relief in the mother and decreases fetal loss
Bile acid synthesis defects	e.g., 8q2.3	3β-$\Delta 5$-C27-hydroxysteroid oxidoreductase; $\Delta 4$-3-oxosteroid-5β reductase; 3β-hydroxy C27 steroid dehydrogenase/isomerase; oxysterol 7alpha-hydroxylase; 24, 25-dihydroxy-cholanoic cleavage enzyme.	High incidence of fetal loss Intrahepatic cholestasis, neonatal giant cell hepatitis. Normal or elevated GGT, low or elevated serum total bile acids	Ursodeoxycholic acid, chenodeoxycholic acid or cholic acid alone or in combination, depending on subtype.

Progressive Familial Intrahepatic Cholestasis Type 2

Genetic studies revealed that the *FIC1* locus is not involved in all patients with a PFIC phenotype and low serum gamma-GT.[63,65,67,78] Moreover, in a large number of non-Amish patients the disease was mapped to a locus on chromosome 2q24 which later proved to be the *ABCB11 (BSEP)* gene.[16,51,79] Antibodies directed against BSEP sequences enabled localization studies and it became clear that this protein not only is liver-specific but is located in the canalicular domain of the hepatocyte's plasma membrane. Liver specimens of patients with PFIC type 2 stain negative for canalicular BSEP on immunohistochemistry using BSEP-antibodies.[1] As in PFIC type 1, the serum gamma-GT activity in these patients is not elevated and bile duct proliferation is absent. However, there are also some differences with PFIC type 1: in PFIC2 the disease frequently starts as nonspecific giant cell hepatitis which is undistinguishable from idiopathic neonatal giant cell hepatitis; patients are usually permanently jaundiced and the disease rapidly progresses to persistent and progressive cholestasis requiring liver transplantation. The liver histology shows more inflammatory activity than in PFIC type 1, with giant cell transformation, lobular and portal fibrosis.[1,50,51,63] The bile of PFIC type 2 patients is amorphous or filamentous on transmission electronmicroscopy. This contrasts with the coarsely granular bile of PFIC type 1 patients. Extrahepatic manifestations are uncommon. In the majority of non-Amish patients, progressive familial intrahepatic cholestasis is type 2 rather than type 1.

Bile acids are not completely absent in the bile of these patients. MRP2, the canalicular transporter of bilirubin and other organic anions also transports glucuronidated or sulfated bile acids.[80] This transporter may act as an escape pathway under conditions of cholestasis. This may also explain why these patients are jaundiced despite an intact bilirubin transporter: bilirubin transport may be inhibited by bile acid conjugates.

Bile Acid Synthesis Defects

Defects of bile acid synthesis phenotypically resemble PFIC type 2. Clayton et al. described a defect of 3β-Δ5-C27-hydroxysteroid oxidoreductase as a cause of giant cell hepatitis.[81] Deficiency of Δ4-3-oxosteroid-5β reductase and 3β-hydroxy C27steroid dehydrogenase/isomerase and mutations of the oxysterol 7alpha-hydroxylase gene may also be causes of neonatal hepatitis and cholestasis.[1,82,83] In these diseases toxic intermediates are formed which cause cholestasis by interaction with the hepatic bile acid transporter.[84] Bile acid synthesis defects are called PFIC type 4 by some authors.

Progressive Familial Intrahepatic Cholestasis Type 3

The third PFIC subtype, PFIC type 3, is quite different from the other PFIC subtypes and is described elsewhere in this book. In patients with PFIC type 3 symptoms present somewhat later in life than in PFIC types 1 and 2 and liver failure also occurs at a later age. Jaundice may be less apparent. The serum gamma-GT activity is usually markedly elevated in these patients and the liver histology shows extensive bile duct proliferation, portal and periportal fibrosis.[85,86] In humans, the *ABCB4 (MDR3)* gene is mutated in this disease.[57,58,85,86] Patients with a partial *ABCB4* defect respond to ursodeoxycholic acid therapy.[87] The majority of patients however has to be transplanted.

Therapy

Therapeutic interventions for PFIC have not been thoroughly studied in relation to the disease genotypes. Thus, therapy is summarized here according to earlier categorization for the low GGT PFIC (types 1 and 2) and high GGT PFIC (type 3). Subgroups of PFIC types 1-3 may respond to ursodeoxycholic acid.[88] However, progression of disease and insufficient control of symptoms may necessitate further intervention. Ursodeoxycholic may also lead to acceleration of disease[88] and even to very high serum bile acid levels (> 1 mmol/l) without any increase of bile salt secretion.[1] PFIC 3 patients are more likely to respond to ursodeoxycholic acid therapy if they carry mild mutations of the *MDR3* gene.[86,87]

For PFIC types 1 and 2 partial external biliary diversion (PEBD) is an accepted mode of therapy.[89,90] The majority of patients respond with a significant improvement of symptoms.[90-92] Error! No bookmark name given.. When performed early, ongoing hepatic injury may be interrupted with resolution of histological abnormalities including reversal of fibrosis.[92] Clinically, the patients may experience long-term amelioration of pruritus and induction of catch-up growth.[90] Why PEBD improves the clinical picture is still a matter of debate.

Liver transplantation has been performed for patients with PFIC types 1-3. In general, these patients have a good prognosis after successful transplantation.[93] Since PFIC type 1 is a multiorgan disease, diarrhea may persist and pancreatitis may occur after transplantation. Diarrhea in these patients is associated with malabsorption, therefore catch-up growth after transplantation may be disappointing. Cholestyramine may be of symptomatic benefit but does not improve the malabsorption

Intrahepatic Cholestasis of Pregnancy

Jacquemin et al reported a high incidence of intrahepatic cholestasis of pregnancy (ICP) in families with PFIC type 3.[58] This suggests that in persons carrying one mutated *ABCB4* gene, cholestasis may occur during pregnancy. Mutations leading to single amino acid substitutions of the MDR3 protein may cause intracellular traffic mutants, that is the protein is synthesized but does not reach its destination in the canalicular membrane.[94] One can hypothesize that in patients, carrying these mutations, the hormones in the third trimester of pregnancy impair the intracellular targeting which causes the disease to become clinically manifest.

ICP has also been described in families with other PFIC subtypes.[95] Thus, ICP may be associated either with elevated serum gamma-glutamyltransferase or with a normal enzyme activity.

Ursodeoxycholic acid has been shown to be of benefit in these patients with also a decrease of fetal loss.[96-98]

Other Forms of Intrahepatic Cholestasis

There are numerous other genetic transport defects affecting the liver (listed in Table 3). For example, Aagenaes syndrome is a combination of severe progressive lympoedema and episodic intrahepatic cholestasis.[99] The locus for this disease has been mapped to chromosome 15q.[100] Recently Morton et al described an Amish kindred with a disease phenotypically characterized by elevated serum bile acids not due to either a *FIC1* or a *ABCB11* mutation. Most likely this disease represents a bile salt uptake defect, however no direct proof for this is available yet.[101]

Acknowledgements

Our research is supported by the Netherlands Organization for Scientific Research, NWO program grant 902-23-191 and NWO project grants 902-23-253, 902-23-257 and 903-39-188.

References

1. Jansen PL, Strautnieks SS, Jacquemin E et al. Hepatocanalicular bile salt export pump deficiency in patients with progressive familial intrahepatic cholestasis. Gastroenterology 1999; 117:1370-1379.
2. Wang R, alem M, Yousef IM et al. Targeted inactivation of sister of P-glycoprotein gene (spgp) in mice results in nonprogressive but persistent intrahepatic cholestasis. Proc Natl Acad Sci USA 2001; 98:2011-2016.
3. Kullak-Ublick GA, Stieger B, Hagenbuch B et al. Hepatic transport of bile salts. Semin Liver Dis 2000; 20:273-292.
4. Stieger B, Hagenbuch B, Landmann L et al. In situ localization of the hepatocytic Na+/Taurocholate cotransporting polypeptide in rat liver. Gastroenterology 1994; 107:1781-1787.
5. Ananthanarayanan M, Ng OG, Boyer JL et al. Characterization of cloned rat liver Na(+)-bile acid cotransporter using peptide and fusion protein antibodies. Am J Physiol 1994; 267:G637-G643.

Table 3. Familial cholestatic diseases with unknown etiology

Disease	Characteristics	Reference
Tubulo-interstitial nephropathy and cholestatic liver disease	Progressive renal failure and elevated liver enzymes, sclerosing cholangitis	102-104
Familial sclerosing cholangitis, adult form	Laboratory and pathology features of primary sclerosing cholangitis	105
Neonatal sclerosing cholangitis	Intra- and extrahepatic cholangitis, cholestasis, hepatosplenomegaly	106,107
Defect in sinusoidal bile acid transport	Pruritus, poor growth, severe bleeding episodes	101
Infantile cirrhosis in Northern Germany	Normal-high GGT, lethargy, hepatosplenomegaly, jaundice, idiopathic copper toxicosis	108
Tyrolean childhood cirrhosis	Idiopathic copper toxicosis, jaundice, irritability, hepatosplenomegaly	109,110
Indian childhood cirrhosis	Defect in handling of dietary copper, irritability, high GGT activity, hepatosplenomegaly, variceal bleeding	111-114
North American Indian Childhood cirrhosis in Ontario	Associated with zinc toxicity, high GGT activity	115
North American Indian Cholestasis in Quebec	Hepatosplenomegaly, pruritus, portal hypertension, high GGT activity	116
Navajo Neuro(hepato)pathy	Jaundice, failure to thrive, neurological dysfunction, high or low GGT	117,118
ARC-Syndrome	Arthrogryposis, renal dysfunction and cholestasis, normal GGT	119
Aagenaes Syndrome	Cholestasis of neonatal onset, lymphedema, locus on chromosome 15	99,100

6. Shi X, Bai S, Ford AC et al. Stable inducible expression of a functional rat liver organic anion transport protein in HeLa cells. J Biol Chem 1995; 270:25591-25595.
7. Hagenbuch B, Scharschmidt BF, Meier PJ. Effect of antisense oligonucleotides on the expression of hepatocellular bile acid and organic anion uptake systems in Xenopus laevis oocytes. Biochem J 1996; 316(Pt3):901-904.
8. Hagenbuch B, Meier PJ. Sinusoidal (basolateral) bile salt uptake systems of hepatocytes. Semin Liver Dis 1996; 16:129-136.
9. Kullak-Ublick GA, Ismair MG, Stieger B et al. Organic anion-transporting polypeptide B (OATP-B) and its functional comparison with three other OATPs of human liver. Gastroenterology 2001; 120:525-533.
10. Noe B, Hagenbuch B, Stieger B et al. Isolation of a multispecific organic anion and cardiac glycoside transporter from rat brain. Proc Natl Acad Sci USA 1997; 94:10346-10350.
11. Guo GL, Choudhuri S, Klaassen CD. Induction Profile of Rat Organic anion transporting Polypeptide 2 (oatp2) by Prototypical Drug-Metabolizing Enzyme Inducers That Activate Gene Expression through Ligand-Activated Transcription factor Pathways. J Pharmacol Exp Ther 2002; 300:206-212.
12. Staudinger JL, Goodwin B, Jones SA et al. The nuclear receptor PXR is a lithocholic acid sensor that protects against liver toxicity. Proc Natl Acad Sci USA 2001; 98:3369-3374.
13. Cattori V, van Montfoort JE, Stieger B et al. Localization of organic anion transporting polypeptide 4 (Oatp4) in rat liver and comparison of its substrate specificity with Oatp1, Oatp2 and Oatp3. Pflugers Arch 2001; 443:188-195.
14. Walters HC, Craddock AL, Fusegawa H et al. Expression, transport properties, and chromosomal location of organic anion transporter subtype 3. Am J Physiol Gastrointest Liver Physiol 2000; 279:G1188-G1200.

15. Gerloff T, Stieger B, Hagenbuch B et al. The sister of P-glycoprotein represents the canalicular bile salt export pump of mammalian liver. J Biol Chem 1998; 273:10046-10050.
16. Strautnieks SS, Bull LN, Knisely AS et al. A gene encoding a liver-specific ABC transporter is mutated in progressive familial intrahepatic cholestasis. Nat Genet 1998; 20:233-238.
17. Lecureur V, Sun D, Hargrove P et al. Cloning and expression of murine sister of P-glycoprotein reveals a more discriminating transporter than MDR1/P-glycoprotein. Mol Pharmacol 2000; 57:4-35.
18. Green RM, Hoda F, Ward KL. Molecular cloning and characterization of the murine bile salt export pump. Gene 2000; 241:117-123.
19. Vos TA, Hooiveld GJ, Koning H et al. Up-regulation of the multidrug resistance genes, Mrp1 and Mdr1b, and down-regulation of the organic anion transporter, Mrp2, and the bile salt transporter, Spgp, in endotoxemic rat liver. Hepatology 1998; 28:1637-1644.
20. Lee JM, Trauner M, Soroka CJ et al. Expression of the bile salt export pump is maintained after chronic cholestasis in the rat. Gastroenterology 2000; 118:163-172.
21. Oude Elferink RP, Jansen PL. The role of the canalicular multispecific organic anion transporter in the disposal of. Pharmacol Ther 1994; 64:77-97.
22. Trauner M, Arrese M, Lee H et al. Endotoxin downregulates rat hepatic ntcp gene expression via decreased activity of critical transcription factors. J Clin Invest 1998; 101:2092-2100.
23. Vos TA, Ros JE, Havinga R et al. Regulation of hepatic transport systems involved in bile secretion during liver regeneration in rats. Hepatology 1999; 29:1833-1839.
24. Trauner M, Meier PJ, Boyer JL. Molecular pathogenesis of cholestasis. N Engl J Med 1998; 339:1217-1227.
25. Gerloff T, Geier A, Stieger B et al. Differential expression of basolateral and canalicular organic anion transporters during regeneration of rat liver. Gastroenterology 1999; 117:1408-1415.
26. Ogawa K, Suzuki H, Hirohashi T et al. Characterization of inducible nature of MRP3 in rat liver. Am J Physiol Gastrointest Liver Physiol 2000; 278:G438-G446.
27. Hirohashi T, Suzuki H, Takikawa H et al. ATP-dependent transport of bile salts by rat multidrug resistance- associated protein 3 (Mrp3). J Biol Chem 2000; 275:2905-2910.
28. Donner MG, Keppler D. Up-regulation of basolateral multidrug resistance protein 3 (Mrp3) in cholestatic rat liver. Hepatology 2001; 34:351-359.
29. Kiuchi Y, Suzuki H, Hirohashi T et al. cDNA cloning and inducible expression of human multidrug resistance associated protein 3 (MRP3). FEBS Lett 1998; 433:149-152.
30. Hirohashi T, Suzuki H, Ito K et al. Hepatic expression of multidrug resistance-associated protein-like proteins maintained in Eisai hyperbilirubinemic rats. Mol Pharmacol 1998; 53:1068-1075.
31. Konig J, Nies AT, Cui Y et al. Conjugate export pumps of the multidrug resistance protein (MRP) family: localization, substrate specificity, and MRP2-mediated drug resistance. Biochim Biophys Acta 1999; 1461:377-394.
32. Konig J, Rost D, Cui Y et al. Characterization of the human multidrug resistance protein isoform MRP3 localized to the basolateral hepatocyte membrane. Hepatology 1999; 29:1156-1163.
33. Stockel B, Konig J, Nies AT et al. Characterization of the 5'-flanking region of the human multidrug resistance protein 2 (MRP2) gene and its regulation in comparison withthe multidrug resistance protein 3 (MRP3) gene. Eur J Biochem 2000; 267:1347-1358.
34. Fromm MF, Leake B, Roden DM et al. Human MRP3 transporter: identification of the 5'-flanking region, genomic organization and alternative splice variants. Biochim Biophys Acta 1999; 1415:369-374.
35. Inokuchi A, Hinoshita E, Iwamoto Y et al. Enhanced Expression of the Human Multidrug resistance protein 3 by Bile salt in Human Enterocytes. A transcriptional control of a plausible bile acid transporter. J Biol Chem 2001; 276:46822-46829.
36. Oelkers P, Kirby LC, Heubi JE et al. Primary bile acid malabsorption caused by mutations in the ileal sodium- dependent bile acid transporter gene (SLC10A2). J Clin Invest 1997; 99:1880-1887.
37. Wong MH, Oelkers P, Dawson PA. Identification of a mutation in the ileal sodium-dependent bile acid transporter gene that abolishes transport activity. J Biol Chem 1995; 270:27228-27234.
38. Kipp H, Pichetshote N, Arias IM. Transporters on demand: intrahepatic pools of canalicular ATP binding cassette transporters in rat liver. J Biol Chem 2001; 276:7218-7224.
39. Ananthanarayanan M, Balasubramanian N, Makishima M et al. Human bile salt export pump promoter is transactivated by the farnesoid X receptor/bile acid receptor. J Biol Chem 2001; 276:28857-28865.
40. Kast HR, Goodwin B, Tarr PT et al. Regulation of multidrug resistance-associated protein 2 (MRP2;ABCC2) by the nuclear receptors PXR, FXR, and CAR. J Biol Chem 2001.
41. Lu TT, Makishima TM, Repa JJ et al. Molecular basis for feedback regulation of bile acid synthesis by nuclear receptors. Mol Cell 2000; 6:507-515.

42. Goodwin B, Jones SA, Price RR et al. A regulatory cascade of the nuclear receptors FXR, SHP-1, and LRH-1 represses bile acid biosynthesis. Mol Cell 2000; 6:517-526.

43. Denson LA, Sturm E, Echevarria W et al. The orphan nuclear receptor, shp, mediates bile acid-induced inhibition of the rat bile acid transporter, ntcp. Gastroenterology 2001; 121:140-147.

44. Schuetz EG, Strom S, Yasuda K et al. Disrupted bile acid homeostasis reveals an unexpected interaction among nuclear hormone receptors, transporters, and cytochrome P450. J Biol Chem 2001; 276:39411-39418.

45. Sinal CJ, Tohkin M, Miyata M et al. Targeted disruption of the nuclear receptor FXR/BAR impairs bile acid and lipid homeostasis. Cell 2000; 102:731-744.

46. Beigneux AP, Moser AH, Shigenaga JK et al. The acute phase response is associated with retinoid X receptor repression in rodent liver. J Biol Chem 2000; 275:16390-16399.

47. Denson LA, Auld KL, Schiek DS et al. Interleukin-1beta suppresses retinoid transactivation of two hepatic transporter genes involved in bile formation. J Biol Chem 2000; 275:8835-8843.

48. Gartung C, Ananthanarayanan M, Rahman MA et al. Down-regulation of expression and function of the rat liver Na+/bile acid cotransporter in extrahepatic cholestasis. Gastroenterology 1996; 110:199-209.

49. Zollner G, Fickert P, Zenz R et al. Hepatobiliary transporter expression in percutaneous liver biopsies of patients with cholestatic liver diseases. Hepatology 2001; 33:633-646.

50. Thompson R, Jansen PL. Genetic defects in hepatocanalicular transport. Semin Liver Dis 2000; 20:365-372.

51. Jacquemin E, Hadchouel M. Genetic basis of progressive familial intrahepatic cholestasis. J Hepatol 1999; 31:377-381.

52. Sinke RJ, Carlton VE, Juijn JA et al. Benign recurrent intrahepatic cholestasis (BRIC): evidence of genetic heterogeneity and delimitation of the BRIC locus to a 7-cM interval between D18S69 and D18S64. Hum Genet 1997; 100:382-387.

53. Bull LN, Juijn JA, Liao M et al. Fine-resolution mapping by haplotype evaluation: the examples of PFIC1 and BRIC. Hum Genet 1999; 104:241-248.

54. Bull LN, van Eijk MJ, Pawlikowska L et al. A gene encoding a P-type ATPase mutated in two forms of hereditary cholestasis. Nat Genet 1998; 18:219-224.

55. Tygstrup N, Steig BA, Juijn JA et al. Recurrent familial intrahepatic cholestasis in the Faeroe Islands. Phenotypic heterogeneity but genetic homogeneity. Hepatology 1999; 29:506-508.

56. Klomp LW, Bull LN, Knisely AS etal. A missense mutation in FIC1 is associated with Greenland familial cholestasis. Hepatology 2000; 32:1337-41.

57. De Vree JM, Jacquemin E, Sturm E et al. Mutations in the MDR3 gene cause progressive familial intrahepatic cholestasis. Proc Natl Acad Sci USA 1998; 95:282-287.

58. Jacquemin E, Cresteil D, Manouvrier S et al. Heterozygous non-sense mutation of the MDR3 gene in familial intrahepatic cholestasis of pregnancy. Lancet 1999; 353:210-211.

59. Zielenski J, Tsui LC. Cystic fibrosis: genotypic and phenotypic variations. Annu Rev Genet 1995; 29:777-807.

60. Mosser J, Lutz Y, Stoeckel ME et al. The gene responsible for adrenoleukodystrophy encodes a peroxisomal membrane protein. Hum Mol Genet 1994; 3:265-271.

61. Paulusma CC, Kool M, Bosma PJ et al. A mutation in the human canalicular multispecific organic anion transporter gene causes the Dubin-Johnson syndrome. Hepatology 1997; 25:1539-1542.

62. Kartenbeck J, Leuschner U, Mayer R et al. Absence of the canalicular isoform of the MRP gene-encoded conjugate export pump from the hepatocytes in Dubin-Johnson syndrome. Hepatology 1996; 23:1061-1066.

63. Bull LN, Carlton VE, Stricker NL et al. Genetic and morphological findings in progressive familial intrahepatic cholestasis (Byler disease [PFIC-1] and Byler syndrome): evidence for heterogeneity. Hepatology 1997; 26:155-164.

64. Bourke B, Goggin N, Walsh D et al. Byler-like familial cholestasis in an extended kindred. Arch Dis Child 1996; 75:223-227.

65. Arnell H, Nemeth A, Anneren G et al. Progressive familial intrahepatic cholestasis (PFIC): evidence for genetic heterogeneity by exclusion of linkage to chromosome 18q21-q22. Hum Genet 1997; 100:378-381.

66. Whitington PF, Freese DK, Alonso EM et al. Clinical and biochemical findings in progressive familial intrahepatic cholestasis. J Paediatr Gastroenterol Nutr 1994; 18:134-41.

67. Van Ooteghem NA, Klomp LW, van Berge-Henegouwen GP et al. Benign recurrent intrahepatic cholestasis progressing to progressive familial intrahepatic cholestasis: low GGT cholestasis is a clinical continuum. J Hepatol 2002; 36:439-43.

68. Oshima T, Ikeda K, Takasaka T. Sensorineural hearing loss associated with Byler disease. Tohoku J Exp Med 1999; 187:83-88.

69. Alonso EM, Snover DC, Montag A et al. Histologic pathology of the liver in progressive familial intrahepatic cholestasis. J Pediatr Gastroenterol Nutr 1994; 18:128-133.
70. Jansen PL, Muller M. Early events in sepsis-associated cholestasis. Gastroenterology 1999; 116:486-488.
71. Naveh Y, Bassan L, Rosenthal E et al. Progressive familial intrahepatic cholestasis among the Arab population in Israel. J Pediatr Gastroenterol Nutr 1997; 24:548-554.
72. Kagalwalla AF, Al Amir AR, Khalifa A et al. Progressive familial intrahepatic cholestasis (Byler's disease) in Arab children. Ann Trop Paediatr 1995; 15:321-327.
73. Ding J, Wu Z, Crider BP et al. Identification and functional expression of four isoforms of AT-Pase II, the putative aminophospholipid translocase. Effect of isoform variation on the ATPase activity and phospholipid specificity. J Biol Chem 2000; 275:23378-23386.
74. Floreani A, Molaro M, Mottes M et al. Autosomal dominant benign recurrent intrahepatic cholestasis (BRIC) unlinked to 18q21 and 2q24. Am J Med Genet 2000; 95:450-453.
75. Brenard R, Geubel AP, Benhamou JP. Benign recurrent intrahepatic cholestasis. A report of 26 cases. J Clin Gastroenterol 1989; 11:546-551.
76. Cancado EL, Leitao RM, Carrilho FJ et al. Unexpected clinical remission of cholestasis after rifampicin therapy in patients with normal or slightly increased levels of gamma-glutamyl transpeptidase. Am J Gastroenterol 1998; 93:1510-1517.
77. Balsells F, Wyllie R, Steffen R et al. Benign recurrent intrahepatic cholestasis: improvement of pruritus and shortening of the symptomatic phase with rifampin therapy: a case report. Clin Pediatr (Phila) 1997; 36:483-485.
78. Strautnieks SS, Kagalwalla AF, Tanner MS et al. Locus heterogeneity in progressive familial intrahepatic cholestasis J Med Genet 1996; 33:833-6.
79. Strautnieks SS, Kagalwalla AF, Tanner MS et al. Identification of a locus for progressive familial intrahepatic cholestasis PFIC2 on chromosome 2q24. Am J Hum Genet 1997; 61:630-3.
80. Oude Elferink RP, de Haan J, Lambert KJ et al. Selective hepatobiliary transport of nordeoxycholate side chain conjugates in mutant rats with a canalicular transport defect. Hepatology 1989; 9:861-865.
81. Clayton PT, Leonard JV, Lawson AM et al. Familial giant cell hepatitis associated with synthesis of 3 beta, 7 alpha-dihydroxy-and 3 beta,7 alpha, 12 alpha-trihydroxy-5-cholenoic acids. J Clin Invest 1987; 79:1031-1038.
82. Jacquemin E, Setchell KD, O'Connell NC et al. A new cause of progressive intrahepatic cholestasis: 3 beta-hydroxy-C27- steroid dehydrogenase/isomerase deficiency. J Pediatr 1994; 125:379-384.
83. Setchell KD, Schwarz M, O'Connell NC et al. Identification of a new inborn error in bile acid synthesis: mutation of the oxysterol 7alpha-hydroxylase gene causes severe neonatal liver disease. J Clin Invest 1998; 102:1690-1703.
84. Stieger B, Zhang J, O'Neill B et al. Differential interaction of bile acids from patients with inborn errors of bile acid synthesis with hepatocellular bile acid transporters. Eur J Biochem 1997; 244:39-44.
85. Deleuze JF, Jacquemin E, Dubuisson C et al. Defect of multidrug-resistance 3 gene expression in a subtype of progressive familial intrahepatic cholestasis. Hepatology 1996; 23:904-908.
86. Jacquemin E, De Vree JM, Cresteil D et al. The wide spectrum of multidrug resistance 3 deficiency: from neonatal cholestasis to cirrhosis of adulthood. Gastroenterology 2001; 120:1448-1458.
87. Jacquemin E, Hermans D,Myara A et al. Ursodeoxycholic acid therapy in pediatric patients with progressive familial intrahepatic cholestasis. Hepatology 1997; 25:519-523.
88. Jacquemin E. Progressive familial intrahepatic cholestasis. J Gastroenterol Hepatol 1999; 14:594-599.
89. Emond JC, Whitington PF. Selective surgical management of progressive familial intrahepatic cholestasis (Byler's disease). J Pediatr Surg 1995; 30:1635-1641.
90. Melter M, Rodeck B, Kardorff R et al. Progressive familial intrahepatic cholestasis: partial biliary diversion normalizes serum lipids and improves growth in noncirrhotic patients. Am J Gastroenterol 2000; 95:3522-3528.
91. Rebhandl W, Felberbauer FX, Turnbull J et al. Biliary diversion by use of the appendix (cholecystoappendicostomy) in progressive familial intrahepatic cholestasis. J Pediatr Gastroenterol Nutr 1999; 28:217-219.
92. Whitington PF, Whitington GL. Partial external diversion of bile for the treatment of intractable pruritus associated with intrahepatic cholestasis. Gastroenterology 1988; 95:130-136.
93. Soubrane O, Gauthier F, Devictor D et al. Orthotopic liver transplantation for Byler disease. Transplantation 1990; 50:804-806.
94. Dixon PH, Weerasekera N, Linton KJ et al. Heterozygous MDR3 missense mutation associated with intrahepatic cholestasis of pregnancy: evidence for a defect in protein trafficking. Hum Mol Genet 2000; 9:1209-1217.

95. de Pagter AG, Berge Henegouwen GP, Bokkel Huinink JA et al. Familial benign recurrent intrahepatic cholestasis. Interrelation with intrahepatic cholestasis of pregnancy and from oral contraceptives? Gastroenterology 1976; 71:202-207.

96. Mazzella G, Nicola R, Francesco A et al. Ursodeoxycholic acid administration in patients with cholestasis of pregnancy: effects on primary bile acids in babies and mothers. Hepatology 2001; 33:504-508.

97. Berkane N, Cocheton JJ, Brehier D et al. Ursodeoxycholic acid in intrahepatic cholestasis of pregnancy. A retrospective study of 19 cases. Acta Obstet Gynecol Scand 2000; 79:941-946.

98. Palma J, Reyes H, Ribalta J et al. Ursodeoxycholic acid in the treatment of cholestasis of pregnancy: a randomized, double-blind study controlled with placebo. J Hepatol 1997; 27:1022-1028.

99. Aagenaes O. Hereditary cholestasis with lymphoedema (Aagenaes syndrome, cholestasis- lymphoedema syndrome). New cases and follow-up from infancy to adult age. Scand J Gastroenterol 1998; 33:335-345.

100. Bull LN, Roche E, Song EJ et al. Mapping of the locus for cholestasis-lymphedema syndrome (Aagenaes syndrome) to a 6.6-cM interval on chromosome 15q. Am J Hum Genet 2000; 67:994-999.

101. Morton DH, Salen G, Batta AK et al. Abnormal hepatic sinusoidal bile acid transport in an Amish kindred is not linked to FIC1 and is improved by ursodiol. Gastroenterology 2000; 119:188-195.

102. Harris HW Jr, Carpenter TO, Shanley P et al. Progressive tubulointerstitial renal disease in infancy with associated hepatic abnormalities. Am J Med 1986; 81:169-176.

103. Neuhaus TJ, Stallmach T, Leumann E et al. Familial progressive tubulo-interstitial nephropathy and cholestatic liver disease—a newly recognized entity? Eur J Pediatr 1997; 156:723-726.

104. Popovic-Rolovic M, Kostic M, Sindjic M et al. Progressive tubulointerstitial nephritis and chronic cholestatic liver disease. Pediatr Nephrol 1993; 7:396-400.

105. Quigley EM, LaRusso NF, Ludwig J et al. Familial occurrence of primary sclerosing cholangitis and ulcerative colitis. Gastroenterology 1983; 85:1160-1165.

106. Debray D, Pariente D, Urvoas E et al. Sclerosing cholangitis in children. J Pediatr 1994; 124:49-56.

107. Isoyama K, Yamada K, Ishikawa K et al. Coincidental cases of primary sclerosing cholangitis and biliary atresia in siblings? Acta Paediatr 1995; 84:1444-1446.

108. Muller T, Schafer H, Rodeck B et al. Familial clustering of infantile cirrhosis in Northern Germany: A clue to the etiology of idiopathic copper toxicosis. J Pediatr 1999; 135:189-196.

109. Muller T, Feichtinger H, Berger H et al. Endemic Tyrolean infantile cirrhosis: an ecogenetic disorder. Lancet 1996; 347:877-880.

110. Wijmenga C, Muller T, Murli IS et al. Endemic Tyrolean infantile cirrhosis is not an allelic variant of Wilson's disease. Eur J Hum Genet 1998; 6:624-628.

111. Bhagwat AG. Indian childhood cirrhosis (ICC): review (Part III) 1980-1982. Indian Pediatr 1983; 20:152-153.

112. Gupta V, Mudgil A, Srivastava A et al. Serum gamma glutamyl transferase in Indian childhood cirrhosis. J Indian Med Assoc 1983; 80:6-7.

113. Tanner MS, Portmann B, Mowat AP et al. Increased hepatic copper concentration in Indian childhood cirrhosis. Lancet 1979; 1:1203-1205.

114. Popper H, Goldfischer S, Sternlieb I et al. Cytoplasmic copper and its toxic effects. Studies in Indian childhood cirrhosis. Lancet 1979; 1:1205-1208.

115. Phillips MJ, Ackerley CA, Superina RA et al. Excess zinc associated with severe progressive cholestasis in Cree and Ojibwa-Cree children. Lancet 1996; 347:866-868.

116. Weber AM, Tuchweber B, Yousef I et al. Severe familial cholestasis in North American Indian children: a clinical model of microfilament dysfunction? Gastroenterology 1981; 81:653-662.

117. Vu TH, Tanji K, Holve SA et al. Navajo neurohepatopathy: a mitochondrial DNA depletion syndrome? Hepatology 2001; 34:116-120.

118. Holve S, Hu D, Shub M et al. Liver disease in Navajo neuropathy. J Pediatr 1999; 135:482-493.

119. Di Rocco M, Callea F, Pollice B et al. Arthrogryposis, renal dysfunction and cholestasis syndrome: report of five patients from three Italian families. Eur J Pediatr , 1995154:835-839.

CHAPTER 14

Genetic Defects in Biliary Lipid Transport

Ronald Oude Elferink

Introduction

In terms of solute mass, lipids are the second most important component of bile. Biliary lipids mainly consist of phospholipid (almost exclusively phosphatidylcholine; PC) and cholesterol. The ratio in which these two lipids are secreted varies considerably between species. In rodents the phospholipid over cholesterol ratio is 5-10 while in man this is considerably lower. The relatively high concentration of cholesterol in human bile represents the major risk factor to gallstone formation. In the past years it has become clear that biliary lipid secretion serves an important function in defending hepatocytes and bile duct epithelial cells against bile salt induced toxicity. The secretion of biliary lipids is a complex process that we only poorly understand. It is clear, however, that it involves translocation of phosphatidylcholine across the canalicular membrane by MDR3 P-glycoprotein (Pgp) and that the actual secretion process is driven by bile salt present in the canalicular lumen. Patients lacking MDR3 P-glycoprotein develop progessive liver disease. It is becoming clear that other lipid translocation processes occur in the canalicular membrane which may play a role in biliary lipid and bile salt secretion.

Hepatobiliary Lipid Secretion

Upon production of a knockout mouse for the Mdr2 gene it was discovered that the transporter encoded by this gene is essential for biliary lipid secretion.[1] Mdr2 P-glycoprotein is normally localized in the canalicular membrane but in its absence, mice do not secrete phospholipid nor cholesterol. In different experimental systems it was subsequently demonstrated that the murine Mdr2 Pgp as well as the human orthologue MDR3 Pgp are phosphatidylcholine translocators.[2-4] Studies with a set of transgenic animals with different expression levels of MDR3 and/or Mdr2 revealed that these lipid translocators are not only essential for biliary phospholipid secretion but also rate-controlling.[5,6] Although cholesterol secretion is also impaired in Mdr2-/- mice, it was found to be a secondary effect of the absence of phospholipid from canalicular bile.[7]

On the basis of available data, we have proposed a model[8] in which PC, after delivery to the inner leaflet of the canalicular plasma membrane, is translocated by Mdr2/MDR3 to the outer leaflet of the membrane. This leaflet of the membrane is thought to have a high content of sphingmyelin and cholesterol. Interaction between sphingomyelin and cholesterol greatly enhances this rigidity and increases resistance towards bile salts.[9] Sphingomyelin has high affinity for cholesterol[10] and these two lipids tend to segregate in lateral domains.[11,12] The presence of these two lipids prevents solubilization of the canalicular membrane by high bile salt concentrations in the canaliculus. The translocated PC resides in fluid PC-rich lipid domains that are laterally separated from the more rigid sphingolipid-rich lipids of the outer canalicular membrane leaflet. The more fluid PC-rich microdomains are destabilized by the canalicular bile salt molecules and, in combination with the continuous pumping of PC into these domains, vesiculation of PC rich membrane bilayer occurs. These vesicles then pinch off to give rise to typical biliary lipid vesicles.[8] Morphological evidence for this model was provided by Crawford et al.[13,14]

It was observed that the expression of the Mdr2 gene is regulated by the bile salt pool size. Enlargement of the bile salt pool by feeding mice a diet containing cholate (0.1%) induced Mdr2 mRNA levels.[15,16] In addition, chronic bile diversion of rats led to a reciprocal reduction of Mdr2 expression.[16] This type of regulation makes sense as phospholipid secretion directly depends on the rate of bile salt secretion. The transcription factors involved in this regulation have not been identified yet.

The mechanism of cholesterol secretion into bile is still largely unknown. As yet, no direct evidence exists for the involvement of a translocator protein which, in analogy with Mdr2/MDR3 Pgp for phospholipids, would catalyze the translocation of cholesterol across the membrane. Controversy exists on the rate of spontaneous flip-flop of cholesterol across biological membranes. Cholesterol is not secreted into bile in the absence of phospholipid secretion (in the Mdr2-/- mouse); this is caused by the fact that simple bile salt micelles have very poor cholesterol solubilizing capacity, especially those of the more hydrophilic bile salt species, such as muricholate (the main murine bile salt) and ursodeoxycholate. Infusion of a more hydrophobic bile salt such as taurodeoxycholate, or feeding of a diet containing cholate induced cholesterol secretion in Mdr2-/- mice while phospholipid secretion remained negligible.[7] This observation does, however, not exclude the possibility that cholesterol translocation needs a transporter protein. Indeed, recent evidence suggests that sterol translocation may also be protein-mediated: the two halftransporters ABCG5 and ABCG8 from the ABC transporter familie appear to be involved in the intestinal and biliary elimination of plant sterols.[17,18] Transport has thus far not been demonstrated directly, but this function can be inferred from accumulation of plant sterols in patients with sitosterolemia, who have a mutation in either of the two genes encoding these proteins (see below). ABCG5 and ABCG8 are expressed in the intestine and the liver.

Recently, yet another ABC transporter, ABCA1 has been implicated in the transport of cholesterol. Several groups simultaneously discovered that ABCA1 is the causative gene for Tangier disease.[19-22] Tangier disease is a rare inherited disorder characterized by the virtual absence of HDL. Fibroblasts from these patients are incapable of donating cholesterol and phospholipid to lipid-poor apoA-I, a step that is essential for maturation of the lipoprotein. When the association between ABCA1 defects and Tangier disease was discovered, the initial suggestion was that ABCA1 transports cholesterol, but recent evidence demonstrates that this is not the case.[23] Most likely ABCA1 functions primarily as an outward phospholipid translocase rather than a cholesterol transporter. Translocation of phospholipid to lipid-poor apoA-I probably secondarily facilitates the transfer of cholesterol. Recently, Vaisman et al[24] reported a study on transgenic mice that overexpress ABCA1. As expected, these mice had increased plasma levels of HDL. The authors also reported that the concentration of cholesterol in bile was increased. This could mean that ABCA1 also fullfills a function in lipid secretion in the canalicular membrane. To investigate this aspect in a more rigorous model, Groen et al[25] measured cholesterol secretion rates into bile in Abca1 (-/-) mice during bile diversion. They observed no difference between Abca1 (-/-) mice and controls. This rules out the possibility that Abca1 contributes to biliary cholesterol efflux, unless this function would be completely taken over by another gene with compensatory expression, which is rather unlikely.

Progressive Familial Intrahepatic Cholestasis (PFIC)

It has been known for a long time that a group of pediatric patients exists, who suffer from an inherited form of progressive intrahepatic cholestasis. This group has been described under various other synonyma such as fatal familial intrahepatic cholestasis[26], or progressive idiopathic cholestasis[27] The first report identified this disease entity in an Amish family (the Byler family), where seven members of four related sibships suffered from the same symptoms.[26] These children presented with steatorrhea, diarrhoea, jaundice with intermittent exacerbations, hepatosplenomegaly and failure to thrive. Because of the patient family history it was clear that this represented a recessively inherited disease which was called Byler's disease after the family. Subsequent studies showed that a similar disease phenotype also existed outside the Amish community.[28-30] Patients with the same disease phenotype that did not belong to the Byler pedigree were generally described as suffering from Byler syndrome, due to the uncertainty of the genetic cause of this form. The outcome of the disease was generally fatal due to liver failure within the first decade of life. The advent of liver transplantation for children provided the opportunity to cure these patients. In further studies biochemical and histological features provided support for heterogeneity of this disease entity, although the clinical development was very similar. First and foremost the group fell into two parts, one subgroup of patients with a high serum γ-glutamyltranspeptidase (GGT) activity and the other with a normal serum GGT activity.[31,32] The latter form was the most frequent and included the patient group from the original Byler pedigree. Further analysis learned that these subgroups differed in more aspects than only the serum GGT. Liver histology of patients with high GGT PFIC revealed prominent bile duct proliferation and cirrhosis which was absent or much less prominent in patients with low GGT. In the last few years considerable progress was made in identifying the genetic background of this group of patients, although it must be stressed that not all involved genes have been identified yet. There are at least three groups which are now commonly designated by PFIC type 1, type 2 and type 3. Type 1 and type 2 patients have a low serum GGT activity, while type 3 patients have a high serum GGT activity.

Progressive Familial Intrahepatic Cholestasis Type 1 (Formerly Called Byler's Disease)

Byler's disease primarily manifests itself as a chronic intrahepatic cholestasis. Bile salt concentrations are high in serum and very low in bile, suggesting that hepatic transport is impaired. The jaundice in these patients is regarded as a secondary consequence of insufficient bile flow, the latter being dependent on bile salt secretion. This also holds for the fat-soluble vitamin deficiency and steatorrhoea which are the consequence of impaired fat absorption, due to the absence of bile salts from the intestine. Liver transplantation is presently the main therapeutic option for this progressive disease that ends in liver failure. It is of importance to note, however, that interruption of the enterohepatic circulation (chronic bile diversion) was reported to improve the clinical picture of these patients dramatically. Whitington et al [33] described a procedure in which chronic external partial bile diversion was achieved through a jejunal stoma in four patients. In these patients pruritus dramatically improved and serum bile salt levels fell from >200 μM to less than 10 μM. Several later studies have reported similar results.[34-36] This did not occur in all patients, however, and it would be of great interest to distinguish the beneficial effect of this procedure in PFIC type 1 vs. type 2 patients (see below).

With the patients from the Byler pedigree and their family members a genetic screen was performed to identify the disease locus.[37] This was located on chromosome 18q21-q22, the same chromosomal region on which a very similar disease type, Benign recurrent intrahepatic cholestasis was localized (see below). A combined search for the two disease loci led to the identification of the mutated gene, FIC1 (gene code ATP8B1), and its mutations in a group of patients, including those from the original Byler pedigree.[38] More recently it became possible to stain the protein by western blotting and immunohistochemistry.[39] It is a 140 kD protein which is localized in the canalicular membrane of the hepatocyte as well as in the apical membrane of bile duct epithelial cells. The protein was virtually absent from a liver specimen of a patient with the common Amish mutation. Many issues concerning this protein remain puzzling, however. Expression of FIC1 in liver is actually quite low, while the gene is highly expressed in intestine and pancreas.[38] Expression is also found in many other tissues. Thus, it is unclear why the absence of FIC1 from many tissues leads primarily to a phenotype in the liver. On the basis of sequence homology this gene must be a member of the P-type ATPase family. Within this family, subgroups of transporters with different functions have been identified.

The subgroup of cation pumps harbours the plasma membrane Na^+/K^+ATPase and the gastric K^+/H^+ ATPase. A second subgroup contains metal ion transporters: such as the hepatic copper transporter, mutated in Wilsons disease[40] and the extrahepatic copper transporter that is defective in Menke's disease.[41] The function of FIC1 as a transporter is as yet unknown, but it has highest homology with the drs2 protein from yeast and with the bovine ATPase II (gene code ATP8A1), which form the third subgroup of P-type ATPases.[38,42] Recently, evidence was presented that FIC1 is an aminophospholipid flippase that is capable of translocating phosphatidylserine.[43] This observation is in line with the finding that drs2 would be an aminophospholipid flippase as well.[42] The latter is a controversial issue, however, because later publications showed that the protein was unable to translocate lipids[44,45] and suggested that it might be involved in metal ion transport.[46] It remains enigmatic how the absence of FIC1 can lead to defective canalicular bile salt transport. Given the fact that BSEP is the main canalicular bile salt transporter, that FIC1 is mainly expressed in the intestine, and that chronic bile diversion leads to disappearance or reduction of PFIC type 1 symptoms, it can be speculated that FIC1 activity plays a role in the enterohepatic circulation of bile salts and that the intrahepatic cholestasis in these patients is a secondary consequence of dysregulation of the enterohepatic circulation.[47] In the line of this hypothesis several possible functions of FIC1 can be envisioned. FIC1 may be an aminophospholipid flippase, as originally suggested; the (inward) flipping of aminophospholipids could be of importance for the regulation of the activity of other transporters such as the bile salt transporters in liver and intestine. It would have to be assumed then, that lipid asymmetry is of major regulatory function for normal bile salt transport. Alternatively, the inward flipping of PS might be important for the binding of auxiliary proteins to, or fusion of vesicles with, the cytosolic leaflet of the plasma membrane of hepatocytes and/or enterocytes. It is known that binding of certain proteins (like annexin) as well as fusion of exocytotic vesicles critically depends on the PS concentration in the cytosolic membrane leaflet.[48] Another possibility is that FIC1 is a bile salt pump itself. FIC1 could have affinity for hydrophobic bile salts and/or cholestatic bile salt metabolites. The latter mostly concern metabolites from the primary bile salts, cholate and chenodeoxycholate, produced by bacteria in the gut. The high expression of FIC1 in the gut might serve as a direct defense against such compounds. In line with this hypothesis it has been observed that bile from PFIC type1 deficient patients contains very low amounts of bile salts with a disproportionately low content of the hydrophobic bile salts.[29,49]

With the identification of the disease locus for PFIC type 1 and the responsible gene, FIC1, it became clear that the same gene is mutated in BRIC. BRIC most likely represents a mild form of PFIC type 1. PFIC type 1 may start with bouts of cholestasis and progressively develops into chronic persistent cholestasis, but in BRIC patients the phenotype remains restricted to periods of cholestasis that resolve after days to months. Importantly, when cholestasis resolves it leaves no detectable liver damage. The main features of BRIC are elevated serum bile salt concentrations, jaundice and pruritus. The milder phenotype of BRIC compared to PFIC type 1, seems to correlate with the mutations found in these two patient groups.[38] DNA sequencing of the FIC1 gene learned that deletions, frame shifts and nonsense mutations appear to lead to the PFIC type 1 phenotype, while in BRIC patients generally missense mutations are found. This allows the hypothesis that the FIC1 protein with BRIC mutations may have residual activity, while the protein is absent or non-functional in PFIC type 1 patients.[38]

Progressive Familial Intrahepatic Cholestasis Type 3

As mentioned above this form of PFIC is fundamentally different from type 1 and 2 in that these patients have a high serum GGT activity. The onset of this disease is somewhat later than in the other two forms, but the histological picture is more severe; there is strong bile duct proliferation and cirrhosis. The most prominent features of the disease are portal hypertension, hepatosplenomegaly, jaundice and pruritus. If untreated the disease develops into liver failure. The genetic background of this subgroup of PFIC patients was elucidated after it was found that mice with a disruption in the Mdr2 gene, the murine orthologue of MDR3, develop a similar phenotype (see below). This led Deleuze et al[56] to investigate the possible involvement of this gene in PFIC. Subsequently, in a group of 31 patients with high GGT PFIC, 17 were found to have a mutation in the MDR3 gene.[57,58] Since the gene was not completely sequenced in all these patients it is not clear whether the remaining 14 patients have as yet unidentified mutations or that another gene might be involved in this form of PFIC. MDR3 P-glycoprotein functions in the translocation of phosphatidylcholine, thereby facilitating the secretion of this phospholipid into bile. The secretion of phospholipid is of crucial importance in the protection of the cellular membranes of the biliary tree against the high concentrations of bile salt detergents.[8]

Treatment of patients with ursodeoxycholate (UDCA) appeared to be beneficial in about half of the cases.[57] Given the fact that the hepatic damage is caused by bile salts, this is a rational treatment, because UDCA is a hydrophilic bile salt that has low cytotoxicity. Indeed, in Mdr2-/- mice with the equivalent defect, feeding of this bile salt also halted the disease process.[59,60] The observation that not all patients improve on UDCA treatment may be explained by the fact that administration of this bile salt insufficiently replaces the endogenous bile salt pool. Thus, in patients who have no MDR3 Pgp activity at all, the reduction of bile salt cytotoxicity by UDCA is insufficient, while in patients with residual phospholipid secretion, UDCA might bring the overall bile salt cytotoxicity below a critical threshold. Indeed, none of the patients with a truncated MDR3 gene improved on UDCA, while some of those with missense mutations did.[57]

Adult Forms of MDR3 Deficiency

Interestingly, it became clear more recently that defects in the MDR3 gene do not only give rise to pediatric liver disease. Jacquemin et al reported[61] that the mother of a patient with PFIC type 3 and several other women from this family suffered from intrahepatic cholestasis of pregnancy. These women turned out to be heterozygotes for the mutation in the MDR3 gene, that caused PFIC type 3 in the homozygous index patient. Moreover, Rosmorduc et al[62] reported on 6 gallstone patients, in whom mutations in the MDR3 gene were found. This included homozygous and heterozygous missense mutations as well as a heterozygous 1 bp insertion leading to frame shift and truncation. None of these mutations were observed in >100 chromosomes from control subjects suggesting that they may be associated with the disease phenotype. These two publications suggest that, apart from the severe phenotype associated with complete or nearly complete absence of MDR3 function, also milder phenotypes exist that are associated with reduction but not complete absence of MDR3 Pgp function. Very recently, Thompson and coworkers[63] also reported, in a preliminary form on three members of a family having heterozygous mutations in the MDR3 gene. These patients had symptoms of Primary biliary cirrhosis but were negative for antimitochondrial antibodies. Two of the three patients were sisters who originally presented with cholestasis of pregnancy. Sequencing of the MDR3 gene revealed that all three patients were compound heterozygotes for 2 mutations in this gene: T175K and T424M. Although the effect of these mutations remains to be established, it strongly suggests that mutations in the MDR3 gene may also cause PBC-like liver pathology in adults.

Serum Cholesterol Level in PFIC Patients

In several studies on patients with low GGT PFIC, it was observed that these patients have a normal plasma cholesterol level as opposed to other pediatric and adult forms of cholestasis.[26,64,65] Very recently, this was also reported for PFIC type 3 patients.[57] The plasma lipoprotein profile of cholestatic patients reveals an abnormal lipoprotein fraction in the LDL region. This abnormal "cholestatic" lipoprotein was characterized as a unilamellar vesicle with an aqeous lumen, and was designated lipoprotein X (LpX).[66] Using Mdr2-/- mice it could be shown that the formation of LpX depends on the function of canalicular secretion machinery.[67]

Wild type mice develop hypercholesterolemia upon bile duct ligation and this is associated with the presence of massive amounts of LpX in the plasma, but LpX was completely absent in plasma of Mdr2-/- mice.[67] Apparently, during cholestasis the formation of biliary vesicles continues and the release of these vesicles is redirected to the plasma compartment. Indeed, it was shown that during bile duct ligation, the expression of Bsep as well as Mdr2 are not down regulated and that the proteins are redistributed from the canalicular membrane into an intracellular, subapical compartment, where they may continue to function in the formation of biliary vesicles. The conclusion from these data is that during obstructive cholestasis the continued formation of lipid vesicles by Mdr2/MDR3 Pgp and the release into plasma leads to a dysregulation of cholesterol homeostasis. In patients with all forms of PFIC the process of biliary lipid vesicle formation is impaired, because both bile salt transport and phospholipid translocation are necessary for this process. The absence of either of these two processes precludes the formation of biliary vesicles and therefore during cholestasis also no LpX will be formed. Thus the normal serum cholesterol level represents an important diagnostic tool to distinguish all forms of PFIC from other inherited or acquired cholestatic syndromes.

Sitosterolemia

Sitosterolemia is a very rare, recessively inherited, disease. The clinical presentation includes tendon xanthomas, accelerated atherosclerosis particularly affecting males at young age, hemolytic episodes, and arthritis and arthralgias.[68] The hallmark biochemical feature of the disease is the elevated concentration of plant sterols in plasma. Because sitosterol (24-ethyl cholesterol) is the most important accumulated sterol in plasma of these patients (as well as the most abundant plant sterol in the diet), this disease is referred to as sitosterolemia. A host of other sterol variants, such as campesterol, stigmasterol and brassicasterol, are present in plants and other compontents of the diet like shell fish. Several studies have indicated that absorption of plant sterols in the intestine is strongly increased in sitosterolemic patients. Control subjects absorb about 50-70% of dietary cholesterol, but only less than 5% of the ingested plant sterols. In stark contrast, sitosterolemic patients absorb plant sterols to about the same extent (30-60%) as cholesterol. As a consequence there is marked accumulation of these sterols in plasma and this leads to similar accumulation in tissues like liver, lung, heart and red blood cells. The brain content of plant sterols is low, indicating that the blood brain barrier is intact with respect to these exogenous sterols.[68] Importantly, in normal subjects the low amounts of absorbed sitosterol is quickly secreted into bile so that only trace amounts of it can be found in blood.[68] However, in sitosterolemic patients the biliary secretion of sitosterol and other phytosterols is impaired.[69-71] Bile and plasma analysis in two patients revealed a more than 30-fold reduced bile/plasma ratio of total plant sterols compared to controls.[71] These data show that phytosterol handling in these patients is impaired at two levels: absorption in the intestine and secretion into bile.

The disease locus had been localized to chromosome 2p21 [72] and more recently, ABCG5 and ABCG8, the genes involved in this disease were identified.[17,18] These encode two ABC halftransporters from the G-subfamily. Separated by only about 150 basepairs they are located in opposite orientation on chromosome 2p21. Lee et al[18] reported mutations in the ABCG5 gene in nine patients, while Berge et al [17] identified mutations in the ABCG8 gene of 8 patients and in the ABCG5 gene of one patient. Both genes are expressed mainly in liver and intestine.[17,18]

ABC halftransporters are thought to dimerize into functional pumps. The fact that mutations in either ABCG5 or ABCG8 cause sitosterolemia suggests that these two halftransporters indeed form a functional (and obligatory) heterodimer. No studies have been reported yet, in which the cDNAs have been transfected in cell lines in order to perform transport studies Also the localization of the proteins in normal human tissues has not been described yet. However, on basis of the description of the increased sterol absorption and the decreased biliary secretion in patients with sitosterolemia, it is tempting to speculate on the localization and function of a potential heterodimer of ABCG5 and ABCG8. The minimum hypothesis would be that this represents an outward pump for plant sterols present in the apical membrane, which in the intestine reduces absorption and in the liver mediates canalicular secretion of these unwanted sterols.

References

1. Smit JJM, Schinkel AH, Oude Elferink RPJ et al. Homozygous disruption of the murine mdr2 P-glycoprotein gene leads to a complete absence of phospholipid from bile and to liver disease. Cell 1993; 75:451-462.
2. Ruetz S, Gros P. Phosphatidylcholine translocase: a physiological role for the mdr2 gene. Cell 1994; 77:1071-1082.
3. Smith AJ, Timmermans-Hereijgers JLPM, Roelofsen B et al. The human MDR3 P-glycoprotein promotes translocation of phosphatidylcholine through the plasma membrane of fibroblasts from transgenic mice. FEBS Lett 1994; 354:263-266.
4. van Helvoort A, Smith AJ, Sprong H et al. MDR1 P-glycoprotein is a lipid translocase of broad specificity, while MDR3 P-glycoprotein specifically translocates phosphatidylcholine. Cell 1996; 87(3):507-17.
5. Groen AK, Oude Elferink RP, Tager JM. Control analysis of biliary lipid secretion. Journal of Theoretical Biology 1996; 182(3):427-36.
6. Smith AJ, de Vree JM, Ottenhoff R et al. Hepatocyte-specific expression of the human MDR3 P-glycoprotein gene restores the biliary phosphatidylcholine excretion absent in Mdr2 (-/-) mice. Hepatology 1998; 28(2):530-6.
7. Oude Elferink RPJ, Ottenhoff R, van Wijland M et al. Uncoupling of biliary phospholipid and cholesterol secretion in mice with reduced expression of mdr2 P-glycoprotein. J Lipid Res 1996; 37(5):1065-1075.
8. Oude Elferink RP, Tytgat GN, Groen AK. Hepatic canalicular membrane 1: The role of mdr2 P-glycoprotein in hepatobiliary lipid transport. FASEB Journal 1997; 11(1):19-28.
9. Eckhardt ERM, Moschetta A, Renooij W et al. Asymmetric distribution of phosphatidylcholine and sphingomyelin between micellar and vesicular phases: potential implications for canalicular bile formation. Journal of Lipid Research 1999; 40(11):2022-2033.
10. Demel RA, Jansen JWCM, van Dijck PWM et al. The preferential interaction of cholesterol with different classes of phospholipids. Biochim Biophys Acta 1977; 465:1-10.
11. Mattjus P, Bittman R, Vilcheze C et al. Lateral domain formation in cholesterol/phospholipid monolayers as affected by the sterol side chain conformation. Biochimica et Biophysica Acta 1995; 1240(2):237-47.
12. Slotte JP. Lateral domain formation in mixed monolayers containing cholesterol and dipalmitoylphosphatidylcholine or N- palmitoylsphingomyelin. Bba-Biomembranes 1995; 1235:419-427.
13. Crawford JM, Möckel G-M, Crawford AR et al. Imaging biliary lipid secretion in the rat: ultrastructural evidence for vesiculation of the hepatocyte canalicular membrane. J.Lipid Res. 1995; 36:2147-2163.
14. Crawford AR, Smith AJ, Hatch VC et al. Hepatic secretion of phospholipid vesicles in the mouse critically depends on mdr2 or MDR3 P-glycoprotein expression. Visualization by electron microscopy. Journal of Clinical Investigation 1997; 100(10):2562-7.
15. Frijters CM, Ottenhoff R, van Wijland MJ et al. Regulation of mdr2 P-glycoprotein expression by bile salts. Biochemical Journal 1997; 321(Pt 2):389-95.
16. Gupta S, Todd Stravitz R, Pandak WM et al. Regulation of multidrug resistance 2 P-glycoprotein expression by bile salts in rats and in primary cultures of rat hepatocytes. Hepatology 2000; 32(2):341-7.
17. Berge KE, Tian H, Graf GA et al. Accumulation of dietary cholesterol in sitosterolemia caused by mutations in adjacent ABC transporters. Science 2000; 290(5497):1771-5.
18. Lee MH, Lu K, Hazard S et al. Identification of a gene, ABCG5, important in the regulation of dietary cholesterol absorption. Nat Genet 2001; 27(1):79-83.
19. Rust S, Rosier M, Funke H et al. Tangier disease is caused by mutations in the gene encoding atp-binding cassette transporter 1. Nature Genetics 1999; 22(4):352-355.
20. Bodzioch M, Orso E, Klucken T et al. The gene encoding atp-binding cassette transporter 1 is mutated in Tangier disease. Nature Genetics 1999; 22(4):347-351.
21. Drobnik W, Liebisch G, Biederer C et al. Growth and cell cycle abnormalities of fibroblasts from Tangier disease patients. Arteriosclerosis Thrombosis & Vascular Biology 1999; 19(1):28-38.
22. Brooks-Wilson A, Marcil M, Clee SM et al. Mutations in abc1 in Tangier disease and familial high-density lipoprotein deficiency. Nature Genetics 1999; 22(4):336-345.

23. Wang N, Silver DL, Thiele C, Tall AR. Atp-binding cassette transporter a1 (abca1) functions as a cholesterol efflux regulatory protein. J Biol Chem 2001; 276(26):23742-7.
24. Vaisman BL, Lambert G, Amar M et al. ABCA1 overexpression leads to hyperalphalipoproteinemia and increased biliary cholesterol excretion in transgenic mice. J Clin Invest 2001; 108(2):303-9.
25. Groen AK BV, Bandsma RH, Ottenhoff R et al. Hepatobiliary cholesterol transport is not impaired in ABCA1 null mice lacking high density lipoproteins. J Clin Invest 2001; 108:843-850.
26. Clayton RJ IF, Ruebner BH. Byler Disease. Fatal Familial Intrahepatic Cholestasis in an Amish kindred. Am J Dis Child 1969; 117:112-124.
27. Chobert MN, Bernard O, Bulle F et al. High hepatic gamma-glutamyltransferase (gamma-GT) activity with normal serum gamma-GT in children with progressive idiopathic cholestasis. J Hepatol 1989; 8(1):22-25.
28. Linarelli LG, Williams CN, Phillips MJ. Byler's disease: fatal intrahepatic cholestasis. J Pediatr 1972; 81(3):484-92.
29. Tazawa Y, Yamada M, Nakagawa M et al. Bile acid profiles in siblings with progressive intrahepatic cholestasis: absence of biliary chenodeoxycholate. J Pediatr Gastroenterol Nutr 1985; 4(1):32-7.
30. Ornvold K NI, Poulsen H. Fatal familial cholestatic cholestatic syndrome in Greenland Eskimo children. Virchows Archiv A Pathol Anat 1989; 415:275-281.
31. Maggiore G, Bernard O, Hadchouel M et al. Diagnostic value of serum gamma-glutamyl transpeptidase activity in liver diseases in children. J Pediatr Gastroenterol Nutr 1991; 12(1):21-6.
32. Maggiore G, Bernard O, Riely CA et al. Normal serum gamma-glutamyl-transpeptidase activity identifies groups of infants with idiopathic cholestasis with poor prognosis. J Pediatr 1987; 111(2):251-2.
33. Whitington PF, Whitington GL. Partial external diversion of bile for the treatment of intractable pruritus associated with intrahepatic cholestasis. Gastroenterology 1988; 95(1):130-6.
34. Emond JC, Whitington PF. Selective surgical management of progressive familial intrahepatic cholestasis (Byler's disease). Journal of Pediatric Surgery 1995; 30(12):1635-41.
35. Ismail H, Kalicinski P, Markiewicz M et al. Treatment of progressive familial intrahepatic cholestasis: liver transplantation or partial external biliary diversion. Pediatr Transplant 1999; 3(3):219-24.
36. Ng VL, Ryckman FC, Porta G et al. Long-term outcome after partial external biliary diversion for intractable pruritus in patients with intrahepatic cholestasis. J Pediatr Gastroenterol Nutr 2000; 30(2):152-6.
37. Carlton VEH, Knisely AS, Freimer NB. Mapping of a locus for progressive familial intrahepatic cholestasis (Byler disease) to 18q21-q22, the benign recurrent intrahepatic cholestasis region. Hum Mol Genet 1995; 4:1049-1053.
38. Bull LN, van Eijk MJ, Pawlikowska L et al. A gene encoding a P-type ATPase mutated in two forms of hereditary cholestasis. Nat Genet 1998; 18(3):219-24.
39. Eppens EF, van Mil SWC, de Vree JM et al. FIC1, the protein affected in two forms of hereditary cholestasis, is localized in the canalicular membrane of the hepatocyte. J Hepatol 2001; in press.
40. Bull PC, Thomas GR, Rommens JM et al. The Wilson disease gene is a putative copper transporting P-type ATPase similar to the Menkes gene. Nat Genet 1993; 5(4):327-37.
41. Vulpe C, Levinson B, Whitney S et al. Isolation of a candidate gene for Menkes disease and evidence that it encodes a copper-transporting ATPase. Nat Genet 1993; 3(1):7-13.
42. Tang X, Halleck MS, Schlegel RA et al. A subfamily of P-type ATPases with aminophospholipid transporting activity. Science 1996; 272(5267):1495-7.
43. Ujhazy P, Ortiz D, Misra S et al. Familial intrahepatic cholestasis 1: studies of localization and function. Hepatology 2001; 34(4 Pt 1):768-75.
44. Marx U, Polakowski T, Pomorski T et al. Rapid transbilayer movement of fluorescent phospholipid analogues in the plasma membrane of endocytosis-deficient yeast cells does not require the Drs2 protein. Eur J Biochem 1999; 263(1):254-63.
45. Siegmund A, Grant A, Angeletti C et al. Loss of Drs2p does not abolish transfer of fluorescence-labeled phospholipids across the plasma membrane of Saccharomyces cerevisiae. J Biol Chem 1998; 273(51):34399-405.
46. Siegmund A, Grant A, Angeletti C et al. Loss of drs2p does not abolish transfer of fluorescence-labeled phospholipids across the plasma membrane of saccharomyces cerevisiae. Journal of Biological Chemistry 1998; 273(51):34399-34405.
47. Oude Elferink RP, van Berge Henegouwen GP. Cracking the genetic code for benign recurrent and progressive familial intrahepatic cholestasis [editorial]. Journal of Hepatology 1998; 29(2):317-20.
48. Zachowski A, Henry JP, Devaux PF. Control of transmembrane lipid asymmetry in chromaffin granules by an ATP-dependent protein. Nature 1989; 340(6228):75-6.

49. Jacquemin E, Dumont M, Bernard O et al. Evidence for defective primary bile acid secretion in children with progressive familial intrahepatic cholestasis (Byler disease). Eur J Pediatr 1994; 153(6):424-8.
50. Prezant TR, Chaltraw WE Jr, Fischel-Ghodsian N. Identification of an overexpressed yeast gene which prevents aminoglycoside toxicity. Microbiology 1996; 142(Pt 12):3407-14.
51. Jones EA, Bergasa NV. The Pruritus of Cholestasis and the Opioid System. JAMA 1992; 268:3359-3362.
52. Jones EA, Bergasa NV. The pruritus of cholestasis. Hepatology 1999; 29(4):1003-6.
53. Gregorio GV, Ball CS, Mowat AP et al. Effect of rifampicin in the treatment of pruritus in hepatic cholestasis. Arch Dis Child 1993; 69(1):141-3.
54. Xie W, Radominska-Pandya A, Shi Y et al. An essential role for nuclear receptors SXR/PXR in detoxification of cholestatic bile acids. Proc Natl Acad Sci USA 2001; 98(6):3375-80.
55. Staudinger JL, Goodwin B, Jones SA et al. The nuclear receptor PXR is a lithocholic acid sensor that protects against liver toxicity. Proc Natl Acad Sci USA 2001; 98(6):3369-3374.
56. Deleuze JF, Jacquemin E, Dubuisson C et al. Defect of multidrug-resistance 3 gene expression in a subtype of progressive familial intrahepatic cholestasis. Hepatology 1996; 23(4):904-908:.
57. Jacquemin E, de Vree JML, Cresteil D et al. The wide spectrum of MDR3 deficiency in patients with progressive familial intrahepatic cholestasis type 3: from neonatal cholestasis to cirrhosis of adulthood. Gastroenterology 2001; 120:1448-1458.
58. de Vree JM, Jacquemin E, Sturm E et al. Mutations in the MDR3 gene cause progressive familial intrahepatic cholestasis. Proceedings of the National Academy of Sciences of the United States of America 1998; 95(1):282-7.
59. Van Nieuwkerk CM, OudeElferink RP, Groen AK et al. Effects of Ursodeoxycholate and cholate feeding on liver disease in FVB mice with a disrupted mdr2 P-glycoprotein gene. Gastroenterology 1996; 111(1):165-71.
60. van Nieuwkerk CM, Groen AK, Ottenhoff R et al. The role of bile salt composition in liver pathology of mdr2 (-/-) mice: differences between males and females. Journal of Hepatology 1997; 26(1):138-45.
61. Jacquemin E, Cresteil D, Manouvrier S et al. Heterozygous non-sense mutation of the MDR3 gene in familial intrahepatic cholestasis of pregnancy. Lancet 1999; 353(9148):210-1.
62. Rosmorduc O, Hermelin B, Poupon R. MDR3 gene defect in adults with symptomatic intrahepatic and gallbladder cholesterol cholelithiasis. Gastroenterology 2001; 120(6):1459-67.
63. Thompson RJ SS, Gerred S, Kniseley A et al. Adult onset cholangiopathy (AMA-ve PBC) due to mutations in ABCB4. 2001:184.
64. Whitington PF, Freese DK, Alonso EM et al. Clinical and biochemical findings in progressive familial intrahepatic cholestasis. Journal of Pediatric Gastroenterology & Nutrition 1994; 18(2):134-41.
65. Alonso EM, Snover DC, Montag A et al. Histologic pathology of the liver in progressive familial intrahepatic cholestasis. Journal of Pediatric Gastroenterology & Nutrition 1994; 18(2):128-33.
66. Hamilton RL, Havel RJ, Kane JP et al. Cholestasis: lamellar structure of the abnormal human serum lipoprotein. Science 1971; 172(982):475-8.
67. Oude Elferink RP, Ottenhoff R, van Marle J et al. Class III P-glycoproteins mediate the formation of lipoprotein X in the mouse. Journal of Clinical Investigation 1998; 102(9):1749-57.
68. Salen G, Shefer S, Nguyen L et al. Sitosterolemia. J Lipid Res 1992; 33(7):945-55.
69. Lutjohann D, Bjorkhem I, Beil UF et al. Sterol absorption and sterol balance in phytosterolemia evaluated by deuterium-labeled sterols: effect of sitostanol treatment. J Lipid Res 1995; 36(8):1763-73.
70. Miettinen TA. Phytosterolaemia, xanthomatosis and premature atherosclerotic arterial disease: a case with high plant sterol absorption, impaired sterol elimination and low cholesterol synthesis. Eur J Clin Invest 1980; 10(1):27-35.
71. Gregg RE, Connor WE, Lin DS et al. Abnormal metabolism of shellfish sterols in a patient with sitosterolemia and xanthomatosis. J Clin Invest 1986; 77(6):1864-72.
72. Patel SB, Salen G, Hidaka H et al. Mapping a gene involved in regulating dietary cholesterol absorption. The sitosterolemia locus is found at chromosome 2p21. J Clin Invest 1998; 102(5):1041-4.

CHAPTER 15

Transport of Bilirubin Conjugates across Hepatocellular Membrane Domains and the Conjugated Hyperbilirubinemia of Dubin-Johnson Syndrome

Anne T. Nies, Yunhai Cui, Jörg König and Dietrich Keppler

Summary

Bilirubin, the end product of heme catabolism, needs to be taken up into hepatocytes and is then glucuronidated within the cells prior to its excretion via bile. Members of the SLC21A family in the sinusoidal membrane of hepatocytes selectively mediate the uptake of unconjugated bilirubin and of bilirubin conjugates. After conjugation within the hepatocyte by UGT1A1, yielding monoglucuronosyl and bisglucuronosyl bilirubin, the conjugates are transported across the canalicular membrane into bile by the apical conjugate export pump MRP2, a member of the ABCC subfamily of ATP-dependent transporters. MRP2 also mediates the export of a number of other amphiphilic anions and anionic substances including xenobiotics conjugated with glutathione, glucuronate, or sulfate. Mutations in the *ABCC2* gene leading to the absence of a functional MRP2 protein from the canalicular membrane, are the molecular basis of Dubin-Johnson syndrome in humans. Two hyperbilirubinemic rat strains with a hereditary Mrp2 deficiency may be considered as animal models for Dubin-Johnson syndrome and have been helpful in the molecular identification and functional characterization of MRP2. A number of single nucleotide polymorphisms has been recently described in the *ABCC2* gene, however, the functional consequences, if any, of each of the polymorphism still await clarification. A transitory MRP2 deficiency with MRP2 being retrieved into subapical vesicles is observed under cholestatic conditions in rats and humans. Under conditions of hereditary or acquired MRP2 deficiency, the isoform MRP3 is upregulated in the sinusoidal membrane of hepatocytes. This mechanism may be more pronounced in rat than in human liver. MRP3 mediates the export of bilirubin conjugates and other anionic conjugates with glucuronate or sulfate into blood. Recent results show that the expression of MRP2 is regulated by several nuclear receptor-mediated pathways including the FXR, the PXR, and the CAR receptor.

Introduction

Bilirubin is the end product of hemoglobin and hemoprotein catabolism.[1] Already in 1847, Virchow isolated bilirubin crystals from hematomes and suggested that bilirubin was derived from blood.[2] The structure of bilirubin was defined by Fischer and Orth in 1937 as that of a tetrapyrrol closely related to hemoglobin,[3] and, in 1950, London et al succeeded in the definitive demonstration that bilirubin is derived from heme.[4] The serum bilirubin concentration is

Molecular Pathogenesis of Cholestasis, edited by Michael Trauner and Peter L.M. Jansen.
©2004 Eurekah.com and Kluwer Academic / Plenum Publishers.

widely used as an indicator for assessing liver dysfunction. In 1916, Hymans van den Bergh developed a method for the measurement of bilirubin separating bilirubin into two chemically distinct fractions, i.e., the ‚direct-reacting‘ and the ‚indirect-reacting‘ bilirubin.[5] But it was not until 1956 and 1957 that the direct-reacting bilirubin was identified as the bisglucuronoside of bilirubin.[6-8]

Approximately 80% of the bilirubin is derived from the degradation of hemoglobin, the rest results from breakdown of other hemoproteins such as cytochromes, catalase, peroxidase, and tryptophan pyrrolase, and from that of free heme.[4] Normal serum bilirubin concentration ranges between 5 – 17 μM, about 4% of which is conjugated.[1] Unconjugated bilirubin is toxic[9,10] and tightly bound to albumin in serum. The liver is the main organ for conversion of bilirubin to the more hydrophilic bilirubin conjugates. Glucuronidation of bilirubin to monoglucuronosyl and bisglucuronosyl bilirubin is catalyzed by the UDP-glucuronosyl trans-ferase 1A1 (UGT1A1), the only isoform that significantly contributes to bilirubin metabo-lism.[11,12] UGT1A1 is also synthesized in extrahepatic tissues such as intestine and colon,[13-15] so that bilirubin conjugates found in serum[16,17] may originate from the intestinal tract.

In this article, we focus on the transport processes involved in uptake and export of biliru-bin and its conjugates by the hepatocyte (Fig. 1). Conjugated and unconjugated bilirubin are taken up into the hepatocyte by members of the SLC21A family[18,19] which are located in the sinusoidal, i.e., basolateral, membrane. After conjugation of bilirubin by UGT1A1, bilirubin conjugates are secreted into bile across the canalicular, i.e., apical, membrane by the isoform 2 of the ABCC subfamily (Multidrug resistance protein 2, MRP2).[20-24] There is no significant amount of unconjugated bilirubin in bile.[25] During conditions of either hereditary (i.e., hu-man Dubin-Johnson syndrome) or acquired (e.g., cholestasis) MRP2 deficiency, bilirubin conju-gates are transported across the sinusoidal membrane into blood via another ABCC isoform, i.e., ABCC3 (Multidrug resistance protein 3, MRP3),[26] leading to conjugated hyperbilirubinemia.

Newborns frequently become jaundiced during the first week of life with serum bilirubin being predominantly unconjugated.[1] A high bilirubin production due to a decreased erythro-cyte half-life, low hepatic uptake of bilirubin, or low UGT1A1 activity may lead to a transient rise in bilirubin concentration.[1] Phototherapy is commonly used as a treatment.[1] Several of the oxidative breakdown products of bilirubin can be eliminated into bile.[27]

Uptake mechanisms of bilirubin into the hepatocyte, conjugation of bilirubin within the hepatocyte, and the ATP-dependent export of bilirubin conjugates into bile are a related se-quence of reactions with major importance for heme catabolism. It is of interest that David Moore and his colleagues (Baylor College of Medicine, Houston, TX) recently showed that expression of SLC21A6, UGT1A1, and ABCC2 are coordinately regulated by the constitutive androstane receptor, CAR.[28] Moreover, CAR is activated by bilirubin.[28]

Hepatocyte Uptake Transporters of the SLC21A Family

The first step in bilirubin elimination is the selective uptake of conjugated and unconjugated bilirubin from blood over the basolateral membrane into hepatocytes. This uptake has been considered to be a process mediated by specific membrane proteins, although passive diffusion has also been proposed. It has been suggested that uptake of bilirubin and structurally related substances like bromosulfophthalein (BSP) are mediated by organic anion uptake transporters. The first transport protein for BSP belonging to the uptake transporter family SLC21A has been cloned from rat liver and designated as Oatp1 (symbol Slc21a1).[29,30] To date, a total of five different rat SLC21A family members have been cloned and characterized and at least four of them (Oatp1 through Oatp4; corresponding to Slc21a1, Slc21a5, Slc21a7, and Slc21a10) are expressed in liver. In humans, the first identified SLC21A family member was OATP-A (SLC21A3), which is expressed at a high level in brain and kidney and at a very low level in human liver.[31] In addition to this low level of hepatic expression, kinetic studies revealed only a moderate affinity of human OATP-A for BSP.[31] In a search for additional SLC21A family members expressed in human liver, two uptake transporters have been identified more recently and they are preferentially, if not exclusively, expressed in human liver. We designated them as

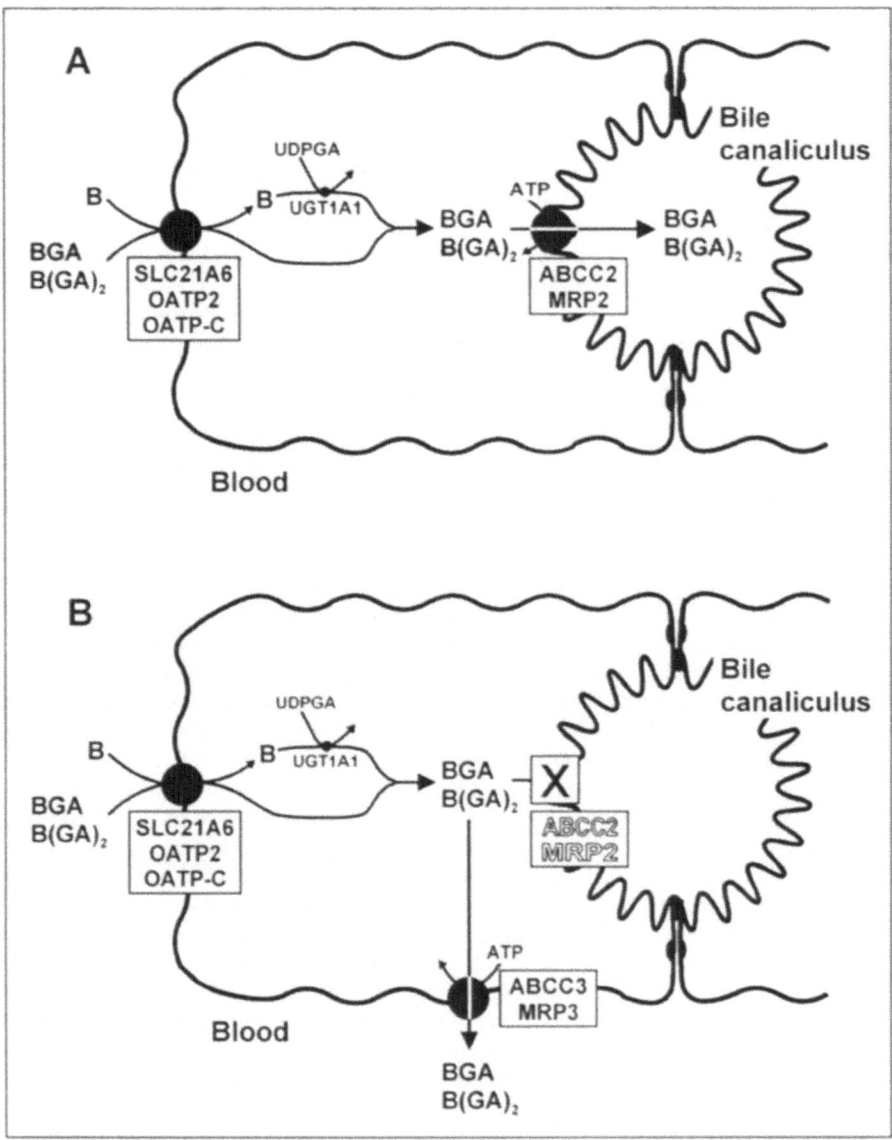

Figure 1. Bilirubin uptake into human hepatocytes, conjugation, and ATP-dependent export into bile. A, Bilirubin (B) is taken up across the basolateral membrane by SLC21A6 and conjugated with glucuronate by UDP-glucuronosyl transferase 1A1 (UGT1A1) resulting in monoglucuronosyl bilirubin (BGA) and bisglucuronosyl bilirubin (B(GA)₂). Bilirubin conjugates are secreted into bile by the ATP-dependent apical conjugate export pump Multidrug resistance protein 2 (MRP2, symbol ABCC2). B, Under conditions of hereditary (i.e., human Dubin-Johnson syndrome) or acquired (e.g., cholestasis) MRP2 deficiency, bilirubin conjugates are exported across the sinusoidal membrane into blood by MRP3 (symbol ABCC3) for subsequent renal elimination.

OATP2 (symbol SLC21A6; also known as OATP-C and LST-1)[18,32-35] and OATP8 (SLC21A8).[36] Most recently, four additional human SLC21A family members have been identified which are expressed in a variety of different tissues.[34] At least three members of the

OATP family are expressed in human hepatocytes: OATP2 (SLC21A6),[18,32,33] OATP-B (SLC21A9),[34,35] and OATP8 (SLC21A8).[36] The SLC21A family members expressed in hepatocytes show an overlapping substrate specificity and for all of them BSP has been identified as a substrate.[35]

The comparison of rat and human SLC21A family members with respect to tissue distribution and substrate specificity demonstrated that the human SLC21A family members have no direct and comparable rat orthologs. The degree of amino acid identity between the human and the rat OATP proteins is consistently below 75%. For example, the rat Oatp family member closest to human OATP2 is rat Oatp4 (Slc21a10) with 64 % amino acid identity. Rat Oatp4 seems to be expressed exclusively in liver and it may be considered at least a functional ortholog of human OATP2. This is in contrast to the export pumps, where one can clearly identify the rat ortholog of human MRP2 being rat Mrp2.

Uptake of Unconjugated Bilirubin by SLC21A6

Bilirubin is the end product of heme catabolism and the major bile pigment in most mammals.[1] In blood circulation, unconjugated bilirubin is bound tightly to serum albumin, which prevents its potential toxicity thought to be caused by the free ligand.[9,10] Despite the high-affinity binding to albumin, unconjugated bilirubin is rapidly and selectively taken up into the liver.[37,38] Competition studies indicated that unconjugated bilirubin shares the same carrier system with organic anions like BSP and indocyanine green (ICG).[38] In human liver, sodium-independent uptake of BSP is mediated by at least three members of the SLC21A family: SLC21A6,[19,35] SLC21A8,[36] and SLC21A9 (OATP-B).[35] Whereas SLC21A9 has been found in various tissues,[34,35] SLC21A6 and SLC21A8 have been detected exclusively in the basolateral membrane of human hepatocytes.[18,36] Among the three SLC21A family members found in human liver, SLC21A6 shows the highest affinity for BSP with a K_m value of 140 nM.[19] Interestingly, SLC21A6, but not SLC21A8, is able to transport BSP in the presence of human serum albumin in a large molar excess,[19,39] although BSP, similar to bilirubin, binds with high-affinity to albumin. Moreover, SLC21A6, but not SLC21A8, is strongly inhibited by nanomolar ICG concentrations.[19] These findings suggest that SLC21A6 mediates the hepatocellular uptake of unconjugated bilirubin. In fact, uptake of unconjugated bilirubin could be measured with *SLC21A6*-transfected HEK293 cells.[19] In addition, BSP and unconjugated bilirubin showed strong mutual inhibition.[19,39] Moreover, rifamycin SV and rifampicin, both of which were shown to be high-affinity substrates for SLC21A6 and SLC21A8,[40,41] interfere with hepatobiliary excretion of bilirubin and BSP.[42,43] Recently, Wang et al reported that SLC21A6 was not able to transport unconjugated bilirubin using *SLC21A6*-transfected HeLa cells.[44] The reason for this discrepancy is not yet known. Nevertheless, SLC21A6 continues to be the best candidate for the hepatic uptake of unconjugated bilirubin.

Uptake of Bilirubin Conjugates by Members of the SLC21A Family

After being taken up into the hepatocyte, unconjugated bilirubin is conjugated with glucuronic acid, catalyzed by UDP-glucuronosyl transferase 1A1 (UGT1A1) which is localized in the endoplasmic reticulum membrane.[11,45] Interestingly, the catalytic site of UGT1A1 is localized in the lumen of the ER.[46] Thus, unconjugated bilirubin has to cross the ER membrane to reach the active site of UGT1A1 and bilirubin glucuronosides have to be transported back into cytosol from the ER lumen. The transport mechanisms for these processes are still unknown.

Although bilirubin glucuronosides are mainly excreted into bile via the apical conjugate export pump ABCC2 (MRP2) localized in the canalicular membrane of hepatocytes (see Members of the ABCC subfamily transporting bilirubin conjugates),[20-24] bilirubin glucuronosides were also found in normal serum.[16,17] The origin of serum bilirubin glucuronosides is unknown. Although UGT1A1 expression was also detected in extrahepatic tissues like intestine and colon,[13-15] extrahepatic UGT1A1 may play only a minor role in bilirubin glucuronidation, because of the selective uptake of bilirubin into hepatocytes.

Uptake of bilirubin glucuronosides from serum into hepatocytes has been shown to be sodium-independent.[47] It shares a common uptake mechanism with BSP and bilirubin.[47] In human hepatocytes, uptake of monoglucuronosyl bilirubin is mediated by SLC21A6 and SLC21A8.[19] Both transporters showed high affinity for monoglucuronosyl bilirubin (K_m 0.1 μM for SLC21A6 and 0.5 μM for SLC21A8).[19] However, only SLC21A6 is able to transport bisglucuronosyl bilirubin with a K_m value of 0.28 μM.[19] Whether other SLC21A family members can transport bilirubin glucuronosides is currently unknown.

Members of the ABCC Subfamily Transporting Bilirubin Conjugates

The ABCC subfamily is comprised of 12 members (ABCC1 – ABCC12) with 9 (ABCC1 – ABCC6 and ABCC10 – ABCC12) of them being members of the MRP family of conjugate export pumps. The first member of this family was identified in 1992 by the cloning of human MRP1 (ABCC1).[48] The analysis of the amino acid sequence of MRP1,[48] followed by the elucidation of the function as a conjugate export pump[49,50] suggested the presence of MRP1-related transporters in other tissues. As a consequence, up to now eight MRP1-related ABC transporters have been identified, cloned, and at least partially characterized. All nine MRP family members are encoded by different genes located on several chromosomes. In human liver, at least MRP2 (ABCC2), MRP3 (ABCC3), MRP4 (ABCC4), MRP5 (ABCC5), and MRP6 (ABCC6) proteins are expressed at substantial amounts and from these, MRP2 and MRP3 are shown to be important in bilirubin conjugate transport. The *ABCC2* gene is located on chromosome 10,[51] contains 32 exons,[52,53] and spans around 69 kb.[52] The comparison of the genomic organization of the *ABCC1* gene, which is located on chromosome 16 and contains 31 exons,[54] and the *ABCC2* gene displays some remarkable similarities as indicated by the number and size of exons, by the high proportion of class 0 introns, and by 21 identical splice junction sites when the exon-intron organization is transferred to the amino acid sequence alignment. The genomic organization of the *ABCC3* gene was elucidated by means of a cosmid clone containing the entire *ABCC3* gene. A comparison of the two experimentally analyzed gene structures of *ABCC1* and *ABCC2* with the *ABCC3* gene again revealed remarkable similarities. The *ABCC3* gene, located on chromosome 17, contains 31 exons, as does the *ABCC1* gene[54] and a high proportion of class 0 introns. Transferring a gene structure alignment to the amino acid alignment of all three proteins indicates that 21 splice junction sites are located at the same position.[26] This similar gene structure demonstrates the close evolutionary relation of MRP1, the founding member of this family, with MRP2 and MRP3, both of which are involved in the hepatic elimination of anionic conjugates including bilirubin conjugates.

The close relationship of MRP2 and MRP3 is also evident from the comparison of the topology of both transporters. Both transporters are predicted to consist of a MDR-like core structure of two ATP-binding domains and two transmembrane regions and, in addition, of a large third aminoproximal transmembrane region. Most computational-based transmembrane organization analyses favor an odd number of transmembrane domains leading to an extracellularly located amino terminus. This extracellular localization was experimentally established for MRP2 by immunofluorescence microscopy studies.[55]

MRP2 is localized in the canalicular membrane of rat[20] and human hepatocytes.[21,56] In addition, MRP2 has been detected in the apical membrane of epithelial cells including those of kidney proximal tubule,[57,58] of gallbladder,[59] and of small intestine.[60] The transport of anionic conjugates across the canalicular membrane has been characterized as an unidirectional, ATP-dependent process.[61-64] But it was not before 1996 that the protein mediating this conjugate export was cloned and identified as Mrp2.[20,22,51] Prior to this molecular identification of the conjugate export pump, ATP-dependent transport measurements using inside-out oriented canalicular membrane vesicles from normal and hyperbilirubinemic mutant rats were carried out.[61-63,65-67] These mutant rats are deficient in the secretion of anionic conjugates into bile.[68-71] These transport measurements retrospectively contributed to our knowledge on the substrate specificity of Mrp2.[21,72] Using inside-out-oriented membrane vesicles from cells stably

Table 1. Selected substrates for the apical conjugate export pump MRP2 and for the basolateral isoform MRP3

Substrate	K_M Value (μM)	Reference
MRP2 (Human recombinant)		
Monoglucuronosyl bilirubin	0.7	Kamisako et al, 1999[24]
Bisglucuronosyl bilirubin	0.9	Kamisako et al, 1999[24]
Leukotriene C_4	1.0	Cui et al, 1999[55]
S-Glutathionyl		
2,4-dinitrobenzene	6.5	Evers et al, 1998[73]
17β-Glucuronosyl estradiol	7.2	Cui et al, 1999[55]
MRP3 (Human recombinant)		
Monoglucuronosyl bilirubin		Keppler et al, 2000[26]
Bisglucuronosyl bilirubin		Lee et al, 2002[a]
Leukotriene C_4	5.3	Zeng et al, 2000[81]
S-Glutathionyl		
2,4-dinitrobenzene	5.7	Zeng et al, 2000[81]
17β-Glucuronosyl estradiol	26	Zeng et al, 2000[81]

[a]Unpublished data.

expressing either rat or human recombinant MRP2, the substrate specificity of MRP2 was studied under defined conditions.[55,73-75] Prototypic high-affinity substrates of MRP2 include monoglucuronosyl bilirubin, bisglucuronosyl bilirubin, and the endogenous glutathione S-conjugate leukotriene C_4 (Table 1).[24,55,73] A large number of glutathione S-conjugates, glucuronosides, and sulfoconjugates of drugs and other xenobiotics are also substrates for MRP2, although the affinity of MRP2 for most of these drug conjugates is much lower than for the prototypic endogenous substrates.[76] In addition, non-conjugated amphiphilic anions are also transported by MRP2. References 72 and 77 give an overview of currently identified MRP2 substrates.

In contrast to monoglucuronosyl and bisglucuronosyl bilirubin, unconjugated bilirubin is found at a very low concentration in bile and probably arises from hydrolysis of secreted bilirubin conjugates.[25] Neither MRP2, nor the bile salt export pump (BSEP), nor MDR1 P-glycoprotein are involved in transport of unconjugated bilirubin across the canalicular membrane.

In contrast to the apical localization of MRP2, the isoform MRP3 has been detected in the basolateral membrane of human hepatocytes and other polarized cells.[59,78-80] Substrates for human MRP3 include monoglucuronosyl bilirubin, bisglucuronosyl bilirubin, and 17β-glucuronosyl estradiol (Table 1).[26,81,82] Sulfated bile salts have also been identified as substrates for human and, in addition, for rat MRP3,[83-85] whereas glutathione conjugates are relatively poor substrates.[84] Human MRP3 also transports glycocholate, taurocholate, and methotrexate.[81,82,85,86] MRP3 is present at a low level under normal conditions in human hepatocytes and is upregulated under cholestatic conditions.[78-80,87,88] Thus, when secretion of anionic conjugates into bile is impaired by either hereditary or acquired MRP2 deficiency, glucuronidated and sulfated conjugates may be transported out of the hepatocyte into blood by MRP3 (Fig. 1).

Hereditary Deficiency of MRP2 in Dubin-Johnson Syndrome

The absence of a functionally active MRP2 protein from the canalicular membrane is the molecular basis of Dubin-Johnson syndrome in humans.[21,52,56,89] This syndrome was first described in 1954 as an autosomal recessively inherited disorder characterized by conjugated hyperbilirubinemia.[90,91] The liver of patients with Dubin-Johnson syndrome appears dark blue

Table 2. *Homozygous mutations in the human* ABCC2 *gene causing*
       Dubin-Johnson syndrome

Type of Mutation and Designation[a]	Nucleotide Change	Predicted Consequence or Amino Acid Change	Functional Consequence
Splice site			
S1	1815 + 2T → A	Splice donor, loss of exon 13 (ref. 53)	Lack of functional MRP2
S2	1967 + 2T → C	Splice donor, loss of exon 15 (ref. 92)	Lack of functional MRP2
S3	2439 + 2T → C	Splice donor, loss of exon 18 (ref. 53)	Lack of functional MRP2
Missense			
M1	2302C → T, exon 18	Arg768Trp (ref. 53)	Deficient maturation and impaired sorting (ref. 101)
M2	3449G → A, exon 25	Arg1150His (ref. 95)	Deficient transport function (ref. 95)
M3	3517A → T, exon 25	Ile1173Phe (ref. 95)	Deficient maturation and impaired sorting, and deficient transport function (refs. 95,100)
M4	4145A → G, exon 30	Gln1382Arg (ref. 53)	Deficient transport function (ref. 101)
Nonsense			
N1	3196C → T, exon 23	Arg1066X (ref. 56)	mRNA degradation due to nonsense-mediated decay (ref. 98)
Deletion			
D1	4175-4180DelGGATGA, exon 30	Arg1392+Met1393del (ref. 52)	Deficient maturation and impaired sorting (ref. 99)

[a]See Figure 2 for location.

or black because of deposition of a dark pigment in the pericanalicular area of hepatocytes.[1] Dubin-Johnson syndrome is a very rare disorder (e.g., approximate incidence of 1:300,000 in Japan[92]) except for Jews of Persian origin among whom the incidence is 1:1,300.[93]

A number of naturally occurring mutations in the *ABCC2* gene has been described (Fig. 2, Table 2) that lead to lack of a functionally active protein in the canalicular membrane and, therefore, to an impaired transport of conjugated bilirubin and other anionic conjugates across the canalicular membrane into bile. Established homozygous mutations identified in patients with Dubin-Johnson syndrome include splice site mutations leading to exon deletions and subsequent premature stop codons,[53,92,94] missense mutations,[53,95] a nonsense mutation resulting in a premature stop codon,[52,56] and a deletion mutation leading to the loss of 2 amino acids from the second nucleotide-binding domain.[52] In addition, three heterozygous mutations—a nonsense mutation in exon 3 (Arg105X),[96] a nonsense mutation in exon 28 (Arg1309X),[97] and a missense mutation in exon 30 (Gln1382Arg)[96]—have recently been identified in Japanese patients with Dubin-Johnson syndrome.

Although all Dubin-Johnson syndrome-associated mutations result in lack of a functionally active MRP2 protein from the canalicular membrane, the consequences of mutations on the cellular itinerary of MRP2 may differ (Table 2). It is likely that a premature stop codon leads to degradation of mRNA due to a mechanism termed 'nonsense-mediated decay' in which a stop

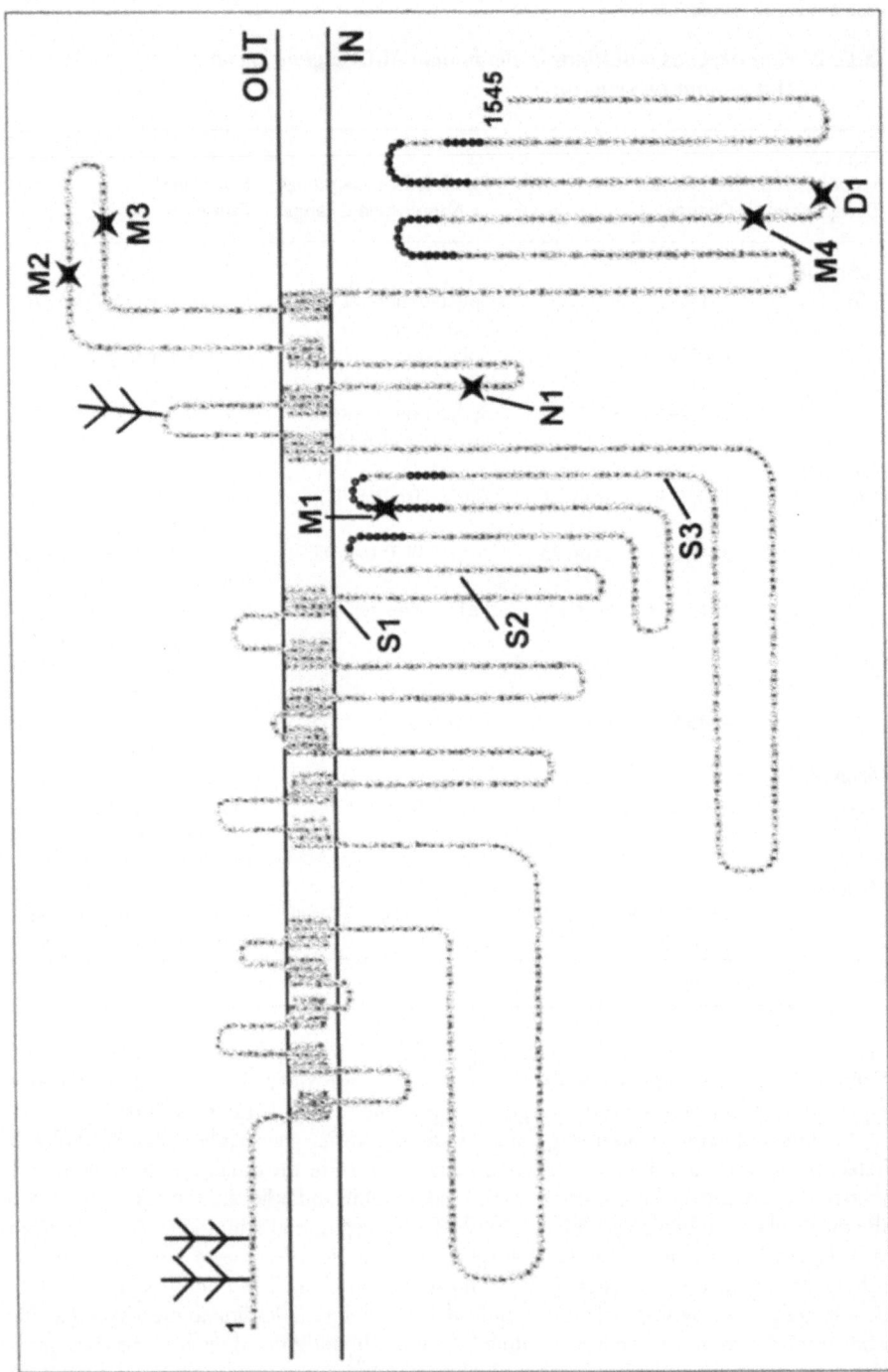

Figure 2. A predicted membrane topology model of the human MRP2 protein. Amino acids in the nucleotide-(ATP)-binding domains are indicated, with the Walker A and B motifs in dark gray and the ABC transporter family signature in black. Mutations causing Dubin-Johnson syndrome are indicated as detailed in Table 2. S = splice site mutation, N = nonsense mutation, M = missense mutation, D = deletion mutation.

codon preceding the last splice site is recognized during translation.[98] Absence of the MRP2 protein from the canalicular membrane in certain cases of Dubin-Johnson syndrome may thus be a consequence of rapid degradation of the mutated *MRP2* mRNA. So far, no mutations in the *ABCC2* gene have been found that lead to a truncated protein localized in the canalicular membrane.[52,56] Other Dubin-Johnson syndrome-associated mutations may lead to the synthesis of a mutated, yet immature MRP2 protein which is recognized by the cellular quality control machinery, retained in the endoplasmic reticulum, and subsequently degraded by proteasomes.[95,99-101] Few mutations apparently lead to the synthesis of mutant MRP2 proteins which are correctly localized to the canalicular membrane but lack transport activity.[95,101] Moreover, certain mutations in the *ABCC2* gene may affect the interaction of MRP2 with chaperone proteins required for correct sorting of MRP2 to the canalicular membrane. A recent study in mice has demonstrated that a deficiency of radixin, a protein linking Mrp2 either directly, or via additional coupling proteins, to actin, leads to loss of the Mrp2 protein from the canalicular membrane and to conjugated hyperbilirubinemia.[102]

Hereditary Mrp2 Deficiency in Animal Models

Two distinct hyperbilirubinemic mutant rat strains, the GY/TR[-68] and the Eisai hyperbilirubinemic (EHBR)[70,71] rats, have been described that are deficient in the secretion of anionic conjugates into bile. After the molecular identification and cloning of Mrp2 in 1996 as the apical conjugate export pump,[20,22] which is defective in these rats,[20,22,103] these two mutant rat strains may now be considered as animal models of the human Dubin-Johnson syndrome. Distinct premature termination codons within the *Abcc2* gene, either at codon 401 (GY/TR$^-$)[22] or at codon 855 (EHBR),[103] were identified and lead to the absence of the Mrp2 protein from the hepatocyte canalicular membrane.[20,22,67] Similar to corresponding human mutations, no truncated Mrp2 protein was detected[20] and the *Mrp2* mRNA was below the detection level of Northern blot analysis.[20,22,103] As described for the Dubin-Johnson syndrome-associated mutations, the introduction of premature termination codons may lead to rapid degradation of the mutated mRNA by ,nonsense-mediated decay'.[98] In addition, a mutant strain of golden lion tamarins (*Leontopithecus rosalia rosalia*),[104] and one of Corriedale sheep[105] with conjugated hyperbilirubinemia similar to that observed in Dubin-Johnson syndrome have been described. Because the two rat strains lacking Mrp2 are readily available, Mrp2 knockout mice have not been generated so far. However, Mrp2 knockout mice may become of interest, especially for cross-breeding with mice deficient in other Mrp isoforms.

The GY/TR$^-$ and EHBR mutant rat strains have been valuable in suggesting the substrate specificity of Mrp2 by using them for hepatobiliary elimination studies,[66,68-70] as well as for measurements of ATP-dependent transport by inside-out oriented canalicular membrane vesicles derived from Mrp2-deficient rat hepatocytes in comparison with those derived from normal hepatocytes.[20,61-63,65-67] The non-invasive assessment by positron emission tomography of hepatobiliary and renal transport of a high-affinity substrate of Mrp2, the endogenous lipophilic leukotriene C_4 metabolite N-acetyl leukotriene E_4, demonstrated that the selective absence of Mrp2 from the canalicular membrane of mutant GY/TR$^-$ rats abolishes transport of anionic conjugates across the canalicular membrane into bile.[106] After a prolonged time of hepatocellular storage and metabolism, as compared to normal rats, anionic conjugates are then secreted into the blood and subsequently eliminated via the kidney into urine of the mutant rat.[106]

The bile salt-independent bile flow is largely driven by Mrp2 as indicated by the markedly decreased bile flow in both mutant rat strains.[68,70] Moreover, an impaired biliary excretion of GSH has been described in the mutant GY/TR$^-$ rats.[107] MRP2-dependent release of GSH either out of rat hepatocytes or out of MDCK cells expressing recombinant human MRP2 correlated with the amount of MRP2 protein in the apical membrane.[107] These studies show that MRP2 also transports GSH across the canalicular membrane, either with low affinity in the absence of a co-substrate or more efficiently together with a lipophilic co-substrate.[107]

Table 3. Polymorphisms in the ABCC2 *gene*

Nucleotide Change	Amino Acid Change	Genomic Localization	Allelic Frequency	Reference
1-24C → T	Promoter	Promoter	0.188	Ito et al[108]
842G → A	Ser281Asp	Exon 7	0.017	Mor-Cohen et al[95]
1249G → A	Val417Ile	Exon 10	0.125	Ito et al[108]
2303C → T	Arg768Trp	Exon 18	0.010	Ito et al[108]
2366C → T	Ser789Phe	Exon 18	0.010	Ito et al[108]
3972C → T	Ile1324Ile	Exon 28	0.219	Ito et al[108]
4348G → A	Ala1450Thr	Exon 31	0.010	Ito et al[108]

Since hepatocellular GSH concentrations are high (5-10 mM) MRP2 may have an important physiological function in the maintenance of GSH-dependent bile flow and in hepatic GSH turnover.[107]

Polymorphisms in the Human *ABCC2* Gene

In addition to mutations in the *ABCC2* gene which are associated with Dubin-Johnson syndrome, a number of nucleotide exchanges have been identified which do not cause Dubin-Johnson syndrome and are, therefore, considered as single nucleotide polymorphisms (SNPs). Table 3 gives an overview of currently known polymorphisms with an allelic frequency of >0.01.[95,108,109] In addition, a number of SNPs exist in the non-coding regions of the *ABCC2* gene, either in the promoter or in introns.[95,110] The consequences of these polymorphisms on the function of MRP2 remain to be elucidated.

Acquired MRP2 Deficiency in Cholestasis

In contrast to the hereditary deficiency of MRP2 in Dubin-Johnson syndrome, a transitory MRP2 deficiency may occur in humans during pathophysiological conditions such as cholestasis. The molecular mechanisms of this acquired MRP2 deficiency have been mostly studied using the rat model. Endotoxin (lipopolysaccharide)-treated rats, estrogen-treated rats, and bile duct-ligated rats are experimental models representing sepsis-associated cholestasis, intrahepatic cholestasis of pregnancy or cholestasis caused by oral contraceptives, and obstructive cholestasis, respectively. These cholestatic conditions are also observed in humans. In all three models, total Mrp2 protein is markedly decreased during cholestasis.[111-113] Mrp2 concentrates near the central (perivenous) area of the liver lobule after bile duct ligation, whereas, in normal liver, Mrp2 is homogeneously distributed throughout the liver lobule.[114] As an early event of cholestasis, selective retrieval of Mrp2 from the canalicular membrane to pericanalicular vesicles of rat hepatocytes has been demonstrated by immunofluorescence microscopy[111-113,115-117] and immunogold electron microscopy.[116,118] Under cholestatic conditions, tauroursodeoxycholate stimulates re-insertion of internalized Mrp2 into the canalicular membrane in a protein kinase C dependent manner,[118,119] however, activation of protein kinase C in HepG2 cells results in relocation of MRP2 to the basolateral membrane.[120] Treatment of isolated rat hepatocytes with cAMP results in enhanced insertion of Mrp2 into the apical membrane.[121] CyclicAMP also partially protects against endocytic retrieval of rat Mrp2 caused by 17β-glucuronosyl estradiol-induced cholestasis.[113] Endocytic retrieval of MRP2 from the canalicular membrane has also been described in human liver samples with cholestatic liver disease.[88,122]

Drug-induced cholestasis is often observed in patients as a side effect of drug treatment[123] and may occur during therapy e.g., with antibiotics[124] or with the immunosuppressant cyclosporin A.[125] The mono-anionic steroidal antibiotic fusidate, which is used to treat infectious diseases caused by multi-resistant *Staphylococcus aureus* strains, frequently induces conjugated hyperbilirubinemia.[126,127] At least two different mechanisms have been identified in the rat which account for the impaired hepatobiliary elimination of anionic conjugates after fusidate administration, i.e., the direct and competitive inhibition of Mrp2- and Bsep-mediated transport, and the impairment of biliary secretion by a decreased level of the Mrp2 protein.[128] The immunosuppressant cyclosporin A also directly and competitively inhibits Mrp2-mediated transport in the rat.[129,130] In addition, cyclosporin A is a potent inhibitor of Bsep- and Mdr1 P-glycoprotein-mediated transport.[129,130] Furthermore, cholestasis is observed as an early effect of acute phalloidin poisoning.[131] Phalloidin is one of the toxic peptides of the mushroom *Amanita phalloides*. In addition to a loss of canalicular contractility due to disruption of the actin organization, phalloidin causes endocytic retrieval of Mrp2 and other membrane proteins from the canalicular membrane into the hepatocyte.[115]

Recent studies indicate that the expression of rat Mrp2 is regulated by three distinct nuclear receptor signaling pathways involving the farnesoid X-activated receptor (FXR), the pregnane X receptor (PXR), and the constitutive androstane receptor (CAR).[132] All three factors bind to the same element in the rat Mrp2 promoter.[132] Because human MRP2 expression is induced by PXR ligands, including rifampicin[132] and peptide mimetic HIV protease inhibitors,[133] and by the CAR ligand phenobarbital,[134] similar mechanisms may be involved in regulation of human MRP2. Four imperfect potential PXR binding sites are present in the human MRP2 promoter.[135,136] Moreover, recent data by David Moore and his colleagues suggest that, in humans, CAR coordinately regulates and induces expression of proteins required for bilirubin uptake (SLC21A6), bilirubin conjugation (UGT1A1), and ATP-dependent export of bilirubin conjugates (MRP2).[28] This observation may become important in understanding the regulation-induced MRP2 deficiency which might occur in newborns who have low levels of CAR.[28] Additional (patho)physiological conditions may be found under which MRP2 deficiency is attributable to a decreased level of CAR or to the presence of CAR antagonists.

Acknowledgements

The studies in the authors' laboratory were supported by the Deutsches Krebsforschungszentrum and by the Deutsche Forschungsgemeinschaft, Bonn, Germany. We acknowledge the contributions to this work from past and present members of our department, especially from Hiroyuki Tsujii, Gabriele Jedlitschky, Verena Keitel, and Young-Min Lee, as well as the collaboration with Jürgen Kartenbeck and Herbert Spring from the Cell Biology Division of the Deutsches Krebsforschungszentrum.

References

1. Roy Chowdhury N, Arias IM, Wolkoff AW et al. Disorders of bilirubin metabolism. In: Arias IM, Boyer JL, Chisari FV et al. The Liver: Biology and Pathobiology. 4th ed. Philadelphia: Lippincott, Williams & Wilkins, 2001:291-309.
2. Virchow R. Die pathologischen Pigmente. Arch Pathol Anat 1847; 1:379-486.
3. Fischer H, Orth H. Die Chemie des Pyrrols. Akademische Verlagsges m b H 1937; Leipzig.
4. London IM, West R, Shemin D et al. On the origin of bile pigment in normal man. J Biol Chem 1950; 184:351-358.
5. Hymans van den Bergh AA, Mueller P. Ueber eine direkte und indirekte Diazoreaktion auf Bilirubin. Biochem Z 1916; 77:90.
6. Billing BH, Cole PG, Lathe GH. The excretion of bilirubin as a diglucuronide giving the direct van den Bergh reaction. Biochem J 1957; 65:774-783.
7. Schmid R. The identification of "direct reacting" bilirubin as bilirubin glucuronide. J Biol Chem 1957; 229:881-888.
8. Talafant E. Properties and composition of the bile pigments giving a direct diazo reaction. Nature 1956; 178:312.
9. Gourley GR. Bilirubin metabolism and kernicterus. Adv Pediatr 1997; 44:173-229.
10. Brodersen R, Stern L. Deposition of bilirubin acid in the central nervous system—a hypothesis for the development of kernicterus. Acta Paediatr Scand 1990; 79:12-19.

11. Senafi SB, Clarke DJ, Burchell B. Investigation of the substrate specificity of a cloned expressed human bilirubin UDP-glucuronosyltranferase: UDP sugar specificity and involvement in steroid and xenobiotic glucuronidation. Biochem J 1994; 303:233-240.

12. Bosma PJ, Seppen J, Goldhoorn B et al. Bilirubin UDP-glucuronosyltranferase 1 is the only relevant bilirubin glucuronidating isoform in man. J Biol Chem 1994; 269:17960-17964.

13. Strassburg CP, Kneip S, Topp J et al. Polymorphic gene regulation and interindividual variation of UDP-glucuronosyltranferase activity in human small intestine. J Biol Chem 2000; 275:36164-36171.

14. Paine MF, Fisher MB. Immunochemical identification of UGT isoforms in human small bowel and in Caco-2 cell monolayers. Biochem Biophys Res Commun 2000; 273:1053-1057.

15. Czernik PJ, Little JM, Barone GW et al. Glucuronidation of estrogens and retinoic acid and expression of UDP-glucuronosyltransferase 2B7 in human intestinal mucosa. Drug Metab Dispos 2000; 28:1210-1216.

16. van Roy FP, Heirwegh KP. Determination of bilirubin glucuronide and assay of glucuronosyltransferase with bilirubin as acceptor. Biochem J 1968; 107:507-518.

17. Muraca M, Blanckaert N. Liquid-chromatographic assay and identification of mono- and diester conjugates of bilirubin in normal serum. Clin Chem 1983; 29:1767-1771.

18. König J, Cui Y, Nies AT et al. A novel human organic anion transporting polypeptide localized to the basolateral hepatocyte membrane. Am J Physiol Gastrointest Liver Physiol 2000; 278:G156-G164.

19. Cui Y, König J, Leier I et al. Hepatic uptake of bilirubin and its conjugates by the human organic anion transporter SLC21A6. J Biol Chem 2001; 276:9626-9630.

20. Büchler M, König J, Brom M et al. cDNA cloning of the hepatocyte canalicular isoform of the multidrug resistance protein, cMRP, reveals a novel conjugate export pump deficient in hyperbilirubinemic mutant rats. J Biol Chem 1996; 271:15091-15098.

21. Keppler D, Kartenbeck J. The canalicular conjugate export pump encoded by the *cmrp/cmoat* gene. In: Boyer JL, Ockner RK eds. Progress in Liver Diseases. Philadelphia: Saunders, 1996:55-67.

22. Paulusma CC, Bosma PJ, Zaman GJR et al. Congenital jaundice in rats with a mutation in a multidrug resistance associated-protein gene. Science 1996; 271:1126-1127.

23. Jedlitschky G, Leier I, Buchholz U et al. ATP-dependent transport of bilirubin glucuronides by the multidrug resistance protein MRP1 and its hepatocyte canalicular isoform MRP2. Biochem J 1997; 327:305-310.

24. Kamisako T, Leier I, Cui Y et al. Transport of monoglucuronosyl and bisglucuronosyl bilirubin by recombinant human and rat multidrug resistance protein 2. Hepatology 1999; 30:485-490.

25. Pascolo L, Bayon EJ, Cupelli F et al. ATP-dependent transport of unconjugated bilirubin by rat liver canalicular plasma membrane vesicles. Biochem J 1998; 331:99-103.

26. Keppler D, Kamisako T, Leier I et al. Localization, substrate specificity, and drug resistance conferred by conjugate export pumps of the MRP family. Advan Enzyme Regul 2000; 40:339-349.

27. McDonagh AF, Palma LA. Hepatic excretion of circulating bilirubin photoproducts in the Gunn rat. J Clin Invest 1980; 66:1182-1185.

28. Huang W, Zhang J, Qatanani M et al. Metabolic regulation by the xenobiotic receptor CAR. Drug metabolism Rev 2002; 34:16 (abstract) suppl.1.

29. Jacquemin E, Hagenbuch B, Stieger B et al. Expression cloning of a rat liver Na(+)-independent organic anion transporter. Proc Natl Acad Sci USA 1994; 91:133-137.

30. Meier PJ, Eckhardt U, Schroeder A et al. Substrate specificity of sinusoidal bile acid and organic anion uptake systems in rat and human liver. Hepatology 1997; 26:1667-1677.

31. Kullak-Ublick GA, Hagenbuch B, Stieger B et al. Molecular and functional characterization of an organic anion transporting polypeptide cloned from human liver. Gastroenterology 1995; 109:1274-1282.

32. Abe T, Kakyo M, Tokui T et al. Identification of a novel gene family encoding human liver-specific organic anion transporter LST-1. J Biol Chem 1999; 274:17159-17163.

33. Hsiang B, Zhu Y, Wang Z et al. A novel human hepatic organic anion transporting polypeptide (OATP2). Identification of a liver-specific human organic anion transporting polypeptide and identification of rat and human hydroxymethylglutaryl-CoA reductase inhibitor transporters. J Biol Chem 1999; 274:37161-37168.

34. Tamai I, Nezu J, Uchino H et al. Molecular identification and characterization of novel members of the human organic anion transporter (OATP) family. Biochem Biophys Res Commun 2001; 273:251-260.

35. Kullak-Ublick GA, Ismair MG, Stieger B et al. Organic anion-transporting polypeptide B (OATP-B) and its functional comparison with three other OATPs of human liver. Gastroenterology 2001; 120:525-533.

36. König J, Cui Y, Nies AT et al. Localization and genomic organization of a new hepatocellular organic anion transporting polypeptide. J Biol Chem 2000; 275:23161-23168.

37. Arias IM, Johnson L, Wolfson S. Am J Physiol 1961; 200:1091-1094.
38. Scharschmidt BF, Waggoner JG, Berk PD. Hepatic organic anion uptake in the rat. J Clin Invest 1975; 56:1280-1292.
39. Cui Y, Walter B. Influence of albumin-binding on the substrate transport mediated by human hepatocyte transporters OATP2 and OATP8. J Gastroenterol 2003; 38:in press
40. Cui Y, König J, Keppler D. Vectorial transport by double-transfected cells expressing the human uptake transporter SLC21A8 and the apical export pump ABCC2. Mol Pharmacol 2001; 60:934-943.
41. Vavricka SR, Van Montfoort J, Ha HR et al. Interactions of rifamycin SV and rifampicin with organic anion uptake systems of human liver. Hepatology 2002; 36:164-172.
42. Capelle P, Dhumeaux D, Mora M et al. Effect of rifampicin on liver function in man. Gut 1972; 13:366-371.
43. Acocella G, Nicolis FB, Tenconi LT. The effect of an intravenous infusion of rifamycin SV on the excretion of bilirubin, bromsulphalein, and indocyanine green in man. Gastroenterology 1965; 49:521-525.
44. Wang P, Roy Chowdhury J, Kim R et al. Expression of the human organic anion transporting protein OATP2 (OATP-C) is not sufficient to produce bilirubin transport. Hepatology 2001; 34:256A
45. Gordon ER, Meier PJ, Goresky CA et al. Mechanism and subcellular site of bilirubin diglucuronide formation in rat liver. J Biol Chem 1984; 259:5500-5506.
46. Vanstapel F, Blanckaert N. Topology and regulation of bilirubin UDP-glucuronosyltranferase in sealed native microsomes from rat liver. Arch Biochem Biophys 1988; 263:216-225.
47. Adachi Y, Kinne R, Chowdhury JR et al. Uptake of bilirubin glucuronides by isolated rat hepatocytes. Gastroenterol Jpn 1991; 26:350-355.
48. Cole SPC, Bhardwaj G, Gerlach JH et al. Overexpression of a transporter gene in a multidrug-resistant human lung cancer cell line. Science 1992; 258:1650-1654.
49. Jedlitschky G, Leier I, Buchholz U et al. ATP-dependent transport of glutathione S-conjugates by the multidrug resistance-associated protein. Cancer Res 1994; 54:4833-4836.
50. Leier I, Jedlitschky G, Buchholz U et al. The *MRP* gene encodes an ATP-dependent export pump for leukotriene C$_4$ and structurally related conjugates. J Biol Chem 1994; 269:27807-27810.
51. Taniguchi K, Wada M, Kohno K et al. A human canalicular multispecific organic anion transporter (cMOAT) overexpressed in cisplatin-resistant human cancer cell lines with decreased drug accumulation. Cancer Res 1996; 56:4124-4129.
52. Tsujii H, König J, Rost D et al. Exon-intron organization of the human multidrug-resistance protein 2 (MRP2) gene mutated in Dubin-Johnson syndrome. Gastroenterology 1999; 117:653-660.
53. Toh S, Wada M, Uchiumi T et al. Genomic structure of the canalicular multispecific organic anion-transporter gene (MRP2/cMOAT) and mutations in the ATP-binding-cassette region in Dubin-Johnson syndrome. Am J Hum Genet 1999; 64:739-746.
54. Grant CE, Kurz EU, Cole SPC et al. Analysis of the intron-exon organization of the human multidrug-resistance protein gene (*MRP*) and alternative splicing of its mRNA. Genomics 1997; 45:368-378.
55. Cui Y, König J, Buchholz U et al. Drug resistance and ATP-dependent conjugate transport mediated by the apical multidrug resistance protein, MRP2, permanently expressed in human and canine cells. Mol Pharmacol 1999; 55:929-937.
56. Paulusma CC, Kool M, Bosma PJ et al. A mutation in the human canalicular multispecific organic anion transporter gene causes the Dubin-Johnson syndrome. Hepatology 1997; 25:1539-1542.
57. Schaub TP, Kartenbeck J, König J et al. Expression of the conjugate export pump encoded by the *mrp2* gene in the apical membrane of kidney proximal tubules. J Am Soc Nephrol 1997; 8:1213-1221.
58. Schaub TP, Kartenbeck J, König J et al. Expression of the MRP2 gene-encoded conjugate export pump in human kidney proximal tubules and in renal-cell carcinoma. J Am Soc Nephrol 1999; 10:1159-1169.
59. Rost D, König J, Weiss G et al. Expression and localization of the multidrug resistance proteins MRP2 and MRP3 in human gallbladder epithelia. Gastroenterology 2001; 121:1203-1208.
60. Fromm MF, Kauffmann HM, Fritz P et al. The effect of rifampin treatment on intestinal expression of human MRP transporters. Am J Pathol 2000; 157:1575-1580.
61. Kitamura T, Jansen P, Hardenbrook C et al. Defective ATP-dependent bile canalicular transport of organic anions in mutant (TR-) rats with conjugated hyperbilirubinemia. Proc Nat Acad Sci USA 1990; 87:3557-3561.
62. Ishikawa T, Müller M, Klünemann C et al. ATP-dependent primary active transport of cysteinyl leukotrienes across liver canalicular membrane: Role of the ATP-dependent transport system for glutathione S-conjugates. J Biol Chem 1990; 265:19279-19286.

63. Fernandez-Checa JC, Takikawa H, Horie T et al. Canalicular transport of reduced glutathione in normal and mutant Eisai hyperbilirubinemic rats. J Biol Chem 1992; 267:1667-1673.
64. Keppler D, König J. Expression and localization of the conjugate export pump encoded by the *MRP2* (*cMRP/cMOAT*) gene in liver. FASEB J 1997; 11:509-516.
65. Akerboom TP, Narayanaswami V, Kunst M et al. ATP-dependent S-(2,4-dinitrophenyl)glutathione transport in canalicular plasma membrane vesicles from rat liver. J Biol Chem 1991; 266:13147-13152.
66. Oude Elferink RPJ, Meijer DKF, Kuipers F et al. Hepatobiliary secretion of organic compounds: Molecular mechanisms of membrane transport. Biochim Biophys Acta 1995; 1241:215-268.
67. Mayer R, Kartenbeck J, Büchler M et al. Expression of the *MRP* gene-encoded conjugate export pump in liver and its selective absence from the canalicular membrane in transport-deficient mutant hepatocytes. J Cell Biol 1995; 131:137-150.
68. Jansen PLM, Peters WHM, Lamers WH. Hereditary chronic conjugated hyperbilirubinemia in mutants rats caused by defective hepatic anion transport. Hepatology 1985; 5:573-579.
69. Huber M, Guhlmann A, Jansen PL et al. Hereditary defect of hepatobiliary cysteinyl leukotriene elimination in mutant rats with defective hepatic anion excretion. Hepatology 1987; 7:224-228.
70. Takikawa H, Sano N, Narita T et al. Biliary excretion of bile acid conjugates in a hyperbilirubinemic mutant Sprague-Dawley rat. Hepatology 1991; 14:352-360.
71. Hosokawa S, Tagaya O, Mikami T et al. A new rat mutant with chronic conjugated hyperbilirubinemia and renal glomerular lesions. Lab Anim Sci 1992; 42:27-34.
72. König J, Nies AT, Cui Y et al. Conjugate export pumps of the multidrug resistance protein (MRP) family: localization, substrate specificity, and MRP2-mediated drug resistance. Biochim Biophys Acta 1999; 1461:377-394.
73. Evers R, Kool M, van Deemter L et al. Drug export activity of the human canalicular multispecific organic anion transporter in polarized kidney MDCK cells expressing *cMOAT* (*MRP2*) cDNA. J Clin Invest 1998; 101:1310-1319.
74. Ito K, Suzuki H, Hirohashi T et al. Functional analysis of a canalicular multispecific organic anion transporter cloned from rat liver. J Biol Chem 1998; 273:1684-1688.
75. Chen ZS, Kawabe T, Ono M et al. Effect of multidrug resistance-reversing agents on transporting activity of human canalicular multispecific organic anion transporter. Mol Pharmacol 1999; 56:1219-1228.
76. Suzuki H, Sugiyama Y. Transporters for bile acids and organic anions. In: Amidon GL, Sadée W eds. Membrane transporters as drug targets. 1999:387-439.
77. Suzuki H, Sugiyama Y. Single nucleotide polymorphisms in multidrug resistance associated protein 2 (MRP2/ABCC2): its impact on drug disposition. Adv Drug Deliv Rev 2002; 54:1311-1331.
78. Kool M, van der Linden M, de Haas M et al. MRP3, an organic anion transporter able to transport anti-cancer drugs. Proc Natl Acad Sci USA 1999; 96:6914-6919.
79. König J, Rost D, Cui Y et al. Characterization of the human multidrug resistance protein isoform MRP3 localized to the basolateral hepatocyte membrane. Hepatology 1999; 29:1156-1163.
80. Scheffer GL, Kool M, de Haas M et al. Tissue distribution and induction of human multidrug resistance protein 3. Lab Invest 2002; 82:193-201.
81. Zeng H, Liu G, Rea PA et al. Transport of amphipathic anions by human multidrug resistance protein 3. Cancer Res 2000; 60:4779-4784.
82. Oleschuk CJ, Deeley RG, Cole SP. Substitution of Trp1242 of transmembrane segment 17 alters substrate specificity of human multidrug resistance protein, MRP3. Am J Physiol Gastrointest Liver Physiol 2003; 284:G280-G289.
83. Hirohashi T, Suzuki H, Takikawa H et al. ATP-dependent transport of bile salts by rat multidrug resistance-associated protein 3 (Mrp3). J Biol Chem 2000; 275:2905-2910.
84. Hirohashi T, Suzuki H, Sugiyama Y. Characterization of the transport properties of cloned rat multidrug resistance-associated protein 3 (MRP3). J Biol Chem 1999; 274:15181-15185.
85. Akita H, Suzuki H, Hirohashi T et al. Transport activity of human MRP3 expressed in Sf9 cells: comparative studies with rat MRP3. Pharm Res 2002; 19:34-41.
86. Zelcer N, Saeki T, Bot I et al. Transport of bile acids in multidrug resistance protein 3 over-expressing cells co-transfected with the ileal sodium-dependent bile acid transporter. Biochem J 2003; 369:23-30.
87. Kiuchi Y, Suzuki H, Hirohashi T et al. cDNA cloning and inducible expression of human multidrug resistance associated protein 3 (MRP3). FEBS Lett 1998; 433:149-152.
88. Shoda J, Kano M, Oda K et al. The expression levels of plasma membrane transporters in the cholestatic liver of patients undergoing biliary drainage and their association with the impairment of biliary secretory function. Am J Gastroenterol 2001; 96:3368-3378.

89. Kartenbeck J, Leuschner U, Mayer R et al. Absence of the canalicular isoform of the *MRP* gene-encoded conjugate export pump from the hepatocytes in Dubin-Johnson syndrome. Hepatology 1996; 23:1061-1066.
90. Dubin IN, Johnson FB. Chronic idiopathic jaundice with unidentified pigment in liver cells; a new clinicopathologic entity with report of 12 cases. Medicine 1954; 33:155-179.
91. Sprinz H, Nelson RS. Persistent nonhemolytic hyperbilirubinemia associated with lipochrome-like pigment in liver cells; report of four cases. Ann Intern Med 1954; 41:952-962.
92. Kajihara S, Hisatomi A, Mizuta T et al. A splice mutation in the human canalicular multispecific organic anion transporter gene causes Dubin-Johnson syndrome. Biochem Biophys Res Comm 1998; 253:454-457.
93. Shani M, Seligsohn U, Gilon E et al. Dubin-Johnson syndrome in Israel. I. Clinical, laboratory, and genetic aspects of 101 cases. Quarterly J Med 1970; 39:549-567.
94. Wada M, Toh S, Taniguchi K et al. Mutations in the canalicular multispecific organic anion transporter (cMOAT) gene, a novel ABC transporter, in patients with hyperbilirubinemia II/ Dubin-Johnson syndrome. Hum Mol Genet 1998; 7:203-207.
95. Mor-Cohen R, Zivelin A, Rosenberg N et al. Identification and functional analysis of two novel mutations in the multidrug resistance protein 2 gene in Israeli patients with Dubin-Johnson syndrome. J Biol Chem 2001; 276:36923-36930.
96. Suzuki H, Shoda J, Kano M et al. Novel mutations identified in the human multidrug resistance-associated protein 2 (MRP2) gene in a Japanese patient with Dubin-Johnson syndrome (DJS). Gastroenterology 2003; in press:
97. Tate G, Li M, Suzuki T et al. A new mutation of the ATP-binding cassette, sub-family C, member 2 (ABCC2) gene in a Japanese patient with Dubin.Johnson syndrome. Genes Genet Syst 2002; 77:117-121.
98. Thermann R, Neu-Yilik G, Deters A et al. Binary specifications of nonsense codons by splicing and cytoplasmic translation. EMBO J 1998; 17:3484-3494.
99. Keitel V, Kartenbeck J, Nies AT et al. Impaired protein maturation of the conjugate export pump MRP2 as a consequence of a deletion mutation in Dubin-Johnson syndrome. Hepatology 2000; 32:1317-1328.
100. Keitel V, Nies AT, Brom M et al. A common Dubin-Johnson syndrome mutation impairs protein maturation and transport activity of MRP2 (ABCC2). Am J Physiol Gastrointest Liver Physiol 2003; 284:G165-G174.
101. Hashimoto K, Uchiumi T, Konno T et al. Trafficking and functional defects by mutations of the ATP-binding domains in MRP2 in patients with Dubin-Johnson syndrome. Hepatology 2002; 36:1236-1245.
102. Kikuchi S, Hata M, Fukumoto K et al. Radixin deficiency causes conjugated hyperbilirubinemia with loss of Mrp2 from bile canalicular membranes. Nature Genet 2002; 31:320-325.
103. Ito K, Suzuki H, Hirohashi T et al. Molecular cloning of canalicular multispecific organic anion transporter defective in EHBR. Am J Physiol Gastrointest Liver Physiol 1997; 272:G16-G22.
104. Schulman FY, Montali RJ, Bush M et al. Dubin-Johnson-like syndrome in golden lion tamarins (Leontopithecus rosalia rosalia). Vet Pathol 1993; 30:491-498.
105. Barnhart JL, Gronwall RR, Combes B. Biliary excretion of sulfobromophthalein compounds in normal and mutant Corriedale sheep. Evidence for a disproportionate transport defect for conjugated sulfobromophthalein. Hepatology 1981; 1:441-447.
106. Guhlmann A, Krauss K, Oberdorfer F et al. Noninvasive assessment of hepatobiliary and renal elimination of cysteinyl leukotrienes by positron emission tomography. Hepatology 1995; 21:1568-1575.
107. Paulusma CC, van Meer MA, Evers R et al. Canalicular multispecific organic anion transporter/ multidrug resistance protein 2 mediates low-affinity transport of reduced glutathione. Biochem J 1999; 338:393-401.
108. Ito S, Ieiri I, Tanabe M et al. Polymorphisms of the ABC transporter genes, MDR1, MRP1 and MRP2/cMOAT, in healthy Japanese subjects. Pharmacogenetics 2001; 11:175-184.
109. Itoda M, Saito Y, Soyama A et al. Polymorphisms in the ABCC2 (CMOAT/MRP2) gene found in 72 established cell lines derived from Japanese indivduals: an association between single nucleotide polymorphisms in the 5'-untranslated region and exon 28. Drug Metab Dispos 2002; 30:363-364.
110. Saito S, Iida A, Sekine A et al. Identification of 779 genetic variations in eight genes encoding members of the ATP-binding cassette, subfamily C (ABCC/MRP/CFTR). J Hum Genet 2002; 47:147-171.
111. Trauner M, Arrese M, Soroka CJ et al. The rat canalicular conjugate export pump (Mrp2) is down-regulated in intrahepatic and obstructive cholestasis. Gastroenterology 1997; 113:255

112. Kubitz R, Wettstein M, Warskulat U et al. Regulation of the multidrug resistance protein 2 in the rat liver by lipopolysaccharide and dexamethasone. Gastroenterology 1999; 116:401-410.
113. Mottino AD, Cao J, Veggi LM et al. Altered localization and activity of canalicular Mrp2 in estradiol-17beta-D-glucuronide-induced cholestasis. Hepatology 2002; 35:1409-1419.
114. Paulusma CC, Kothe MJ, Bakker CT et al. Zonal down-regulation and redistribution of the multidrug resistance protein 2 during bile duct ligation in rat liver. Hepatology 2000; 31:684-693.
115. Rost D, Kartenbeck J, Keppler D. Changes in the localization of the rat canalicular conjugate export pump Mrp2 in phalloidin-induced cholestasis. Hepatology 1999; 29:814-821.
116. Dombrowski F, Kubitz R, Chittattu A et al. Electron-microscopic demonstration of multidrug resistance protein 2 (Mrp2) retrieval from the canalicular membrane in response to hyperosmolarity and lipopolysaccharide. Biochem J 2000; 348:183-188.
117. Schmitt M, Kubitz R, Wettstein M et al. Retrieval of the mrp2 gene encoded conjugate export pump from the canalicular membrane contributes to cholestasis induced by tert-butyl hydroperoxide and chloro-dinitrobenzene. Biol Chem 2000; 381:487-495.
118. Beuers U, Bilzer M, Chittattu A et al. Tauroursodeoxycholic acid inserts the apical conjugate export pump, Mrp2, into canalicular membranes and stimulates organic anion secretion by protein kinase C-dependent mechanisms in cholestatic rat liver. Hepatology 2001; 33:1206-1216.
119. Fickert P, Zollner G, Fuchsbichler A et al. Effects of ursodeoxycholic and cholic acid feeding on hepatocellular transporter expression in mouse liver. Gastroenterology 2001; 121:170-183.
120. Kubitz R, Huth C, Schmitt M et al. Protein kinase C-dependent distribution of the multidrug resistance protein 2 from the canalicular to the basolateral membrane in human HepG2 cells. Hepatology 2001; 34:340-350.
121. Roelofsen H, Soroka CJ, Keppler D et al. Cyclic AMP stimulates sorting of the canalicular organic anion transporter (Mrp2/cMoat) to the apical domain in hepatocyte couplets. J Cell Sci 1998; 111:1137-1145.
122. Zollner G, Fickert P, Zenz R et al. Hepatobiliary transporter expression in percutaneous liver biopsies of patients with cholestatic liver diseases. Hepatology 2001; 33:633-646.
123. Bohan A, Boyer JL. Mechanisms of hepatic transport of drugs: implications for cholestatic drug reactions. Semin Liver Dis 2002; 22:123-136.
124. Westphal JF, Vetter D, Brogard JM. Hepatic side-effects of antibiotics. J Antimicrob Chemother 1994; 33:387-401.
125. Gulbis B, Adler M, Ooms HA et al. Liver-function studies in heart-transplant recipients treated with cyclosporin A. Clin Chem 1988; 34:1772-1774.
126. Menday AP, Marsh BT. Intravenous fusidic acid ('Fucidin') in the management of severe staphylococcal infections: a review of 46 cases. Curr Med Res Opin 1976; 4:132-138.
127. Kutty KP, Nath IV, Kothandaraman KR et al. Fusidic acid-induced hyperbilirubinemia. Dig Dis Sci 1987; 32:933-938.
128. Bode KA, Donner MG, Leier I et al. Inhibition of transport across the hepatocyte canalicular membrane by the antibiotic fusidate. Biochem Pharmacol 2002; 64:151-158.
129. Böhme M, Büchler M, Müller M et al. Differential inhibition by cyclosporins of primary-active ATP-dependent transporters in the hepatocyte canalicular membrane. FEBS Letters 1993; 333:193-196.
130. Böhme M, Jedlitschky G, Leier I et al. ATP-dependent export pumps and their inhibition by cyclosporins. Advan Enzyme Regul 1994; 34:371-380.
131. Frimmer M. What we have learned from phalloidin. Toxicol Lett 1987; 35:169-182.
132. Kast HR, Goodwin B, Tarr PT et al. Regulation of multidrug resistance-associated protein 2 (ABCC2) by the nuclear receptors pregnane X receptor, farnesoid X-activated receptor, and constitutive androstane receptor. J Biol Chem 2002; 277:2908-2915.
133. Dussault I, Lin M, Hollister K et al. Peptide mimetic HIV protease inhibitors are ligands for the orphan receptor SXR. J Biol Chem 2001; 276:33309-33312.
134. Courtois A, Payen L, Le Ferrec E et al. Differential regulation of multidrug resistance-associated protein 2 (MRP2) and cytochromes P450 2B1/2 and 3A1/2 in phenobarbital-treated hepatocytes. Biochem Pharmacol 2002; 63:333-341.
135. Tanaka T, Uchiumi T, Hinoshita E et al. The human multidrug resistance protein 2 gene: functional characterization of the 5'-flanking region and expression in hepatic cells. Hepatology 1999; 30:1507-1512.
136. Stöckel B, König J, Nies AT et al. Characterization of the 5'-flanking region of the human multidrug resistance protein 2 (MRP2) gene and its regulation in comparison with the multidrug resistance protein 3 (MRP3). Eur J Biochem 2000; 267:1347-1358.

CHAPTER 16

Hepatic Copper Transport

Iqbal Hamza and Jonathan D. Gitlin

Summary

Copper is an essential nutrient that is required in a number of critical metabolic path ways. This metal is absorbed in the stomach and duodenum, stored in the liver and excreted in the bile. The liver functions to maintain copper balance as the amount of copper excreted in the bile is directly proportional to the size of the hepatic copper pool. Biliary copper excretion increases rapidly as the hepatic copper pool expands, providing a mechanism that normally prevents systemic copper overload. However, as the biliary system provides the only route for copper excretion, any interference with this process, as may occur in cholestasis, will result in hepatic copper accumulation. The molecular mechanisms of biliary copper excretion have begun to be elucidated with identification of the genetic defect in Wilson disease, an inherited disorder resulting in hepatic copper accumulation. The Wilson disease gene encodes a copper-transporting P-type ATPase (ATP7B) localized to the trans-Golgi network of hepatocytes and required for biliary copper excretion. In this chapter we discuss what is currently known about the molecular basis of hepatic copper transport and overview the molecular pathogenesis of hepatic copper accumulation in Wilson disease and the cholestatic syndromes.

Introduction

A diverse but limited number of cuproproteins including cytochrome oxidase, ceruloplasmin, tyrosinase, dopamine β-hydroxylase, superoxide dismuatse, peptidyl _-amidating monoxygenase and lysl oxidase utilize copper as a cofactor in essential electron transfer reactions.[1] The signs and symptoms of copper deficiency result from the loss of activity of these copper-dependent enzymes and include impairment in cellular respiration, iron oxidation, pigment formation, neurotransmitter biosynthesis, antioxidant defense, peptide amidation, and connective tissue formation. The chemical reactivity that makes copper a useful cofactor in these metabolic pathways can also result in considerable cellular injury in those circumstances where the metabolism of this metal is disturbed.[2] Therefore, unique cellular mechanisms have evolved that permit the intracellular trafficking and compartmentalization of copper, ensuring an adequate tissue supply of this metal and avoiding cellular toxicity.

Numerous food sources are rich in dietary copper and about half of the total daily intake of copper is absorbed, predominately in the stomach and duodenum. Biliary excretion serves as the only route for copper elimination and as there is no enterohepatic circulation of this metal, each day an amount of copper equivalent to that absorbed is excreted via the biliary tract (Fig. 1). Therefore, this pathway of absorption and excretion, which is the only physiologically relevant mechanism for maintaining copper homeostasis, is critically dependent upon normal liver function.[3]

Copper is transported in the blood complexed to amino acids such as histidine and thus renal filtration from the plasma can also provide a mechanism for copper excretion, but only under circumstances of marked copper overload where tubular reabsorption is overwhelmed.

Molecular Pathogenesis of Cholestasis, edited by Michael Trauner and Peter L.M. Jansen.
©2004 Eurekah.com and Kluwer Academic / Plenum Publishers.

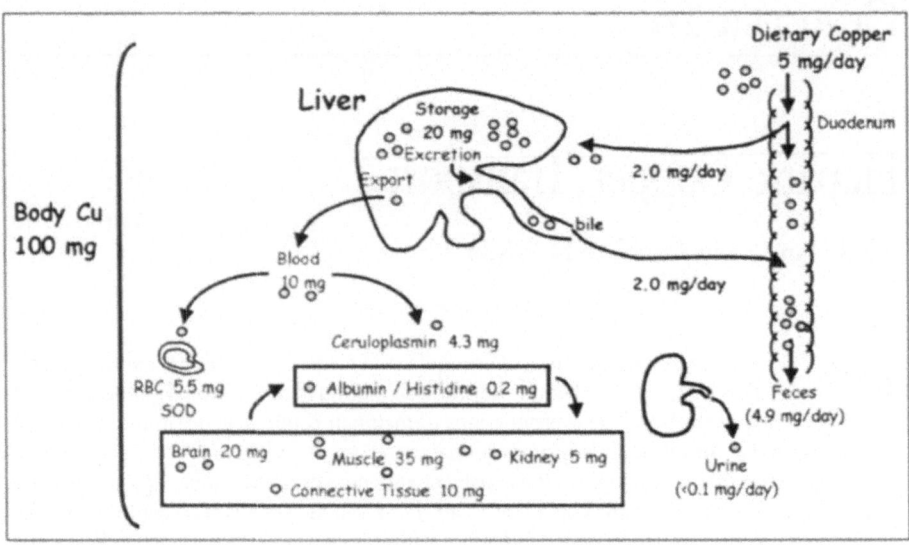

Figure 1. Physiology of human copper metabolism. The liver is the central organ of copper homeostasis. Copper balance is determined by biliary excretion which is the only physiologic mechanism of excretion. There is no enterohepatic circulation of copper. (Modified and reproduced with permission from Harris ZL, Gitlin JD. Am J Clin Nutr 1996; 63: 836S-841S).

Hepatic Copper Metabolism

Overview

As noted in Fig. 1, the liver plays a critical role in copper metabolism, serving as both the central site of storage for this metal as well as the primary determinant regulating biliary excretion. Tracer kinetic studies utilizing copper isotopes reveal rapid hepatic clearance of this metal from the portal circulation. Approximately 24 hours following a single dose of radioactive copper, 10% of the isotope will reappear in the serum bound to the plasma protein ceruloplasmin.[4] This protein is a ferroxidase containing 95% of the copper present in plasma. Despite the abundance of copper in this protein, ceruloplasmin has no essential role in copper transport or distribution.[4] The remaining copper present in the plasma is bound to amino acids and it is believed that these complexes provide the mechanism for transport of this metal to various tissues.

In the liver, hepatocytes are responsible for the uptake and storage of copper, as well as the regulation of excretion of this metal into the bile. The critical role of these cells in systemic copper homeostasis is illustrated by the normalization of copper metabolism in the severely copper-overloaded patient with Wilson disease following hepatic transplantation.[5] Any increase in copper intake results in a similar increase in the hepatocyte copper pool and metabolic studies reveal that under normal conditions the amount of copper excreted in the bile will be directly proportional to the size of this pool.[6] Given this capability to rapidly increase biliary copper excretion, hepatic copper excess is an unusual finding under normal physiological conditions. The form of copper appearing in the bile is unknown; however, several studies suggest that, once excreted into the bile, copper exists as an unabsorbable complex that is then eliminated in the stool. Although trace amounts of most plasma proteins are detected in bile, ceruloplasmin is not required for biliary copper excretion as patients with aceruloplasminemia reveal no evidence of impaired hepatic copper metabolism.[4] Similarly, the lack of detectable ceruloplasmin in bile samples from patients with Wilson disease (vide infra) is simply a consequence of the decreased serum concentration of this protein in affected patients and does not indicate a role for ceruloplasmin in the pathogenesis of this disease.

The human fetus acquires copper by placental transport. Biliary excretion is markedly decreased in the fetus and reaches adult capacity only after the first postnatal year.[7] Consistent with this developmental physiology, copper accumulates in the fetal liver such that the hepatic copper content at birth is relatively increased. This stored hepatic copper only becomes available to the secretory pathway of the hepatocyte once bile flow increases after birth. As a result, the fetal and newborn liver synthesizes and secretes ceruloplasmin devoid of copper. The slow increase in postnatal biliary copper excretion is paralleled by an increase in the serum ceruloplasmin concentration, reflecting copper movement into the secretory pathway of the hepatocyte with resulting maturation in the capacity for holoceruloplasmin biosynthesis.[4,8] The diminished serum ceruloplasmin content of newborn plasma abrogates the use of this protein as a marker for newborn screening in patients with Wilson disease or aceruloplasminemia.

Metallothionein

Metallothioneins are cysteine-rich, cytosolic proteins capable of binding several metal ions, including copper, under physiologic conditions.[9] Although the human genome contains at least sixteen distinct genes encoding metallothioneins all clustered on chromosome 16, there are currently four well-characterized, highly homologous metallothioneins that have been extensively studied in mice and man. Two of these proteins, MT I and MT II, are ubiquitously expressed in all cell types including hepatocytes. While genetic experiments in mice suggest no essential role for MT I or MT II in copper metabolism, such studies do reveal a critical role for these proteins when the homeostasis of this metal is perturbed.[10-12] These findings indicate that these proteins protect against the toxicity of copper, presumably by binding and sequestration, and suggest the possibility of an essential role for metallothioneins in situations of hepatic copper excess.

Ctr1

Genetic studies in *Saccharomyces cerevisiae* identified a protein, termed ctr1, required for high-affinity copper uptake.[13] Complementation experiments in ctr1Δ yeast resulted in characterization of a functional human homologue termed hCtr1.[14] Genetic ablation of ctr1 function in mice causes early fetal demise revealing an essential role for this protein in embryonic development and suggesting a critical role for ctr1 in mammalian copper homeostasis.[15,16] hCtr1 is a multimeric plasma membrane protein expressed in multiple cell types, which transports copper with high-affinity in a metal-specific and saturable manner dependent upon a series of critical methionine residues clustered in the extracellular domain.[17-19] These findings suggest that ctr1 functions as the high-affinity copper transporter at the basolateral surface of hepatocytes.

Ceruloplasmin

Ceruloplasmin is a multicopper oxidase that is required for the oxidation and movement of iron out of cells with mobilizable iron stores.[4] In hepatocytes this protein is synthesized and secreted into the plasma following the incorporation of six atoms of copper in the secretory pathway. Any situation that prevents copper incorporation into ceruloplasmin will result in the secretion of an apoprotein that lacks enzymatic activity and is rapidly catabolized. In copper deficient states, as the size of the hepatocyte copper pool falls, the movement of copper into the secretory pathway decreases and biliary copper excretion and holoceruloplasmin secretion are diminished. As noted above, ceruloplasmin contains greater than 95% of the copper found in human plasma and therefore the serum concentration of this protein is a useful and sensitive marker of hepatic copper metabolism. In patients with Wilson disease, the lack of functional ATP7b activity (vide infra) abrogates copper transport into the hepatocyte secretory pathway, resulting in secretion of apoceruloplasmin and resulting in the decrease in serum ceruloplasmin

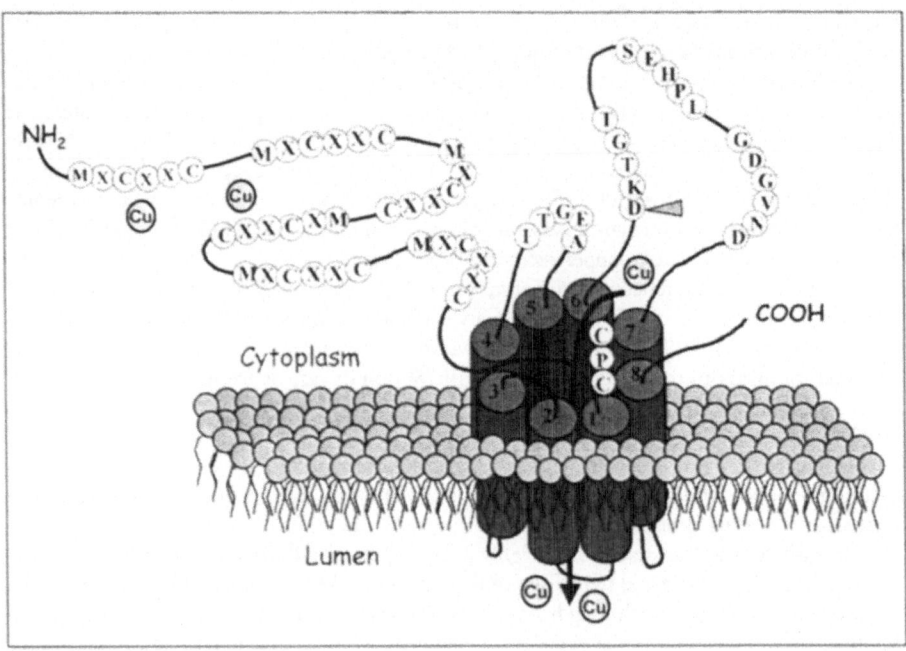

Figure 2. Topological model of ATP7b. Specific amino acids are noted in the conserved motifs discussed in the text. The proposed mechanism of energy-dependent ATP-driven cation transport across the membrane is illustrated. (Modified and reproduced with permission from Payne A, Gitlin JD J Biol Chem 1998; 273:3765-3770).

concentration diagnostic of this disease. As revealed by earlier metabolic studies and more recent work in patients and mice with aceruloplasminemia, ceruloplasmin has no essential role in copper transport or metabolism.[4]

ATP7b

ATP7b is a copper-transporting P-type ATPase expressed within the secretory pathway of hepatocytes.[20] This ATPase plays a critical role in copper homeostasis, as inherited loss-of-function mutations in the gene encoding human ATP7b result in the disorder of hepatic copper overload termed Wilson disease (vide infra). ATP7b is localized to the trans-Golgi network of hepatocytes and is required for the movement of copper into the secretory pathway for both incorporation into apoceruloplasmin and excretion into the bile.[20-23] With an increase in the hepatocyte cytosolic copper concentration, ATP7b localizes to a vesicular compartment near the canalicular membrane where copper is accumulated by this ATPase for subsequent excretion into the bile. As bile is the only route for copper excretion, this copper-dependent trafficking of ATP7b appears to provide a sensitive post-translational mechanism for maintaining copper homeostasis. The molecular mechanisms determining recycling of ATP7b have not been determined, but studies of an homologous copper-transporting ATPase, ATP7a, suggest that specific motifs within the carboxyl terminus may be required for this response.[24,25] The mechanisms involved in subsequent copper movement across the canalicular membrane are also unknown.

ATP7b is a polytopic membrane protein with features characteristic of known P-type ATPases, including a consensus motif with an invariant aspartate reside (DKTGT) (Fig. 2).[26] Phosphorylation of this aspartate residue results in a β-aspartyl phosphoryl intermediate that is required for ATP-dependent transfer of copper across the lipid bilayer. Recent experiments

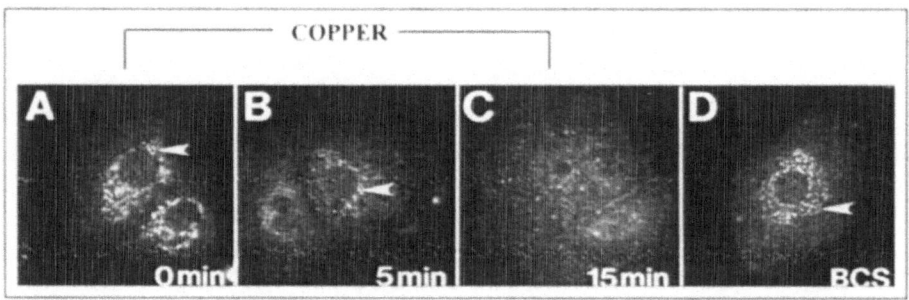

Figure 3. Intracellular localization of the Wilson disease copper-transporting ATPase (ATP7B). Primary rat hepatocytes were incubated for indicated times (A-C) in 50 μM copper or (D) 40 μM bathocuproine disulfonate (BCS) following copper incubation for 1 hr. ATP7b was detected by immunofluoresence and is shown in the trans-Golgi network with arrowhead. (Reproduced with permission from Schaefer M, Hopkins R, Failla M et al. Am J Physiol 1999; 276:G639-G646).

with ATP7b have demonstrated copper-dependent formation of this phosphorylated interme-diate suggesting that the catalytic cycle of copper transport begins with the binding of copper to high affinity binding sites in the transmembrane channel, followed by ATP binding and transient aspartate phosphorylation.[27] Site-directed mutagenesis studies suggest that the CPC sequence within the 6th transmembrane domain, highly conserved in all heavy metal transport P-type ATPases,[28] is the site of copper binding during the catalytic cycle of transmembrane transport of this metal by ATP7b.[20,29]

The amino terminus of ATP7b consists of six highly homologous domains, each of which contains the copper-binding motif MXCXXC (Fig. 2). These domains are critical for copper binding and transport and are the site of direct interaction with the copper chaperone atox1.[30-33] Structural analysis of an homologous domain in ATP7a reveals a linear bicoordinate copper-binding environment dependent upon the conserved cysteine residues in the MXCXXC motif.[34] Recent studies suggest that this region interacts with the largest cytoplasmic loop of ATP7b, perhaps regulating copper transfer from these amino terminal domains to the CPC in the transport channel.[35] The histidine residue in the sequence SEHPL located within this cytoplasmic loop is the site of the most common mutation (H1069Q) in Northern European populations with Wilson disease accounting for up to 40% of disease alleles. This mutation results in impaired trafficking of ATP7b indicating a role for this region in the intracellular localization.[36]

As noted above, upon increasing intracellular copper concentration, ATP7b traffics from the trans-Golgi network to a cytoplasmic location (Fig. 3). This vesicular compartment in mammalian cells into which copper is transported by ATP7b has not been well characterized. Copper transport into the homologous compartment in *Saccharomyces cerevisiae* is dependent upon the function of both the H^+ transporting V-type ATPase[37] and the CLC chloride channel Gef1.[38,39] These proteins presumably provide the acidic milieu and the charge balance required to maintain active vectorial copper transport. Recent studies also indicate that the provision of chloride ions by the CLC chloride channel in yeast is required for the allosteric assembly of copper into the ceruloplasmin homologue Fet3.[40]

Atox1

Under physiological circumstances intracellular copper availability is restricted by the pres-ence of intracellular chelators.[41] For this reason, copper delivery to specific pathways within the cell is mediated by a family of proteins termed copper chaperones.[42] These metallochaperones function to provide copper directly to target proteins while protecting this metal from intracel-lular scavenging (Fig. 4). The copper chaperone atox1 is required for copper delivery to ATP7b

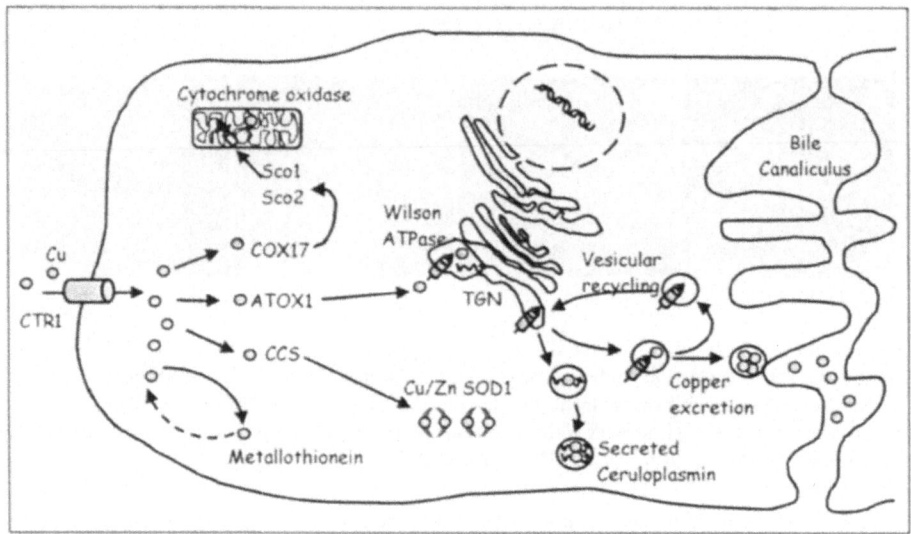

Figure 4. Model of the proposed pathways of intracellular copper trafficking within the human hepatocyte. The copper chaperones, recycling of ATP7b and pathway of copper excretion into bile are illustrated. In this model, copper movement to the canalicular membrane is predicted to be a final stage which may be altered in cholestasis or idiopathic copper toxicosis. (Modified and reproduced with permission from Harris ZL, Gitlin JD. Am J Clin Nutr 1996; 63: 836S-841S).

in the secretory pathway and genetic disruption of the atox1 locus in mice reveals that this protein plays a critical role in perinatal copper homeostasis.[43] Wilson disease-associated mutations in the amino terminus of ATP7b have been shown to result in a marked diminution in atox1 binding indicating that impaired copper delivery by this chaperone constitutes the molecular basis of Wilson disease in patients harboring these mutations.[31] Atox1 contains a single copy of the MXCXXC copper-binding motif present in the amino-terminus of ATP7b and in vitro and in vivo studies indicate that these cysteines are required for copper binding and transport to this ATPase.[30,31] The crystal structure of atox1 has been resolved and this structural data suggests a mechanism for copper transfer between atox1 and ATP7b dependent upon direct protein-protein interaction and copper binding at the MXCXXC motifs.[44,45]

Murr1

The presence in Bedlington terriers of an inherited disorder of copper homeostasis prompted studies to map and identify the involved locus. These animals have impaired copper excretion into bile but no abnormality in copper incorporation into ceruloplasmin suggesting that the defect occurs distal to the function of ATP7b in intracellular copper transport (vide infra). This disorder has recently been shown to result from deletion of a gene on dog chromosome 10q26 encoding a small cytosolic protein termed murr1.[46] The gene encoding murr1 is abundantly expressed in human liver suggesting that this protein plays a role in hepatic copper transport and biliary copper excretion in man.

Hepatic Copper Disorders

Wilson Disease

Wilson disease is an autosomal recessive disorder of copper metabolism resulting in hepatic cirrhosis and neurodegeneration. Recognition of this genetic disorder of hepatic copper homeostasis indicated that specific mechanisms are involved in copper trafficking within hepatocytes, a concept confirmed with the cloning of the Wilson disease gene, (ATP7b).[47] Loss of function of ATP7b in the hepatocyte results in a marked decrease in both holoceruloplasmin biosynthesis and biliary copper excretion with intracellular copper accumulation and eventual copper overload in most tissues (Fig. 4). Although ATP7b is expressed in extrahepatic tissues, the multi-organ copper overload observed in this disease is the result of impaired ATPase function in the hepatocyte, as this is reversed following hepatic transplantation.[5,47] More than two hundred distinct mutations have been identified in affected patients, approximately 50% of which are missense, most within well-defined consensus motifs or predicted transmembrane domains.[47,48] Biochemical analysis of these mutations has revealed specific molecular mechanisms accounting for Wilson disease including abnormalities in chaperone interaction, copper transport, subcellular localization and copper-induced trafficking of ATP7b.[20,29,31,49,50]

The model of cellular pathogenesis revealed by analysis of ATP7b mutations also provides a starting point for defining additional genetic and environmental factors affecting hepatic copper metabolism that may contribute to the clinical heterogeneity observed in individuals with Wilson disease. Such factors include proteins determining the rate of copper-delivery to the secretory pathway such as atox1, as well as potential homologues of the V-ATPase and gef1 chloride channel shown to be required for vesicular copper accumulation in yeast.[37-40] Metallothioneins are essential when copper homeostasis is perturbed[12] and allelic variability or loss-of-function of these proteins might also contribute to clinical outcome in any given patient. Although little is known about the specific mechanisms resulting in hepatocyte injury following copper accumulation, recent studies implicating specific apoptotic pathways suggest additional proteins that may influence disease outcome.[51] Although heterozygous loss of function of ATP7b is not associated with clinical abnormalities, the presence of such mutations might serve as risk factors promoting copper-mediated injury in more common liver disorders such as alcoholic cirrhosis.[52]

Animal models of Wilson disease have also provided insight into hepatic copper metabolism. Long Evans Cinnamon (LEC) rats have a marked impairment in biliary copper excretion with resulting hepatitis secondary to a deletion in the rat orthologue of ATP7b.[53,54] Hepatic copper accumulation in these animals results in hepatocellular carcinoma and abnormalities in hepatic iron metabolism.[55] As these findings are not observed in humans with Wilson disease, the data suggest significant species differences in the response of hepatocytes to copper accumulation. Toxic milk mice contain spontaneous missense mutations (M1356V or (G712D) in the murine orthologue of ATP7b.[56,57] Newborn mice suckled by the affected mother develop severe copper deficiency indicating a critical role for this ATPase in perinatal copper metabolism. Interestingly, although adult mice demonstrate significant hepatic copper overload these animals do not develop cirrhosis again suggesting species specific differences in factors which determine the outcome of copper-mediated hepatocyte injury. Consistent with this concept, deletion of a portion of the murine ATP7b gene by homologous recombination also results in mice with significant hepatic copper-overload by 8 weeks of age but no evidence of hepatic cirrhosis.[58]

Although Wilson disease is the most common disorder resulting in hepatic copper overload, any process that interferes with biliary excretion will eventually result in hepatic copper accumulation.[59] Accordingly, hepatic copper content is frequently elevated in cholestatic syndromes secondary to intrahepatic and extrahepatic bile duct injury.[60] In such cases, the serum ceruloplasmin will be normal or elevated indicating that the defect in copper excretion is distal to the function of ATP7b in this pathway (Fig. 4). Although accumulated copper may play a role in the eventual hepatic injury observed in these conditions, this does not appear to be a major factor in such injury as chelation with D-penicillamine is not effective in reversing this process.[60]

Idiopathic Childhood Cirrhosis

A severe form of rapidly progressive cirrhosis associated with a marked increase in hepatic copper has been described in children from rural, middle class Hindu families in India.[61]

Originally termed Indian childhood cirrhosis, similar clinical cases have now been reported worldwide and this disorder is now referred to as idiopathic childhood cirrhosis.[62] Affected children are diagnosed by two years of age with hepatosplenomegaly, elevation of serum aminotransferases, cirrhosis and elevated liver copper. Interestingly, the serum ceruloplasmin in these patients is normal or elevated, suggesting that the defect in biliary copper excretion is distal to the role of ATP7b in this process (Fig. 4). Epidemiological investigations of idiopathic childhood cirrhosis indicate that both genetic and environmental factors may play a role in this disease. These studies have revealed an increase in the copper content of the diet of affected children while analysis of some families suggests autosomal recessive inheritance with incomplete penetrance.[63] In support of an underlying defect in hepatic copper excretion, D-penicillamine is effective in many cases and hepatic transplantation can be curative.

A similar form of copper-associated cirrhosis is observed as an autosomal recessive disorder in inbred Bedlington terriers.[64] In these animals radioisotope studies reveal impaired biliary copper excretion but not holoceruloplasmin synthesis, once again suggesting a defect distal to the role of ATP7b in biliary copper excretion (Fig. 4). As noted above, genetic analysis has now localized the affected gene in these dogs and this should permit a more careful molecular analysis in affected patients. North Ronaldsay sheep also accumulate significant amounts of hepatic copper with concomitant liver injury suggesting that these animal models may allow for a detailed evaluation of the molecular mechanisms of canalicular bile excretion in children with idiopathic copper toxicosis.[65]

Acknowledgments

Work from the author's laboratory reported in this chapter was supported in part by National Institute of Health Grants DK44464, DK61763, and HD39952. Jonathan D. Gitlin is a recipient of a Burroughs-Welcome Scholar Award in Experimental Therapeutics.

References

1. Culotta VC, Gitlin JD. Disorders of copper transport. In: Scriver CR, Beaudet AL, Sly WS et al, ed. The Molecular and Metabolic Basis of Inherited Disease. New York: McGraw-Hill, 2001:3105-3136.
2. Harris ZL, Gitlin JD. Genetic and molecular basis for copper toxicity. Am J Clin Nutr 1996; 63:836S-841S.
3. Hamza I, Gitlin JD. Copper metabolism and the liver. In: Arias IM, Boyer JL, Chisari FV et al, ed. The Liver: Biology and Pathology. Philadelphia: Lippincott Williams & Wilkins, 2001:331-343.
4. Hellman N, Gitlin JD. Ceruloplasmin metabolism and function. Ann Rev Nutrit 2002; 22:439-458.
5. Loudianos G, Gitlin JD. Wilson disease. Sem Liv Dis 2000; 20:353-354.
6. Gollan JL, Gollan TJ. Wilson disease in 1998: genetic, diagnostic and therapeutic aspects. J Hepatol 1998; 28:28-36.
7. Arrese M, Ananthananarayanan M, Suchy FJ. Hepatobiliary transport: molecular mechanisms of development and cholestasis. 1998; Ped Res 44:141-147.
8. Gitlin D, Biasucci A. Development of gamma G, gamma A, Gamma M, beta 1C, beta 1A, Cl esterase inhibitor, ceruloplasmin, transferrin, hemopexin, haptoglobin, fibrinogen, plasminogen, alpha 1-antritrypsin, orosomucoid, beta-lipoprotein, alpha 2 macroglobulin and prealbumin in the human conceptus. J Clin Invest 1969; 48:1433-1446.
9. Palmiter RD. The elusive function of metallothioneins. Proc Natl Acad Sci 1998; 95:8428-8430.
10. Michalska AE, Choo AKH Targeting and germ-line transmission of a null mutation at the metallothionein I and II loci in mouse. 1993; Proc Natl Acad Sci 90:8088-8092.
11. Masters BA, Kelley EJ, Quaife CJ, et al. Targeted disruption of metallothionein I and II genes increases sensitivity to cadmium Proc Natl Acad Sci 1994; 91:584-588.
12. Kelley EJ, Palmiter RJ. A murine model of Menkes disease reveals a physiological function of metallothionein. 1996; Nat Genet 13:219-222.
13. Dancis A, Yuan DS, Haile D et al. Molecular characterization of a copper transport protein in S. cerevisiae: an unexpected role for copper in iron transport. Cell 1994; 76:393-402.
14. Zhou B, Gitschier J. hCTR1: A human gene for copper uptake identified by complementation in yeast. Proc Natl Acad Sci 1997, 94:7481-7486.
15. Kuo YM, Zhou B, Cosco D et al. The copper transporter CTR1 provides an essential function in mammalian embryonic development. Proc Natl Acad Sci 2001; 98:6836-6841.

16. Lee J, Prohaska JR, Thiele DJ. Essential role of mammalian copper transporter Ctr1 in copper homeostasis and embryonic development. Proc Natl Acad Sci 2001; 98:6842-6847.

17. Lee J, Pena MM, Nose Y et al. Biochemical Characterization of the human copper transporter Ctr1. J Biol Chem 2002; 277:4380-4387.

18. Puig S, Lee J, Lau M et al. Biochemical and genetic analyses of yeast and human high-affinity copper transporters suggest a conserved mechanism for copper uptake. J Biol Chem 2002; 277: in press.

19. Klomp AE, Tops BB, Van Denberg IE et al. Biochemical characterization and subcellular localization of human copper transporter 1 (hCTR1). Biochem J 2002; 364:497-505.

20. Hung IH, Suzuki M, Yamaguchi Y et al. Biochemical characterization of the Wilson disease protein and functional expression in the yeast Saccharomyces cerevisiae. J Biol Chem 1997; 272:21461-21466.

21. Schaefer M, Hopkins R, Failla M et al. Hepatocyte-specific localization and copper-dependent trafficking of the Wilson's disease protein in the liver. Am J Physiol 1999; 276:G639-G646.

22. Schaefer M, Roelofsen H, Wolters H et al. Localization of the Wilson's disease protein in human liver. Gastroenterology 1999; 117:1380-1385.

23. Roelofsen H, Wolters H, Van Luyn MJ et al. Copper-induced apical trafficking of ATP7B in polarized hepatoma cells provides a mechanism of biliary copper excretion. Gastroenterology 2000; 119:782-793.

24. Francis M, Jones E, Levy E et al. Identification of a di-leucine motif within the C terminus domain of the menkes disease protein that mediates endocytosis from the plasma membrane. J Cell Sci 1999; 112:1721-1732.

25. Petris MJ, Mercer JF. The Menkes protein (ATP7A; MNK) cycles via the plasma membrane both in basal and elevated extracellular copper using a C-terminal di-leucine endocytic signal. Hum Mol Genet 1999; 8:2107-2115.

26. Moller JV, Juul B, LeMaire M. Structural organization, ion transport, and energy transduction of P-type ATPases. Biochim Biophys Acta 1996; 1286:1-51.

27. Vanderwerf SM, Cooper MJ, Stetsenko et al. Copper specifically regulates intracellular phosphorylation of the Wilson's disease protein, a human copper-transporting ATPase. J Biol Chem 2001; 276:36289-36294.

28. Solioz M, Vulpe C. CPx-type ATPases: a class of P-type ATPases that pump heavy metals. Trends Biochem Sci 1996; 21:237-241.

29. Forbes JR, Cox DW. Functional characterization of missense mutations in ATP7B: Wilson disease mutation or normal variant? Am J Hum Genet 1998; 63:1663-1674.

30. Larin D, Mekios C, Das K et al. Characterization of the interaction between the Wilson and Menkes disease proteins and the cytoplasmic copper chaperone, HAH1p. J Biol Chem 1999; 274:28497-28504.

31. Hamza I, Schaefer M, Klomp LW et al. Interaction of the copper chaperone HAH1 with the Wilson disease protein is essential for copper homeostasis. Proc Natl Acad Sci 1999; 96:13363-13368.

32. Walker JM, Tsivkovskii R, Lutsenko S. Metallochaperone Atox1 transfers copper to the N-terminal domain of the Wilson's disease protein and regulates its catalytic activity. J Biol Chem 2002; 277: in press.

33. Forbes J, Hsi G, Cox D. Role of the copper-binding domain in the copper transporter function of ATP7B, the P-type ATPase defective in Wilson disease. J Biol Chem 1999; 274:12408-12413.

34. Gitschier J, Moffat B, Reilly D et al. Solution structure of the fourth metal-binding domain from the Menkes copper-transporting ATPase. Nat Struct Biol 1998; 5:47-54.

35. Tsivkovskii R, MacArthur BC, Lutsenko S. The Lys1010-Lys1325 fragment of the Wilson's disease protein binds nucleotides and interacts with the N-terminal domain of this protein in a copper-dependent manner. J Biol Chem 2001; 276:2234-2242.

36. Payne AS, Kelly EJ, Gitlin JD. Functional expression of the Wilson disease protein reveals mislocalization and impaired copper-dependent trafficking of the common H1069Q mutation. Proc Natl Acad Sci 1998; 95:10854-10859.

37. Eide D, Bridgham JT, Zhao Z et al. The vacuolar H+ ATPase of Saccharomyces cerevisiae is required for efficient copper detoxification, mitochondrial function and iron metabolism. Mol Gen Genet 1993; 241:447-456.

38. Gaxiola RA, Yuan DS, Klausner RD et al. The yeast CLC chloride channel functions in cation homeostasis. Proc Natl Acad Sci 1998; 95: 4046-4050.

39. Schwappach B, Strobrawa S, Hechenberger M et al. Golgi localization and functionally important domains in the NH2 and COOH terminus of the yeast CLC putative chloride channel Geflp. J Biol Chem 1998; 273:15110-15118.

40. Davis-Kaplan SR, Askwith CC, Bengtzen AC et al. Chloride is an allosteric effector of copper assembly for the yeast multicopper oxidase fet3p: An unexpected role for intracellular chloride channels. Proc Natl Acad Sci 1998; 95:13641-13645.
41. Rae T, Schmidt P, Pufahl R et al. Undetectable intracellular free copper: the requirement of a copper chaperone for superoxide dismutase. Science 199; 284:805-808.
42. Huffman DL, O'Halloran TV. Function, structure, and mechanism of intracellular copper trafficking proteins. Annu Rev Biochem 2001; 10:677-701.
43. Hamza I, Faisst A, Prohaska J et al. The metallochaperone Atox1 plays a critical role in perinatal copper homeostasis. Proc Natl Acad Sci 2001; 98:6848-6852.
44. Wernimont AK, Huffman DL, Lamb AL et al. Structural basis for copper transfer by the metallochaperone for Menkes/Wilson disease proteins. Nat Struct Biol 2000; 7:766-771.
45. Huffman DL, O'Halloran TV. Energetics of copper trafficking between the atx1 metallochaperone and the intracellular copper-transporter, Ccc2. J Biol Chem 2000; 275:18611-18614.
46. van De Sluis B, Rothuizen J, Pearson PL et al. Identification of a new copper metabolism gene by positional cloning in a purebred dog population. Hum Mol Genet 2002; 11:165-173.
47. Schilsky ML. Diagnosis and treatment of Wilson's disease. Pediatr Transplant 2002; 6:15-19.
48. Sternlieb I. Wilson's disease. Clin Liver Dis 2000; 4:229-239.
49. Forbes JR, Cox DW. Copper-dependent trafficking of Wilson disease mutant ATP7B proteins. Human Molec Genet 2000; 13:1927-1935.
50. Payne A, Gitlin JD. Functional expression of the Menkes disease protein reveals common biochemical mechanisms among the copper-transporting P-type ATPases. J Biol Chem 1998; 273:3765-3770.
51. Strand S, Hofmann WJ, Grambihler A et al. Hepatic failure and liver cell damage in acute Wilson's disease involve CD95 (APO-1/Fas) mediated apoptosis. Nat Med 1998; 4:588-593.
52. Pyeritz RE. Genetic heterogeneity in Wilson disease: lessons from rare alleles. Ann Intern Med 1997; 127:70-72.
53. Wu J, Forbes JR, Chen HS et al. The LEC rat has a deletion in the copper transporting ATPase gene homologous to the Wilson disease gene. Nature Gen 1994; 7:541-545.
54. Li Y, Togahsi Y, Sato S et al. Spontaneous hepatic copper accumulation in Long Evans Cinnamon rats with hereditary hepatitis. A model of Wilson's disease. J Clin Invest. 1991; 87:1858-1861.
55. Kato J, Kobune M, Kohgo Y et al. Hepatic iron deprivation prevents spontaneous development of fulminant hepatitis and liver cancer in Long Evans Cinnamon rats. J Clin Invest 1996; 98:923-929.
56. Theophanous MB, Cox DW, Mercer JF. The toxic milk mouse is a murine model of Wilson disease. Hum Mol Genet 1996; 5:1619-1624.
57. Coronado V, Nanji M, Cox DW. The Jackson toxic milk mouse as a model for copper loading. Mam Genome 2001; 12:793-795.
58. Buiakova OI, Xu J, Lutsenko S et al. Null mutation of the murine ATP7B (Wilson disease) gene results in intracellular copper accumulation and late-onset hepatic nodular transformation. Hum Mol Genet 1999; 8:1665-1671.
59. Ferenci P, Zollner G, Trauner M. Hepatic transport systems. J Gastroenterol Hepatol 2002;17:S105-S112.
60. Gossard AA, Lindor KD. Medical management of chronic cholestatic liver diseases. Can J Gastroenterol 2000; 14:93D-98D.
61. Pandit A, Bhave S. Present interpretation of the role of copper in Indian childhood cirrhosis. Am J Clin Nutr 1996; 63:830S-835S.
62. Muller T, Muller W, Feichtinger H. Idiopathic copper toxicosis. Am J Clin Nutrit 1998; 67:1082S-1086S.
63. Scheinberg IH, Sternlieb I. Wilson disease and idiopathic copper toxicosis. Am J Clin Nutrit 1996; 63:842S-845S.
64. Hultgren BD, Stevens JB, Hardy RM. Inherited, chronic progressive hepatic degeneration in Bedlington terriers with increased liver copper concentrations: clinical and pathologic observations and comparison with other copper-associated liver diseases. Am J Vet Res 1986; 47:365-377.
65. Haywood S, Muller T, Muller W et al. Copper-associated liver disease in North Ronaldsay sheep: a possible animal model for non-Wilsonian hepatic copper toxicosis on infancy and childhood. J Pathol 2001; 195:264-269.

CHAPTER 17

Molecular Basis of Primary Biliary Cirrhosis

Margaret F. Bassendine

Introduction

Primary biliary cirrhosis (PBC) can be defined by the triad of positive PBC-specific autoantibodies (antimitochondrial antibodies (AMA) in >95%), cholestatic liver function tests and diagnostic or compatible liver histology. PBC is generally considered to be an autoimmune disease. To be accepted as an autoimmune disease, tissue damaging lesions must be the consequence of either an antibody or T cell orchestrated response to one or more host components. Tissue damaging autoreactivity is a rarity requiring a complex series of events to occur. This complex induction requires multiple factors. In humans, as well as in animal models, there is a strong genetic component that always involves multiple genes. Non-genetic factors must also be involved and are assumed to include environmental, possibly dietary components and infectious agents. It is not clear if this autoimmune condition should be regarded as a single disease with a unique aetiology or a syndrome with a range of aetiological factors initiating similar pathogenic processes and culminating in the clinical picture that we recognise as PBC. How these factors lead to the development of primary biliary cirrhosis is not understood (Fig. 1) but important insights into the underlying molecular mechanisms have been made and parts of the intriguing puzzle are now clearer.

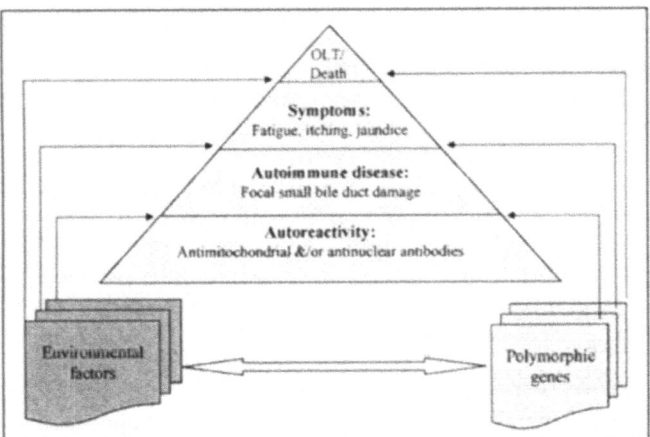

Figure 1. Putative natural history of primary biliary cirrhosis indicating that the development of disease specific antimitochondrial and/or antinuclear antibodies is an early event requiring complex interaction of environmental factors with multiple host genes. Progression of the disease may be regarded as a pyramid in which some patients with autoreactivity develop tissue damage, some of whom progress with time to symptomatic disease and a minority of these require an orthotopic liver transplant (OLT) or die of liver failure. The ongoing role of environmental factors and host genes in disease progression is not understood.

Molecular Pathogenesis of Cholestasis, edited by Michael Trauner and Peter L.M. Jansen.
©2004 Eurekah.com and Kluwer Academic / Plenum Publishers.

Tissue Damage

Primary biliary cirrhosis is characterised histologically by necroinflammation of bile ducts less than 100 μm in diameter, whilst sparing medium- and large-sized bile ducts. The terminal bile ductules (canals of Hering), which connect bile canaliculi to the interlobular bile ducts and represent the most proximal portion of the bile drainage pathway with a cholangiocyte lining are also destroyed in PBC in concert with destruction of the small bile ducts. This destruction appears to be an early event.[1] Ultrastructural changes in the lateral membranes of the hepatocyte have also been reported in PBC livers using transmission electron microscopy, with appearance of a large number of microvilli and localisation of the Golgi apparatus in front of these microvilli.[2] The inflammatory infiltrate around the bile ducts/ductules includes lymphocytes, plasma cells, eosinophils and mast cells.[3]

Early Bile Duct Lesion

In early disease (histological stage 1) the T cells (CD3+) occur predominantly in the portal areas with some found in the bile ducts. CD11+ cells (predominantly monocytes) have a similar distribution to T cells. T cells in the inflammatory infiltrate include both CD4+ and CD8+ cells but in early disease cytotoxic CD8+ T cells predominate within and around the bile duct epithelium.[4] There is also evidence for an enhanced type1 memory T cell response around damaged bile ducts of early PBC shown by expression of CXR chemokine receptor 3.[5]

Bile duct epithelial cells (BEC) aberrantly express HLA-DR, DP and DQ antigens in early disease and less frequently in the late cirrhotic stages of the disease (Stage 4)[4] HLA class II antigens are similarly expressed on terminal bile ductules in PBC but not normal controls.[1] Ectopic expression of MHC class II antigens may be induced in response to proinflammatory cytokines.[6] This has led to the proposal that such class II antigen expression allows BEC to present antigen to infiltrating CD4+ T cells, thereby amplifying and perpetuating immune responses within the liver. Functional activation of lymphocytes requires a second co-stimulatory signal via interaction between CD28 and B7-1 (CD80) or B7-2 (CD86) in addition to the interaction between peptide-bound MHC molecules and their cognate T cell receptor (TCR).[7] Comprehensive in vitro studies have failed to demonstrate CD80 or CD86 protein and mRNA in cultured BECs.[8,9] It is therefore unlikely that BECs stimulate efficient primary T cell activation. However immunohistochemical studies have shown expression of B7-1 and B7-2 on BECs in PBC[10] and this area remains controversial. It may be possible that these non co-stimulatory class II MHC complex positive BECs play a role in modulating immune responses (Fig. 2). It is noteworthy that MHC class II expression on bile duct epithelium has sometimes been observed without bile duct lesions[11-13] In addition it has recently been shown that MHC class II genes can be activated under situations of environmental "stress", using the probe pregnenolone 16α-carbonitrile, an agonist for the pregnane X receptor.[14] Thus exposure to "stressful" toxic insults could not only switch on a set of genes which stimulate components of the liver's xenobiotic metabolizing system but also genes which may be involved in immune surveillance or in down-regulation of autoimmune responses.

Activated CD83-positive dendritic cells have been identified within the infiltrate[15] in PBC, supporting a role for professional antigen-presenting cells in the pathogenesis of PBC.

The capacity for cytokine-stimulated human BEC to form high affinity adhesive bonds with T cells has been demonstrated using a sensitive flow cytometric assay.[16] Combination of this system with antibody blockade has allowed demonstration of the major contributions made by ICAM-1 and, to a lesser extent, LFA-3 to the adhesion of T cells to cultured BEC. These adhesions are essential for effective induction of BEC cytolysis by activated lymphocytes. More recent data shows up-regulation of LFA-1 (CD11a) expression on peripheral CD4+ T cells in PBC, supporting Th1 predominance in the autoimmune process.[17]

Epithelioid granulomas closely apposed to injured bile ducts are a characteristic, if not pathognomic, finding of PBC. Although they may arise as part of an immune mediated injury to the ducts, some contain foamy histiocytes, suggesting that rupture of the ducts with release

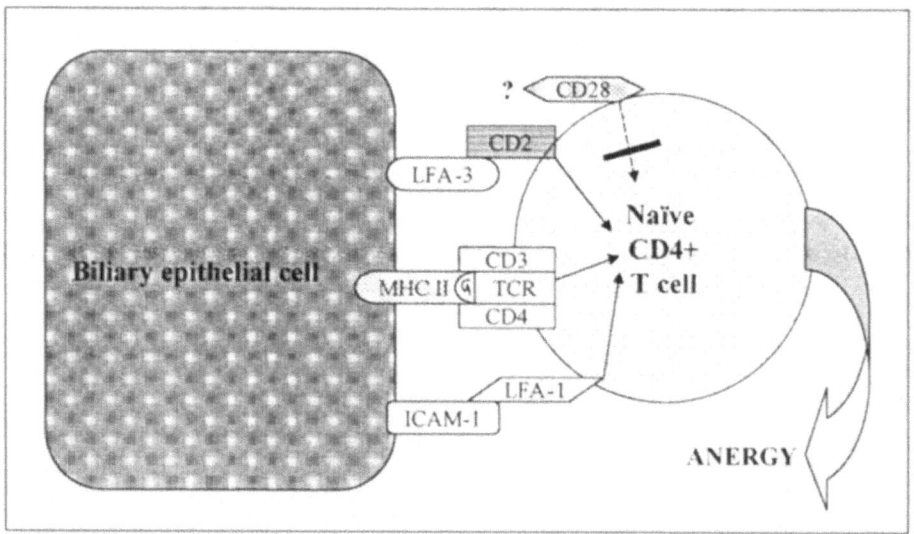

Figure 2. T cell activation is provided by cells that present antigenic peptides in the context of MHC class II antigens to the T cell receptor (TCR) in the presence of the required adhesion and co-stimulatory molecules. CD28 is particularly important for the activation of naïve T cells and lack of CD28 ligands on biliary epithelial cells means that the likely outcome will be anergy.

of bile acids and phospholipids may contribute to their development. CD1d, an MHC class I-like molecule that can present microbial non-peptide lipid antigen to T cells, has been shown to be expressed in epithelioid granuloma cells in PBC livers aswell as focally in small bile duct epithelial cells in early-stage disease.[18] Monocyte chemotactic protein-2 and -3 appear to be involved in recruitment of monocytes into the epithelioid granulomata and portal tracts,[19] implicating bacterial materials derived from bile in the overall pathogenesis of PBC. Other evidence supporting a role for bacteria in tissue damage is the finding of gram-positive bacterial lipotechoic acid in portal monocytes expressing the scavenger receptor class B type1 in PBC livers.[20]

Progression of Bile Duct Damage

The modality of cell death in PBC remains controversial. Some data suggests that lytic necrosis and not apoptosis accounts for most of the bile duct loss.[21,22] Apoptosis of BEC appears to be secondary to the invasion of inflammatory cells[23] and involve CD40-Fas interactions[24] and granzyme B.[25] Many autoantigens are selectively modified during apoptotsis, which has focused attention on apoptotic cells as a potential source of "neo-antigens" responsible for activating autoreactive lymphocytes. Apoptotic cholangiocytes have been shown to be a potential source of immunogenic mitochondrial autoantigen (pyruvate dehydrogenase complex-E2— see below), possibly due to depletion of glutathione.[26] Recent data suggests that following apoptosis the caspase family of proteolytic enzymes have the potential to generate immunogenic fragments of the mitochondrial autoantigen that could contribute to the autoantigen reservoir and the production of AMA.[27]

As the disease progresses, injured bile ducts disappear. A lymphoid aggregate may be present at the site of a previous duct and tissue damage may become "non-specific" and secondary to the initial bile duct injury.[28] A ductular reaction is seen and is characterised by anastomosing duct like structures at the interface between the portal tract and the parenchyma; the ductules express cytokeratins 7 and 19 and integrins normally expressed by bile ducts.[29] The stimulus for this reaction remains uncertain but it is likely that bile salt accumulation plays a role. There are features of cholate stasis with copper-associated protein accumulation, progressive necroinflammatory changes at the edge of the portal tracts and fibrosis.

Differential Gene Expression in PBC Liver

Analysis of differential gene expression using complementary DNA (cDNA) arrays is an important new approach to understanding the molecular basis of PBC. This has been applied to tissue from end-stage cirrhotic PBC obtained from liver explants[30] and to isolated biliary epithelial cells from the same source.[31] In the former the mixture of cell-types, with contributions from hepatocytes, biliary epithelial cells, hepatic stellate cells, endothelial cells, Kupffer cells, T cells and macrophages, generates an enormous breadth of data. The later approach may appear more specific but caution is required in interpretation of results as isolation methods may not only activate cells but remove important interactions that are present in vivo such as those with the extracellular matrix. Array analysis identified many differentially expressed genes that are important in inflammation such as the chemokine CXCR4 , Th2 associated cytokines Interleukin 4 (IL-4), IL-5, and IL-13, and Th1 associated molecules including lymphotoxin-beta and interferon-γ receptor beta subunit. Other genes involved in processes including fibrosis, apoptosis, cell proliferation, intracellular signalling and the stress response were identified and point to molecular pathways that may be involved in PBC pathogenesis. Some of these pathways may be "non-specific" and secondary to the initial bile duct injury. However this approach led to novel observations of differential expression of many genes of the Wnt and notch pathways in PBC; both pathways were initially identified in *Drosophila* development and differentiation and their physiological function in PBC is unknown. Future studies will require localisation of gene expression, particularly focusing on early disease to throw more light on disease induction.

Other Changes in Biliary Epithelial Cells

Reduced gene expression may also play a role in disease pathogenesis. Decreased hepatic levels of anion exchanger 2 mRNA and protein have been reported in PBC.[32] This transporter protein is normally expressed in canaliculi and the luminal membrane of terminal and interlobular bile ducts. Impaired biliary secretion of bicarbonate has been shown in patients with PBC[33] and cholangiocytes from PBC patients exhibit a widespread failure in the regulation of carriers involved in transepithelial $H(+)/HCO3(-)$ transport.[34] In contrast no change in expression of transporters for bile salts (NTCP, BSEP), organic anions (OATP2, MRP2, MRP3), organic cations (MDR1), phospholipids (MDR3) and aminophospholids (FIC1) has been found in early anicteric PBC.[35]

Glutathione-S-transferase-pi expression has also recently been shown to be markedly reduced in PBC, reflecting reduction of intracellular glutathione. As glutathione is a defensive substance against oxidative cell damage, this may suggest a role for lipid peroxidation in bile duct damage.[36] Ursodeoxycholic acid appears to enhance hepatic glutathione levels and this may provide an additional explanation of its effect in PBC.[37]

Another change that has been reported in all PBC livers is the expression of Met-enkephalin immunoreactivity in hepatocytes and bile ducts, suggesting that the liver may be a source of endogenous opioids in this disease.[38]

Expression of Autoantigen(s)

Targeting of the biliary epithelium in PBC may be explained by abnormal expression of mitochondrial autoantigens on the surface of biliary epithelial cells.[39],[40],[41,42] This is found early in the natural history of PBC and is localised predominantly around the biliary lumen.[43,44] Data suggests that the mitochondrial autoantigen found at the BEC apical surface spans the entire inner lipoyl domain of pyruvate dehydrogenase complex (PDC) E2 (see below).[45] Autoantigens relevant to PBC are also found on the luminal surface of ductal epithelial cells in

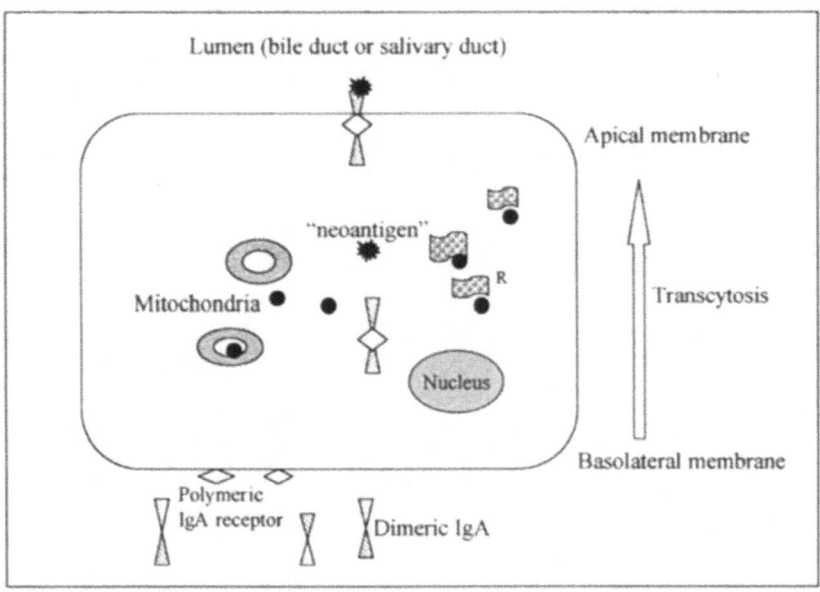

Figure 3. A model for the hypothetical role played by high titre anti-PDC IgA in epithelial damage in PBC. Dimeric IgA binds to the polymeric IgA receptor, undergoes transcytosis and may bind to components of mitochondrial autoantigens following translation in ribosomes (R) or "neoantigens" prior to their uptake by mitochondria, thus transporting them to the apical surface where they can be detected prior to being shed into the lumen of the bile or salivary ducts. This could lead to depletion of 2-oxo acid dehydrogenase complexes in the mitochondria and metabolic consequences for the epithelial cell.

the salivary gland,[46] independent of the presence of Sjoegren's syndrome,[47] suggesting that the same pathological process is occurring in both the salivary gland and the liver. Increased synthesis of relevant autoantigens in BEC seems unlikely in patients with PBC.[48] One physiological process that is common to both biliary and salivary epithelium in humans is the capacity to export IgA and IgM by the process of transcytosis. The normal passage of these immunoglobulins through secretory epithelial cells involves binding to the polymeric Ig receptor (pIgR) on the basolateral cell membrane, followed by internalisation (endocytosis) and vesicular transport to the apical surface. Here the pIgR is proteolytically cleaved between its external and intramembranous domains, thereby releasing IgA, still bound to the external domain of the pIgR (termed secretory component), into the secretions. This IgA provides the first line of defence against mucosal pathogens by helping to prevent them adhering to and penetrating the mucosal epithelium. Recently studies have shown the presence of IgA autoantibodies (antimitochondrial antibodies—AMA) in bile[49] and saliva in PBC.[50,51,52] In addition to the traditional concept of IgA action in the mucosal secretions, evidence has emerged in recent years to support a potentially important locus of IgA action inside mucosal epithelial cells. Studies in viral infection models have shown that IgA specific for viral proteins can, during transcytosis neutralise viral particles.[53] These results suggest that IgA antibody has the opportunity to interact with intracellular antigens and raise the possibility that IgA autoantibodies may interact with autoantigens (or cross reactive epitopes) within BECs. This could both explain the aberrant apical expression (Fig. 3) and lead to metabolic consequences with subsequent cell damage. A pathogenic role for autoantibodies in PBC is supported by studies showing that the level of AMA (antibodies to PDC) correlates with both disease progression[54,55] and several histological parameters, including fibrosis and inflammatory infiltrate.[56]

Autoreactivity

Autoimmune diseases do not occur under normal circumstances since lymphocytes with high affinity receptors for self-antigens are removed from the immune repertoire during central and peripheral tolerance induction. However, in reality, potentially autoreactive lymphocytes are present in normal individuals, but normally these cells are either not activated or are tolerised. Furthermore, it has become apparent that lymphocyte receptors may react with multiple epitopes, many of which may not be closely related in primary amino acid sequence with each other.[57,58] Indeed by some estimates a single T cell receptor may react with up to 10^6 different ligands. Such degeneracy means that a T cell clone expanded following antigen triggering could react with numerous other epitopes including self peptides, presented by self MHC.

The most widely accepted hypothesis for development of PBC is that some environmental trigger(s) acts in a genetically predisposed host to lead to the development of autoantibodies, usually AMA. The development of these autoantibodies is an early event in the disease pathway; AMA can be detected in serum before abnormalities in liver function and long before the onset of symptoms of PBC.[59-61] The close association of significant titres of AMA with focal non-suppurative destructive cholangitis is not only useful in diagnosis of PBC but may also be an important clue to disease aetiology.

Antimitochondrial Antibodies (AMA)

The mitochondrial autoantigens reactive with AMA are all components of the 2-oxo acid dehydrogenase multienzyme complexes; pyruvate, 2-oxoglutarate and branched-chain 2-oxo acid dehydrogenase complexes (PDC, OGDC and BCOADC, respectively). These three complexes are among the best studied examples of multifunctional proteins catalysing a set of sequential chemical reactions. Each complex occupies a key position in energy metabolism in a cell. PDC links glycolysis to the Krebs cycle, OGDC is in the Krebs cycle itself and BCOADC is involved in the regulation of the oxidation of the branched-chain amino acids. Each complex consists of multiple copies of at least three enzymes (E1, E2, and E3), which are encoded by genes in the nucleus and separately imported into mitochondria for assembly into high molecular weight multimers on the inner membrane. E3 is common to all three complexes whereas E1 and E2 are unique to each complex. In addition PDC contains a fourth polypeptide, protein X, which plays a structural role as an E3-binding protein and so now has the functional designation E3BP.[62] Mammalian and *Saccharomyces cerevisiae* PDC has a polypeptide chain ratio of 60E1α:60E1β:60E2:12E3BP:24E3[63] and is slightly larger than a ribosome. The activity of each mammalian complex located within the mitochondrion is under stringent control by hormones and dietary factors.[64,65]

The pivotal development in the study of AMA was the cloning of the major 70kD mitochondrial antigen,[66] which is recognised by antibodies in the sera of more than 90% of patients with PBC. This lead to its identification as the E2 component of PDC,[67,68] and the demonstration that all sera that reacted with PDC-E2 also reacted with protein X (PDC-E3BP).[67] It was rapidly shown that AMA often also react with the E2 components of the other two 2-oxo acid dehydrogenase complexes, OGDC[69] and BCOADC.[69,70] The frequency of AMA reacting with OGDC-E2 and BCOADC-E2 was lower than with PDC-E2 and E3BP[71] and AMA reacting with PDC-E2 did not cross-react with the E2 components of OGDC and BCOADC.[69,72] Subsequently PDC-E1α and β components have also been found to react with AMA in the sera of a minority of PBC patients.[73,74] All these studies utilised purified protein derived from animal tissue or recombinant human fusion proteins expressed in *E.coli.*, but subsequently purified native human PDC has been shown to have an identical pattern of immunoreactivity with AMA.[75] More recently autoantibodies reactive with BCOADC-E1α have been identified in approximately 50% of sera from PBC patients using highly purified human BCOADC as the antigen source[76] (Table 1). These autoantibodies do not cross react with BCOADC-E2 or PDC-E1α and seem to occur subsequent to antibodies to BCOADC-E2, supporting the concept of intermolecular determinant spreading.

Table 1. Mitochondrial and nuclear antigens reacting with PBC-specific autoantibodies

Antigen	% of PBC Sera Containing Reactive Autoantibodies
Mitochondrial	
E2 component of pyruvate dehydrogenase complex (PDC) (lipoyl domain)	90-95
E3 binding protein of PDC (lipoyl domain)	90-95
E2 component of 2-oxoglutarate dehydrogenase complex (lipoyl domain)	40-65
E2 component of branched chain oxo-acid dehydrogenase complex (lipoyl domain)	50-55
E1-alpha component of PDC	40-65
E1-beta component of PDC	<10
E1-alpha component of BCOADC	50
Nuclear	
Nucleoporin p62	32
Nuclear pore glycoprotein, gp210	10-25
Lamin B receptor	<5
2 nuclear dot-associated proteins: Sp100 & promyelocytic leukaemia antigen	10-30

The frequency and pattern of autoantibodies to PDC-E2 + E3BP, OGDC-E2 and/or BCOADC-E2 does not correlate with markers of disease progression.[71] Despite earlier reports to the contrary, data also suggests that other AMA profiles do not predict prognosis in PBC.[77]

Epitope Mapping

The E2 chains of the three 2-oxo acid dehydrogenase complexes are highly segmented structures that have been conserved in evolution: they comprise, from the N terminus, one to three lipoyl domains, a peripheral subunit-binding domain and a large core-forming acyltransferase catalytic domain, all linked together by long (25-30 residues) segments of flexible polypeptide chain rich in alanine and proline[78] (Fig. 4). Similarly E3BP has a multidomain substructure consisting of an amino-terminal lipoyl domain, followed by an E3-binding domain and then a carboxyl-terminal E2-binding domain.[62] Each lipoyl domain consists of approximately 80 amino acids with the lipoyl-lysine being found within a DKA motif. Attachment of the lipoate to the susceptible lysine is an enzyme catalysed process which, in mammalian cells occurs within mitochondria following import of the E2 precursor. All experimental data indicates that the immunodominant B-cell epitopes in PBC are located within these lipoyl domains,[72,79-88] and that the lysine residue (K) to which lipoic acid is attached is critical. The three-dimensional structure of lipoyl domains from both human PDC-E2 and BCOADC E2 has been determined by means of nuclear magnetic resonance spectroscopy.[89,90] They are similar to the lipoyl domains of bacterial PDC-E2s[91] comprising a flattened beta-barrel formed by two antiparallel β sheets, each of which contains three major strands and one minor strand. The lipoylation site (lysine residue) is physically exposed at the tip of a tight turn in one of the β-sheets, and the residues surrounding the lipoyl-lysine are highly flexible. It would appear that AMA in PBC patients' sera recognise this physically exposed lipoylation site and other distinctive surface conformational features in the lipoyl moiety. This conclusion is strengthened by the observation that monoclonal antibodies derived from fusions of circulating B lymphocytes from patients with PBC show significantly stronger binding to lipoyl-containing domains than their unlipoylated couterparts.[92]

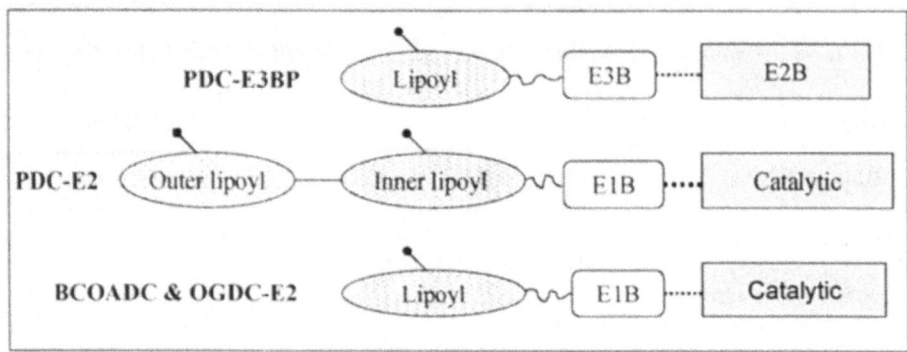

Figure 4. Diagrammatic representation of the structural and functional domains of the E2 polypeptides of the 2-oxo acid dehydrogenase complexes (PDC, BCOADC & OGDC), compared to the E3 binding protein of PDC [B = binding domain, ⌐ = lipoic acid co-factor].

Enzyme Inhibitory Antimitochondrial Antibodies

A striking property of AMA in PBC sera is their capacity to rapidly inactivate the catalytic function of PDC in vitro.[93] This assay has been utilised to demonstrate a population of autoantibodies in PBC sera that inhibit PDC function but do not show reactivity in an immunoblot with recombinant human PDC-E2.[94] These non-blotting inhibitory AMA are presumed to react with exclusively conformational determinant(s) perhaps presented by the tertiary structure of the entire multienzyme complex. Interestingly AMA in PBC sera have been shown to be highly inhibitory of the mammalian PDC (99%), but only moderately inhibitory for yeast PDC (70%), and weakly inhibitory for E.coli PDC (26%).[95]

Reactivity of AMA with Bacterial Antigens

The first observation that PBC sera cross-react with bacterial antigens was made in 1976 when Sayers and Baum demonstrated cross-reactivity with membrane vesicles of *Paracoccus denitrificans*.[96] This observation was confirmed and cross-reactivity was demonstrated with a number of other bacteria,[97] including with rough mutants of *Enterobacteriaceae*,[98] leading to the hypothesis that PBC may have a bacterial aetiology. The explanation for these cross-reactions with bacterial antigens came, of course, with the identification of PDC-E2 and PDC-E3BP as the dominant M2 antigens. E2 polypeptides are highly conserved in evolution and indeed, a gene encoding PDC-E2 has been found in the smallest known genome of any free-living organism, *Mycoplasma genitalium*.[99] In addition, comparison of the deduced amino acid sequences of E2 and E3BP from *S.cervisiae* indicates that there is strong homology between the polypeptides in the region that corresponds to the lipoyl domain and that they evolved from a common ancestor.[100] PBC sera do indeed react with E.coli PDC-E2[81] and OGDC-E2[79] but distinct antibodies against these prokaryotic antigens are present in PBC sera and the antibody titre against mammalian PDC-E2 is approximately 100-fold higher than against bacterial PDC-E2[72] Furthermore PBC sera containing AMA to mammalian PDC-E1 subunits do not recognise E.coli PDC-E1.[73] These results, taken together with the subsequent observation of marked differences between the inhibitory effects of PBC sera on mammalian PDC as opposed to yeast and bacterial PDC[101] do not appear to support the concept of molecular mimicry/cross-reactivity as an explanation for the origin of AMA. These earlier observations are supported by a recent study analysing changes in specificity of human anti-PDC antibodies during affinity maturation; the demonstration of intermolecular epitope shift suggests that bacterial molecular mimicry is not involved in initiating PBC.[102]

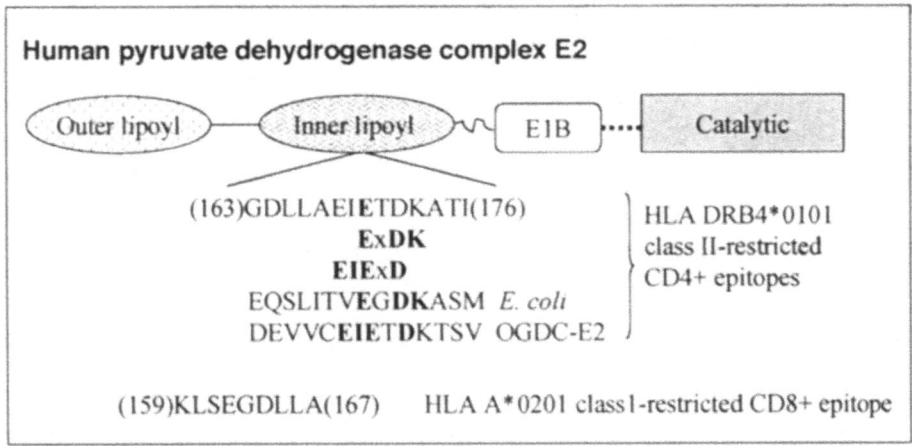

Figure 5. Diagrammatic representation of the structural and functional domains of the E2 component of human pyruvate dehydrogenase complex. The amino acid sequences of HLA class I (amino acid 159-167) and class II (amino acid 163-176) restricted autoepitopes within the inner lipoyl domain are shown, and compared with cross reactive HLA class II restricted epitopes found in the E2 component of OGDC and a microbial peptide from *E. coli*. The most critical T cell receptor contact residues are indicated in bold.

Autoreactive T Cells

Autoreactive T cells play a central role in the development of various autoimmune diseases.[103] Autoreactive peripheral blood T cells responses to biochemically purified PDC-E2 + E3BP have been shown to largely restricted to PBC patients.[104,105] The peripheral blood T cell response appears to be heterogeneous; one study reported response to the inner and/or the outer lipoyl domains[106] whilst another showed that some T cell autoepitopes within PDC-E2 + E3BP lie outside the inner lipoyl domain in PBC patients.[107] Self-reactivity is universal in peripheral blood T cells from PBC patients when co-cultured with PDC pulsed autologous dendritic cells and this method may help to further localize T cell autoepitopes within PDC.[108]

Early studies showed that CD4+ T cell lines from liver biopsies of patients with PBC produce Interleukin-2 specifically in response to PDC-E2 or BCOADC-E2.[109] Subsequently work has focused on identifying T cell autoepitopes. One study identified an HLA class II (DRB4*0101) restricted epitope within the inner lipoyl domain of PDC-E2 (peptide 163-176).[110] This study also demonstrated that these CD4+ T cell clones cross-reacted with the corresponding peptide from *E.coli*, raising the possibility of molecular mimicry being involved in the induction of PBC at a T cell level. Indeed T cell clones reactive to peptide 163-176 (the lipoic acid binding site in the inner lipoyl domain) were subsequently shown to be activated by mimicry peptides derived from several microbial proteins.[111] Further studies were undertaken aimed at addressing molecular mimicry and cross-recognition among mitochondrial autoantigens using cloned T cell lines. Fine specificities were unique to every single T cell clone, but the clones could be categorised into two distinct groups based on recognition motifs of the T cell receptor (TCR).[112] Data suggested [170]E is the most critical TCR contact residue for both groups, with one group recognising [170]ExDK[173] and the other [168]EIExD[172] (Fig. 5). Some T cell clones cross-reacted with E3BP residues 34-47, OGDC-E2 100-113, as well as several peptides derived from various microbrial proteins.

HLA class I (A*0201) restricted epitopes recognised by CD8+ T cells in PBC have also been identified recently. A tenfold increase in the frequency of PDC-E2 [159-167] -specific CD8+ T cells has been found in the liver, compared with the blood in PBC, with the precursor frequency of the cytotoxic T lymphocytes (CTLs) in the blood being significantly higher in early-stage PBC.[113] Again data suggests that molecular mimicry may be implicated in the initiation of these autoreactive T cells, with a peptide from *Psuedomonas aeruginosa* showing not only a higher affinity binding to the HLA A*0201 than the prototype peptide from human PDC-E2 but also recognition by the CTLs specific for the prototype peptide.[114] Furthermore, alanine substitution at position 5 of the epitope significantly reduced peptide-specific effector functions of CTLs, providing preliminary evidence that peptides may have potential in immunotherapy. The same group screened 79 overlapping 15mer peptides, spanning the entire PDC-E2 molecule and identified a 10mer peptide, PDC-E2 amino acids 165-174, as another CD8 epitope restricted by HLA-A*0201.[115] Analysis of the variable T cell receptor alpha-chain and beta-chain of T cells specific to PDC suggests that few relevant epitopes are recognised,[116] supporting the possibility that altered peptide ligands may ultimately be used to modulate disease activity.

The peptide 159-167 of PDC-E2 has also been shown to induce specific MHC class I-restricted CD8+ CTL lines from 10/12 HLA-A2(+) PBC patients, but not controls after in vitro stimulation with antigen-pulsed dendritic cells (DCs). Furthermore, using soluble PDC-E2 complexed with either affinity purified antibodies or human monoclonal antibody against PDC-E2, the generation of PDC-E2 specific CTLs occurred at 10-fold and 100-fold less concentration, respectively, compared with soluble antigen alone. Thus autoantigen-autoantibody immune-complexes can be taken up by professional antigen-presenting cells and present autoantigen at a higher relative efficiency implicating a role for autoantibodies in pathogenesis of PBC.[117] This observation is particularly interesting in the light of earlier reports that high concentrations of large immune complexes are present in the circulation of patients with PBC;[118,119] this finding led to the hypothesis that immune complexes could be formed in the vicinity of the bile ducts by antigen absorbed from the bile or biliary epithelium.[118] It is also in keeping with the concept that prolonged antigen/immune complex delivery via DCs is a crucial factor for the conversion of transient autoimmunity to manifest autoimmune disease.[120]

The liver is a unique lymphoid organ with an enhanced proportion of natural killer T (NKT) cells compared to the peripheral blood.[121] This subset of lymphocytes is thought to play a role in the modulation of the innate immune response. A recent study has quantitated NKT cells in PBC using a human CD1d tetramer produced by a baculovirus expressing CD1d protein complexed with alpha-galactosylceramide (alpha-GalCer). The frequency of CD1d-alphaGalCer-restricted NKT cells in the liver was higher in PBC patients than controls[122] and, as noted above, CD1 is expressed in epithelioid granuloma cells in PBC livers as well as focally in small bile duct epithelial cells in early-stage disease.[18]

AMA Negative PBC

It has been suggested that the "syndrome" of PBC is made up of a spectrum of disorders in which the common finding is intrahepatic bile duct destruction in the presence of any of several serum autoantibodies[123] This spectrum would include not only classic AMA-positive PBC but also AMA-negative PBC in which disease-specific autoantibodies react with nuclear autoantigens. AMA-negative PBC has also been called "autoimmune cholangitis".[124] Two distinct antinuclear antibodies (ANA), one giving a membrane-like pattern of positivity (M-ANA)[125] and the other reacting with multiple nuclear dots (MND-ANA)[126] by immunofluorescence have been described in AMA-negative PBC.

Membrane ANA

Several investigators showed that most of the autoantibodies from patients with PBC that label the nuclear envelope recognise a protein with a molecular mass of about 200 kD.[125,127] This protein has been identified as the nuclear pore membrane glycoprotein gp210,[128,129] and autoantibodies to gp210 are a sensitive and specific serological marker for PBC.[129,130] Subsequent work focused on identification of the epitope(s) of gp210 recognised by M-ANA; one study using a recombinant gp210 fusion protein has shown that autoantibodies in PBC react with a stretch of 15 amino acids in the cytoplasmic carboxy-terminal domain.[131] Another study, using biochemically purified gp210 has demonstrated that autoantibodies against gp210 recognise at least two different epitopes, some reacting as above with the short cytoplasmic tail whilst other sera react with a novel epitope within the large glycosylated luminal amino-terminal domain.[132] Two studies have used a recombinant protein or a synthetic polypeptide containing one immunodominant epitope of gp210 in ELISA's for the diagnosis of PBC;[133],[134] both ELISA's were highly specific for the diagnosis of PBC (96% and 99% respectively) and one study found these gp210 autoantibodies in 47% of AMA-negative PBC patients.

Another glycoprotein of the nuclear pore complex, p62, has recently also been shown to be an autoantigen in about a third of PBC patients[135](Table 1). Anti-p62 antibodies are highly specific for PBC but do not co-localise with anti-gp210 autoantibodies, so patients positive for one or other M-ANA appear to represent immunologically distinct subsets. At present it is not clear how these M-ANA relate to the pathological process of intrahepatic bile duct destruction, but autoantibodies to nuclear pore complexes have been shown to be associated with more active and severe PBC.[136]

A very small subset of patients with PBC have M-ANA against an integral protein of the inner nuclear membrane, the lamin B receptor[137](Table 1). Autoantibodies from four of these rare PBC patients have been shown to recognise a conformational epitope within amino acids 1-60 of the nucleoplasmic amino-terminal domain[138] So far autoantibodies against the lamin B receptor also appear to be PBC-specific and are more often present in the sera of patients who are AMA negative.

Multiple Nuclear Dot ANA

PBC sera which give the immunofluorescence pattern of multiple nuclear dots have been reported to bind to proteins of 95kD[139] and 78-92 and 96-100kD[140] in immunoblots. Two nuclear proteins have now been shown to react with these MND-ANA; the first to be cloned was Sp100.[141] This cDNA codes for a protein of unknown function with sequence similarities to several transcriptional transactivating proteins, including HIV-1 nef proteins. Screening for anti-Sp100 autoantibodies in sera from patients with various diagnoses has been performed by ELISA using recombinant fusion proteins; anti-Sp100 antibodies were found in 34% of PBC patients but not in patients with other liver diseases.[142] Anti-Sp100 positivity was shown to be a highly specific serological marker of PBC and to be associated with higher gamma-globulin levels; MND-ANA were found in rheumatology patients but data suggested the antigenic target is other than Sp100. Epitope mapping has shown that one domain of Sp100, which contains the sequence similarity with HIV nef proteins is recognised by all anti-Sp100 sera.[143]

A second protein that co-localizes to dot-like nuclear domains, and is aberrantly expressed in promyelocytic leukaemia cells (PML) has also been shown to react with MND-ANA in PBC sera.[144] Similar to Sp100, PML sequence similarities with several transcription factors have been identified, suggesting a transcription-regulatory function.[145] Interferons have been shown to strongly increase the levels of both Sp100 and PML mRNA and protein[146,147] and thus Sp100 and PML belong to the growing family of interferon-stimulated genes.. Autoantibodies against PML are as highly prevalent and specific for patients with PBC as those against Sp100.[144]

Like AMA, both autoantibodies to gp210 and Sp100 have been shown to persist in the sera of PBC patients after liver transplantation, but this does not reflect recurrent disease activity in the graft.[148] These ANA may indicate differing environmental triggers to AMA positive disease although recent data comparing antibody expression in PBC and autoimmune cholangitis using phage display suggests that there is similar autoimmune targeting in these disease variants.[149]

Genetic Factors

The prevalence of PBC is increased in the close relatives of affected patients suggesting that genetic factors play a significant role in determining PBC susceptibility.[150,151] A single report has suggested discordant disease in monozygotic twins,[152] confirming the view that increased individual susceptibility and environmental factors are both necessary for the development of PBC.[153]

The methods available for the genetic dissection of complex traits in humans have been frequently reviewed.[154-156] These fall into three main categories of concept and methodology: linkage analysis, allele sharing methods and association studies. All the published data on genetic predisposition to primary biliary cirrhosis has been obtained from association studies, largely because of the paucity of genetically informative families. The age of presentation in late middle age, together with the relative rarity of PBC in the population means that there are very few suitable simplex (index case and both parents), let alone multiplex families available for genetic studies.

Major Histocompatibility Complex (MHC) Associations in PBC

The human major histocompatibility complex, occupying a 4000Kb segment on the short arm of chromosome 6, is highly polymorphic. It comprises three distinct groups of genes, termed human leucocyte antigens (HLA) class I, class II and class III.[157] This group of genes encode various products involved in cell-cell interactions with antigen specific cell surface receptors of T cells, related to the functioning of the immune response to foreign antigen presentation.

Early studies identified various associations with serologically defined HLA DR specificity's including; DR2,[158] DR3,[159] DR4[160] and later with DR8.[161] However, only the association with DR8 has been widely confirmed.[162-167] Molecular genotyping studies suggest that the primary susceptibility allele is *DRB1*0801* in Caucasian patients and *DRB1*0803* in Japanese patients and, though there is linkage with specific *DQA1*, *DQB1* and to a lesser extent *DPB1* alleles these associations are most likely due to linkage disequilibrium with the *DRB1*08* alleles (Table 2). It is of interest that a recent study has shown that the frequency of *DRB1*08* was significantly lower in AMA negative PBC, compared to AMA positive patients.[168] This supports the concept that autoimmune cholangitis/AMA negative PBC results from a similar triggering event in a host with different genetic susceptibility, rather than shared genetic risk with varying environmental trigger.[169]

The antigen processing genes, *TAP1* and *TAP2*, situated between the DR and DP regions, are of interest as candidate susceptibility loci because of their role in endogenous antigen presentation but no disease association has been found in a preliminary study of over 100 PBC patients.[170]

There are some published data for the class III region; early studies of the complement genes identified strong associations with *C4B*2*[171] and *C4A*Q0*[172] though neither association has been confirmed. A more recent study has suggested that HLA-DR8 and C4B2 are in linkage disequilibrium and that C4B2 is not an independent susceptibility locus; indeed a genetic marker located midway between HLA-DR8 and C4B2 termed G91 appears to be the telomeric limit for the gene associated with HLA-DR8 that encodes susceptibility to PBC.[173] Studies of the *TNFA* promoter A/G single nucleotide polymorphisms (SNPs) at *-238* and *-308* produced conflicting data and remain controversial.[174-176]

Other Candidate PBC Susceptibility Loci

The interleukin 1 (*IL1*) family represent good functional candidates and recent micro-array analysis of PBC livers reported a four-fold increase in *IL1A* mRNA transcripts.[30] The *IL1* gene family on chromosome 2q includes genes for the two agonist forms of IL-1 (*IL1A* and *IL1B*) and the receptor antagonist (*IL1RN*) , and also two of the IL-1 receptors; *IL1R1* and *IL1R*. Two polymorphisms have been investigated in PBC, the *IL1B* SNP at +3953 and the 86bp VNTR in *IL1RN* and a strong association has been reported with both disease susceptibility and progression.[177]

Table 2. Some association studies in PBC performed using genotyping techniques

Candidate Gene	References
Human leucocyte antigen class II - HLA DRB1 0801 (Caucasian) & 0803 (Japanese)	161-169
Antigen processing genes (TAP1/2)	170
Tumour necrosis factor alpha (TNFα)	174-176
Interleukin 1 beta (IL1β) &Interleukin 1 receptor antagonist (IL1RN)	177
Interleukin 10 (IL10)	178
Cytotoxic T lymphocyte-associated antigen-4(CTLA-4)	183
Mannose binding lectin (MBL)	185
Vitamin D receptor (VDR)	186
Apolipoprotein E	187

Interleukin-10 (IL10) is another candidate cytokine as it plays a role in controlling the balance between potentially autoreactive Th-1 and immunomodulatory Th-2 CD4+ T-cell subsets. A potentially functional biallelic polymorphism at position −592 in the promoter region of *IL10* has recently been studied but no significant disease association with PBC was identified.[178]

Polymorphisms in the gene encoding Cytotoxic T lymphocyte-associated antigen-4 (CTLA-4, CD152) have been associated with susceptibility to other autoimmune diseases.[179] CTLA-4 is expressed exclusively on activated CD4+ and CD8+ T cells, and it binds to the same ligands (B7.1 or CD80 and B7.2 or CD86) as CD28. It appears to be critical in the regulation of CD4 T cell homeostasis,[180] and may preferentially dampen pathological immune responses to self proteins whilst permitting protective immunity to foreign agents.[181] The *CTLA4* gene on chromosome 2q33 includes 12 SNP's, one of which involves substitution of a guanine (G) for an adenine (A) at position 49. This substitution leads to an alanine to threonine amino acid exchange at codon 17 (A17T) of the leader peptide, which affects the CTLA-4 down regulation of T cell activation.[182] Recent studies have suggested that the G allele at position 49 is associated with susceptibility to PBC.[183]

Another candidate gene that is located near the *CTLA-4* gene is that encoding the divalent cation transporter, NRAMP1 (correct gene name: solute carrier family *SLC11a1*). NRAMP1 has an important role in early phase macrophage activation and therefore in host innate immunity and a preliminary study has suggested association of a microsatellite polymorphism in the promoter region of the *NRAMP1* gene with PBC.[184]

Mannose binding lectin (MBL) is another key factor in innate mucosal defenses and 4 SNPs have been examined in another small case-control study; individuals homozygous for a haplotype leading to hyper-production of MBL were found to have a significantly increased risk for PBC.[185]

Other genetic polymorphisms have been reported to be associated with PBC; the Bsm1 polymorphism of the vitamin D receptor with disease susceptibility[186] whilst an apolipoprotein E polymorphism may influence progression and response to ursodeoxycholic acid treatment.[187] Clearly the complex genetic determinants of PBC are still far from being understood. The challenge of modern molecular research is to dissect out the contribution of genetic from non-genetic factors.

Non-Genetic Factors

The incidence of PBC not only appears to be increasing[188] but also varies widely between regions.[189] Recent data shows that it is also unevenly distributed within one defined region

(Northeast England) with significant clustering of cases,[190] implicating one or more environmental risk factors in the aetiology of PBC. Further data pointing to the importance of environmental factors is provided by the higher prevalence of PBC in atomic bomb survivors in Nagasaki, estimated to be at least 615 cases per million (792 per million women), compared to the general population in Japan.[191] A population based case-control study aimed at elucidating possible environmental factors has drawn a blank, but found an unexpected association with smoking, with an odds ratio of 3.5 for PBC in individuals who have smoked for 20 years or more.[192]

Xenobiotics

Many xenobiotics are metabolised in the liver and in a small minority of cases have been associated with autoimmunity; for example tienilic acid induced autoimmune hepatitis is associated with antibodies to cytochrome P450(CYP)2C9 whilst dihydralazine hepatitis is associated with autoantibodies to CYP1A2.[193] In dihydralazine immunoallergic hepatitis it is suggested that dihydralazine is biotransformed into a reactive metabolite which covalently binds to CYP1A2, which is involved in its metabolism, and triggers an immunological response as a "neoantigen".[194] Similarly halothane hepatitis occurs because susceptible individuals mount immune responses to trifluoroacetylated protein antigens, formed following cytochrome P450-mediated bioactivation of halothane to trifluroacetyl chloride. Indeed it is thought that interindividual variability in the balance between metabolic bioactivation of halothane by CYP2E1 and detoxification of reactive metabolites is important.[195]It is of interest that the trifluoroacetyl (TFA) metabolite covalently links to a lysine residue on CYP2E1 and that the autoantibodies to CYP2E1 cross-react with lipoylated PDC-E2. However the epitope on PDC-E2 recognised by antibodies to the TFA-modified CYP2E1 differs from that recognised by AMA in PBC sera.[196] Another example of xenobiotic induced autoimmunity is provided by the accidental exposure of 9 workers to hydrochlorofluorocarbons; these were metabolised to form reactive trifluoroacetyl halide intermediates and resulted in autoantibodies to CYP2E1 in 5 affected workers and hepatocellular necrosis in one severely affected worker.[197] It is clear from these examples that an initiator of an autoimmune response can be exposure to a xenobiotic which is biotransformed into a reactive intermediate which bind to proteins and results in neoantigen formation.

The issue of whether PBC may be induced by exposure to a xenobiotic has recently been addressed by synthesis of the inner lipoyl domain of PDC-E2 and replacement of the lipoic acid moiety with synthetic structures designed to mimic a xenobiotically-modified lipoyl hapten. Three of the 18 modified autoepitopes were recognised by AMA from PBC sera better than the native PDC-E2 peptide; thus the substitution of the lipoic acid side chain with a phenyl ring containing a halide increased AMA binding.[198] It would have been of interest to see if these modified autoepitopes could absorb the PDC enzyme inhibitory activity demonstrated by AMA but this was not reported. However the data are consistent with the hypothesis that the initiating event in PBC could be exposure to a xenobiotically modified self-antigen in a genetically susceptible host. In this context the genes involved in biotransformation of xenobiotics become candidates for predisposition to PBC.

It may be relevant that ursodeoxycholic acid acts on the pregnane X receptor which induces expression of genes encoding cytochrome P450 enzymes and drug transporters by binding to the PXR response elements found within the promoter regions of these target genes. Recently activation of PXR has been shown to regulate xenobiotic sulfonation and proposed that PXR serves as the master regulator of the phase1 and 11 responses to facilitate rapid and efficient detoxification and elimination of foreign chemicals. Thus one action of ursodeoxycholic acid (UDCA) in PBC could be not only to detoxify lithocholic acid but also to hasten detoxification of an initiating xenobiotic[199] (Fig. 6). This is consistent with the observations that UDCA appears to be of most benefit when instituted in early stage disease[200] and UDCA treatment results in a reduction in AMA titres.[201]

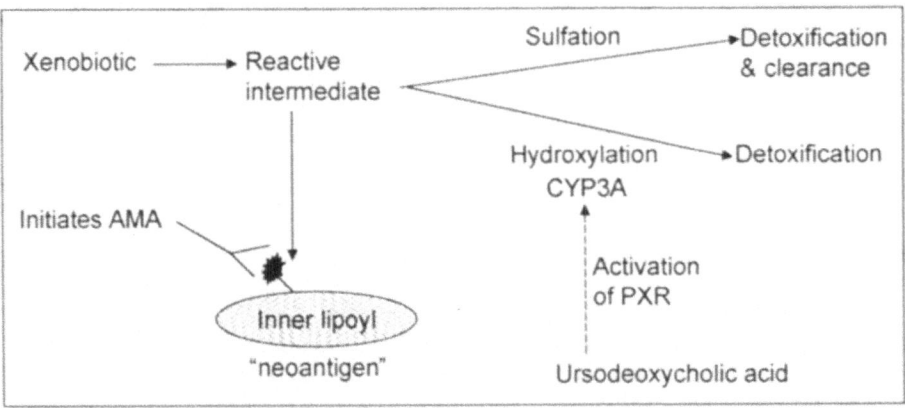

Figure 6. Hypothetical schema suggesting a mechanism whereby a xenobiotic metabolised in the liver to a reactive intermediate may covalently bind to the lysine residue or lipoic acid co-factor of the lipoyl domains of the 2-oxo acid dehydrogenase complexes, generating a "neoantigen" which could elicit an immune response. One action of ursodeoxycholic acid is activation of the pregnane X receptor (PXR) which regulates enzymes involved in rapid and efficient detoxification and elimination of foreign chemicals, as well as toxic bile acids.

Infectious Agents

Viruses have been implicated as initiators, perpetuators and most recently as terminators of autoimmune diseases, but even the best evidence is only circumstantial.[202] Preliminary studies in PBC were stimulated by reports suggesting that Sjogren's syndrome might be triggered by a retroviral infection,[203] with antibodies detectable to retroviral proteins detected in a subset of patients.[204] Antibodies to these retroviral proteins were subsequently discovered in a subset of patients with systemic lupus erythematosus[205] and primary biliary cirrhosis.[206] Evidence has also been presented that raises the possibility of a transmissible agent in PBC lymph nodes. BEC isolated from normal individuals incubated with homogenised lymph node from a PBC patient and then cultured for seven days develop aberrant mitochondrial antigen staining on the plasma membrane, a "phenotype" found only in PBC.[39,40,41,42] If supernatant from these "infected" BEC is incubated with normal BEC they too develop the PBC phenotype, an effect which can be abolished by irradiation of the supernatant.[207] Subsequent attempts to detect, clone and treat a putative retrovirus associated with PBC have recently been reviewed.[208] From the data presented it is probably a safe bet that virus infection is not the principle cause of PBC. However the involvement of retroviral sequences in the complex pathway of breakdown of tolerance to autoantigens, possibly by infection unveiling host antigens or genetic alterations caused by transposon insertion, can not yet be excluded.

As noted above the early observations that AMA in PBC sera cross-reacted with a number of bacteria,[97] including with rough mutants of *Enterobacteriaceae*,[98] lead to the hypothesis that PBC may have a bacterial aetiology. This hypothesis remains controversial with a number of studies reporting conflicting data. The observation that sera from patients with tuberculosis recognise PDC-E2 raised the possibility that mycobacteria may play a role in PBC.[209] Subsequently PBC patients were found to have antibodies in their serum that cross-react with antigens of the atypical mycobacterium *M. gordonae*.[210,211] However there was no evidence of excess T cell responses to mycobacterial proteins[212] and mycobacterial DNA was not detected in liver sections from patients with PBC.[213] Similarly sera from patients with recurrent urinary tract infections (UTIs) recognise mitochondrial antigens,[214] and it has been suggested, but not confirmed, that there is an increased incidence of UTIs in patients with PBC.[215,216] However T cell cross-reactivity has now been confirmed between human and bacterial PDC in patients with PBC.[110-112,114] It seems that the explanation for these observations lies in the fact that the autoepitopes recognised by AMA in PBC sera are evolutionarily conserved, and thus this data should not be used to implicate any particular microorganism in the aetiology of the disease. However studies in mice have shown that self-tolerance can be broken by means of immunisation with foreign (xenogeneic) cytochrome c that has sequence homologies with the self protein.[217] The foreign proteins initiate a T cell response, resulting in activated B cells that can then present the self-proteins to T-cells, thus allowing the generation of autoreactive T cells.[218] On this basis, infection with any microorganism containing PDC-E2 +E3BP and/or OGDCE2 and or BCOADCE2 could conceivably play a role in the immunopathogenesis of PBC. This hypothesis has recently been explored more fully in PBC on the basis of data from a murine model of breakdown of tolerance to PDC.[219-221] In this murine model it has also been shown that bacterial DNA containing unmethylated CpG dinucleotide motifs skews the immune response towards the Th1 phenotype, augmenting the autoimmune reaction.[222]

Other studies have attempted to confirm or refute bacterial involvement in the pathogenesis of PBC. Initial studies using polymerase chain reaction (PCR) techniques and primers specific for the 16s rRNA gene of Eubacteria, Archaeabacteria, Mycobacteria and Helicobacter in liver tissue from patients with PBC found no evidence of an ongoing chronic infectious process.[223] Subsequent studies using PCR have found evidence of gram-positive bacteria in bile,[224] evidence of Helicobacter species in liver tissue[225] and evidence of Propionibacterium acnes in granulomatous lesions in PBC.[226] Genomes of P.acnes have also been found in and around sarcoid granulomatous lesions and there is no suggestion that the finding of these infectious agents is PBC-specific.[227] It could however be that any infectious agent that is found in the liver, bile or at any mucosal surface has the capacity, in an immunogenetically susceptible individual, to generate a transient immune response that is cross-reactive with self-PDC. A second "hit" in biliary epithelial cells, such as unveiling of host PDC lipoyl domains by xenobiotic modification or viral infection resulting in apoptosis and sustained presentation of self peptides via dendritic cells, may then be sufficient to convert this transient autoimmunity into manifest autoimmune disease[120] (Fig. 7).

Summary and Conclusions

In PBC an early event is the development of disease specific autoantibodies, usually AMA. The reactivity of these autoantibodies is now well characterised; they react predominantly with conformational epitopes around the lipoic-acid binding lysine of the lipoyl domains of components of the human 2 oxo acid dehydrogenase complexes. AMA have been shown to cross-react not only with microbial proteins but also with synthetic structures designed to mimic a xenobiotically-modified lipoyl hapten. It seems likely that, in an immunogenetically susceptible individual, exposure to one or other of these immunogens could result in the development of antibodies that cross react with self. Prolonged delivery via dendritic cells of antigen/ AMA immune complexes formed in the vicinity of the bile ducts by antigen absorbed from the bile or unveiled in biliary epithelium could then convert these autoimmune responses into overt disease. Changes in the biliary epithelial cells involving secretion of IgA AMA, viral infection or focal apoptosis could result in abnormal membrane expression of mitochondrial autoantigens, and lead to cytotoxic T cell recognition of this target tissue. Recurrent mucosal bacterial infection could exert an adjuvant effect, and influence progression. As the maintenance of self-tolerance is so vital, it is likely that a multi-step process will be involved, similar in complexity to those occurring in carcinogenesis. The pieces of the jigsaw involved in the conversion of transient autoimmunity into non-suppurative destructive cholangitis are becoming clearer, largely through detailed study of the autoantigen/antibody interaction. Insight into the molecular basis of the female predominance of PBC however remains elusive. It is hoped that improved understanding of the molecular pathways within the target biliary epithelial cell, and of mucosal tolerance (which may not be identical in all cases), will eventually lead to a cure for this enigmatic disease. In addition further molecular epidemiological studies may identify

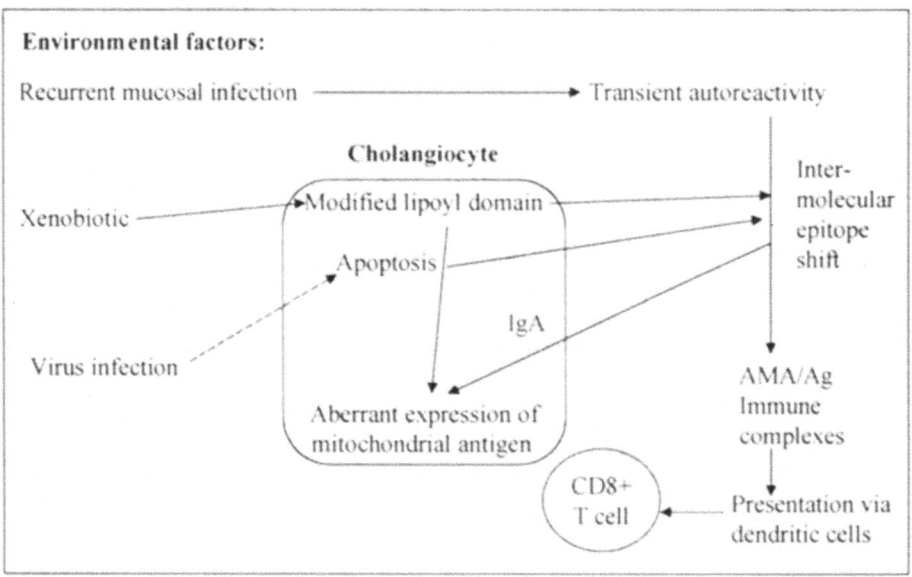

Figure 7. A model suggesting that complex exposure to assorted environmental factors may break tolerance in PBC. Autoantigen-antibody immune complexes may be taken up by dendritic cells and efficiently presented, converting transient autoimmunity to cholangiocyte damage mediated by cytotoxic T lymphocytes.

environmental triggers. Clinicians should take heart however from the use of UDCA in PBC, where an understanding of its molecular mechanism of action on the pregnane X receptor followed clinical observations of its benefit!

References

1. Saxena R, Hytiroglou P, Thung SN et al. Destruction of canals of Hering in primary biliary cirrhosis. Hum Pathol 2002; 33:983-988.
2. Jorge AD, Gutierrez LS, Jorge O et al. Ultrastructural and cytochemical changes in the liver of primary biliary cirrhosis patients. Biocell 2002; 26:253-62.
3. Nakamura A, Yamazaki K, Suzuki K et al. Increased portal tract infiltration of mast cells and eosinophils in primary biliary cirrhosis. Am J Gastroenterol 1997; 92:2245-9.
4. Floreani A, Bennett MK, Mitchison HC et al. Progression of autoimmune damage in primary biliary cirrhosis: An immunohistochemical study. Autoimmunity 1989; 2:311-21.
5. Harada K, Tsuneyama K, Yasoshima M et al. Type1 and type2 memory T cells imbalance shown by expression of intrahepatic chemokine receptors relates to pathogenesis of primary bilairy cirrhosis. Hepatol Res 2002; 24:290.
6. Ayres RC, Neuberger JM, Shaw J et al. Intercellular adhesion molecule-1 and MHC antigens on human intrahepatic bile duct cells: effect of pro-inflammatory cytokines. Gut 1993; 34:1245-9.
7. Janaway CAJ, Bottomly K. Signals and signs for lymphocyte responses. Cell 1994; 76:275-85.
8. Leon MP, Kirby JA, Gibbs P et al. Immunogenicity of biliary epithelial cells: study of the expression of B7 molecules. J Hepatol 1995; 22:591-95.
9. Leon MP, Bassendine MF, Wilson JL et al. Immunogenicity of biliary epithelium: Investigation of antigen presentation to CD4+ T cells. Hepatology 1996; 24:561-67.
10. Tsuneyama K, Harada K, Yasoshima M et al. Expression of co-stimulatory factor B7-2 on the intrahepatic ducts in primary biliary cirrhosis and primary sclerosing cholangitis: an immunohistochemical study. J Pathol 1998; 186:126-30.
11. Van Den Oord JJ, Sciot R, Desmett VJ. Expression of MHC products by normal and abnormal bile duct epithelium. J Hepatol 1986; 3:310-317.
12. Ballardini G, Bianchi FB, Mirakian R et al. HLA-A, BC, HLA-D/DR and HLA-D/DQ expression on unfixed liver biopsy sections from patients with chronic liver damage. Clin Exp Immunol 1987; 70:35-46.

13. Spengler U, Pape GR, Hoffman RM et al. Differential expression of MHC class II subregion products on bile duct epithelial cells and hepatocytes in patients with primary biliary cirrhosis. Hepatology 1988; 8:459-62.
14. Jiminez BD, Maldonado L, Dahl RH et al. Ectopic expression of MHC class II genes (RT1.B(1)β/α) in rat hepatocytes in vivo and in culture can be elicited by treatment with the pregnane X receptor agonists pregnenolone 16α-carbonitrile and dexamethasone. Life Sci 2002; 71:311-323.
15. Tanimoto K, Akbar SM, Michitaka K et al. Immunohistochemical localization of antigen presenting cells in liver from patients with primary biliary cirrhosis; highly restricted distribution of CD83-positive activated dendritic cells. Pathol Res Pract 1999; 195:157-62.
16. Leon MP, Bassendine MF, Gibbs P et al. Immunogenicity of biliary epithelium: Study of the adhesive interaction with lymphocytes. Gastroenterology 1997; 112:968-77.
17. Shiina M, Kobayashi K, Mano Y et al. Up-regulation of CD11a (LFA-1) expression on peripheral CD4+ T cells in primary biliary cirrhosis. Dig Dis Sci 2002; 47:1209-15.
18. Tsuneyama K, Yasoshima M, Harada K et al. Increased CD1d expression on small bile duct epithelium and epithelioid granuloma in livers in primary biliary cirrhosis. Hepatology 1998; 28:620-3.
19. Tsuneyama K, Harada K, Yasoshima M et al. Monocyte chemotactic protein-1, -2,and -3 are distinctly expressed in portal tracts and granulomata in primary biliary cirrhosis: implications for pathogenesis. J Pathol 2001; 193:102-9.
20. Tsuneyama K, Harada K, Kono N et al. Scavenger cells with gram-positive bacterial lipotechoic acid infiltrate around the damaged interlobular bile ducts of primary biliary cirrhosis. J Hepatol 2001; 35:156-63.
21. Ballardini G, Guidi M, Susca M et al. Bile duct cell apoptosis is a rare event in primary bilairy cirrhosis. Dig Liver Dis 2001; 33:122-4.
22. Marucci L, Ugili L, Macarri G et al. Primary biliary cirrhosis: modalities of injury and death in biliary epithelium. Dig Liver Dis 2001; 33:576-83.
23. Tinmouth J, Lee M, Wanless IR et al. Apoptosis of biliary epithelial cells in primary biliary cirrhosis and primary sclerosing cholangitis. Liver 2002; 22:228-34.
24. Afford SC, Ahmed-Choudhury J, Randhawa S et al. CD40 activation-induced, Fas-dependent apoptosis and NF-kappB/AP-1 signalling in human intrahepatic biliary epithelial cells. FASEB J 2001; 15:2345-54.
25. Fox CK, Furtwaengler A, Nepomuceno RR et al. Apoptotic pathways in primary biliary cirrhosis and autoimmune hepatitis. Liver 2001; 21:272-279.
26. Odin JA, Huebert RC, Casciola-Rosen L et al. Bcl-2-dependent oxidation of pyruvate dehydrogenase-E2, a primary biliary cirrhosis autoantigen, during apoptosis. J Clin Invest 2001; 108:223-32.
27. Matsumura S, Van de Water J, Kita H et al. Contribution to antimitochindrial antibody production: cleavage of pyruvate dehydrogenase complex-E2 by opoptosis related proteases. Hepatology 2002; 35:14-22.
28. Burt AD. Primary biliary cirrhosis and other ductopaenic diseases. Clinics in Liver Disease 2002; 6:363-380.
29. Roskams T, Desmet V. Ductular reaction and its diagnostic significance. Semin Diagn Pathol 1998; 15:259-69.
30. Shackel NA, P.H. M, Abbott CA et al. Identification of novel molecules and pathogenic pathways in primary biliary cirrhosis: cDNA array analysis of intrahepatic differential gene expression. Gut 2001; 49:565-76.
31. Tanaka A, Leung PSC, Kenny TP et al. Genomic analysis of differentially expressed genes in liver and biliary cells of patients with primary biliary cirrhosis. J Autoimmun 2001; 17:89-98.
32. Medina JF, Martinex-Anso, Vazquez JJ et al. Decreased anion exchanger 2 immunoreactivity in the liver of patients with primary biliary cirrhosis. Hepatology 1997; 25:12-17.
33. Prieto J, Garcia N, Marti-Climent JM et al. Assessment of biliary bicarbonate secretion in humans by positron emission tomography. Gastroenterology 1999; 117:167-72.
34. Melero S, Spirli C, Zsembery A et al. Defective regulation of cholangiocyte Cl-/HCO3(-) and Na+/H+ exchanger activities in primary biliary cirrhosis. Hepatology 2002; 35:1513-21.
35. Zollner G, Fickert P, Zenz R et al. Hepatobiliary transporter expression in percutaneous liver biopsies of patients with cholestatic liver diseases. Hepatology 2001; 33:633-46.
36. Tsuneyama K, Harada K, Kono N et al. Damaged interlobular bile ducts in primary biliary cirrhosis show reduced expression of glutathione-S-transferase-pi and aberrant expression of 4-hydroxynonenal. J Hepatol 2002; 37:176-83.
37. Rodriguez-Ortigosa CM, Cincu RN, Sanz S et al. Effect of ursodeoxycholic acid on methionine adenosyltransferase activity and hepatic glutathione metabolism in rats. Gut 2002; 50:701-6.

38. Bergasa NV, Liau S, Homel P et al. Hepatic Met-enkephalin immunoreactivity is enhanced in primary biliary cirrhosis. Liver 2002; 22:107-13.
39. Joplin R, Gordon Lindsay J, Johnson GD et al. Membrane dihydrolipoamide acetyltransferase (E2) on human biliary epithelial cells in primary biliary cirrhosis. Lancet 1992; 339:93-94.
40. Van de Water J, Turchany J, Leung PS et al. Molecular mimicry in primary biliary cirrhosis. Evidence for biliary epithelial expression of a molecule cross-reactive with pyruvate dehydrogenase complex-E2. J Clin Invest 1993; 91(6):2653-64.
41. Joplin R, Wallace LL, Lindsay JG et al. Pyruvate dehydrogenase-X (PDH-X) is associated with the biliary epithelial cell (BEC) plasma membrane in primary biliary cirrhosis (PBC). Hepatology 1996; 24:165A.
42. Joplin RE, Wallace LL, Lindsay JG et al. The human biliary epithelial cell plasma membrane antigen in primary biliary cirrhosis: pyruvate dehydrogenase X? Gastroenterology 1997; 113:1727-33.
43. Joplin R, Wallace LL, Johnson GD et al. Subcellular localisation of pyruvate dehydrogenase dihydrolipoamide acetyltransferase in human intrahepatic biliary epithelial cells. J Pathol 1995; 176:381-390.
44. Nakanuma Y, Tsuneyama K, Kono N et al. Biliary epithelial expression of pyruvate dehydrogenase complex in primary biliary cirrhosis:an immunohistochemical and immunoelectron microscopic study. Hum Pathol 1995; 26(1):92-98.
45. Migliaccio C, Van de Water J, Ansari AA et al. Heterogeneous response of antimitochondrial autoantibodies and bile duct apical staining monoclonal antibodies to pyruvate dehydrogenase complex E2: the molecule versus the mimic. Hepatology 2001; 33:792-801.
46. Joplin RE, Johnson GD, Matthews JB et al. Distribution of pyruvate dehydrogenase dihydrolipoamide acetyltransferase (PDC-E2) and another mitochondrial marker in salivary gland and biliary epithelium from patients with primary biliary cirrhosis. Hepatology 1994; 19:1375-1380.
47. Tsuneyama K, Van de Water J, Yamazaki K et al. Primary biliary cirrhosis an epithelitis: evidence of abnormal salivary gland immunohistochemistry. Autoimmunity 1997; 26:23-31.
48. Harada K, Sudo Y, Kono N et al. In situ nucleic acid detection of PDC-E2, BCOADC-E2, OGDC-E2, PDC-E1alpha, OGDC-E1, and the E3 binding protein (protein X) in primary biliary cirrhosis. Hepatology 1999; 30:36-45.
49. Nishio A, Van de Water J, Leung PS et al. Comparative studies of antimitochondrial autoantibodies in sera and bile in primary biliary cirrhosis. Hepatology 1997; 25:1085-89.
50. Reynoso-Paz S, Leung PSC, Van de Water J et al. Evidence for a locally driven mucosal response and the presence of mitochondrial antigens in saliva in primary biliary cirrhosis. Hepatology 2000; 31:24-29.
51. Palmer JM, Doshi M, Kirby JA et al. Secretory autoantibodies in primary biliary cirrhosis. Clin Exp Immunol 2000; 122:1-7.
52. Ikuno N, Mackay IR, Jois J et al. Antimitochondrial autoantibodies in saliva and sera from patients with primary biliary cirrhosis. J Gastroenterol Hepatol 2001; 16:1390-4.
53. Fujioka H, Emancipator SN, Aikawa M et al. Immunocytochemical colocalization of specific immunogloublin A with Sendai virus protein in infected polarized epithelium. J Exp Med 1998; 188:1223-29.
54. Christensen E, Crowe J, Doniach D et al. Clinical pattern and course of disease in primary biliary cirrhosis based on analysis of 236 patients. Gastroenterology 1980; 78:236-46.
55. Heseltine L, Turner IB, Fussey SP et al. Primary biliary cirrhosis: quantitation of autoantibodies to purified mitochondrial enzymes and correlation with disease progression. Gastroenterology 1990; 99(6):1786-1792.
56. Kisand KE, Kisand KV, Karvonen AL et al. Antibodies to pyruvate dehydrogenase in primary biliary cirrhosis: correlation with histology. APMIS 1998; 106:884-92.
57. Mason D. A very high level of cross-reactivity is an essential feature of the T-cell receptor. Immunol Today 1998; 19:395-404.
58. Wilson DB, Pinilla C, Wilson DH et al. Immunogenicity. 1. Use of peptide libraries to identify epitopes that activate clonotypic CD4+ T cells and induce T cell responses to native peptide ligands. J Immunol 1999; 163:6424-6234.
59. Mitchison HC, Bassendine MF, Hendrick AM et al. Positive antimitochondrial antibody but normal liver function tests: is this primary biliary cirrhosis? Hepatology 1986; 6:1279-1284.
60. Metcalf JV, Mitchison HC, Palmer JM et al. Natural history of early primary biliary cirrhosis. Lancet 1996; 348:1399-1402.
61. Kisand KE, Metskula K, Kisand KV et al. The follow-up of asymptomatic persons with antibodies to pyruvate dehydrogenase in adult population samples. J Gastroenterol 2001; 36:248-54.

62. Maeng CY, Yazdi MA, Reed LJ. Stoichiometry of binding of mature and truncated forms of the dihydrolipoamide dehydrogenase-binding protein to the dihydrolipoamide acetyltransferase core of the pyruvate dehydrogenase complex from *Saccharomyces cerevisiae*. Biochemistry 1996; 35:5879-5882.

63. Maeng C-Y, Yazdi MA, Niu X-D et al. Expression, purification, and characterisation of the dihydrolipoamide dehydrogenase-binding protein of the pyruvate dehydrogenase complex from *Saccharomyces cerevisiae*. Biochemistry 1994; 33:13801-13807.

64. Yeaman SJ. The 2-oxo acid dehydrogenase complexes: recent advances. Biochem J 1989; 257:625-632.

65. Denton RM, McCormack JG, Rutter GA et al. The hormonal regulation of pyruvate dehydrogenase complex. Adv Enzyme Regul 1996; 36:183-98.

66. Gershwin ME, Mackay IR, Sturgess A et al. Identification and specificity of a cDNA encoding the 70 kD mitochondrial antigen recognized in primary biliary cirrhosis. J Immunol 1987; 138:3525-3531.

67. Yeaman SJ, Fussey SP, Danner DJ et al. Primary biliary cirrhosis: identification of two major M2 mitochondrial autoantigens. Lancet 1988(i):1067-1070.

68. Van de Water J, Gershwin ME, Leung P et al. The autoepitope of the 74-kD mitochondrial autoantigen of primary biliary cirrhosis corresponds to the functional site of dihydrolipoamide acetyltransferase. J Exp Med 1988; 167:1791-1799.

69. Fussey SPM, Guest JR, James OFW et al. Identification and analysis of the major M2 autoantigens in primary biliary cirrhosis. Proc Natl Aca Sci USA 1988; 85:8654-8658.

70. Surh CD, Danner DJ, Ahmed A et al. Reactivity of primary biliary cirrhosis sera with a human fetal liver cDNA clone of branched-chain α-keto acid dehydrogenase dihydrolipoamide acyltransferase, the 52 kDa mitochondrial autoantigen. Hepatology 1989; 9:63-68.

71. Mutimer DJ, Fussey SP, Yeaman SJ et al. Frequency of IgG and IgM autoantibodies to four specific M2 mitochondrial autoantigens in primary biliary cirrhosis. Hepatology 1989; 10:403-407.

72. Fussey SP, Lindsay JG, Fuller C et al. Autoantibodies in primary biliary cirrhosis: analysis of reactivity against eukaryotic and prokaryotic 2-oxo acid dehydrogenase complexes. Hepatology 1991; 13(3):467-74.

73. Fussey SPM, Bassendine MF, Fittes D et al. The E1α and β subunits of the pyruvate dehydrogenase complex are M2"d" and M2"e" autoantigens in primary biliary cirrhosis. Clin Sci 1989; 77:365-368.

74. Fussey SP, West SM, Lindsay JG et al. Clarification of the identity of the major autoantigen in primary biliary cirrhosis. Clin Sci 1991; 80(5):451-5.

75. Palmer JM, Bassendine MF, James OFW et al. Human pyruvate dehydrogenase complex as an autoantigen in primary biliary cirrhosis. Clin Sci 1993; 85(3):289-93.

76. Mori T, Ono K, Hakozaki M et al. Autoantibodies of sera from patients with primary biliary cirrhosis recognize the alpha subunit of the decarboxylase component of human branched-chain 2-oxo acid dehydrogenase complex. J Hepatol 2001; 34:799-804.

77. Joshi S, Cauch-Dudek K, Heathcote EJ et al. Antimitochondrial antibody profiles: are they valid prognostic indicators in primary biliary cirrhosis. Am J Gastroenterol 2002; 97:999-1002.

78. Perham RN. Domains,motifs and linkers in 2-oxo acid dehydrogenase multienzyme complexes: a paradigm in the design of a multifunctional protein. Biochemistry 1991; 30:8501-8512.

79. Fussey SP, Bassendine MF, James OFW et al. Characterisation of the reactivity of autoantibodies in primary biliary cirrhosis. FEBS Letters 1989; 246(1-2):49-53.

80. Fussey SPM, Bassendine MF, James OFW et al. The lipoate-containing domain of PDC E2 contains the main immunogenic region of the 70 kD M2 autoantigen in primary biliary cirrhosis. Ann N Y Acad Sci 1989; 573:444-6.

81. Fussey SP, Ali ST, Guest JR et al. Reactivity of primary biliary cirrhosis sera with *Escherichia coli* dihydrolipoamide acetyltransferase (E2p): characterisation of the main immunogenic region. Proc Natl Aca Sci USA 1990; 87(10):3987-3991.

82. Surh CD, Coppel R, Gershwin ME. Structural requirement for autoreactivity on human pyruvate dehydrogenase-E2, the major autoantigen of primary biliary cirrhosis. J Immunol 1990; 144:3367-3374.

83. Tuaillon N, Andre C, Briand JP et al. A lipoyl synthetic octapeptide of dihydrolipoamide acetyltransferase specifically recognised by anti-M2 autoantibodies in primary biliary cirrhosis. J Immunol 1992; 148:445-450.

84. Quinn J, Diamond AG, Palmer JM et al. Lipoylated and unlipoylated domains of human PDC-E2 as autoantigens in primary biliary cirrhosis: significance of lipoate attachment. Hepatology 1993; 18(6):1384-91.

85. Leung PS, Chuang DT, Wynn RM et al. Autoantibodies to BCOADC-E2 in patients with primary biliary cirrhosis recognize a conformational epitope. Hepatology 1995; 22(2):505-513.

86. Moteki S, Leung PS, Dickson ER et al. Epitope mapping and reactivity of autoantibodies to the E2 component of 2-oxoglutarate dehydrogenase complex in primary biliary cirrhosis using recombinant 2-oxoglutarate dehydrogenase complex. Hepatology 1996; 23(3):436-444.
87. Dubel L, Tanak A, Leung PS et al. Autoepitope mapping and reactivity of autoantibodies to the dihydrolipoamide dehydrogenase-binding protein (E3BP) and the glycine cleavage proteins in primary biliary cirrhosis. Hepatology 1999; 29:1013-8.
88. Palmer JM, Jones DEJ, Quinn J et al. Characterisation of the autantibody response to recombinant E3 binding protein of pyruvate dehydrogenase complex in primary biliary cirrhosis. Hepatology 1999; 30:21-26.
89. Howard MJ, Fuller C, Broadhurst W et al. Three-dimensional structure of the major autoantigen in primary biliary cirrhosis. Gastroenterology 1998; 115:1-9.
90. Chang CF, Chou HT, Chuang JL et al. Solution structure and dynamics of the lipoic acid-bearing domain of human mitochondrial branched-chain alpha-keto-acid dehydrogenase complex. J Biol Chem 2002; 277:15865-73.
91. Berg A, Vervoort J, de Kok A. Three-dimensional structure in solution of the N-terminal lipoyl domain of the pyruvate dehydrogenase complex from *Azotobacter vinelandii*. Eur J Biochem 1997; 244:352-360.
92. Thomson RK, Davis Z, Palmer JM et al. Immunogenetic analysis of a panel of monoclonal IgG and IgM anti-pyruvate dehydrogenase complex antibodies derived from patients with primary biliary cirrhosis. J Hepatol 1998; 28:582-94.
93. Van de Water J, Fregeau D, Davis P et al. Autoantibodies of primary biliary cirrhosis recognize dihydrolipoamide acetyltransferase and inhibit enzyme function. J Immunol 1988; 141:2321-4.
94. Rowley MJ, McNeilage JL, Armstrong JM et al. Inhibitory autoantibody to a conformational epitope of the pyruvate dehydrogenase complex, the major autoantigen in primary biliary cirrhosis. Clin Immunol Immunophthol 1991; 60:356-370.
95. Teoh KL, Mackay IR, Rowley MJ et al. Enzyme inhibitory autoantibodies to pyruvate dehydrogenase complex in primary biliary cirrhosis differ for mammalian, yeast and bacterial enzymes: implications for molecular mimicry. Hepatology 1994; 19(4):1029-1033.
96. Sayers T, Baum H. Possible cross-reaction of human anti-mitochondrial antibody with membrane vesicles of *Paracoccus dinitrificans*. Biochem Soc Trans 1976; 4:138-9.
97. Lindenborn-Fotinos J, Baum H, Berg PA. Mitochondrial antigens in primary biliary cirrhosis: further characterisation of the M2 antigen by immunoblotting, revealing species and non-species specific determinants. Hepatology 1985; 5:763-9.
98. Stemerowicz R, Hopf U, Moller B et al. Are antimitochondrial antibodies in primary biliary cirrhosis induced by R(rough)-mutants of enterobacteriaceae? Lancet 1988; ii:1166-69.
99. Fraser CM, Gocayne JD, White O et al. The minimal gene complement of *Mycoplasma genitalium*. Science 1995; 270:397-403.
100. Behal RH, Browning KS, Hall TB et al. Cloning and nucleotide sequence of the gene for protein X from Saccharomyces cerevisiae. Proc Natl Aca Sci USA 1989; 86(22):8732-36.
101. Teoh KL, Rowley MJ, Zafirakis H et al. Enzyme inhibitory autoantibodies to pyruvate dehydrogenase complex in primary biliary cirrhosis: applications of a semiautomated assay. Hepatology 1994; 20(5):1220-1224.
102. Potter KN, Thomson RK, Hamblin A et al. Immunogenetic analysis reveals that epitope shifting occurs during B-cell affinity maturation in primary biliary cirrhosis. J Mol Biol 2001; 306:37-46.
103. Oldstone MB. Molecular mimicry and immune-mediated diseases. FASEB J 1998; 12:1255-65.
104. Jones DEJ, Palmer JM, Yeaman SJ et al. T-cell responses to components of pyruvate dehydrogenase complex in primary biliary cirrhosis. Hepatology 1995; 21:995-1002.
105. Jones DEJ, Palmer JM, Yeaman SJ et al. T-cell responses to native human proteins in primary biliary cirrhosis. Clin Exp Immunol 1997; 107:562-68.
106. Van de Water J, Ansari A, Prindiville T et al. Heterogeneity of autoreactive T cell clones specific for the E2 component of the pyruvate dehydrogenase complex in primary biliary cirrhosis. J Exp Med 1995; 181(2):723-33.
107. Palmer JM, Diamond A, Yeaman SJ et al. T-cell responses to the putative dominant autoepitope in primary biliary cirrhosis. Clin Exp Immunol 1999; 116:133-139.
108. Akbar SM, Yamamoto K, Miyakawa H et al. Peripheral blood T cell responses to pyruvate dehydrogenase complex in primary biliary cirrhosis: role of antigen presenting dendritic cells. Eur J Clin Invest 2001; 31:639-46.
109. Van de Water J, Ansari AA, Surh CD et al. Evidence for the targeting by 2-oxo-dehydrogenase enzymes in the T cell response of primary biliary cirrhosis. J Immunol 1991; 146:89-94.

110. Shimoda S, Nakamura M, Ishibishi H et al. HLA DRB4 0101-restricted immunodominant T-cell autoepitope of pyruvate dehydrogenase complex in primary biliary cirrhosis: evidence of molecular mimicry in human autoimmune disease. J Exp Med 1995; 181:1835-45.

111. Shimoda S, Nakamura M, Shigematsu H et al. Mimicry peptides of human PDC-E2 163-176 peptide, the immundominant T-cell epitope of primary biliary cirrhosis. Hepatology 2000; 31:1212-16.

112. Shigematsu H, Shimoda S, Nakamura M et al. Fine specificity of T cells reactive to human PDC-E2 163-176 peptide, the immunodominant autoantigen in primary biliary cirrhosis: implications for molecular mimicry and cross-recognition among mitochondrial autoantigens. Hepatology 2000; 32:901-9.

113. Kita H, Matsumura S, He XS et al. Quantitative and functional analysis of PDC-E2 specific autoreactive cytotoxic T lymphocytes in primary bilisry cirrhosis. J Clin Invest 2002; 109:1231-40.

114. Kita H, Matsumura S, He XS et al. Analysis of TCR antagonism and molecular mimicry of an HLA A0201-restricted CTL epitope in primary biliary cirrhosis. Hepatology 2002; 36:918-26.

115. Matsumura S, Kita H, He.X.S. et al. Comprehensive mapping of HLA-A-0201-restricted CD8 T-cell epitopes on PDC-E2 in primary biliary cirrhosis. Hepatology 2002; 36:1125-34.

116. Pingel S, Arenz M, Meyer Zum Buschenfelde KH et al. Pyruvate dehydrogenase specific T cells in primary biliary cirrhosis show restricted antigen recognition sites. Liver 2002; 22:308-16.

117. Kita H, Lian ZX, Van de Water J et al. Identification of HLA-A2-restricted CD8(+) cytotoxic T cell responses in primary biliary cirrhosis: T cell activation is augmented by immune complexes cross-presented by dendritic cells. J Exp Med 2002; 195:113-23.

118. Thomas HC, Potter BJ, Sherlock S. Is primary biliary cirrhosis an immune complex disease? Lancet 1977; (ii):1261-3.

119. Wands JR, Dienstag JL, Bhan AK et al. Circulating immune complexes and complement activation in primary biliary cirrhosis. N Engl J Med 1978; 298:233-7.

120. Ludwig B, Junt T, Hengartner H, Zinkernagel RM. Dendritic cells in autoimmune diseases. Curr Opin Immunol 2001; 13:657-662.

121. Mackay IR. Hepatoimmunology: A perspective. Immunology and Cell Biology 2002; 80:36-44.

122. Kita H, Naidenko OV, Kronenberg M et al. Quantitation and phenotypic analysis of natural killer T cells in primary biliary cirrhosis using a human Cd1d tetramer. Gastroenterology 2002; 123:1031-43.

123. Lacerda MA, Ludwig J, Dickson ER et al. Antimitochondrial antibody-negative primary biliary cirrhosis. Am J Gastroenterol 1995; 90(2):247-249.

124. Michieletti P, Wanless IR, Katz A et al. Antimitochondrial antibody negative primary biliary cirrhosis: a distinct syndrome of autoimmune cholangitis. Gut 1994; 35(2):260-265.

125. Lozano F, Pares A, Borche L et al. Autoantibodies against nuclear envelope-associated proteins in primary biliary cirrhosis. Hepatology 1988; 8:930-8.

126. Powell F, Schoeter AL, Dickson ER. Antinuclear antibodies in primary biliary cirrhosis. Lancet 1984(i):288-9.

127. Lassoued K, Guilly MN, Andre C et al. Autoantibodies to 200 kD polypeptide(s) of the nuclear envelope: a new serologic marker of primary biliary cirrhosis. Clin Exp Immunol 1988; 74(2):283-288.

128. Courvalin JC, Lassoued K, Bartnik E et al. The 210-kD nuclear envelope polypeptide recognized by human autoantibodies in primary biliary cirrhosis is the major glycoprotein of the nuclear pore. J Clin Invest 1990; 86(1):279-285.

129. Nickowitz RE, Wozniak RW, Schaffner F et al. Autoantibodies against integral membrane proteins of the nuclear envelope in patients with primary biliary cirrhosis. Gastroenterology 1994; 106(1):193-9.

130. Lassoued K, Brenard R, Degos F et al. Antinuclear antibodies directed to a 200-kilodalton polypeptide of the nuclear envelope in primary biliary cirrhosis. A clinical and immunological study of a series of 150 patients with primary biliary cirrhosis. Gastroenterology 1990; 99(1):181-186.

131. Nickowitz RE, Worman HJ. Autoantibodies from patients with primary biliary cirrhosis recognize a restricted region within the cytoplasmic tail of the nuclear pore membrane glycoprotein Gp210. J Exp Med 1993; 178(6):2237-42.

132. Wesierska-Gadek J, Hohenauer H, Hitchman E et al. Autoantibodies from patients with primary biliary cirrhosis preferentially react with the amino-terminal domain of the nuclear pore complex glycoprotein gp210. J Exp Med 1995; 182:1159-62.

133. Tartakovsky F, Worman HJ. Detection of Gp210 autoantibodies in primary biliary cirrhosis using a recombinant protein containing the predominant autoepitope. Hepatology 1995; 21(2):495-500.

134. Bandin O, Courvalin JC, Poupon R et al. Specificity and sensitivity of gp210 autoantibodies detected using an enzyme-linked immunoabsorbent assay and a synthetic polypeptide in the diagnosis of primary biliary cirrhosis. Hepatology 1996; 23(5):1020-4.
135. Wesierska-Gadek J, Honenauer H, Hitchman E et al. Autoantibodies against nucleoporin p62 constitute a novel marker of primary biliary cirrhosis. Gastroenterology 1996; 110:840-847.
136. Invernizzi P, Podda M, Battezzati PM et al. Autoantibodies against nuclear pore complexes are associated with more active and severe liver disease in primary biliary cirrhosis. J Hepatol 2001; 34:366-72.
137. Courvalin JC, Lassoued K, Worman HJ et al. Identification and characterisation of autoantibodies against the nuclear envelope lamin B receptor from patients with primary biliary cirrhosis. J Exp Med 1990; 172(3):961-967.
138. Lin F, Noyer CM, Ye Q et al. Autoantibodies from patients with primary biliary cirrhosis recognize a region within the nucleoplasmic domain of inner nuclear membrane protein LBR. Hepatology 1996; 23(1):57-61.
139. Evans J, Rueben A, Craft C. PBC95K, a 95 kilodalton nuclear autoantigen in primary biliary cirrhosis. Arthritis Rheum 1991; 34:731-6.
140. Fusconi M, Cassani F, Govoni M et al. Anti-nuclear antibodies of primary biliary cirrhosis recognise 78-92-kD and 96-100-kD proteins of nuclear bodies. Clin Exp Immunol 1991; 83:291-297.
141. Szostecki C, Guldner HH, Netter HJ et al. Isolation and characterisation of cDNA encoding a human nuclear antigen predominantly recognised by autoantibodies from patients with primary biliary cirrhosis. J Immunol 1990; 145:4338-4347.
142. Muratori P, Muratori L, Cassani F et al. Anti-multiple nuclear dots (anti-MND) and anti-SP100 antibodies in hepatic and rheumatological disorders. Clin Exp Immunol 2002; 127:172-5.
143. Szostecki C, Will H, Netter HJ et al. Autoantibodies to the nuclear Sp100 protein in primary biliary cirrhosis and associated diseases: epitope specificity and immunoglobulin class distribution. Scand J Immunol 1992; 36(4):555-564.
144. Sternsdorf T, Gulder HH, Szostecki C et al. Two nuclear dot associated proteins,PML and Sp100, are often co-autoimmunogenic in patients with primary biliary cirrhosis. Scand J Immunol 1995; 42(2):257-268.
145. Kakizuka A, Miller WHJ, Umesono K et al. Chromosomal translocation t(15:17) in human promyelocytic leukaemia fuses RARα with a novel putative transcription factor, PML. Cell 1991; 66:663-674.
146. Gulder HH, Szostecki C, Grotzinger T et al. IFN enhance expression of Sp100, an autoantigen in primary biliary cirrhosis. J Immunol 1992; 149:4067-73.
147. Grotzinger H, Sternsdorf T, Jensen K. Interferon-modulated expression of genes encoding the nuclear-dot-associated proteins Sp100 and promyelocytic leukemia protein (PML). Eur J Biochem 1996; 238(2):554-560.
148. Luettig B, Boeker KH, Schoessler W et al. The antinuclear autoantibodies Sp100 and gp210 persist after orthotopic liver transplantation in patients with primary biliary cirrhosis. J Hepatol 1998; 28:824-8.
149. Ikuno N, Scealy M, Davies JM et al. A comparitive study of antibody expression in primary biliary cirrhosis and autoimmune cholangitis using phage display. Hepatology 2001; 34:478-86.
150. Bach N, Schaffner F. Familial primary biliary cirrhosis. J Hepatol 1994; 20:698-701.
151. Brind AM, Bray GP, Portmann BC et al. Prevalence and pattern of familial disease in primary biliary cirrhosis. Gut 1995; 36:615-17.
152. Kaplan MM, Rabson AR, Lee YM et al. Discordant occurence of primary biliary cirrhosis in monozygotic twins. N Eng J Med 1994; 331:952.
153. Douglas JG, Finlayson NDC. Are increased individual susceptiblity and environmental factors both necessary for the development of primary biliary cirrhosis? Br Med J 1979(2):419-20.
154. Vyse TJ, Todd JA. Genetic analysis of autoimmune disease. Cell 1996; 85:311-8.
155. Lander ES, Schork NJ. Genetic dissection of complex traits. Science 1994; 265:2037-48.
156. Lander E, Kruglyak L. Genetic dissection of complex traits: guidelines for interpreting and reporting linkage results. Nature Genet 1995; 11:241-47.
157. Bodmer JG, Marsh SG, Albert ED et al. Nomenclature for factors of the HLA system. Human Immunlogy 1997; 53(1):98-128.
158. Miyamori H, Kato Y, Kobayashi K et al. HLA antigens in Japanese patients with primary biliary cirrhosis and autoimmune hepatitis. Digestion 1983; 26:213-7.
159. Ercilla G, Pares A, Arriaga F et al. Primary biliary cirrhosis associated with HLA-DRw3. Tissue Antigens 1979; 14:449-52.
160. Johnston DE, Kaplan MM, Milleer KB et al. Histocompatibility antigens in primary biliary cirrhosis. Am J Gastroenterol 1987; 82:1127-9.

161. Gores GJ, Moore SB, Fisher LD et al. Primary biliary cirrhosis: associations with class II major histocompatibility antigens. Hepatology 1987; 7:889-92.
162. Underhill J, Donaldson P, Bray G et al. Susceptiblity to primary biliary cirrhosis is associated with the HLA-DR8-DQB1*0402 haplotype. Hepatology 1992; 16:1404-8.
163. Maeda T, Onishi S, Saibara T et al. HLA DRw8 and primary biliary cirrhosis. Gastroenterology 1992; 103:1118.
164. Gregory WL, Mehal W, Dunn AN et al. Primary biliary cirrhosis: contribution of HLA class II allele DR8. Q J Med 1993; 86:393-99.
165. Begovich AB, Klitz W, Moonsamy PV et al. Genes within HLA class II region confer both predisposition and resistance to primary biliary cirrhosis. Tissue Antigens 1994; 43:71-7.
166. Underhill JA, Donaldson PT, Doherty DG et al. HLA DPB polymorphism in primary sclerosing cholangitis and primary biliary cirrhosis. Hepatology 1995; 21:959-62.
167. Tanaka A, Borchers AT, Ishibashi H et al. Genetic and familial considerations of primary biliary cirrhosis. Am J Gastroenterol 2001; 96:8-15.
168. Stone J, Wade JA, Cauch-Dudek K et al. Human leukocyte antigen Class II associations in serum antimitochondrial antibodies (AMA)-positive and AMA-negative primary biliary cirrhosis. J Hepatol 2002; 36:8-13.
169. Agarwal K, Jones DE, Watt FE et al. Familial primary biliary cirrhosis and autoimmune cholangitis. Dig Liver Dis 2002; 34:50-2.
170. Gregory WL, Daly AK, Dunn AN et al. Analysis of HLA-class-II-encoded antigen-processing genes *TAP1* and *TAP2* in primary biliary cirrhosis. Q J Med 1994; 87:237-244.
171. Briggs DC, Donaldson PT, Hayes P et al. A major histocompatibility complex class III allotype (C4B2) associated with primary biliary cirrhosis (PBC). Tissue Antigens 1987; 29:141-45.
172. Manns MP, Bremm A, Scheider PM et al. HLA-DRw8 and complement C4 deficiency as risk factors in primary biliary cirrhosis. Gastroenterology 1991; 101:1367-73.
173. Mehal WZ, Gregory WL, Lo Y-MD et al. Defining the Immunogenetic Suceptibility to Primary biliary cirrhosis. Hepatology 1994; 20:1213-19.
174. Gordon MA, Oppenheim E, Camp NJ et al. Primary biliary cirrhosis shows associations with genetic polymorphisms of tumour necrosis factor alpha promoter region. J Hepatol 1998; 31:242-7.
175. Jones DEJ, Watt FE, Grove J et al. Tumor necrosis factor-alpha promoter polymorphisms in primary biliary cirrhosis. J Hepatol 1999; 30:232-236.
176. Tanaka A, Quaranta S, Mattalia a et al. The tumor necrosis factor-alpha promoter correlates with progression of primary biliary cirrhosis. J Hepatol 1999; 30:826-9.
177. Donaldson P, Agarwal K, Craggs A et al. HLA and interleukin-I gene polymorphisms in primary biliary cirrhosis; associations with disease progression and disease susceptibility. Gut 2001; 48:397-402.
178. Zappala F, Grove J, Watt FE et al. No evidence for involvement of the interleukin-10 -592 promoter polymorphism in genetic suscepbility to primary biliary cirrhosis. J Hepatol 1998; 28:820-23.
179. Donner H, Rau H, Walfish PG et al. CTLA-4 alanine-17 confers genetic susceptiblity to Graves' Disease and to Type 1 Diabetes Mellitus. J Clin Endocrinol Metab 1997; 82:143-46.
180. Kristiansen OP, Larsen ZM, Pocoit F. *CTLA-4* in autoimmune diseases—a general susceptibility gene to autoimmunity? Gen Immunity 2000; 1:170-184.
181. Walker LS, Ausebel LJ, Chodos A et al. Ctla-4 differentially regulates T cell responses to endogenous tissue protein versus exogenous immunogen. J Immunol 2002; 169:6202-9.
182. Maurer M, Loserth S, Kolb-Maurer A et al. A polymorphism in the human cytotoxic T-lymphocyte antigen 4 (CTLA4) gene (exon1 +49) alters T-cell activation. Immunogenetics 2002; 54:1-8.
183. Agarwal K, Jones DEJ, Daly A et al. CTLA-4 gene polymorphism confers susceptibility to primary biliary cirrhosis. J Hepatol 2000; 32:538-541.
184. Wyllie S, Seu P, Goss JA. The natural resistance-associated macrophage protein 1 slc11a1 (formerly Nramp1) and iron metabolism in macrophages. Microbes Infect 2002; 4:351-9.
185. Matsushita M, Miyakawa H, Tanaka A et al. Single nucleotide polymorphisms of the mannose-binding lectin are associated with susceptibility to primary biliary cirrhosis. J Autoimmun 2001; 17:251-7.
186. Vogel A, Strassburg CP, Manns MP. Genetic association of vitamin D receptor polymorphisms with primary biliary cirrhosis and autoimmune hepatitis. Hepatology 2002; 35:126-31.
187. Corpechot C, Benlian P, Barbu V et al. Apolipoprotein E polymorphism, a marker of disease severity in primary biliary cirrhosis? J Hepatol 2001; 35:324-8.
188. James OF, Bhopal R, Howel D et al. Primary biliary cirrhosis once rare, now common in the United Kingdom? Hepatology 1999; 30:390-4.
189. Metcalf JV, James OFW. The geoepidemiology of primary biliary cirrhosis. Seminars in Liver Disease 1997; 17:13-22.

190. Prince MI, Chetwynd A, Diggle P et al. The geographical distribution of primary biliary cirrhosis is a well-defined cohort. Hepatology 2001; 34:1083-8.
191. Ohba K, Omagari K, Kinoshita H et al. Primary biliary cirrhosis among atomin bomb survivors in Nagasaki, Japan. J Clin Epidemiol 2001; 54:845-50.
192. Howel D, Fischbacher CM, Bhopal RS et al. An exploratory population-based case-control study of primary biliary cirrhosis. Hepatology 2000; 31:1055-60.
193. Obermayer-Straub P, Strassburg CP, Manns MP. Autoimmune hepatitis. J Hepatol 2000; 32(1 Suppl):181-97.
194. Belloc C, Gauffre A, Andre C et al. Epitope mapping of human CYP1A2 in dihydralazine-induced autoimmune hepatitis. Pharmacogenetics 1997; 7:181-6.
195. Eliasson E, Gardner I, Hume-Smith H et al. Interindividual variability in P450-dependent generation of neoantigens in halothane hepatitis. Chem Biol Interact 1998; 6:123-41.
196. Sasaki M, Ansari A, Pumford N et al. Comparative immunoreactivity of anti-trifluoroacetyl(TFA) antibody and anti-lipoic acid antibody in primary biliary cirrhosis: searching for a mimic. J Autoimmun 2000; 15:51-60.
197. Hoet P, Graf ML, Bourdi M et al. Epidemic of liver disease caused by hydrochlorofluorocarbons used as ozone-sparing substitutes of chlorofluorcarbons. Lancet 1997; 350:556-9.
198. Long SA, Quan C, Van de Water J et al. Immunoreactivity of organic mimeotopes of the E2 component of pyruvate dehydrogenase connecting xenobiotics with primary biliary cirrhosis. J Immunol 2001; 167:2956-63.
199. Sonoda J, Xie W, Rosenfeld JM et al. Regulation of a xenobiotic sulfonation cascade by nuclear pregnane X receptor (PXR). Proc Nat Acad Sci USA 2002; 99:13801-6.
200. Jorgensen R, Angulo P, Dickson ER et al. Results of long-term ursodiol treatment for patients with primary biliary cirrhosis. Am J Gastroenterol 2002; 97:2647-50.
201. Poupon RE, Balkau B, Eschwege E et al. A multicenter controlled trial of ursodiol for the treatment of primary biliary cirrhosis. N Eng J Med 1991; 324:1548-54.
202. Olson JK, Croxford JL, Miller SD. Virus-induced autoimmunity: potential role of viruses in initiation, perpetuation and progression of T cell mediated autoimmune disease. Viral Immunol 2001; 14:227-50.
203. Garry RF, Fermin CD, Hart DJ et al. Detection of a human intracisternal A-type retroviral particle antigenically related to HIV. Science 1990; 250:1127-9.
204. Talal N, Dauphinee MJ, Dang H et al. Detection of antibodies to retroviral proteins in patients with primary Sjogren's syndrome (autoimmune exocrinopathy). Arthritis Rheum 1990; 33:774-81.
205. Talal N, Garry RF, Schur PH et al. A conserved idiotype and antibodies to retroviral proteins in systemic erythematosus. J Clin Invest 1990; 85:1866-71.
206. Mason AL, Xu L, Guo L et al. Detection of retroviral antibodies in primary biliary cirrhosis and other idiopathic biliary disorders. Lancet 1998; 351:1620-4.
207. Sadamoto T, Joplin R, Keogh A et al. Expression of pyruvate dehydrogenase complex PDC-E2 on biliary epithelial cells induced by lymph nodes from primary biliary cirrhosis. Lancet 1998; 352:1595-96.
208. Mason A, Nair S. Primary biliary cirrhosis: new thoughts on pathophysiology and treatment. Curr Gastroenterol Rep 2002; 4:45-51.
209. Klein R, Wiebel M, Engelhart S et al. Sera from patients with tuberculosis recognise the M2a-epitope (E2-subunit of pyruvate dehydrogenase) specific for primary biliary cirrhosis. Clin Exp Immunol 1993; 92:308-316.
210. Vilagut L, Vila J, Vinas O et al. Cross-reactivity of anti-*Mycobacterium gordonae* antibodies with the major mitochondrial autoantigens in primary bilary cirrhosis. J Hepatol 1994; 21:673-677.
211. Vilagut L, Pares A, Vinas O et al. Antibodies to mycobacterial 65-kD heat shock protein cross-react with the main mitochondrial antigens in patients with primary biliary cirrhosis. Eur J Clin Invest 1997; 27:667-72.
212. Jones DEJ, Palmer JM, Leon MP et al. T-cell responses to Tuberculin-purified protein derivative in primary biliary cirrhosis: evidence for defective T-cell function. Gut 1997; 40:277-83.
213. O'Donohue J, Fidler H, Garcia-Barcelo M et al. Mycobacterial DNA not detected in liver sections from patients with primary biliary cirrhosis. J Hepatol 1998; 28:433-8.
214. Butler P, Hamilton-Miller J, Baum H et al. Detection of M2 antibodies in patients with recurrent urinary tract infection using an ELISA and purified PBC specific antigens. Evidence for a molecular mimicry mechanism in the pathogenesis of primary biliary cirrhosis. Biochem Mol Biol Int 1995; 35:473-85.
215. Burroughs AK, Rosenstein IJ, Epstein O et al. Bacteriuria and primary biliary cirrhosis. Gut 1984; 25:133-7.

216. Floreani A, Bassendine MF, Mitchison HC et al. No specific association between primary biliary cirrhosis and bacteriuria. J Hepatol 1989; 8:201-7.
217. Lin R-H, Mamula MJ, Hardin JA et al. Induction of autoreactive B cells allows priming of autoreactive T cells. J Exp Med 1991; 173:1433-.
218. Mamula MJ, Lin RH, Janeway CA et al. Breaking T cell tolerance with foreign and self co-immunogens: a study of autoimmune T and B cell epitopes of cytochrome c. J Immunol 1992; 149:789-795.
219. Jones DEJ, Palmer JM, Yeaman SJ et al. Breakdown of tolerance to pyruvate dehydrogenase complex in experimental autoimmune cholangitis: a mouse model of primary biliary cirrhosis. Hepatology 1999; 30:65-70.
220. Jones DE, Palmer JM, Bennett K et al. Investigation of a mechanism for accelerated breakdown of immune tolerance to the primary biliary cirrhosis-associated autoantigen, pyruvate dehydrogenase complex. Lab Invest 2002; 82:211-9.
221. Palmer JM, Kirby JA, Jones DE. The immunology of primary biliary cirrhosis: the end of the beginning? Clin Exp Immunol 2002; 129:191-7.
222. Jones DE, Palmer JM, Burt AD et al. Bacterial motif DNA as an adjuvant for the breakdown of immune self-tolerance to pyruvate dehydrogenase complex. Hepatology 2002; 36:679-86.
223. Tanaka A, Prindiville TP, Gish R et al. Are infectious agents involved in primary biliary cirrhosis? A PCR approach. J Hepatol 1999; 31:664-71.
224. Hiramatsu K, Harada K, Tsuneyama K et al. Amplification and sequence analysis of partial bacterial 16S ribosomal RNA gene in gallbladder bile from patients with primary biliary cirrhosis. J Hepatol 2000; 33:9-18.
225. Nilsson HO, Taneera J, Castedal M et al. Identification of Helicobacter pylori and other Helicobacter species by PCR, hybridization, and partial sequencing in human liver samples from patients with primary sclerosing cholangitis or primary bilairy cirrhosis. J Clin Microbiol 2000; 38:1072-6.
226. Harada K, Tsuneyama K, Sudo Y et al. Molecular identification of bacterial 16S ribosomal RNA gene in liver tissue of primary biliary cirrhosis: is Propionibacterium acnes involved in granuloma formation? Hepatology 2001; 33:530-6.
227. Yamada T, Eishi Y, Ikeda S et al. In situ localization of Propionibacterium acnes DNA in lymph nodes from sarcoidosis patients by signal amplification with catalysed reporter deposition. J Pathol 2002; 198:541-7.

CHAPTER 18

Primary Sclerosing Cholangitis

Einar Björnsson and Roger W. Chapman

Introduction

Primary sclerosing cholangitis (PSC) is a chronic cholestatic liver disorder of the intrahepatic and/or extrahepatic bile ducts. PSC is characterized by concentric obliterative fibrosis and bile duct strictures. Its course is very variable and can in individual cases follow a benign course but for the most part, PSC is a progressive disease and leads eventually to cirrhosis and can also lead to cholangiocarcinoma. It is associated with inflammatory bowel disease (IBD) in the majority of cases. The etiology and the pathogenesis of PSC is still unknown. Autoimmune mediated liver injury has been proposed due to association with HLA haplotypes, detection of autoantibodies and the association with other autoimmune diseases. Other non-immune factors such as molecular alterations of tight junctions of the biliary epithelial cells, bacterial infections, ischemia and toxicity have also been proposed to be of significance in the etiopathogenesis of PSC.

Clinical, Biochemical and Serological Findings

The clinical course of PSC is highly variable. Patients with IBD may have cholestatic liver function tests, without symptoms or patients may present with attacks of jaundice or upper abdominal pain and fever and /or pruritus.[1-3] The median survival to death or liver transplantation has been estimated to be 12 years in studies from Sweden,[1] as well as from the United States[2] and England.[3] In the largest study,[1] 66% of patients were symptomatic at diagnosis but 22% of those originally asymptomatic developed symptoms during the follow-up period. In a long term follow-up study of the natural history of PSC in 305 patients with PSC, 26% died of cholangiocarcinoma which was diagnosed after a median time of 32 months after diagnosis. The biochemical profile of PSC patients is cholestatic with alkaline phosphatase (ALP) activity elevation of at least three times the upper limit.[4-6] The ALP characteristically fluctuates during the course of the disease and can return to normal periodically in some patients.[7] In most cases moderate activity of ALT and AST is observed, approximately 2-3 times the upper limit.[4-6] However, synthetic function of the liver as estimated by serum albumin and coagulations factors (INR) is normal in most patients at the time of diagnosis.[4-6] Moreover, in most cases bilirubin is normal at the diagnosis, especially in those who present with IBD.[5] Perinuclear antineutrophilic antibodies are present in 26-85% of patients.[8] These are however not related to disease activity. Moreover, antinuclear antibodies (ANA) and smooth muscle cell antibodies (SMA) are present in low titres in a minority of PSC patients.[4] Elevated levels of circulating immunoglobulins, both IgG and IgM are detected in almost half of patients.[9,10] No specific disease marker has however been observed and the diagnosis is based on typical cholangiographic findings with exclusion of secondary sclerosing cholangitis and typical histological features.[11]

Molecular Pathogenesis of Cholestasis, edited by Michael Trauner and Peter L.M. Jansen.
©2004 Eurekah.com and Kluwer Academic / Plenum Publishers.

Etiopathogenesis

Any proposed hypothesis for the etiology and the pathogenesis of PSC needs to take into account the diverse and focal involvement of the biliary tree. In most of the cases both extra- and intrahepatic bile ducts are involved as identified by cholangiography. However, in some cases only changes in the small bile ducts are identified by histology and in other cases only the large extrahepatic bile ducts can be affected. The association with IBD as well as the strong association with HLA haplotypes and the high prevalence of autoantibodies have led to the hypothesis that PSC is an autoimmune disease. Furthermore, association with other classical autoimmune diseases and altered T cell function have led to suggestion of an immune medi- ated cause of PSC. However, at least in one third of the patients no association is found with HLA haplotypes. Moreover, the autoantigen and the immune mechanism for the disorder have not been identified.[12,13] Thus, it remains controversial whether PSC should be regarded an autoimmune disease.[14] It is clear that immunological factors seem to play a part in the pathogenesis of PSC[14-16] but the primary cause of the activation of the immune system is unknown.

Any proposed mechanism for the induction of the immune system in PSC has also to take into consideration the strong association with IBD. The association with IBD has been used to support an autoimmune hypothesis for PSC but recent observations of importance of the colonic flora in the etiology of ulcerative colitis and the interplay between bacterial antigens and the immune system has put forward an alternative hypothesis of the initial event in the pathogenesis of PSC.[13] According to this hypothesis, the primary event is the reaction of parts of the immune system in the immunogenitically susceptible host to bacterial cell products, brought to the liver via diseased colon either in the IBD patient or during gastroenteritis. Animal experiments support a hypothesis of a possible bacterial etiology of PSC.[17-19] The evidence will be reviewed below.

Molecular Alterations and Defects in PSC

Tight junctions (TJ) of biliary epithelial cells (BEC) have previously been shown to be defective in many types of cholestasis[20] such as in primary biliary cirrhosis (PBC).[21] These changes increase the likelihood of biliary regurgitation from the biliary tract to the periductal area. Sakisaka et al have recently investigated two proteins of the tight junction of both BEC and hepatocytes in frozen sections from livers of patients with PBC and PSC.[22] Significant alterations were demonstrated in these TJ proteins, ZO-1 and 7H6 in both PBC and PSC. Changes were present predominantly in bile ducts in PBC but in contrast in hepatocytes in PSC. These findings are suggestive of increased paracellular permeability through different pathways in PBC and PSC. However, it cannot be excluded that changes of TJ in hepatocytes from PSC patients might be secondary changes due to impaired flow of bile through narrowed and strictured biliary tree in these patients.

Trinitrobenzene sulphonic acid (TNB)-induced colitis is the only rat model of IBD that gives concomitant intrahepatic cholestasis.[23] TJ proteins were examined in a rat model of TNB- induced colitis, with focus on portal endotoxins and proinflammatory cytokines, in order to elucidate the importance of these factors in the etiology of cholestasis.[24] The levels of portal endotoxins but not those of proinflammatory peptides were increased in the early stages of cholestasis. Furthermore, pretreatment with polymyxin B which binds to and neutralizes en- dotoxins, could prevent alterations in TJ proteins associated with intrahepatic cholestasis. These findings are suggestive of the importance of gut derived endotoxins, from inflamed colon, in the induction of cholestasis in this model of IBD and cholestasis. However, this animal model of IBD is only available in the acute setting of IBD and a chronic IBD model with cholestasis would be important for future studies to reveal the pathogenesis of the association with cholestasis in IBD.

In another study in TNB-induced colitis, a low level of portal endotoxins were demon- strated during the early phase of colitis.[25] Bile acid secretion and cytochrome P-450 metabolic capacity were reduced in the TNB-induced colitis model.[25]

Role of Cystic Fibrosis-Mediated Chloride Channel Function

Some patients with cystic fibrosis (CF) develop liver disease with histological and radiological changes similar to those observed in PSC. Patients with CF have defective chloride secretion and sodium reuptake in epithelial cells because of mutations in the cystic fibrosis transmembrane conductance regulator (CFTR) gene. Potential difference (PD) across nasal epithelial cells can be used as a measure of CFTR regulated ion transport in humans. Sheth et al found evidence of CFTR-mediated ion transport dysfunction with decreased nasal PD in 8 out of 16 patients with PSC.[26] Altered ion transport in the bile duct epithelial cells could therefore play a role in some patients with PSC. The association between PSC and CF carrier state has been further characterized by the same group, with sophisticated genotypic and phenotypic analysis.[27] CFTR mutations were demonstrated in 6 of 18 (33%) PSC patients. All of those 6, had mutations on one allele and all had as well IBD. Furthermore, nasal PD response was reduced in half of PSC patients studied. Combining genotypic and phenotypic data revealed CFTR abnormalities in 67% of PSC patients compared with 17% of disease controls (27). CF carrier state could therefore predispose to the development of PSC, as mutations were observed in one allele and the findings might explain why a subset of IBD patients develops PSC. Based on the hypothesis that carrier state of CF could predispose to development of PSC in IBD patients, a CF knockout mice model has recently been used to elucidate this association (28). It was hypothesized that the induction of colitis in homozygous (CFTR -/-) and heterozygous (CFTR +/-) CF knockout mice would lead to bile duct injury. Colitis was induced by dextran sodium sulphate and biochemical and histological features of intrahepatic cholestasis examined. Both the homozygous and heterozygous knockout mice with colitis had significantly more severe bile duct injury, compared with those without colitis. Moreover, portal tract injury characterized by acute and chronic inflammation was noted only in the CFTR (-/-) mice. Furthermore, alkaline phosphatase as a measure of biochemical cholestasis was significantly higher in homozygous knockout mice compared with the control animals. Thus, colitis in the setting of single CFTR mutation is enough to evoke bile duct injury in mice. These findings further support the human data, with CFTR mutations on one allele that were shown to predispose to the development of PSC in patients with IBD.[27] However, a study from Italy could not demonstrate any association between PSC and CFTR genes in 64 patients from the North of Italy.[28] HLA-DRB1 and tumor necrosis factor were found to contribute to the genetic susceptibility of primary sclerosing cholangitis.[28] Further studies on the association between the CFTR abnormalities and PSC are clearly needed.

Animal Models of PSC

Experimental Induction of PSC-Like Pathology

A number of animal models have suggested that bacterial products from the inflamed gut can lead to immune response in the liver leading to PSC-like pathology. Recent animal experiments have shown that proinflammatory peptide synthesized by colonic bacteria in rats with chemically induced colitis induced histological changes similar to those seen in patients with PSC.[17] Moreover, non-pathogenic *E. coli* infused into the portal blood in rabbits caused portal fibrosis.[18] Furthermore, a model of chemically induced colitis in rabbits is also associated with histological changes similar to the pathological lesions in PSC patients.[30] Similar histological features as those reported in PSC are also observed in a rat model of E. coli chemotactic peptide-induced colitis.[31] Small bowel bacterial overgrowth induced in rats has recently been shown to lead to strictures in the biliary tract and portal inflammation which are the hallmarks of the pathology of PSC.[19] In this particular model, phagocytosis of bacterial wall parts were considered to be the initial event leading to bile duct injury as the blockage of liver macrophage phagocytosis could prevent liver injury. Moreover, cleaving of the bacterial cell wall peptidoglycan-polysaccharide by the enzyme mutanolysin could also prevent the inflammation and liver injury. The liver injury was directly correlated to tumor necrosis factor TNF-α production by the hepatic macrophages. Furthermore, hepatobiliary inflammation was only present in the genetically susceptible strain of rats which is in line with human data showing association with HLA haplotypes.[12] Recently, in vitro data from studies of white blood cells from PSC patients have demonstrated higher levels of TNF-α[32] which is in accordance with the rat model with bacterial overgrowth.[19] A novel rat model of fibrosing cholangitis, has been described in which 2, 4, 6-trinitrobenzenesulphonic acid, injected in bile ducts caused cholangiographic abnormalities similar to those seen in PSC.[33] Furthermore, pathological infiltration of T-cells and an upregulation of MHC class II in bile ducts was observed as expressed in human PSC.

Experimental Bile Ductular Obstruction

Biliary strictures in patients with PSC are by definition associated with cholestasis and portal inflammation and concentric obliterative fibrosis also contribute to biliary obstruction. Cholestasis in PSC has been shown to lead to derangement in the cellular immune system, with an increase in activated, autoreactive T-lymphocytes.[34] Immunoregulation is thus clearly disturbed in PSC and animal models of biliary obstruction could therefore increase the understanding of the immunological features present in PSC. Bile obstruction induced experimentally, is associated with a number of changes in immunological expression as well as of all kinds of molecular alterations.[35] Among the most important events are activation of hepatic macrophages by LPS and increased TNF-α expression as well as, IL-1, IL-6 and leukotrienes. Furthermore, the biliary obstruction is associated with changes of tight junctions in biliary epithelial cells as seen in hepatocytes in livers from PSC patients.[21] The alterations in tight junctions were induced by TNF-α which seems to be important for liver injury of biliary obstruction in experimental model of PSC[19] and by T lymphoctes from PSC patients.[32] Furthermore, biliary epithelial cells are stimulated to secrete chemokines and cytokines which attract different populations of white blood cells. This could lead to activation of myofibroblasts (stellate cells) which have been shown to enzymatically digest extracellular matrix and lead to increased synthesis of collagen resulting in peribiliary cirrhosis.[36] In a rat model of biliary obstruction, increased secretion of platelet-derived growth factor by cholangiocytes can promote activation of hepatic stellate cells, leading to fibrosis and eventually cirrhosis.[37]

Role of Bacteria in Human Disease

Because of the well-known association between PSC and ulcerative colitis, many investigators have speculated that portal bacteremia or bacterial products or toxins from the inflamed colon could be of importance in the pathogenesis of PSC.[38,39]

Bacterial cultures obtained from portal venous blood and from liver biopsies from patients with IBD have revealed a number of different bacteria only present in a minority of cultures which failed to support the hypothesis that bacteria play a role in the etiopathogenesis of the liver damage observed in IBD.[40,41] However, it is unclear whether the patients involved in these studies really had PSC as these were performed before endoscopic retrograde cholangiography (ERC) was developed as a clinical tool. Recently, study of explanted livers revealed striking difference in bacterial cultures in patients with PSC compared with PBC patients.[43] Bile from bile duct walls and bile collected from explanted liver at the time of liver transplantation revealed positive bacterial cultures from 21 of 36 PSC patients compared with none of 14 PBC patients. The number of bacterial strains was inversely related to the time after the last ERC, suggesting that the bacterial load was due to contamination from cannulation of the bile ducts at a previous ERC. Interestingly, alpha-haemolytic streptococci were the commonest type of bacteria obtained in the PSC patients. In patients who had an ERC without prophylactic antibiotic treatment, and therefore probably a significantly higher number of bacterial isolates than those patients who had antibiotic treatment in connection with previous ERC(s), the duration of PSC before liver transplantation was shorter, which is suggestive of a deleterious effect of the bacterial contamination on liver function. A high positivity rate of bacterial cultures of alpha-haemolytic streptococci in PSC patients, suggested an etiopathogenic of that particular bacteria in PSC. A study of bile duct bacterial isolates obtained during a diagnostic ERC was performed by the same group in order to elucidate the role of alpha-haemolytic streptococci in this disorder.[44] A comparison between "naive" PSC patients with those patients that had previously undergone ERC, revealed that most "naive" patients were found to have negative bacterial cultures but bacterial cultures were positive in 60% of those patients previously investigated with ERC. However, the occurrence of bacteria in the bile ducts in these patients, with a possible defective clearance of bacteria in the bile, might have negative effects on liver function, whilst not playing a primary role in the progression of PSC. In the original study of the explanted livers,[43] the shorter duration of PSC before liver transplantation is suggestive of deleterious effect of the presence of bacteria in the bile ducts of these patients.

In a rat model of chronic active hepatitis, a novel helicobacter species was observed (*Helicobacter hepaticus*).[45] Human studies have also demonstrated that various helicobacter species can infect both intrahepatic and extrahepatic bile ducts.[46] Recently, Nilsson et al[47] reported that most patients with PSC and Primary bilary cirrhosis (PBC) were positive for *H. pylori* in liver tissue by polymerase chain reaction (PCR). Gene sequences of *Helicobacter* species were found in 20 of 24 liver tissue samples of patients from PSC and PBC but not in non cholestatic livers. The similar positivity rate of *H. pylori* in both PSC and PBC argues against a specific etiopathogenic role of *H. pylori* of these disorders. However, *H. pylori* might have a triggering effect where the response might be modified by host factors. Theoretically, it might be difficult for patients with PSC to clear bacteria that are brought to the liver from their bile due to narrowing of the biliary tract and and biliary strictures.

Role of Autoimmunity

Immunogenetics

A number of genes seem to be involved in the etiopathogenesis of PSC. Several studies have demonstrated association with the genes of the major histocompatibility complex which are involved in the regulation of the immune response in comparing self and non-self.[12,14,48,49] Early studies which were based on HLA serotyping demonstrated increased frequency of HLA-B8 and HLA-DR3 in PSC compared with a control population. Furthermore, a strong negative association with HLA-B44-DR4 was reported.[48] The HLA B8-DR3 haplotype is also associated with a number of classical autoimmune disorders such as diabetes mellitus, thyrotoxicosis and autoimmune hepatitis. The strong association of PSC and HLA, does support the hypothesis that immunological factors play an important role in the etiopathogenesis of PSC.[14] More recent studies, applying molecular genotyping, have confirmed the positive association with DR3 and the negative association with DR4 and identified two susceptibility haplotypes, A1-B8-DRB3*0101DRB1*0103-DQB1*0603 and DRB*0101-DRB1*1301-DQA1*0103-DQB1*0603.[50,51] As the emphasis of this book is more on the metabolic than on the immunological aspects of cholestasis, a detailed discussion on the immunogenteics is referred to elsewhere.[12]

Non-Immune Genes

HLA associations do only account for genetic susceptibility in 50-55% of cases of PSC, and non-HLA genes have also attracted research interest in this area.[52] No association was found between the interleukin 1 family and the susceptibility or resistance to PSC.[52] Association has been reported between tumor necrosis factor polymorphism and PSC.[53] Recently, TNF-α 2 allele was reported to be significantly more common in PSC than in healthy controls.[29] This polymorphism is associated with a functional increase in TNF-α production. The association of HLA DR3 with increased TNF-α production is probably due to linkage disequilibrium.[54]

Humoral-Mediated Immune Abnormalities-Autoantibodies

PSC is associated with a number of humoral immune abnormalities. Perinuclear antineutrophilic antibodies (ANCA) have been reported to be present in 26-85% of patients with PSC, with or without IBD and up to 68% in ulcerative colitis alone.[8,55] ANCA does not correlate with disease activity and does not fall after colectomy or liver transplantation[56] but has been significantly associated with extensive involvement of the biliary tree.[57] However, ANCA is probably of little significance in the etiopathogenesis of PSC. Although a number of other autoantibodies have been reported in PSC in a higher frequency than in healthy controls,[4,10] the significance of these antibodies is unclear. A recent study examined the profile and significance of several autoantibodies in PSC.[58] PSC patients had a greater rate of positivity than controls for antinuclear, anticardiolipin, ANCA, antithyroperoxidase antibodies as well as rheumatoid factor.[58] Anticardiolipins were the only antibodies that had significant correlation with clinical and histological stage. Thus, most autoantibodies are probably nonspecific markers of disturbed immune regulation in PSC. Some anticolon antibodies have been shown to cross-react with antigens in biliary epithelium, suggestive of a shared antigen in colonic and biliary epithelia, although the importance of this finding remains unclear.[59]

Cellular-Mediated Immune Abnormalities

Infiltrates in the portal tracts in PSC consist mostly of CD4 T cells as well as neutrophils.[60] Studies in vitro of lymphocytes from patients with PSC, have suggested that there may be oligoclonal restriction of of T cells.[61] Recent studies have indicated that hepatic T cells in PSC patients have preferentially V beta 3 T cell repertoires which was not seen in peripheral blood.[61] Oligoclonal T cell receptors were observed in PSC patients and in cultures with enterocytes and they were found to be cytotoxic for enterocyte cell lines.[62] These findings give support for the hypothesis of a shared antigen in biliary and colonic epithelium,[59] recognized by the CD4 T cells.

Summary and Conclusions

There is substantial evidence in the literature to support the role of immune mechanisms in the pathogenesis of PSC. Support for the autoimmune pathogenesis has its origin in its HLA associations, high frequency of autoantibodies and associations with other autoimmune diseases. However at least in one third of the patients no association is found with HLA haplotypes and neither the autoantigen nor the immune mechanisms for the disorder have been identified. The unanswered question is whether the immunological factors are the cause or the consequence of the liver injury. A hypothesis of a portal bacteremia that could be involved in the pathogenesis of PSC has been put forward (Fig. 1).[13] According to this hypothesis the primary event is the reaction of the parts of the immune system in the immunogenitically susceptible host to bacterial cell products, brought to the liver via diseased colon either in the IBD patient or during gastroenteritis. Animal models support a hypothesis of a possible bacterial etiology of PSC. Increased levels of portal endotoxins but not those of proinflammatory peptides have been observed in the early stages of cholestasis in an experimental colitis with concomitant cholestasis. However, the relevance of the different animal models of experimental colitis and small bowel bacterial overgrowth, with the associated portal inflammation and bile duct strictures, for the human disease can be questioned. Moreover, chronic portal bacteremia has not been demonstrated in PSC patients, although they have been found to have a high frequency of positive bacterial cultures in liver explants at the time of liver transplantation and in aspirates of bile. Recently, a high frequency of the cystic fibrosis transmembrane conductance regulator (CFTR) gene has been reported in PSC. Mutations in the CFTR gene might predispose to the development of PSC. Animal models of CFTR knock-out mice with experimentally induced colitis support that hypothesis. However, none of the above mentioned immune and non-immune mechanisms seem to be involved in all patients. The starting point of the portal inflammation is still obscure and more work is clearly needed to reveal the etiopathogenesis of PSC.

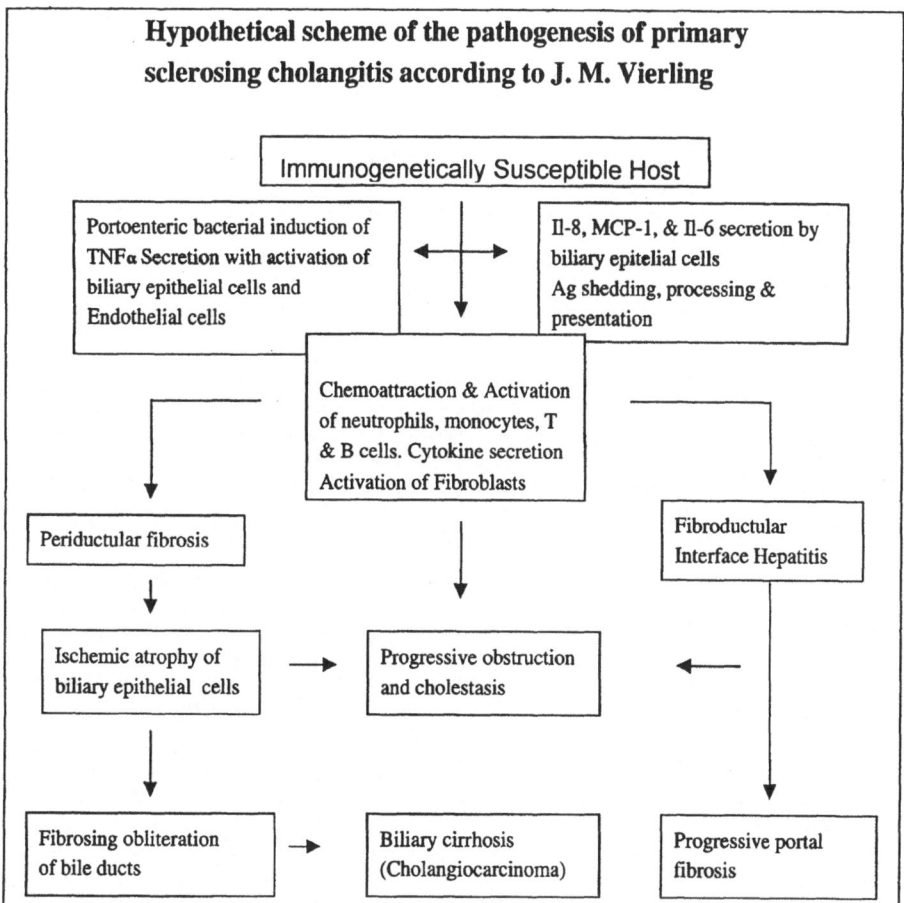

Figure 1. Hypothetical scheme of the pathogenesis of primary sclerosing cholangitis according to J. M. Vierling.

References

1. Broomé U, Olsson R, Lööf L et al. Natural history and prognostic factors in 305 patients with primary sclerosing cholangitis. Gut 1996; 38: 610-615.
2. Wiesner TH, Grambsch PB, Dickson ER et al. Primary sclerosing cholangitis: natural history, prognostic factors, and survival analysis. Hepatology 1989; 10: 430-6.
3. Farrant JM, Hayllar KM, Wikinson ML et al. Natural history and prognostic variables in primary sclerosing cholangitis. Gastroenterology 1991; 100: 1710-1717.
4. Chapman RW, Arborgh BM, Rhodes JM et al. Primary sclerosing cholangitis: a review of its clinical features, cholangiography and hepatic histology. Gut 1980; 21: 1870-1877.
5. Aadland E, Schrumpf E, Fausa et al. Primary sclerosing cholangitis: a long term follow-up study. Scand J Gastroenterol 1987; 22: 655-664.
6. Stockbrügger RW, Olsson R, Jaup B et al. Forty-six cases patients with Primary sclerosing cholangitis: radiological bile duct changes in relationship to clinical course and concomitant inflammatory bowel disease. Hepatogastroenterology 1988; 35: 289-294.
7. Balasubramaniam K, Wiesner RH, Larusso NF. Primary sclerosing cholangiti with normal serum alkaline phosphatase activity. Gastroenterology 1988; 95: 1395-1398.
8. Lo SK, Fleming KA, Chapman RW. Prevalence of anti-neutrophil antibodies in primary sclerosing cholangitis and ulcerative colitis using an alkaline phosphatase technique. Gut 1992; 33:1370-1375.
9. Schrumpf E, Fausa O, Kolmannskog F et al. Sclerosing cholangitis in ulcerative colitis. A follow-up study. Scand J Gastroenterol 1982; 17: 33-39.
10. Boberg KM, Fausa O, Haaland T et al. Features of autoimmune hepatitis in primary sclerosing cholangitis: an evaluation of 114 primary sclerosing cholangitis patients according to a scoring system for the diagnosis of autoimmune hepatitis. Hepatology 1996; 23: 1369-1376.

11. Luwig J. Histopathology of primary sclerosing cholangitis. In: Manns M, Chapman RW, Stiehl A, Wiesner R, Eds. Primary sclerosing cholangitis.London. kluwer Academic publishers 1998; 14-21.
12. Donaldsson PT. Immunogenetics and epidemiology of primary sclerosing cholangitis. In: Manns M, Chapman RW, Stiehl A, Wiesner R, Eds. Primary sclerosing cholangitis.London. kluwer Academic publishers 1998; 22-36.
13. Vierling J. Aetiopathogenesis of primary sclerosing cholangitis. In: Manns M, Chapman RW, Stiehl A, Wiesner R, eds. Primary sclerosing cholangitis. London: Kluwer Academic Publishers, 1998:22-36.
14. Chapman RW. Primary sclerosing cholangitis as an autoimmune disease: pros and cons. In: Manns MP, Paumgartner G, Leuschner U, eds. Immunology and Liver. London: Kluwer Academic Publishers, 2000:279-287.
15. MacSween RNM, Anthony PP, Scheuer PJ. In: Pathology of the liver. London: Churchill Livingstone, 1987; 870-877.
16. Crippin JS, Lindor KD. Primary sclerosing cholangitis; etiology and immunology. Eur J Gastroenterol Hepatol 1992; 4:261-265.
17. Hobson CH, Butt TJ, Ferry DM, Hunter J, Chadwick VS, Broom MF. Enterohepatic circulation of bacterial chemotactic peptide in rats with experimental colitis. Gastroenterology 1988; 94:1006-1013.
18. Kono K, Ohnishi K, Omata et al. Experimental portal fibrosis produced by intrportal injection of killed non-pathogenic Escherichia Coli in rabbits. Gastroenterology 1988; 94: 787-796.
19. Lichtman SN, Okoruwa EE, Keku J et al. Degradation of endogenous bacterial cell wall polymers by the muralytic enzyme mutanalysis prevents hepatobiliary injury in the genetically susceptible rats with experimental intestinal bacterial overgrowth. J Clin Invest 1992; 90: 1313-1322.
20. Trauner M, Meier PJ, Boyer JL. Molecular pathogenesis of cholestasis. N Engl J Med 1998; 339:1217-1227.
21. Nakanuma Y, Tsuneyama, Gershwin ME, Yasoshima M. Pathology and immunopathology of primary biliary cirrhosis with emphasis on bile duct lesions: recent progress. Semin Liver Dis 1995; 15:313-328.
22. Sakisaka S, Kawaguchi T, Taniguchi et al. Alterations in tight junctions differ between primary biliary cirrhosis and primary sclerosing cholangitis. Hepatology 2001; 33: 1460-1468.
23. Lora L, Mazzon E, Martines D, Fries W, Muraca M, Martin A, d'Odorico A et al. Hepatocyte tight-junctional permeability is increased in rat experimental colitis. Gastroenterology 1997; 113:1347-1354.
24. Kawaguchi T, Shotoaro S, Mitsuyama K, Harada M et al. Cholestasis with altered structure and function of hepatocytes tight junction and decreased expression of canalicular multispecific organic anion transporter in a rat model of colitis. Hepatology 2000; 31: 1285-1295.
25. Weidenbach H, Leiz S, Nussler AK, Dikopoulos N et al. Disturbed bile secretion and cytochrome P-450 function during the acute state of experimental colitis in rats. J Hepatol 2000; 32: 708-717.
26. Sheth SG, Bishop MD, Shea JC, Hopper IK et al. Primary sclerosing cholangitis is associated with abnormalities in cystic fibrosis-mediated chloride channel function. Gastroenterology 2000; 118:A-187.
27. Sheth SG, Bishop MD, Shea JC, Hopper IK et al. Association of primary sclerosing cholangitis with the cystic fibrosis carrier state: Results of extensive genotypic and phenotypic analysis. Gastroenterology 2001; 119: A-405.
28. Zaman MM, Blanco PG, Yantiss RY, Nasser IA et al. Bile duct injury in CF knockout mice with colitis: Further evidence of an association between CFTR mutations and primary sclerosing cholangitis. Gastroenterology 2001; 119. A-339.
29. Cavestro GM, Seghini P, Neri TM, Zanelli PF et al. HLA-DRB1 and tumor necrosis factor A but not cystic fibrosis transmembrane conductance regulator genes contribute to the genetic susceptibility of primary sclerosing cholangitis. Gastroenterology 2001; 119. A-2912.
30. Kuroe K, Haga Y, Funakoshi O et al. Pericholangitis in a rabbit colitis model induced by injection of muramyl dipeptide emulsified with a long-chain fatty acid. J Gastroenterol 1996; 31: 347-352.
31. Yamada S, Ishii M, Liang LS et al. Small-duct cholangitis induced by N-formyl-L-methionine- L-leucine L-tyrosine in rats. J Gastroenterol 1994; 29: 631-636.
32. Bo X, Broome U, Remberger M, Sumitran-Holgersson S. Tumour necrosis factor alpha impairs function of liver derived T lymphocytes and natural killer cells in patients with primary sclerosing cholangitis. Gut 2001; 49: 131-141.
33. Orth T, Neurath M, Schirmacher P et al. A novel rat model of chronic fibrosing cholangitis induced by local administration of a hapten reagent into the dilated bile duct is associated with increased TNF-alpha production and autoantibodies. J Hepatol 2000; 33: 862-872.
34. Lindor KD, Wiesner RH, LaRusso NF, Homburger HA. Enhanced autoreactivity of T-lymphocytes in primary sclerosing cholangitis. Hepatology 1987; 7: 884-888.
35. Desmet V, Roskams T, Van Eyken P. Ductular reaction in the liver. Pathol Res Pract 1995; 191:513-524.
36. Arthur MJP. Fibrogenesis II. Metalloproteinases and their inhibitors in liver fibrosis. Am J Physiol 2000; 279:G245-G249.

37. Grappone C, Pinzani M, Parola M et al. Expression of platelet-derived growth factor in newly formed cholangiocytes during experimental biliary fibrosis in rats. J Hepatol. 1999; 31:100-109.
38. Vinnik IE, Kern F, Struthers JE et al. Experimental chronic portal vein bacteremia. Proc Soc Exp Biol Med 1964; 115: 311-314.
39. Palmer KR, Duerden BI, Holdsworth CD. Bacteriological and endotoxin studies of severe cases of ulcerative colitis submitted to surgery. Gut 1980; 21: 851-854.
40. Eade MN, Brooke BN. Portal bacteraemia in cases of ulcerative colitis submitted to colectomy. Lancet 1969; i: 1008-1009.
41. Perret AD, Higgins G, Johnston HH, Massarella GR, Truelove SC, Wright R. Q J Med 1971; 40:211-238.
42. Dew MJ, Thompson H, Allan RJ. The spectrum of hepatic dysfunction in inflammatory bowel disease. Q J Med 1979; 48: 113-135.
43. Olsson R, Björnsson E, Bäckman L, Frimann S, Höckerstedt K, Kajser B, Olausson M. Bile duct bacterial isolates in primary sclerosing cholangitis: a study of explanted livers. J Hepatol 1998; 28:426-432.
44. Björnsson E, Kilander A, Olsson R. Bile duct bacterial isolates in primary sclerosing cholangitis and certain other forms of cholestasis-A study of bile cultures from ERCP. Hepatogastroenterology 2000; 47:1504-1508.
45. Ward JM, Fox JG, Anver et al. Chronic active hepatitis and associated liver tumours in mice caused by a persistent bacterial infection with a novel Helicobacter species. J Natl Cancer Inst 1994; 86:1222-1227.
46. Lin TT, Yeh CT, Wu CS. Detection and partial sequence analysis of Helicobacter pylori DNA in the bile samples. Dig Dis Sci 1995; 40:2214-2219.
47. Nilsson H-O, Taneera J, Castedal M, Glatz E, Olsson R, Wadström T. Ide129-133. Identification of Helicobacter pylori and other Helicobacter species by PCR, hybridization abd partial DNA sequencing in human liver samples from patients with primary sclerosing cholangitis or primary biliary cirrhosis. J Clin Microbiol 2000; 38: 1072-1076.
48. Donaldsson PT, Farrant JM, Wilkinson ML, Hayllar K, Portmann BC, Williams R. Dual association of HLA DR2 and DR3 with primary sclerosing cholangitis. Hepatology 1991; 15:129-133.
49. Olerup O, Olsson R, Hultcrantz R, Broome U. HLA-DR and HLA-DQ are not markers for rapid disease progression in primary sclerosing cholangitis. Gastroenterology 1995; 108: 870-878.
50. Farrant JM, Doherty DG, Donaldsson PT et al. Amino acid substitution at position 38 of the DR beta polypeptide confer susceptibility to and protection from primary sclerosing cholangitis. Hepatology 1992; 16: 390-395.
51. Mehal WZ, LO YM, Wordsworth BP et al. HLA DR4 is a marker for rapid disease progression in primary sclerosing cholangitis. Gastroenterology 1994; 106: 160-167.
52. Donaldsson PT, Norris S, Constanti PK, Bernal W, Harrison P, Williams R. The interleukin-1 and interleukin-10 gene polymorphisms in primary sclerosing cholangitis; no association with disease susceptibility/resistance. J Hepatology 2000; 32: 882-886.
53. Bernal W, Mahoney M, Underhill J, Donaldsson PT. Association of tumour necrosis factor polymorpism with primary sclerosing cholangitis. J Hepatology 1999;30: 237-241.
54. Vasalli P. The pathophysiology of tumor necrosis factor. Ann Rev Immunol 1992; 10: 411-452.
55. Duerr RH, Targan SR, Landers CJ et al. Neutrophil cytoplasmic antibodies: a link between primary sclerosing cholangitis and ulcerative colitis. Gastroenterology 1991; 100: 1385-1391.
56. Lo SK, Fleming KA, Chapman RW. A 2-year follow-up study of anti-neutrophil antibody in primary sclerosing cholangitis: relationship to clinical activity, liver biochemistry and ursodeoxycholic acid. J Hepatology 1994; 21: 974-978.
57. Bansi DS, Fleming KA, Chapman RW. Importance of antrineutrophil cytoplasmic antibodies in primary sclerosing cholangitis and ulcerative colitis: Prevalence, titre, and IgG class. Gut 1996; 38:384-389.
58. Angulo P, Peter JB, Gershwin E, DeSotel CK, Shoenfeld Y, Ahmed AEE, Lindor KD. Serum autoantibodies in patients with primary sclerosing cholangitis. J Hepatology 2000; 32: 182-187.
59. Mandal A, Dasgupta A, Jeffers L, Squillante L, Hyder S, Reddy R, Schiff E, Das KM. Autoantibodies in primary sclerosing cholangitis against a shared peptide in biliary and colonic epithelium. Gastroenterology 1994; 106: 185 192.
60. Hashimoto E, Lindor KD, Homburger HA et al. Immunohistochemical chracterization of hepatic lymphocytes in PBC in comparison with PSC and autoimmune chronic active hepatitis. Mayo Clin Proc 1993; 68: 1049-1055.
61. Broome U, Grunewald J, Scheynius A et al. Preferential V beta 3 usage by hepatic T lymphocytes in patients with primary sclerosing cholangitis. J Hepatology 1997; 26:527-534.
62. Probert CS, Christ AD, Saubermann LJ et al. Analysis of human common bile duct-associated T-cells: Evidence of oligoclonality, T cell persistence, and epithelial cell recognition. J Immunol 1997; 158:1941-1948.

CHAPTER 19

Drug-Induced Cholestatic Liver Disease

Gerd A. Kullak-Ublick

Summary

D rug induced cholestatic liver disease is a subtype of liver injury that is characterized by predominant elevations of alkaline phosphatase and bilirubin secondary to the administration of a hepatotoxic agent. It can manifest itself as a cholestatic hepatitis or as bland cholestasis, depending upon the causative agent and the mechanism of injury. Drugs that typically cause cholestasis with hepatitis include psychotropic agents, antibiotics and nonsteroidal antiinflammatory drugs (NSAIDs). The mechanism is immunoallergic and results from hypersensitivity. Pure cholestasis without hepatitis is observed most frequently with contraceptive and 17α-alkylated androgenic steroids and the mechanism most likely involves interference with hepatocyte canalicular efflux systems for bile salts, organic anions and phospholipids. The rate-limiting step in bile formation is considered to be the bile salt export pump (BSEP) mediated translocation of bile salts across the canalicular hepatocyte membrane. Inhibition of BSEP function by metabolites of cyclosporine A, troglitazone, bosentan, rifampicin and sex steroids is an important cause of drug induced cholestasis. Appropriate screening systems for inhibition of BSEP by drug metabolites have been established in membrane vesicles from Sf9 insect cells overexpressing the BSEP protein. A newly recognized mechanism of drug interactions that could cause cholestasis involves the activation of nuclear receptor signaling cascades which affect the transcription of hepatocyte transporter genes critical for bile formation. The multiple factors that regulate transcription of, for example, the *BSEP*, multidrug resistance protein 2 (*MRP2*) and multidrug resistance gene product 3 (*MDR3*) genes are likely targets for untoward effects of xenobiotics on hepatocyte transport function and disposal of toxic metabolites from the liver.

Introduction

The liver plays a predominant role in drug biotransformation and disposition from the body. In view of its barrier function between the gastrointestinal tract and systemic blood, it is constantly exposed to ingested xenobiotics entering the portal circulation. Drug-induced liver injury accounts for up to 7% of all reports of adverse drug effects voluntarily reported to pharmacovigilance registries. Drugs cause direct damage to hepatocytes, bile ducts or vascular structures or may interfere with bile flow. The phenotypes commonly encountered thus include hepatitis, cholestasis, steatosis, cirrhosis, vascular and neoplastic lesions and even fulminant hepatic failure.

Almost every drug has the potential to cause hepatic injury, be it through direct toxicity of the agent or through an idiosyncratic response of the individual. The susceptibility of the liver to injury by drugs is influenced by various factors such as age, sex, pregnancy, comedication, renal function and genetic factors. The latter variable is exemplified by slow metabolizers of debrisoquine, who exhibit a gene polymorphism in the cytochrome P450 2D6 gene and are more susceptible than fast metabolizers to perhexiline maleate toxicity. Other genetically determined enzyme deficiencies that predispose to drug toxicity have been described for cytochrome

Molecular Pathogenesis of Cholestasis, edited by Michael Trauner and Peter L.M. Jansen.
©2004 Eurekah.com and Kluwer Academic / Plenum Publishers.

P450 2C9 (phenytoin toxicity), *N*-acetyltransferase 2 (sulfonamides, dihydralazine), epoxide hydrolase (phenytoin, halothane), glutathione synthetase (paracetamol) and glutathione *S* transferase type T (tacrin).[1,2]

Epidemiology and Taxonomy

The importance of drugs as hepatotoxins lies not in the overall number of cases, which is relatively small, but in the severity of some reactions and in their potential reversibility provided the drug etiology is promptly recognized. Whereas transient elevations in liver function tests are frequent with many drugs, the incidence of severe liver injury is about 1:100,000 for the most common causative agents including NSAIDs, antibiotics, methyldopa, newer antihypertensive agents and H_2-receptor blockers, >10:100,000 for amoxicillin-clavulanic acid, 14:100,000 for erythromycin esters, and ~1:100 for chlorpromazine and isoniazid.[3,4] According to a consensus conference of the Council for International Organizations of Medical Sciences (CIOMS), drug induced hepatotoxicity can be divided into the three categories: cholestatic, hepatocellular or mixed type injury, depending upon serum biochemistry.[5] Cholestasis with hepatitis is seen with many drugs, notably chlorpromazine, psychotropic agents, erythromycins, clavulanic acid and NSAIDs. Pure cholestasis without hepatitis is observed most frequently with estrogens, oral contraceptive steroids and 17α-alkylated androgenic steroids, and less frequently with cyclosporine A, tamoxifen, griseofulvin, glibenclamide and others. Steroid jaundice caused by methyltestosterone and other C17-alkylated anabolic steroids is dose-related but is also dependent upon the individual susceptibility of the recipient. Whereas hepatic dysfunction is seen in most recipients of steroids, jaundice is seen in only few. A minor degree of hepatic dysfunction in women taking oral contraceptives which contain C-17 ethinyl estrogen and progesterone derivatives is relatively frequent. Women with a personal or family history of cholestatic jaundice of pregnancy are particularly prone to develop jaundice when taking oral contraceptives. In addition to the "cholestatic hepatitis" of the chlorpromazine type and "bland cholestasis" caused by anabolic and contraceptive steroids, a third category of cholestasis that results from injury to bile ducts should be defined. This "cholangiodestructive cholestasis" has been observed following exposure to rapeseed oil contaminated with aniline,[6] paraquat,[7] α-naphthyl-isothiocyanate[8] and intraarterial pump infusions of floxuridine,[9] the latter leading to vascular injury.

Although the prognosis of drug-induced cholestasis is generally good with reversibility of symptoms, a chronic cholestatic lesion resembling primary biliary cirrhosis may ensue.[10] The worst case involves development of a ductopenic or vanishing bile duct syndrome associated with chlorpromazine and phenothiazines, tricyclic antidepressants, flucloxacillin, thiabendazole, tolbutamide, carbamazepine and organic arsenicals.[11] The injury to bile ducts caused by floxuridine may progress to biliary sclerosis.[9,12] Other complications of typically cholestatic drugs include the onset of peliosis hepatis in individuals taking contraceptive or anabolic steroids,[13-15] tamoxifen[16] or azathioprine,[17] as well as tumorigenesis.[18]

From a histologic viewpoint, drug-induced cholestasis can be divided into three categories: hepatocanalicular, canalicular or mixed type injury, depending also upon serum biochemistry. Hepatocanalicular cholestasis is associated with 3- to 10 fold increases of alkaline phosphatase and high serum cholesterol, and is seen in conjunction with phenothiazines and erythromycin estolate. In canalicular cholestasis, typically caused by contraceptive and anabolic steroids, alkaline phosphatase levels are only mildly elevated (up to 3 fold the upper limit of normal) and cholesterol levels are normal or elevated. Mixed-type cholestasis is seen with phenylbutazone, para-aminosalicylic acid and sulfonamides.

Mechanisms of Cholestasis

The following basic mechanisms can be held responsible for the development of drug- or toxin-induced cholestasis (Table 1). The first involves the formation of reactive metabolites[19-22] leading to two types of hepatitis:

Table 1. Prototypic cholestatic hepatotoxins and mechanism of injury

Clinical Manifestation	Causative Agents	Mechanism of Injury
Cholestatic hepatitis	Chlorpromazine Phenothiazines Tricyclic antidepressants Erythromycins Clavulanic acid NSAIDs	Idiosyncrasy / hypersensitivity
Bland cholestasis	Estrogens Oral contraceptive steroids 17α-alkylated androgenic steroids Cyclosporine A Tamoxifen Griseofulvin Glibenclamide	Selective interference with bile excretory mechanisms
Cholangiodestructive cholestasis	Aniline-contaminated rapeseed oil α-naphthyl isothiocyanate Paraquat Floxuridine Sporidesmin	Injury to bile ducts
Unconjugated hyperbilirubinemia/ hypercholanemia	Rifamycin SV / rifampicin Cholecystographic dyes	Selective interference with sinusoidal uptake

    (i) toxic hepatitis, typically occurring after overdoses of a given drug (e.g., paracetamol) and,

    (ii) immunoallergic hepatitis, in which an adverse immune response that is directed against the liver is triggered.

This latter form of hepatitis is an idiosyncratic reaction resulting from hypersensitivity and may be associated with the presence of serum autoantibodies (ANA in the case of nitrofuran-toin, methyldopa, chlorpromazine, diclofenac, sulfonamides and nimesulide; AMA against the E6 subunit following iproniazide; antibodies against the E2 subunit of the pyruvate dehydro-genase complex following halothane).[22-25] The second mechanism of cholestasis involves direct injury to bile ducts, typically seen in conjunction with α-naphthyl-isothiocyanate, aniline-contaminated rapeseed oil, paraquat, 5-floxuridine and sporidesmin. The third and possibly the most important mechanism of cholestasis involves the selective interference of a drug with a bile excretory mechanism. Many cholestatic drugs are substrates for the transport proteins localized at the basolateral and canalicular hepatocyte membrane, the latter class gen-erally belonging to the superfamily of ATP-binding cassette transporters.[26] A variation in trans-porter structure or function may render an individual uniquely susceptible to drug-mediated impairment of bile formation. In analogy to variations in cytochrome P450 function that predispose to altered drug metabolism, variations in transporter function represent a major field of research within the rapidly expanding field of pharmacogenomics.[27] Examples for se-quence variants of hepatocyte canalicular efflux systems include the mutations of the MDR3 phospholipid export pump that predispose to intrahepatic cholestasis of pregnancy as a result of high circulating levels of progesterone metabolites,[28,29] as well as the recently described heterozygous mutations of the bile salt export pump in an adolescent patient suffering from recurrent intrahepatic cholestasis, possibly triggered by the intake of certain NSAIDs.[30]

Hypersensitivity Associated and Toxic Cholestatic Injury

In this form of liver injury the liver becomes the target of an immune reaction directed against an immunogenic drug or drug component. Various mechanisms may trigger an immune reaction. First, the drug itself may produce the initial liver injury, as exemplified by halothane. The drug complexes with a liver-specific (membrane) protein, yielding an antigenic moiety. This moiety is presented by an antigen presenting cell (e.g., Kupffer cell), leading to a CD4+ helper T cell induced immune response. The prerequisite is surface expression of an alkylated peptide derived from the drug-protein complex together with a class II MHC molecule. This pathway will provide for a B cell response to the drug. Alternatively the drug may be metabolized by a P450 enzyme with the formation of a reactive metabolite complexed with a P450 peptide. This complex can be expressed at the cell surface in association with a class I MHC molecule (endogenous antigen presentation). MHC class I associated antigens are recognized by CD8+ T cells, leading to immune induction for a cytotoxic T cell response upon reexposure to the drug. Fortunately MHC molecules have a low capacity to interact with drugs or to bind epitopes derived from drug-protein complexes. In addition, reactive T cells tend to undergo a tolerance rather than an immune response to the drug-protein complex, resulting in clonal anergy. This may explain the low overall frequency of immunoallergic drug reactions.

Hypersensitivity associated liver injury is usually a mixed type "cholestatic hepatitis" as exemplified by chlorpromazine. Chlorpromazine causes hepatocanalicular jaundice in 1% of patients, usually within 1-5 weeks of the initiation of treatment.[18] Severe pruritus is common, and alkaline phosphatase values are elevated 3-10 fold, with transaminases 1-8 fold of the upper limit of normal. Cholestasis is typically seen in zone 3 hepatocytes surrounding the central vein. Eosinophilia occurs in 60% of cases and a rechallenge will lead to a positive recurrence in about 50% of cases.[18] This suggests a hypersensitivity reaction as the underlying mechanism, although a toxic component also appears to be involved. Furthermore, chlorpromazine can lead to inhibition of bile flow in the isolated perfused rat liver,[31-33] to an inhibition of Na^+-K^+-ATPase function and an alteration of membrane fluidity.[34] Although the prognosis is generally good, a small percentage of patients may develop a prolonged cholestatic syndrome strongly resembling PBC, but without the occurrence of AMA.[11,35-37] Risk factors for chlorpromazine-induced liver injury include genetically determined ineffective sulfoxidation [38] and drug interactions with sodium valproate.[39] Treatment should include UDCA as well as the supplementation of vitamins in prolonged cholestasis. Intractable pruritus may necessitate plasmapheresis.

Numerous medications may induce cholestatic or mixed type injury through hypersensitivity reactions including all neuroleptics and the cholestatic NSAIDs sulindac, phenylbutazone, indomethacin, fenoprofen and ticlopidine. Two other classes of drugs are commonly associated with cholestatic injury through immunological idiosyncrasy: erythromycin and amoxicillin-clavulanic acid. All erythromycin esters can lead to cholestatic jaundice, which occurs in 1-2% of adult recipients but only rarely in children.[18] The pattern of injury is also hepatocanalicular and histology shows bile casts with prominent portal inflammation that is often rich in eosinophils. Cholestasis usually develops within 3 weeks after initiation of therapy and may persist for up to 4 months after cessation.

In the case of amoxicillin-clavulanic acid, the causative agent appears to be the clavulanic acid component.[18,40] Cholestasis occurs within 2 weeks of initiation of treatment, although it may only appear several weeks after withdrawal. Another class of antibiotics that can cause severe cholestasis are the oxacillins.[18,40] Flucloxacillin-induced cholestasis can persist for years after withdrawal of the drug and cases of vanishing bile duct syndrome have been reported.[18,40-43]

Interference with Bile Excretory Mechanisms

A simplified term to characterize this form of "bland" cholestasis could be "steroid jaundice", since the most important causative agents are the C17-alkylated and the contraceptive steroids. Steroids that possess an alkyl or ethinyl group on carbon atom 17 can lead to hepatic dysfunction in almost all recipients at high doses. However, at the doses normally implemented the overall incidence is low. The duration of intake that precedes the onset of cholestasis is usually 1-6 months. The development of a prolonged cholestatic syndrome that resembles PBC has been reported.[44]

Contraceptive steroid associated cholestasis exhibits similarities with intrahepatic cholestasis of pregnancy, in which the plasma levels of steroid sulfates, in particular sulfated progesterone metabolites, are markedly elevated.[45] Intrahepatic cholestasis of pregnancy has even been postulated to result from a selective biliary excretory defect for sulfated steroids.[46] Interestingly, treatment with ursodeoxy cholic acid (UDCA) reduces plasma concentrations of sulfated steroid metabolites.[47]

The rate-limiting step for bile formation in man is the efflux of bile salts across the hepatocyte canalicular membrane via the bile salt export pump or Bsep.[48] This protein, which has been isolated and functionally characterized in rat and mouse liver,[49-51] can be expressed at high levels in Sf9 insect cells following baculovirus-mediated gene transfer. By isolating plasma membrane vesicles from Bsep-expressing insect cells, Bsep transport function can be studied directly. In this assay, it was found that cyclosporine A, rifamycin SV, rifampicin, glibenclamide and the endothelin antagonist bosentan *cis*-inhibited Bsep-mediated [³H]taurocholate transport, providing a potential mechanism for intrahepatic cholestasis caused by these agents (Table 2).[27,52] Moreover, the thiazolidinedione insulin sensitizer drug troglitazone, which was withdrawn from the market in March 2000 due to its considerable hepatotoxic potential,[53-55] was shown to competitively inhibit rat Bsep with Ki values of 1.3 μM for troglitazone and 0.23 μM for troglitazone sulfate.[56] In one study, the substrate specificity of Bsep was shown to extend not only to bile salts but also to vinblastine, calcein-acetoxymethyl ester and the linear hexapeptide ditekiren.[50] Interaction of these compounds with the canalicular efflux of bile salts is an important mechanism of drug-induced cholestasis.

Estradiol-17β-D-glucuronide (E$_2$17G) has also been found to inhibit Bsep function in vesicles from transfected Sf9 cells, but only in double transfectants that also expressed the canalicular conjugate export pump Mrp2.[27] This suggests that E$_2$17G first needs to be exported into the canalicular lumen, from where it exerts a trans-inhibitory effect on Bsep, a mechanism of inhibition that has also been postulated for lithocholate. This also explains why E$_2$17G is not cholestatic in GY/TR⁻ rats that lack Mrp2 expression.[57] In contrast to E$_2$17G induced trans-inhibition of Bsep, the mechanism of ethinyl estradiol (EE) induced cholestasis is also a reduction of ATP-dependent canalicular taurocholate transport, however with kinetic parameters that suggest a reduction in the number of ATP-dependent bile salt carriers at the canalicular membrane.[58] Accordingly, in rats treated with EE, Bsep protein levels were decreased to 53% and Mrp2 protein levels to 20% of controls.[59]

The nonsteroidal anti-inflammatory agent sulindac, an established hepatotoxin,[60] may also cause cholestasis by interference with the canalicular excretion of bile salts. Sulindac has been shown to follow the cholehepatic shunt pathway and induce choleresis.[61] However, when coinfused with taurocholate in the isolated perfused rat liver, sulindac causes cholestasis by reducing taurocholate secretion. Sulindac appears to be secreted into the bile canaliculus in unconjugated form via a canalicular bile salt export system and is passively absorbed by the bile duct epithelium, thereby inducing a bicarbonate-rich choleresis. Due to continuous cycling within the cholehepatic shunt pathway, high local concentrations of sulindac could be reached within the hepatocyte that cause cholestasis by inhibition of canalicular bile salt efflux.

An important form of intrahepatic cholestasis is the cholestasis of sepsis, which is caused by the effect of endotoxin on hepatocellular bile formation. In experimental sepsis secondary to lipopolysaccharide (LPS) administration, ATP-dependent uptake of 5 μM taurocholate

Table 2. Inhibitors of hepatocellular bile salt efflux: comparison of inhibitory constants in hepatocyte canalicular plasma membrane vesicles and Bsep-overexpressing Sf9 insect cell membrane vesicles

Bsep Inhibitor	Ki in cLPM Vesicles (µmol/l)	Ki in Sf9 Vesicles (µmol/l)	Reference
Cyclosporine A	0.3	0.3	27
Glibenclamide	6.1	5.7	27
Bosentan		12.0	52
Rifamycin SV	0.9	3.8	27
Rifampicin	8.4	11.9	27
Troglitazone	1.3		56

cLPM vesicles, hepatocyte canalicular plasma membrane vesicles; Sf9 vesicles, membrane vesicles from Bsep-overexpressing Sf9 insect cells.

in canalicular plasma membrane vesicles was reduced to 53% of controls without an apparent change in Km, suggesting a decrease in the number of Bsep molecules in the canalicular membrane.[62] A reduction of Bsep protein levels to 52% of controls following LPS administration was confirmed by western blot analyses.[59] The expression of the canalicular organic anion transporter Mrp2 was decreased even more profoundly, amounting to 11% of controls in the LPS model.[59] LPS was shown to induce an early and selective but reversible retrieval of Mrp2 from the canalicular membrane,[63] whereas phalloidin induced an unselective and irreversible loss of Mrp2 and other proteins from the canalicular membrane.[64]

Selective interference with the sinusoidal uptake of substances destined for biliary secretion such as bilirubin and bromosulphophthalein has been shown for the tuberculostatic agents rifamycin SV and rifampicin.[65] Both are mainly eliminated by hepatic uptake, metabolism and excretion into bile. Rifampicin increases serum bile salt concentrations in 72% of patients after the first dose,[66] suggesting acute interference with sinusoidal uptake of bile salts. The major bile salt uptake system of rat liver is the Na$^+$-taurocholate cotransporting polypeptide or Ntcp, whereas Na$^+$-independent uptake of bile salts is mediated by several organic anion transporting polypeptides including Oatp1 and Oatp2.[48] When selectively expressed in *Xenopus laevis* oocytes by injection of cRNA, rifampicin potently inhibited Oatp2 mediated taurocholate uptake, but did not interfere with Oatp1 mediated taurocholate uptake. Both Oatp1 and Oatp2 were inhibited by 10 µmol/l rifamycin SV, whereas significantly higher concentrations of rifamycin SV and rifampicin were required to inhibit Ntcp.[67] For the human liver OATPs, it was shown that 10 µmol/l rifampicin inhibited OATP8 mediated bromosulphophthalein transport by 50%, whereas inhibition of OATP-A, OATP-B and OATP-C was below 15%. In contrast, all human OATPs were inhibited by more than 50% in the presence of 10 µmol/l rifamycin SV.[68] Inhibition of OATPs can partly explain the known effects of rifamycin SV and rifampicin on hepatic organic anion elimination.

Role of Drug Induced Activation of Nuclear Regulatory Cascades

The role of drug mediated changes in cytochrome P450 expression levels has long been recognized as an important mechanism of drug-drug interactions that can cause enhanced metabolism of P450 substrates following gene induction by phenytoin, carbamazepine, rifampicin and others. It is now becoming increasingly evident that the regulatory cascades that affect P450 gene expression can coordinately affect expression of transporter genes involved in bile formation. Numerous drugs are ligands for "orphan" nuclear receptors such as the pregnane X receptor (PXR) which has recently been shown to increase transcription of the human *MDR1* gene.[69] Increased expression of MDR1 is not a mechanism of drug induced cholestasis, however it can have a major impact on the bioavailability of drugs that are MDR1 substrates such as digoxin or cyclosporine A.[70,71] A second hepatocellular transporter gene that is activated by PXR is the rodent organic anion transporting polypeptide Oatp2 *(Slc21a5)*. Hepatic Oatp2 expression is increased in rats treated with the PXR ligand phenobarbital.[72] Moreover, the PXR ligand taurolithocholate upregulates Oatp2 expression in mouse liver, an effect that is not observed in PXR[-/-] mice.[73] Oatp2 is an uptake system for bile salts, steroid conjugates and numerous drugs including digoxin.[48] Typical PXR ligands that could increase Oatp2 expression include rifampin, RU486, St. John's wort extract, clotrimazole, troglitazone and phenobarbital.[70,74-76]

Decreased expression of Mrp2 in rats treated with bacterial lipopolysaccharide is also attributable to a direct effect on *Mrp2* gene transcription. Lipopolysaccharide administration to rats reproduces the cholestasis which is a common clinical feature of sepsis as well as toxic liver diseases.[62] It has been shown that the inflammatory cytokine interleukin-1β inhibits retinoid mediated activation of the rat *Ntcp* and *Mrp2* promoters by reducing nuclear levels of the retinoic acid receptor (RAR) and retinoid X receptor (RXR) and consequently decreasing the binding of RARα/RXRα heterodimers to the *Ntcp* and *Mrp2* retinoid response elements.[77] In addition, interleukin-1β decreases nuclear levels of hepatocyte nuclear factor 1 alpha (HNF1α),[78] a critical factor required for the transcriptional activation of rat *Ntcp*,[79] human *MRP2*,[80,81] and human *OATP-C (SLC21A6)*.[82] In view of these findings it is reasonable to speculate that any form of xenobiotic induced hepatitis is likely to affect the expression of transporter genes via cytokine mediated changes in the nuclear levels of critical transcription factors.

A nuclear receptor that is likely to come into action during cholestatic liver injury is the farnesoid X receptor (FXR), since its natural ligands are hydrophobic bile salts such as chenodexycholic acid (CDCA).[83,84] Increased intracellular bile salt levels secondary to cholestatic liver injury will activate FXR and consequently affect the expression of genes regulated by FXR. Bile salt induced activation of gene transcription through direct binding of FXR to the corresponding promoter element has been shown for the human *BSEP*[85,86] and *OATP8*[87] genes. To exert a cholestatic effect through a nuclear receptor mediated mechanism, the drug would need to inhibit gene transcription of an efflux pump such as BSEP, MRP2 or MDR3. A first example for such a mechanism is the bile salt induced activation of the short heterodimeric partner (SHP-1), which leads to reduced binding of the RARα/RXRα heterodimer to the gene promoter of the chief hepatocellular uptake system Ntcp.[88] The activation of SHP-1 is also FXR mediated and could contribute to the decreased expression of Ntcp in cholestatic liver injury. Because the Mrp2 gene has also been shown to bind the RARα/RXRα heterodimer,[77] it is conceivable that decreased Mrp2 expression in cholestasis also involves a SHP-1 mediated reduction of RARα/RXRα binding, although this remains to be investigated.

Conclusions

In summary, drug-induced cholestatic liver injury can result from direct damage to the hepatic parenchyma by immunoallergic or toxic mechanisms or from impaired transmembrane transport of cholephilic compounds destined for biliary secretion. The rate-limiting step in bile formation is the canalicular excretion of bile salts via the bile salt export pump, which thus represents an especially vulnerable target for drug or toxin mediated injury. The earliest clinical parameter in these patients, which precedes the onset of symptomatic cholestasis, is a rise in serum bile salts. It is likely that individual susceptibility to drug- and toxin-induced cholestasis is conferred by as yet unknown polymorphisms in hepatic transporter genes, that could affect the regulation or secondary structure of the corresponding protein.

References

1. Edwards IR, Aronson JK. Adverse drug reactions: definitions, diagnosis, and management. Lancet 2000; 356:1255-9.
2. Meyer UA. Pharmacogenetics and adverse drug reactions. Lancet 2000; 356:1667-71.

3. Perez Gutthann S, Garcia Rodriguez LA. The increased risk of hospitalizations for acute liver injury in a population with exposure to multiple drugs. Epidemiology 1993; 4:496-501.
4. Garcia Rodriguez LA, Ruigomez A, Jick H. A review of epidemiologic research on drug-induced acute liver injury using the general practice research data base in the United Kingdom. Pharmacotherapy 1997; 17:721-8.
5. Benichou C. Criteria of drug-induced liver disorders. Report of an international consensus meeting. J Hepatol 1990; 11:272-6.
6. Solis Herruzo JA, Castellano G, Colina F et al. Hepatic injury in the toxic epidemic syndrome caused by ingestion of adulterated cooking oil (Spain, 1981). Hepatology 1984; 4:131-9.
7. Mullick FG, Ishak KG, Mahabir R et al. Hepatic injury associated with paraquat toxicity in humans. Liver 1981; 1:209-21.
8. Plaa GL. The Snider Address. A four-decade adventure in experimental liver injury. Drug Metab Rev 1997; 29:1-37.
9. Ludwig J, Kim CH, Wiesner RH et al. Floxuridine-induced sclerosing cholangitis: an ischemic cholangiopathy? Hepatology 1989; 9:215-8.
10. Zimmerman HJ, Lewis JH. Drug-induced cholestasis. Med Toxicol 1987; 2:112-60.
11. Desmet VJ. Vanishing bile duct syndrome in drug-induced liver disease. J Hepatol 1997; 26:131-5.
12. Hohn DC, Rayner AA, Economou JS et al. Toxicities and complications of implanted pump hepatic arterial and intravenous floxuridine infusion. Cancer 1986; 57:465-70.
13. Bagheri SA, Boyer JL. Peliosis hepatis associated with androgenic-anabolic steroid therapy. A severe form of hepatic injury. Ann Intern Med 1974; 81:610-8.
14. Winkler K, Poulsen H. Liver disease with periportal sinusoidal dilatation. A possible complication to contraceptive steroids. Scand J Gastroenterol 1975; 10:699-704.
15. Schonberg LA. Peliosis hepatis and oral contraceptives. A case report. J Reprod Med 1982; 27:753-6.
16. Loomus GN, Aneja P, Bota RA. A case of peliosis hepatis in association with tamoxifen therapy. Am J Clin Pathol 1983; 80:881-3.
17. Degott C, Rueff B, Kreis H et al. Peliosis hepatis in recipients of renal transplants. Gut 1978; 19:748-53.
18. Zimmerman HJ. Drug-induced liver disease. In: Schiff ER, Sorrell MF, Maddrey WC, eds. Schiff's Diseases of the Liver. Philadelphia: Lippincott-Raven, 1999:973-1064.
19. Larrey D. Drug-induced liver diseases. J Hepatol 2000; 32 (Suppl. 1):77-88.
20. Farrell GC. Drug-induced liver disease. London: Churchill Livingstone, 1994.
21. Pessayre D, Larrey D, Biour M. Drug-induced liver injury. In: Bircher J, Benhamou JP, McIntyre N, Rizzetto M, Rodés J, eds. Oxford Textbook of Clinical Hepatology. Oxford: Oxford University Press, 1999:1261-1315.
22. Dansette PM, Bonierbale E, Minoletti C et al. Drug-induced immunotoxicity. Eur J Drug Metab Pharmacokinet 1998; 23:443-51.
23. Homberg JC, Abuaf N, Bernard O et al. Chronic active hepatitis associated with antiliver/kidney microsome antibody type 1: a second type of „autoimmune" hepatitis. Hepatology 1987; 7:1333-9.
24. Bourdi M, Larrey D, Nataf J et al. Anti-liver endoplasmic reticulum autoantibodies are directed against human cytochrome P-450IA2. A specific marker of dihydralazine-induced hepatitis. J Clin Invest 1990; 85:1967-73.
25. Beaune P, Dansette PM, Mansuy D et al. Human anti-endoplasmic reticulum autoantibodies appearing in a drug-induced hepatitis are directed against a human liver cytochrome P-450 that hydroxylates the drug. Proc Natl Acad Sci U S A 1987; 84:551-5.
26. Trauner M, Meier PJ, Boyer JL. Molecular pathogenesis of cholestasis. N Engl J Med 1998; 339:1217-1227.
27. Stieger B, Fattinger K, Madon J et al. Drug- and estrogen-induced cholestasis through inhibition of the hepatocellular bile salt export pump (Bsep) of rat liver. Gastroenterology 2000; 118:422-430.
28. Jacquemin E, Cresteil D, Manouvrier S et al. Heterozygous non-sense mutation of the MDR3 gene in familial intrahepatic cholestasis of pregnancy [letter]. Lancet 1999; 353:210-1.
29. Lammert F, Marschall HU, Glantz A et al. Intrahepatic cholestasis of pregnancy: molecular pathogenesis, diagnosis and management. J Hepatol 2000; 33:1012-1021.
30. Kullak-Ublick GA, Kerb R, Müllhaupt B et al. A novel R432T mutation in the bile salt export pump gene (BSEP; ABCB11) is associated with recurrent intrahepatic cholestasis in an adolescent patient [abstract]. Hepatology 2001; 34:216A.
31. Boyer JL. Mechanisms of chlorpromazine cholestasis: hypersensitivity or toxic metabolite? Gastroenterology 1978; 74:1331-3.
32. Elias E, Boyer JL. Chlorpromazine and its metabolites alter polymerization and gelation of actin. Science 1979; 206:1404-6.

33. Ros E, Small DM, Carey MC. Effects of chlorpromazine hydrochloride on bile salt synthesis, bile formation and biliary lipid secretion in the rhesus monkey: a model for chlorpromazine-induced cholestasis. Eur J Clin Invest 1979; 9:29-41.
34. Schachter D. Fluidity and function of hepatocyte plasma membranes. Hepatology 1984; 4:140-51.
35. Kohn NN, Myerson RM. Xanthomatous biliary cirrhosis following chlorpromazine. Am J Med 1961; 31:665-667.
36. Lewis JH, Tice HL, Zimmerman HJ. Budd-Chiari syndrome associated with oral contraceptive steroids. Review of treatment of 47 cases. Dig Dis Sci 1983; 28:673-83.
37. Degott C, Feldmann G, Larrey D et al. Drug-induced prolonged cholestasis in adults: a histological semiquantitative study demonstrating progressive ductopenia. Hepatology 1992; 15:244-51.
38. Watson RG, Olomu A, Clements D et al. A proposed mechanism for chlorpromazine jaundice-defective hepatic sulphoxidation combined with rapid hydroxylation. J Hepatol 1988; 7:72-8.
39. Bach N, Thung SN, Schaffner F et al. Exaggerated cholestasis and hepatic fibrosis following simultaneous administration of chlorpromazine and sodium valproate. Dig Dis Sci 1989; 34:1303-7.
40. Reddy KR, Schiff ER. Hepatotoxicity of antimicrobial, antifungal and antiparasitic agents. Gastroenterol Clin N Am 1995; 24:923-936.
41. Hebbard GS, Smith KG, Gibson PR et al. Augmentin-induced jaundice with a fatal outcome. Med J Aust 1992; 156:285-6.
42. Eckstein RP, Dowsett JF, Lunzer MR. Flucloxacillin induced liver disease: histopathological findings at biopsy and autopsy. Pathology 1993; 25:223-8.
43. Munoz SJ, Martinez Hernandez A, Maddrey WC. Intrahepatic cholestasis and phospholipidosis associated with the use of trimethoprim-sulfamethoxazole. Hepatology 1990; 12:342-7.
44. Glober GA, Wilkerson JA. Biliary cirrhosis following the administration of methyltestosterone. Jama 1968; 204:170-3.
45. Sjovall J, Sjovall K. Steroid sulphates in plasma from pregnant women with pruritus and elevated plasma bile acid levels. Ann Clin Res 1970; 2:321-37.
46. Reyes H, Sjövall J. Bile acids and progesterone metabolites in intrahepatic cholestasis of pregnancy. Ann Med 2000; 32:94-106.
47. Meng LJ, Reyes H, Axelson M et al. Progesterone metabolites and bile acids in serum of patients with intrahepatic cholestasis of pregnancy: effect of ursodeoxycholic acid therapy. Hepatology 1997; 26:1573-9.
48. Kullak-Ublick GA, Stieger B, Hagenbuch B et al. Hepatic transport of bile salts. Semin Liver Dis 2000; 20:273-292.
49. Gerloff T, Stieger B, Hagenbuch B et al. The sister of P-glycoprotein represents the canalicular bile salt export pump of mammalian liver. J Biol Chem 1998; 273:10046-50.
50. Lecureur V, Sun D, Hargrove P et al. Cloning and expression of murine sister of P-glycoprotein reveals a more discriminating transporter than *MDR1*/P-glycoprotein. Mol Pharmacol 2000; 57:24-35.
51. Green RM, Hoda F, Ward KL. Molecular cloning and characterization of the murine bile salt export pump. Gene 2000; 241:117-123.
52. Fattinger K, Funk C, Pantze M et al. The endothelin antagonist bosentan inhibits the canalicular bile salt export pump: a potential mechanism for hepatic adverse reactions. Clin Pharmacol Ther 2001; 69:223-31.
53. Gitlin N, Julie NL, Spurr CL et al. Two cases of severe clinical and histologic hepatotoxicity associated with troglitazone. Ann Intern Med 1998; 129:36-8.
54. Neuschwander Tetri BA, Isley WL, Oki JC et al. Troglitazone-induced hepatic failure leading to liver transplantation. A case report. Ann Intern Med 1998; 129:38-41.
55. Herrine SK, Choudhary C. Severe hepatotoxicity associated with troglitazone. Ann Intern Med 1999; 130:163-4.
56. Funk C, Ponelle C, Scheuermann G et al. Cholestatic potential of troglitazone as a possible factor contributing to troglitazone-induced hepatotoxicity: in vivo and in vitro interaction at the canalicular bile salt export pump (Bsep) in the rat. Mol Pharmacol 2001; 59:627-35.
57. Sano N, Takikawa H, Yamanaka M. Estradiol-17 beta-glucuronide-induced cholestasis. Effects of ursodeoxycholate-3-O-glucuronide and 3,7-disulfate. J Hepatol 1993; 17:241-6.
58. Bossard R, Stieger B, O'Neill B et al. Ethinylestradiol treatment induces multiple canalicular membrane transport alterations in rat liver. J Clin Invest 1993; 91:2714-2720.
59. Lee JM, Trauner M, Soroka CJ et al. Expression of the bile salt export pump is maintained after chronic cholestasis in the rat. Gastroenterology 2000; 118:163-172.
60. Rodriguez LAG, Williams R, Derby LE et al. Acute liver injury associated with nonsteroidal anti-inflammatory drugs and the role of risk factors. Arch Intern Med 1994; 154:311-316.
61. Bolder U, Trang NV, Hagey LR et al. Sulindac is excreted into bile by a canalicular bile salt pump and undergoes a cholehepatic circulation in rats. Gastroenterology 1999; 117:962-71.

62. Bolder U, Ton-Nu H-T, Schteingart CD et al. Hepatocyte transport of bile acids and organic anions in endotoxemic rats: impaired uptake and secretion. Gastroenterology 1997; 112:214-225.
63. Kubitz R, Wettstein M, Warskulat U et al. Regulation of the multidrug resistance protein 2 in the rat liver by lipopolysaccharide and dexamethasone. Gastroenterology 1999; 116:401-10.
64. Rost D, Kartenbeck J, Keppler D. Changes in the localization of the rat canalicular conjugate export pump Mrp2 in phalloidin-induced cholestasis. Hepatology 1999; 29:814-21.
65. Acocella G, Nicolis FB, Tenconi LT. The effect of an intravenous infusion of rifamycin SV on the excretion of bilirubin, bromsulphalein, and indocyanine green in man. Gastroenterology 1965; 49:521-5.
66. Galeazzi R, Lorenzini I, Orlandi F. Rifampicin-induced elevation of serum bile acids in man. Dig Dis Sci 1980; 25:108-12.
67. Fattinger K, Cattori V, Hagenbuch B et al. Rifamycin SV and rifampicin exhibit differential inhibition of the hepatic rat organic anion transporting polypeptides, Oatp1 and Oatp2. Hepatology 2000; 32:82-86.
68. Vavricka SR, van Montfoort J, Ha HR et al. Interactions of rifamycin SV and rifampicin with OATP-C, OATP8, OATP-B and OATP-A of human liver. Hepatology 2002; 36:164-72.
69. Geick A, Eichelbaum M, Burk O. Nuclear receptor response elements mediate induction of intestinal MDR1 by rifampin. J Biol Chem 2001; 276:14581-7.
70. Greiner B, Eichelbaum M, Fritz P et al. The role of intestinal P-glycoprotein in the interaction of digoxin and rifampin. J Clin Invest 1999; 104:147-53.
71. Dürr D, Stieger B, Kullak-Ublick GA et al. St John's Wort induces intestinal P-glycoprotein/MDR1 and intestinal and hepatic CYP3A4. Clin Pharmacol Ther 2000; 68:598-604.
72. Hagenbuch N, Reichel C, Stieger B et al. Effect of phenobarbital on the expression of bile salt and organic anion transporters of rat liver. J Hepatol 2001; 34:881-887.
73. Staudinger JL, Goodwin B, Jones SA et al. The nuclear receptor PXR is a lithocholic acid sensor that protects against liver toxicity. Proc Natl Acad Sci U S A 2001; 98:3369-3374.
74. Jones SA, Moore LB, Shenk JL et al. The pregnane X receptor: a promiscuous xenobiotic receptor that has diverged during evolution. Mol Endocrinol 2000; 14:27-39.
75. Moore LB, Goodwin B, Jones SA et al. St. John's wort induces hepatic drug metabolism through activation of the pregnane X receptor. Proc Natl Acad Sci U S A 2000; 97:7500-2.
76. Kullak-Ublick GA, Jung D, Hagenbuch B et al. Organic anion transporting polypeptides, cholestasis and nuclear receptors. Hepatology 2002; 35:732-3.
77. Denson LA, Auld KL, Schiek DS et al. Interleukin-1beta suppresses retinoid transactivation of two hepatic transporter genes involved in bile formation. J Biol Chem 2000; 275:8835-43.
78. Wang B, Cai SR, Gao C et al. Lipopolysaccharide results in a marked decrease in hepatocyte nuclear factor 4α in rat liver. Hepatology 2001; 34:979-989.
79. Karpen SJ, Sun A-Q, Kudish B et al. Multiple factors regulate the rat liver basolateral sodium-dependent bile acid cotransporter gene promoter. J Biol Chem 1996; 271:15211-15221.
80. Tanaka T, Uchiumi T, Hinoshita E et al. The human multidrug resistance protein 2 gene: functional characterization of the 5'-flanking region and expression in hepatic cells. Hepatology 1999; 30:1507-1512.
81. Stockel B, Konig J, Nies AT et al. Characterization of the 5'-flanking region of the human multidrug resistance protein 2 (MRP2) gene and its regulation in comparison with the multidrug resistance protein 3 (MRP3) gene. Eur J Biochem 2000; 267:1347-58.
82. Jung D, Hagenbuch B, Gresh L et al. Characterization of the human OATP-C (SLC21A6) gene promoter and regulation of liver-specific OATP genes by hepatocyte nuclear factor 1α. J Biol Chem 2001; 276:37206-37214.
83. Makishima M, Okamoto AY, Repa JJ et al. Identification of a nuclear receptor for bile acids. Science 1999; 284:1362-1365.
84. Parks DJ, Blanchard SG, Bledsoe RK et al. Bile acids: natural ligands for an orphan nuclear receptor. Science 1999; 284:1365-8.
85. Ananthanarayanan M, Balasubramanian N, Makishima M et al. Human bile salt export pump (BSEP) promoter is transactivated by the farnesoid X receptor/bile acid receptor (FXR/BAR). J Biol Chem 2001; 276:28857-28865.
86. Schuetz EG, Strom S, Yasuda K et al. Disrupted bile acid homeostasis reveals an unexpected interaction among nuclear hormone receptors, transporters, and cytochrome P450. J Biol Chem 2001; 276:39411-39418.
87. Jung D, Podvinec M, Meyer UA et al. Human organic anion transporting polypeptide 8 promoter is transactivated by the farnesoid X receptor/bile acid receptor. Gastroenterology 2002; 122:1954-66.
88. Denson LA, Sturm E, Echevarria W et al. The orphan nuclear receptor, shp, mediates bile acid-induced inhibition of the rat bile acid transporter, ntcp. Gastroenterology 2001; 121:140-147.

CHAPTER 20

Acquired Alterations of Transporter Expression and Function in Cholestasis

Michael Trauner, Peter Fickert and Gernot Zollner

Summary

Exposure to cholestatic injury (e.g., drugs, hormones, proinflammatory cytokines, biliary obstruction/destruction), hereditary mutations of transporter genes, or the combination of both result in decreased expression and function of hepatobiliary transport systems. These molecular changes may contribute to impaired hepatic uptake and excretion of bile salts and other organic anions (e.g., bilirubin) in cholestasis. In addition, alterations in transporter expression may represent secondary and adaptive changes, limiting the accumulation of potentially toxic biliary constituents in the cholestatic liver by providing alternative excretory routes. The mediators and molecular mechanisms responsible for changes in transporter expression are being increasingly understood at transcriptional and post-transcriptional levels. The molecular changes of hepatobiliary transport systems in cholestasis may represent a potential target for specific therapeutic interventions aimed at restoration of defective hepatobilary transporter expression and stimulation of adaptive rescue pathways.

Abbreviations

AE, Cl^-/HCO_3^- anion exchanger; ASBT/Asbt apical Na^+-dependent bile salt transporter; ATP, adenosine triphosphate; cAMP, adenosine 3',5'-cyclic monophosphate; BSEP/Bsep, bile salt export pump; CAR, constitutive androstane receptor; CFTR/Cftr, cystic fibrosis transmembrane conductance regulator; CYP/Cyp, cytochrome P450 enzyme; FIC1/Fic1, putative aminophospholipid transporter (mutated in familial intrahepatic cholestasis); FXR, farnesoid X receptor/bile salt receptor; HNF, hepatocyte nuclear factor; IL-1β, interleukin-1β; LPS, lipopolysaccharide; MDR/Mdr, multidrug resistance P-glycoprotein; MDR1/Mdr1, multidrug export pump; MDR3/Mdr2, phsopholipid export pump; mRNA, messenger ribonucleic acid; MRP/Mrp, multidrug resistance-associated protein; MRP2/Mrp2, conjugate export pump; NTCP/Ntcp, Na^+/taurocholate cotransporter; OATPs/Oatps, organic anion transporting proteins; PBC, primary biliary cirrhosis; PKC, protein kinase C; PSC, primary sclerosing cholangitis; PXR, pregnane X receptor; RAR, retinoic acid receptor; RXR, retinoid X receptor; TNF-α, tumor necrosis factor-a; UDCA, ursodeoxycholic acid.

Note

Please note that by convention human genes and their products are capitalized, whereas rodent genes and their products are written in lower case throughout this article. Transporter genes are set in italic whereas gene products (transporter proteins) are set in normal font.

Molecular Pathogenesis of Cholestasis, edited by Michael Trauner and Peter L.M. Jansen.
©2004 Eurekah.com and Kluwer Academic / Plenum Publishers.

Introduction

Cholestasis may result either from a functional defect in bile formation at the level of the hepatocyte ('hepatocellular cholestasis') or from an impairment in bile secretion and flow at the level of bile ductules or ducts ('ductular/ductal cholestasis').[1-3] By using molecular probes for hepatobiliary transport systems in experimental animal models of cholestasis and human cholestatic liver diseases, it has become evident that decreased expression of transport systems may at least in part explain the impairment of transport function resulting in or maintaining cholestasis.[1,2,4,5] Clinically, primarily hepatocellular forms of acquired cholestasis (e.g., drug-induced cholestasis resulting from direct inhibition of transporter expression and function by drugs) are quite rare, and most clinical forms of cholestasis are caused at a ductular/ductal level by obstruction/destruction of bile ducts (e.g., extrahepatic biliary obstruction, vanishing bile duct syndromes).[6] Therefore, most of the changes in hepatocellular and cholangiocellular transporter expression encountered in these conditions are secondary (rather than primary, causative) changes, which, nevertheless, may explain and maintain the ongoing functional impairment of bile secretion in cholestasis. It is important to keep in mind that not all of the encountered changes in transporter expression are 'negative' from a teleological point of view. While some of these alterations may contribute to cholestasis, some of the changes in transporter expression in liver and extrahepatic tissues (e.g., renal tubules) may represent compensatory (anti-cholestatic) changes, which provide alternative excretory routes for accumulating cholephiles/biliary constituents in cholestasis. Therefore, the transporter changes encountered in cholestasis seem to represent a mixture of pro-cholestatic and anti-cholestatic alterations. The functional implications (i.e., accumulation of cholephiles) will depend on the net-balance of these changes.

This chapter is focused on acquired molecular alterations of transport systems in liver and extrahepatic tissues and a discussion of potential molecular (transcriptional and post-transcriptional) mechanisms responsibel for these changes. Other molecular mechanisms (not discussed in this chapter) which may also contribute to acquired cholestasis include: disruption of the cytoskeleton[3,7-11,12-14] (interfering with transcellular vesicle movement, tight junctional permeability and contraction of bile canaliculi); defects in tight junctional structures[15-25] (leading to dissipation of osmotic gradients via "leaky" paracellular pathways); impairment of gap junction communication[26] and signal transduction pathways normally regulating various steps in bile formation[27-30] (e.g., transporter activity, vesicular trafficking, tight junction permeability) (see related chapters by Anwer & Webster, Echevarria & Nathanson in this book). Most of our knowledge is derived from animal models of cholestasis, where mediators and molecular mechanisms leading to these changes are increasingly being understood. However, the changes observed in experimental models cannot unequivocally be applied to human cholestatic liver diseases, where the picture is still far from being complete and most of the work still needs to be done.[31]

Insights from Experimental Animal Models of Cholestasis

Experimental animal models of cholestasis allow to study the mechanisms of acquired human cholestatic liver diseases including inflammation-induced cholestasis (endotoxin [LPS]-treated rats), oral contraceptive-induced cholestasis/cholestasis of pregnancy (ethinylestradiol-treated rats), and extrahepatic biliary obstruction (common bile duct ligated [CBDL] rats).[1,2,4,5] Since elimination of biliary constituents from the body in cholestasis is not exclusively determined by the liver, it is important to view the hepatic changes (at the level of hepatocytes and cholangiocytes) in context with extrahepatic changes in kidney (proximal renal tubules), intestine (ileum), and—in cholestasis of pregnancy—also in placenta (trophoblast).

Alterations in Transporter Expression in Hepatic and Extrahepatic Tissues (Hepatocytes, Cholangiocytes, Renal Tubular Cells, Enterocytes, Placenta)

In order to understand the functional implications of the alterations in transporter expression encountered in cholestasis, it is important to keep in mind that most biliary constituents (e.g., bile salts, cholesterol, phosphopilids, bilirubin) undergo an efficient enterohepatic circulation between the liver and the intestine.[32] In contrast to rats (which lack a gallbladder), the human bile salt pool is mostly stored in the gallbladder during the fasting state.[33,34] Bile salts are reabsorbed with high efficiency in the small intestine (mainly in the terminal ileum) and circulate via the portal blood back to the liver, completing a full enterohepatic cycle. Of note, the majority of bile salts excreted into bile are derived from the enterohepatic circulation and must first be taken up by the liver, before becoming excreted into bile.[35] As such, the (human) bile salt pool circulates 6 to 10 times per day. As a result of the high degree of interstinal bile salt conservation, only about 3-5% of bile salts excreted into bile are lost through fecal excretion and need to be replaced by de novo bile salt synthesis.[34]

Bile salts also undergo to some extent a cholehepatic shunting/cycling from the bile duct lumen, through cholangiocytes, via the periductular capillary plexus (draining into the portal vein radicles) back to the liver.[32,35] Although probably playing a minor role under normal conditions, cholehepatic shunting of bile salts may become an important escape route for bile salts from obstructed bile ducts under cholestatic conditions. Apart from cholehepatic shunting, bile salt uptake into cholangiocytes may also play an important role for cell signaling in the regulation of secretory and proliferative events within the biliary tree.[36-38]

Biliary constituents (bile salts, conjugated bilirubin) may also be excreted via the kidney into urine, an alternative pathway which may be increasingly utilized under cholestatic conditions with impaired biliary excretion and disruption of the normal enterohepatic circulation.[32,35] Therefore, the degree of the systemic and intrahepatic retention of biliary constituents in cholestasis will depend on the alterations in transporter expression encountered in liver (hepatocytes, cholangiocytes), kidney (proximal renal tubular cells) and intestine (enterocytes). Under the specific condition of maternal cholestasis (e.g., intrahepatic cholestasis of pregancy) placental changes may also come into play in determining the accumulation of cholephiles in the fetus, since the fetal liver is immature and most transport systems are not expressed until shortly before/after birth.[39] The following section will discuss the changes in transporter expression in these cells/tissues which play a role in the handling of bile salts and other biliary constituents.

Hepatocellular Changes

Transport studies in basolateral and canalicular membrane vesicles isolated from cholestatic animals have revealed changes in the maximal transport velocity (V_{max}) of bile salts and non-bile salts organic anions without significant changes in the relative affinity (Michaelis constant, K_m).[40-42] These changes in V_{max} would be consistent with a decrease in the number of functional transport proteins as a consequence of either decreased overall gene expression or impaired targeting of the protein to the cell membrane. Indeed, cholestatic injury (e.g., LPS, ethinylestradiol, or CBDL) in rats results in a marked reduction of messenger ribonucleic acid (mRNA) and protein levels of the basolateral Na^+/taurocholate cotransporter (Ntcp; official gene symbol *Slc10a1*) and organic anion transporting proteins (Oatp1/*Slc21a1*, Lst-1 [but not it's full length isoform Oatp4/*Slc21a10*]), as well as the canalicular conjugate export pump (Mrp2/*Abcc2*) and—to a lesser extent—the bile salt export pump (Bsep/*Abcb11*) (Fig. 1).[40-51] Decreased expression of these transporters may explain impaired hepatocellular uptake and canalicular excretion of bile salts and other (non-bile salt) organic anions (e.g., bilirubin diglucuronide) in cholestasis. Expression of the canalicular Bsep initially decreases and partially recovers (reaching approximately 80% of baseline levels) with prolonged cholestasis.[47,49] Thus, Bsep is more stably expressed than the other transporters (Ntcp, Oatp1, and Mrp2) and may continue—although at reduced levels—to excrete bile salts into bile even under cholestatic

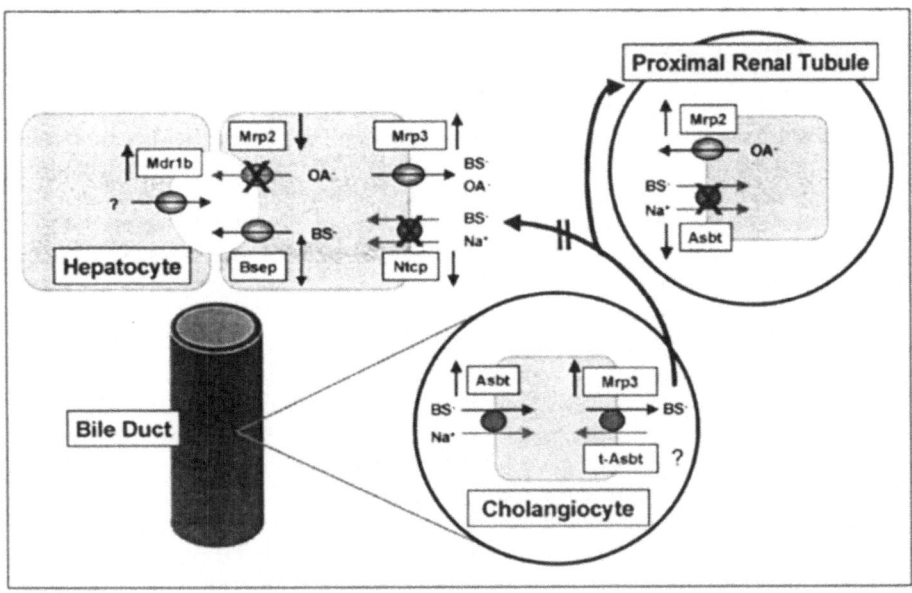

Figure 1. Alterations of bile salt and organic anion transport systems in experimental animal models of cholestasis. Down-regulation of the basolateral Na$^+$/taurocholate cotransporter (Ntcp) together with partial recovery of the canalicular bile salt export pump (Bsep) reduces the hepatocellular retention of bile salts (BS$^-$). In contrast to down-regulation of the canalicular conjugate export pump (Mrp2), expression of another isoform of the multidrug-resistance-associated protein (Mrp3) increases at the (baso)lateral membrane of hepatocytes, providing an alternative route for the elimination of bile salts (BS$^-$) and non-bile salt organic anions (OA$^-$) from cholestatic hepatocytes into the systemic circulation. Increased expression of the apical Na$^+$-dependent bile salt transporter (Asbt) and Mrp3 (and possibly a truncated basolateral Asbt isoform; t-Asbt) facilitates the return of bile salts from obstructed bile ducts to the systemic circulation. In contrast to hepatic downregulation of Mrp2 and upregulation of Asbt, renal Mrp2 is upregulated while Asbt is downregulated in the kidney. These reciprocal changes may help the renal elimination of bile salts and conjugated bilirubin in cholestasis.

conditions.[47,49,52] This response, together with the down-regulation of Ntcp would tend to reduce the hepatocellular retention of bile salts and the degree of bile salt-mediated liver injury (Fig. 1). Since canalicular transport represents the rate limiting step in bile secretion, alterations of canalicular transporter expression may be considered the primary events in hepatocellular forms of cholestasis, while alterations of basolateral transport systems may be secondary, in attempt to limit further uptake and protect the hepatocyte from overload with bile salts and other potentially toxic biliary constituents.

In contrast to the down-regulation of these transporters, expression of the multidrug export pump (Mdr1/*Abcb1*) at the canalicular membrane and isoforms of the multidrug-resistance-associated protein (Mrp1/*Abcc1* and Mrp3/*Abcc3*) at the (baso)lateral membrane increases following cholestatic injury.[47,48,53-59] However, reduced expression of Mrp2 is not necessarily accompanied by compensatory up-regulation of Mrp3 as exemplified by down-regulation of both Mrp2 and Mrp3 in pregnant rats.[60] Up-regulation of Mrp1 and Mrp3 in cholestasis could be a compensatory mechanism that prevents further accumulation of potentially toxic biliary constituents (in particular bile salts) within cholestatic hepatocytes and may explain the shift towards renal excretion of bile salts as a major mechanism for bile salt elimination in chronic, longstanding cholestasis (Fig. 1).[61] Normally, rat Mrp3 is only found on the basolateral membrane of cholangiocytes (where it may be involved in basolateral bile salt efflux from cholangiocytes and the cholehepatic cycling of bile salts) and a single layer of hepatocytes surrounding the central vein in rat liver.[56,57] With progressive cholestasis (e.g., CBDL), Mrp3 expression reaches from the pericentral region throughout the liver lobule and is finally also found in the periportal region, while proliferating bile ducts continue to stain positive.[56,57] Of note, Mrp3 transports not only typical Mrp2 substrates (including sulfated bile salts which are increasingly formed under cholestatic conditions to enhance their renal excretion) but also glycin- and taurin-conjugated bile salts.[62-64] Thus, upregulation of basolateral Mrp3 could not only explain the appearance of conjugated bilirubin in plasma, but could also provide an alternative efflux route for bile salts from hepatocytes into the sinusoidal blood during cholestasis and other conditions with impaired canalicular excretory function (e.g., congenital MRP2/Mrp2 deficiency).[65]

In contrast to down-regulation of Oatp1 and Lst-1, two other liver Oatp isoforms Oatp2/*Slc21a5* and Oatp4/*Slc21a10* are relatively well preserved under cholestatic conditions (e,g, CBDL) (Carsten Gartung, personal communication). The molecular mechanisms for this observation are becoming increasingly clear with the discovery of the role of nuclear orphan receptors in the regulation of hepatobiliary transport systems (see below). Since Oatps function as anion exchangers (with GSH, HCO_3^-),[66] their transport direction may reverse depending on the concentration gradients of their substrates. Therefore it is attractive to speculate that maintained Oatp2 and Oatp4 expression could facilitate efflux of bile salts under cholestatic conditions into the opposite direction (i.e., fom the hepatocyte into the sinusoidal blood) in analogy to conserved OATP1/*SLC21A3* expression in humans (see below). On the other hand, it must be considered that the failure to down-regulate these Oatps could also contribute to continued bile salt loading of cholestatic hepatocytes, resulting in ongoing bile salt accumulation in the cholestatic liver.[67,68]

Expression of Fic1 /FIC1 (*ATP8B1*) remains unchanged in experimental cholestasis (e.g, CBDL)[69] and humans with inflammatory cholestasis,[70] suggesting that alterations of this transport system (mutated in herediary cholestasis; see related chapter by Peter Jansen et al in this book) do not play a major role in acquired cholestasis. However, the functional implications of maintained Fic1/FIC1 expression remain unclear as long as its function is unresolved.[71,72]

Cholangiocellular Changes

Similar to hepatocytes, cholangiocellular transport systems may undergo compensatory changes or become the target of the initial cholestatic injury.[38,73] Bile ducts proliferate in response to various types of liver injury.[37,74] Of note, hepatic progenitor (stem/oval) cells localized within cholangioles (canals of Hering)[75] express ABC transporters of the Mdr and Mrp family (e.g., Mdr1b, Mrp1 and 3) which may protect these cells in conditions of toxic liver injury.[76] Under normal conditions the apical Na^+/dependent bile salt transporter (Asbt/*Slc10a2*) is only expressed in the apical domain of large but not small cholangiocytes.[36,77,78] Bile salt feeding to rats increases proliferative activity and Asbt expression in both small and large rat cholangiocytes in a protein kinase C (PKC) dependent fashion.[79,80] These findings suggest that accumulating bile salts retained during cholestasis, together with increased biliary pressure in biliary obstruction,[81] may be responsible for bile duct proliferation and increased Asbt expression in CBDL rats.[82,83] In addition, Mrp3 expression is maintained at the basolateral membrane of proliferated bile ducts in CBDL rats[56,57] (Fig. 1). The expression of other potential basolateral efflux systems in cholangiocytes such as alternatively spliced/truncated Asbt[84] is unknown under cholestatic conditions. The increases in Asbt and Mrp3 expression result from an increased cholangiocellular cell mass (as a result of bile duct proliferation) while the expression per cell (cholangiocyte) is maintained (Mrp3) or even decreased (Asbt).[83] Since the relative stability of hepatocellular Bsep expression during cholestasis enables some bile salts to continue to be excreted into bile (see above), increased functional capacity of cholangiocyte Asbt and Mrp3 could facilitate the return of bile salts from the obstructed bile ducts to the systemic circulation (Fig. 1). Reabsorption from bile salts from the bile duct lumen may thus prevent/limit bile salt-induced bile duct injury in obstructive cholestasis. On the other hand, increased Asbt activity as a result of bile duct proliferation has also been implicated in the intrahepatic bile salt recirculation contributing to intrahepatic retention of bile salts and, by spillover, to increased systemic bile salt concentrations in cholestasis.[85] Enhanced secretin receptor gene expression, secretin-induced cAMP synthesis, Cl^-/HCO_3^- anion exchanger (AE2/ *SLC4A2*) activity and apical insertion of AQP1 water channels in cholangiocytes of bile duct-ligated rats,[86-89] could account for the well-known clinical observation of "white bile" that occurs during prolonged obstruction of bile ducts (also see related chapter by LeSage et al in this book).

Proinflammatory cytokines profoundly inhibit cholangiocellular Cl^- and HCO_3^- secretion without reducing expression of AE2, the expression of the cystic fibrosis transmembrane conductance regulator (Cftr/*Abcc7*) and secretin receptor being even enhanced probably in a compensatory fashion.[90-92] Rather these effects seem to be mediated via impaired cAMP formation, although the molecular mechanims for this finding remain to be resolved.[91] Of note, gluocorticoids have been identified as potent modulators of cholangiocellular transporter expression and function, which puts these findings into a therapeutic perspective.[93]

Intestinal Changes

The effects of cholestasis on ileal expression of Asbt are controversial, with some studies having shown reduced[94,95] and others unchanged[96] expression and function. Part of these discrepancies may be due to differential effects of cholestasis on the dimeric (reduced) and monomeric (unchanged) form of Asbt in rat ileum.[95] Teleologically, reduced Asbt expression and function would be expected to protect the organisms from bile salt reabsorption and overload under cholestatic conditions with partially preserved enterohepatic circulation.

Bile salt feeding also induces expression of a 14 kD cytosolic intestinal bile acid-binding protein (I-Babp) which is cytoplasmatically attached to Asbt and probably represents the most important protein for transcellular bile salt movement through enterocytes.[34] Similar ontogenic expression patterns and reactions of Asbt and I-Babp to bile salts and dexamethasone suggest that both transport systems might be controlled by similar if not identical regulatory mechanisms.[34] Alternatively, but less likely, induction of I-Babp could also result in cytoplasmatic trapping of bile salts within enterocytes and, thereby, reduced bile salt reabsorption, although this remains speculative and stands in contrast to the currently proposed function of I-Babp.

The effects of cholestasis on other bile salt transport systems such as apical Oatp3 (*Slc21a7*) and basolateral Mrp3 and, possibly, t-Asbt in enterocytes[97,98] remains to be determined. Similarly it would be interesting to learn more about potential alterations of apical efflux pumps for defense against xenobiotics and their metabolites such as Mdr1 and Mrp2 in cholestasis.

Renal Changes

Bile salts which escape the first-pass clearance by the liver are filtered at the glomerulus from plasma into urine, where they are reabsorbed by the apical Na^+-dependent bile salt transporter Asbt in the proximal convoluted tubule.[98] Thus, under normal conditions, urinary bile salt losses are minimal. In addition to Asbt, Mrp2, Oatp1 and Oatp3 have also been localized to the apical brush border membrane, while their basolateral couterparts (e.g., Mrp3, t-Asbt) have not yet been identified in kidney.[98] Under normal conditions Mrp2 could be involved in tubular excretion of organic anions (e.g., para-aminohippuric acid).[99] Increased urinary excretion of bile salts (in particular of sulfated/glucuronidated bile salts which are increasingly formed under cholestatic conditions to enhance renal bile salt excretion) provides an important overflow pathway and partially reduces the accumulating bile salt pool in cholestasis. The potential molecular mechanisms for increased renal bile salt excretion during cholestasis are increasingly becoming clear. In contrast to hepatic down-regulation of Mrp2 and up-regulation of Asbt (see above), renal Mrp2 is up-regulated in proximal tubules, while Asbt in the kidney is down-regulated,[59,83] suggesting organ-specific differences in their gene regulation. These reciprocal changes in liver and kidney expression of both Mrp2 and Asbt may help the renal elimination of biliary constituents (e.g., bile salts, conjugated bilirubin) during cholestasis (Fig. 1). However, the relative contributions of Mrp2-mediated active transport vs. passive glomerular filtration of sulfated / glucuronated bile acids and bilirubin into urine during cholestasis remain to be determined. Other members of the Mrp family also appear to be expressed in kidney and may undego changes in cholestasis. As such, Mrp1 expression increases in renal tubular epithelial cells following CBDL in rats,[58] while renal expression and potential alterations of Mrp3 remain to be determined.

Placental Changes

Bile salts undergo minimal biliary excretion by the fetal liver, since the fetal liver is immature and ontogenic expression of hepatobiliary transporters is not detectable until shortly before birth.[39,100,101] Instead they are eliminated by the maternal liver after vectorial translocation from the fetal to the maternal circulation via the placenta ("placenta-maternal liver tandem").[98] Distinct transport systems have been identified at a functional level at the basolateral (fetal-facing) and apical (maternal-facing) membrane of the trophoblast.[98] The molecular identity of these transport systems is not yet entirely clear. Recent studies have identified partial transcripts of OATPs and BSEP/Bsep in human and rat placenta, as well as MRP3 in the fetal blood endothelia and the syncytiotrophoblast layer, suggesting a role of MRP3 in the extrusion of fetal bile salts.[98,102,103] Maternal cholestasis (e.g., CBDL in pregnant rats) impairs bile salt transfer from the fetus to the mother by reducing bile salt transport at both the basolateral (fetal-facing) and apical (maternal-facing) membrane.[104,105] The molecular basis (e.g., reduced expression of hepatobiliary transport systems in placenta) remains to be clarified. Some of these functional changes, however, appear to be related to major structural changes of the chorionic tissue (e.g., reduction of trophoblast tissue).[104] Of note, maternal cholestasis does not affect the ontogenic expression pattern of hepatobiliary transporters in fetal and neonatal rat liver.[100]

Molecular Mechanisms Leading to Altered Transporter Expression

Although much information has been gathered on the molecular alterations of hepatobiliary transport systems at the mRNA and protein levels, only limited information is currently available regarding the molecular mechanisms and potential mediators leading to these changes. Such effects can be mediated by changes in the rate of gene transcription regulated by the promoter region of each gene, posttranscriptional changes, as well as posttranslational mechanisms.[4,106,107] Evidence existing so far suggests that bile salts, proinflammatory cytokines, oxidative stress, retinoids, drugs, hormones and changes in cell osmolarity may be potent modulators of transporter expression during cholestasis.[4,107,108] Considering the long half-life of most hepatic transport proteins in the range of several (approximately 5) days,[109] posttranscriptional/posttranslational effects may be revelant in the early time period of cholestasis (early events initiating cholestasis), while transcriptional mechanisms may be more relevant in the later course of cholestasis (maintaining alterations of transporter expression and function).[107] For example in LPS-induced cholestasis, endocytotic retrieval of Mrp2 from the canalicular membrane appears to be an early event (first 6 hours after LPS),[153] whereas down-regulation of Mrp2 mRNA may be a later event (occuring beyond 12 hours after LPS).[47-49,110]

Transcriptional Mechanisms

Down-regulation of Ntcp and Mrp2 expression in CBDL and LPS-treated rats may at least in part be due to reduced de-novo gene transcription.[40,45,111] Gene transcription rates of other transport systems have not yet been assessed systematically in cholestasis, but it may be assumed that the observed changes in steady-state mRNA levels also reflect changes in de novo gene transcription. Reduced transcription of transporter genes may be due to alterations of nuclear transcription factors binding to the promoter region and regulating the transcriptional activity of the respective genes.[106] Recent discovery of the role of orphan nuclear receptors / ligand-activated nuclear receptors in regulating hepatobiliary transporter expression has shed some light on the mechanisms of these changes[112-114] (see chapter by Karpen in this book).

Role of Retinoids

The liver is rich in vitamin A (retinol) which is mainly stored as retinyl esters in stellate cells located in the space of Disse. Naturally occuring retinoids serve as ligands that activate specific ligand-activated transcription factors including the retinoid X receptor (RXR; official nuclear receptor name NR2B) activated by 9-*cis* retinoic acid and retinoid acid receptor (RAR; NR1B1) activated by all-trans retinoic acid.[115,116] Stellate cells interact with hepatocytes in a stellate cell-to-hepatocyte retinoid signaling pathway. Several nuclear receptors involved in the regulation of hepatobiliary transporter gene transcription (e.g., FXR, PXR) require RXR as the obligate heterodimeric partner.[115,116] An RXR:RAR heterodimer is essential for constitutive expression of *Ntcp* and *Mrp2*.[117] Modulation of transporter gene expression by other signals (e.g., bile acids, cytokines) may also involve alterations of retinoid signaling (see below). In addition, decreased retinol/retinoid levels in cholestatic livers may contribute to reduced retinoid signaling[118,119] and, thereby, to diminished transporter expression (Fig. 2).

Role of Bile Salts

Hepatic Ntcp steady-state mRNA levels are inversely related to serum bile salt levels in rats subjected to CBDL and choledochocaval fistulas (both resulting in elevated serum bile salt levels).[40] Similarly, hepatic levels of human NTCP and OATP2 also correlate inversely with serum bile salt levels in patients with cholestasis,[70] suggesting that endogenously accumulating bile salts may suppress expression of these genes. Bile salts bind to a nuclear bile salt receptor (farnesoid X receptor, FXR; official nuclear receptor name NR1H4)[120,121] which in turn stimulates expression of a transcriptional repressor (short heterodimer partner, SHP-1; NR0B2) resulting in reduced *Ntcp* expression[112,122,123] (Fig. 2). Induction of SHP-1 parallels elevated serum bile salt levels in CBDL mice and precedes the reduction of Ntcp mRNA levels by several hours, suggesting that this mechanisms also plays a role under cholestatic conditions with retention of endogenous bile salts.[124] SHP-1 may suppress *Ntcp* transcription by competing with coactivators for binding to ligand-activated RXR,[125] resulting in reduced RXR:RAR activity which is required for constitutive *Ntcp* expression.[117] Interestingly, SHP-1 also mediates bile salt-induced negative feed back inhibition of cholesterol-7α-hydroxylase, Cyp7a1, the rate limiting enzyme of bile salt synthesis, via interactions with the essential competence factor 'liver receptor homologue' (LRH-1; NR5A2).[112,126,127] In contrast to these inhibitory effects, bile salts stimulate transcription of *Bsep*[123,128,129] and possibly to some extent also *Mrp2*[130] via RXR:FXR binding elements in the promoters of these genes (Fig. 2). These new insights provide the molecular mechanisms how bile salts may suppress their uptake and synthesis while promoting their own excretion, mechanisms which may serve to protect the hepatocyte from bile salt toxicity. Recent evidence also sugests that bile salts may also stimulate the *Oatp4* promoter via FXR in vitro (Gerd Kullak-Ublick, personal communication), effects which could explain the relatively well preserved expression of Oatp4 during cholestasis. Bile salts have also been identified as ligands for the pregnane X receptor (PXR; NHR1I2) which in turn stimulates expression of Oatp2 (involved in bile salt and organic anion/cation transport) and Cyp3a (involved in bile salt hydroxylation and detoxification).[131,132] Again, maintenance/up-regualtion of these genes in cholestasis could be explained through bile salt effects mediated via RXR:PXR. Recently, LRH-1 has been shown to stimulate Mrp3 expression in enterocytes.[133] Whether this mechanism is also responsible for up-regulation of hepatocellular and cholangiocellular Mrp3 in cholestasis remains to be explored. In addition to bile salts, bilirubin has also been implicated in the induction of Mrp3,[54] which could explain elevated Mrp3/MRP3 levels in non-cholestatic hyperbilirubinemic conditions (e.g., transport deficient TR⁻/GY/EHBR rats, Dubin-Johnson syndrome).[53,54] Of note, recent studies have also implicated the constitutive androstane receptor (CAR; NR1I3), another major xenobiotic receptor,[113] in the regulation of hepatic Mrp3 expression.[134] In addition, CAR also appears to play a role in regulating Mrp2 expression in concert with PXR and FXR.[130] Taken together, many of the changes in transporter expression during cholestasis appear to be mediated by bile salts and now can be explained by bile salt-induced activation of nuclear hormone receptors FXR and PXR.

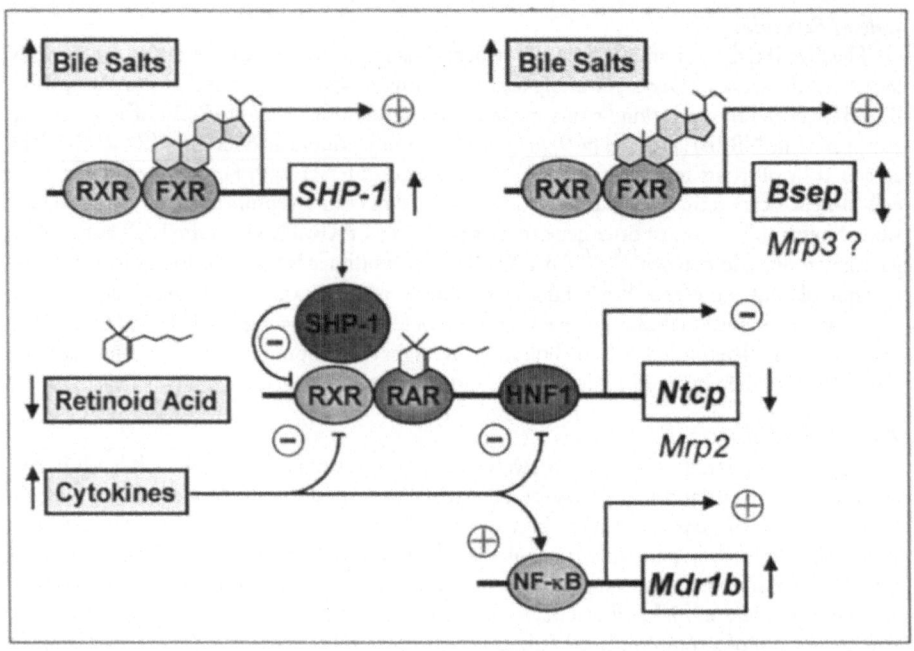

Figure 2. Transcriptional regulation of hepatocellular transport systems in cholestasis. Bile salts (accumulating in cholestasis) activate the nuclear bile salt receptor (farnesoid X receptor FXR) resulting in stimulation of expression of short heterodimer partner 1 (SHP-1) and partial recovery/stabilization of expression of the gene encoding the bile salt export pump (*Bsep*). The molecular mechanisms for induction of *Mrp3* gene expression are still unclear, but appear to be FXR-independent. The transcriptional suppressor SHP-1, together with reduced hepatic retinoid levels in cholestasis and inhibitory effects of proinflammatory cytokines, inhibits constitutive retinoid activation (via the retinoid X receptor RXR and all-trans retinoic acid receptor RAR) of the genes encoding the Na+/taurocholate cotransporter (*Ntcp*) and canalicular conjugate export pump (*Mrp2*). Inhibition of hepatocyte nuclear factor 1 (HNF1) activity by cytokines also results in reduced expression of basolateral transporters such as Ntcp. Cytokines induce expression of the multidrug export pump (Mdr1b) via activation of nuclear factor-kappa B (NF-κB).

Role of Cytokines

Cytokines may reach the liver from extrahepatic inflammatory sites or may be produced within the liver mainly by Kupffer cells.[135,136] Kupffer cell-depletion prevents LPS-induced down-regulation of Mrp2 steady state mRNA levels,[137] suggesting a role for Kupffer cells as intrahepatic source of proinflammatory cytokines.[137,138] In addition, cytokines are also produced by cholangiocytes (e.g., TNF-α, IL-6) and hepatocytes (e.g., TNF-α, IL-8).[135] In vivo administration of TNF-α in mice decreases steady-state mRNA levels of Ntcp, while IL-6 has no effect, despite inhibiting Na+-dependent uptake of conjugated bile acids.[41,139]

Proinflammatory cytokines such as TNF-α and IL-1β inhibit *Ntcp* and *Mrp2* gene transcription at least in part via inhibition of RXR:RAR (previously known as footprint B binding protein, FpB BP).[45,117] LPS reduces RXR α,β,γ mRNA levels and RXRα transcription in rats in vivo.[140] Stress pathway activation induces phosphorylation of RXR via both mitogen-activated protein kinase kinase-4 and its downstream mediator c-Jun N-terminal kinase, resulting in reduced RXR activity.[141] These data indicate that part of the cytokine effects on Ntcp and Mrp2 expression may be mediated at least in part through inhibitory effects of cytokines on retinoid signaling (Fig. 2).[45,117] Of note, some classical bile salts effects on gene transcription (e.g., negative feedback inhibition of Cyp7a1) may by mediated by cytokine and their signaling pathways,[142,143] although this remains to be tested for the specific situation of hepatobiliary transporter regulation.

LPS administration to rats in vivo also inhibits activity of hepatocyte nuclear factor-1α (HNF1α, also known as tcf1)[45] which has been identified as a major transcription factor regulating basolateral transporter expression in hepatocytes (e.g., Ntcp, Oatps).[144-146] However, the hierachy of these changes in relation to changes in the activity of ligand-activated transcription factors is not entirely clear. While reduced HNF1α-activity contributes to reduced *Ntcp* gene transcription in LPS-induced cholestasis in rats[45] (Fig. 2), it appears that stimulation of transporter expression via bile acid-activated transcription factors such as FXR and PXR is able to overcome some effects of reduced HNF1α-activity on Oatp2 and Oatp4 expression, resulting in maintained or even stimulated instead of reduced expression of these transporters in cholestasis[147] (see above). The LPS-induced reduction of HNF1α-activity in inflammatory cholestasis may be secondary to reduced activity of HNF4 which in turn regulates HNF1 transcription.[148] Recent evidence suggests, that the inflammation-inducible transcription factor YB-1 (rat EFI_A) could be involved in suppression of *Mrp2* gene suppression during CBDL.[149]

Proinflammatory cytokines do not always inhibit hepatocellular transporter expression, but can also have stimulatory effects. As such, TNF-α induces Mdr1b (*Abcb1b*) expression in primary rat hepatocyte cultures and rat hepatoma cells, effects which are mediated by the inflammation-inducible nuclear transcription factor nuclear factor-kappa B (NF-κB).[150] In addition, TNF-α increases Mrp1 expression in vitro,[151] but the detailed molecular mechanisms remain unclear.

Taken together, the individual effects of bile acids, cytokines and retinoids on transporter expression may be additive. Therefore, the changes of transporter expression observed during cholestasis may be the result of combined effects of bile acid retention, induction of proinflammatory cytokines, and reduced retinoid levels/signaling (Fig. 2).

Posttranscriptional Mechanisms

In addition to transcriptional events, posttranscriptional changes (affecting mRNA processing, steady-state mRNA stability, translational efficacy) and/or posttranslational changes (such as impaired targeting and sorting, transporter redistribution, enhanced transporter protein degradation, direct protein modifications (e.g., (de-) phosphorylation, (de-) glycosylation), changes in membrane fluidity or cis-/trans-inhibition of transport systems by drugs and other cholestatic agents) can also play an important role in the pathogenesis of cholestasis (Fig. 3).

Translational Efficacy

Estrogen-induced reductions of Mrp2 protein expression are not associated with changes of Mrp2 mRNA levels[48] and may be mediated via changes in translational efficacy of Mrp2 mRNA.[60] So far, potential changes in transporter mRNA stability during cholestasis have not yet been explored in detail.

Impaired Targeting (Reduced Insertion, Increased Retrieval) and Redistribution of Transport Proteins

Insertion / retrieval of transport proteins into / from the plasma membrane via tubulovesicular structures ("vesicular trafficking") is a rapid means of regulating the number of transport proteins and, thereby, transport capacity[107,152-156] (also see related chapter by Kipp & Arias in this book). These processes have been best studied for the canalicular membrane, while similar mechanisms may also apply to the basolateral membrane of hepatocytes, as well as to cholangiocytes.[107] Disturbance of this delicate balance of insertion and retrieval will contribute to cholestasis (Fig. 3). Cholestasis is associated with profound alterations of the cytoskeleton of hepatocytes, including disruption of microtubular system and actin microfilament network, as well as increases in cytokeratin intermediate filaments.[3] These changes may interfere with transcellular vesicle movement, increase tight junctional permeability, and impair contraction

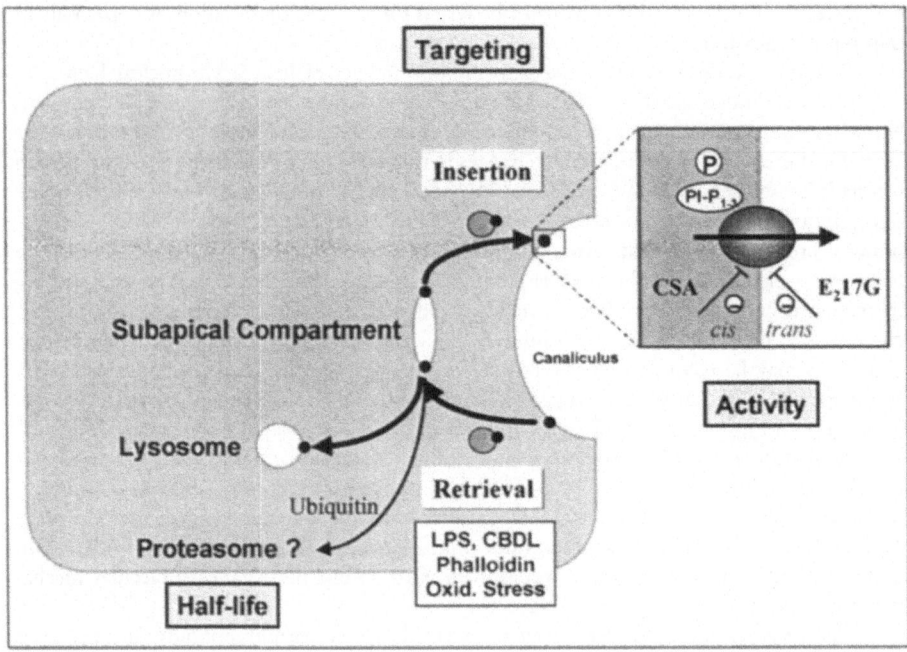

Figure 3. Post-transcriptional regulation of hepatocellular transport systems in cholestasis. Impaired targeting with reduced insertion of transport proteins into to the canalicular membrane will result in a reduced number of functional transporters in the canalicular membrane. Increased retrieval of membrane transporters in response to osmotic stress, endotoxin (LPS) /cytokines, biliary obstruction (CBDL), phalloidin and oxidative stress may also contribute to cholestasis. The retrieved membrane transporters may initially undergo re-insertion from a subapical vesicular compartment, but later are degraded by the lysosomal or ubiquitin-proteasome pathway. Posttranslational modifications of transport proteins (e.g., phosphorylation (P), phosphoinositide (PI)-3-kinase products (PI-P$_{1-3}$)) may rapidly modify transport activity. Cis-inhibition (from cytoplasmatic side; e.g., cylosporin A (CSA)) and trans-inhibition (from canalicular side; e.g., estradiol 17-β-glucuronide (E$_2$17G)) of efflux pumps may be particulary relevant for drug-induced cholestasis.

of bile canaliculi. Impaired vesicular trafficking of transport proteins to the canalicular membrane will result in a reduced number of functional transporters in the canalicular membrane and thus contributes to cholestasis (Fig. 3). It has to be kept in mind that some of the changes in transcellular vesicle movement / vesicular trafficking may be the consequence rather than cause of cholestasis.[107,157,158] Cholestatic concentrations of hydrophobic bile salts such as chenodeoxycholate inhibit microtubular motors such as kinesin and dynein in vitro,[159] which could be a critical factor in the impairment of vesicle movement during mechanical (obstructive) and non-mechanical forms of cholestasis.[160] Phalloidin inhibits actin-dependent insertion of Ntcp into the basolateral membrane,[9] and thus reduces hepatocellular uptake capacity for bile salts.

In addition to impaired targeting to the plasma membrane, increased retrieval of membrane transporters in response to osmotic stress, endotoxin/cytokines, oxidative stress and biliary obstruction may be an important early event in the pathogenesis of cholestasis (Fig. 3).[108,110,161-163] Phalloidin disrupts microfilaments and destabilizes the canalicular membrane with subsequent loss of canalicular membrane proteins through membrane invaginations.[161] Retrieval of transport proteins may be fairly specific for certain transport systems, suggesting a role for specific cytoplasmatic linker proteins (e.g., cortactin binding protein Hax-1), which may group transporters into functonal units and link them to sorting and trafficking networks.[164] The retrieved membrane transporters may initially undergo re-insertion, but later are degraded by the lysosomal or ubiquitin-proteasome pathway.[107,110,152,162]

Cholestasis is also associated with profound alterations (sometimes even loss) of cell polarity and redistribution of canalicular markers to the basolateral membrane.[157,158] It is unclear, whether these changes in cell polarity involve direct mistargeting or retrieval followed by redistribution of canalicular proteins. More recently, lateral diffusion of canalicular transporter (e.g., Mrp2) via leaky tight junctions to the basolateral membrane has been postulated as a potential mechanism.[165] This mechanism could also explain the appearance of basolateral MDR3 staining in human cholestasis (e.g., PBC)[166] (see below).

Changes in Half Life, Protein Degradation (Lysosomal, Proteasomal)

The rapid changes in transporter protein levels cannot sufficiently be explained by changes in mRNA levels, both quantitatively (discrepancies between the changes in protein and mRNA levels) and temporally (dissociation of time courses). Therefore, the rapid reduction of protein levels frequently occuring within the first 24 hours of cholestasis could only be sufficiently explained by premature degradation of transport proteins. Recent evidence suggests, that defective/mutated transport proteins may undergo ubiquitination and end up in the proteasomal degradation pathway,[167,168] although this mechanisms remains to be demonstrated for transport proteins under cholestatic conditions. Moreover, the relative contributions of lysosomal versus proteasomal degradation pathways also remains to be determined. Induction of heat shock proteins may interfere with protein degradation, since heat shock proteins can act as chaperones protecting the transport proteins.[169]

Posttranslational Protein Modifications

Posttranslational modifications of transport proteins (e.g., phosphorylation, glycosylation) potently and rapidly modify transport activity under physiologic conditions (see related chapter by Anwer & Webster in this book). As such, transport activity of mouse Bsep is increased by phosphorylation by PKC in vitro.[170] These findings raise the exciting possibility that Bsep transport activity is regulated by its own substrate, since bile salts are capable of activating specific PKC isoforms.[28,171-174] Moreover, a growing amount of data suggests a major role for phosphoinositide 3-kinase in regulating canalicular transporter targeting and transport activity[153,156,175,176] (see related chapter by Kipp & Arias in this book). Oatp1 looses its transport activity via serine phosphorylation by extracellular ATP.[177] However, the role of posttranslational modifications of transport systems under cholestatic conditions remain to be explored.

Changes on Membrane Fluidity

Cholestasis may also result from altered membrane fluidity with subsequent reduction in the activity of the embedded transporters or enzymes such as Na^+-K^+-ATPase (which maintains the membrane potential and drives Na^+-dependent bile salt uptake via Ntcp). Alterations of membrane fluidity may play an important role in estrogen-induced cholestasis,[170] since estrogen (via modulation of pituitary growth hormone secretion) may increase hepatic low-density lipoprotein receptor expression, resulting in increased cholesterol incorporation into the cell membrane with decreased membrane lipid fluidity and Na^+-K^+-ATPase activity.[178] Decreased basolateral and canalicular membrane fluidity has also been observed after CDBL and may result from changes in phospholipid content.[179]

Direct Inhibition of Transport Activity

Cis-inhibition (from cytoplasmatic side) and trans-inhibition (from canalicular side) of canalicular efflux pumps (e.g., Bsep) may be particulary important for drug-induced cholestasis[180-183] (see related chapter by Kullak-Ublick in this book). Similar mechanisms may also apply to inhibition of basolateral transport systems (e.g., Oatp1 and 2) by drugs (e.g., rifamycin, rifampicin).[184]

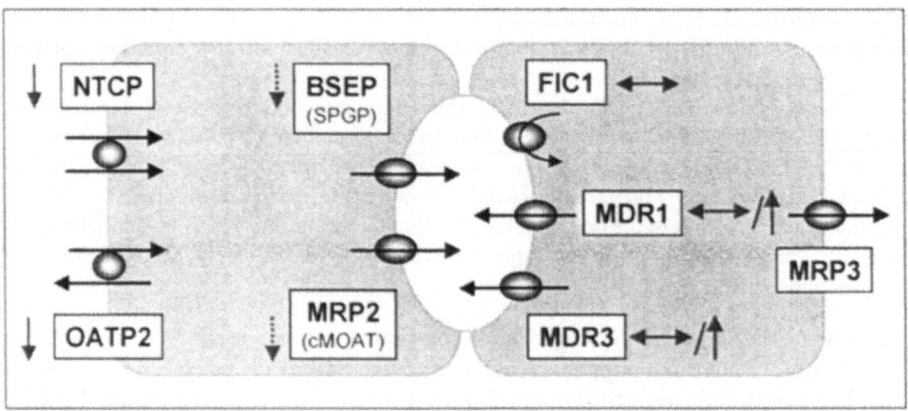

Figure 4. Changes in transporter expression in acute inflammatory cholestasis in humans. Down-regulation of basolateral Na$^+$/taurocholate cotransporter (NTCP) and organic anion transporting protein 2 (OATP2), as well as – to a lesser degree – the canalicular bile salt export pump (BSEP) and (bilirubin) conjugate export pump (MRP2) contributes to inflammation-induced cholestasis caused by drugs or alcohol. Maintainance or even up-regulation of canalicular multidug export pump (MDR1) and phospholipid flippase (MDR3), as well as basolateral MRP3 may help to limit the cholestatic liver injury by providing alternative routes for elimination of bile salts and other biliary constituents.

Acquired Cholestatic Liver Diseases in Humans

While several monogenetic, hereditary cholestatic syndromes now can be attributed to specific mutations of individual hepatobiliary transporter genes[185-188] (see chapters by Jansen et al, Oude Elferink and Nies et al in this book), the information on transporter alterations in acquired cholestatic liver diseases in humans is still fragmentary. Sometimes hereditary and acquired factors causing cholestasis may overlap, since incomplete or heterozygous transport defects/mutations may predispose to acquired cholestatic liver injury (e.g., cholestasis of pregnancy, drug-induced cholestasis). Generally, the acquired changes in transporter expression in human cholestatic liver diseases are consistent with concepts derived from the findings in experimental animal models of cholestasis. As such, expression of the basolateral bile salt transporter NTCP (*SLC10A1*) and organic anion uptake system OATP2 also known as OATP-C (*SLC21A6*), as well as canalicular bile salt export pump BSEP (*ABCB11*) and conjugate export pump MRP2 (*ABCC2*) is reduced in patients with acute inflammation-induced cholestasis (e.g., caused by alcohol, drugs, etc.)[70] (Fig. 4), consistent with findings in endotoxin-treated rats. Reduced MRP2 mRNA expression was reported in patients with chronic hepatitis C and attributed to the action of proinflammatory cytokines.[189] However, the implications of these findings are questionable since these patients were not cholestatic and previous studies showed unchanged MRP2 expression in chronic hepatitis C.[70,190,191] Some transport systems (MRP3/*ABCC3*, MDRs) are well preserved or even up-regulated, which can be interpreted as an attempt to resist the cholestatic liver injury.[70] Similar to the findings in bile duct-ligated rats, NTCP mRNA levels are decreased in patients with extrahepatic biliary atresia and subsequently increase if complete biliary drainage by portoenterostomy (Kasai procedure) is achieved,[192] while BSEP expression appears to be relatively well preserved.[193] Canalicular MDR1 (*ABCB1*) and MDR3 (*ABCB4*) mRNA, as well as hepatocellular and cholangiocellular MRP3 expression are up-regulated in patients with obstructive cholestasis,[194,195] again changes which may help to limit cholestatic liver injury by providing alternative elimination routes for the excretion of bile salts and other (potentially) toxic biliary constituents. The staining pattern for canalicular BSEP and MRP2 appears fuzzy (while overall proteins levels do not change to a significant extent), consistent with a pericanalicular (mis-)localization of transport proteins (e.g., as a result of impaired trageting/increased retrieval under cholestatic conditions; see above) which in turn could contribute (at least in part) to impaired function of these export pumps in obstructive cholestasis.[195]

Alterations in transporter gene expression in primary biliary cirrhosis (PBC), a prototypic chronic cholestatic liver disease, evolve in a disease stage-dependent fashion. In early (anicteric) PBC stages (Ludwig stages I and II) no changes of bile salt and organic anion transporters are seen,[70] indicating that alterations of hepatobiliary transport systems (e.g., down-regulation by proinflammatory cytokines produced in the inflamed portal tracts during the early course of the disease) may not play a central role in the initial pathogenesis of PBC. However, expression and function of the Cl^-/HCO_3^- anion exchanger AE2 (*SLC4A2*) is reduced in cholangiocytes from patients with both early and late stage PBC.[196-198] Since Cl^-/HCO_3^- exchanger activity contributes to secretion of ductular and—to a lesser degree—canalicular bile, decreased hepatic AE2 expression could result in impaired bile flow and cholestasis. Moreover, decreased AE2 expression in salivary and lacrimal glands could explain the frequently associated sicca syndrome in these patients.[199] With disease progression, OATP2 and—to a lesser degree—NTCP expression is down-regulated in later PBC stages (III and IV), while MRP3 and MDR P-glycoprotein expression is up-regulated[200] (Fig. 5). The inverse changes in NTCP (down-regulation) and MRP3 expression (up-regulation) are most predominant in the periportal area (zone 1) of cholate-stasis[166,200,201] where hepatocytes are exposed to the highest bile salt levels under normal and cholestatic conditions. Expression of BSEP and MRP2 even increases in stage III PBC before returning to normal levels in stage IV.[200] This observation is consistent with the presence of canalicular bile plugs in longstanding cholestasis, indicating that bilirubin still is entering the canalicular space. Regurgitation of biliary constituents through leaky tight canalicular tight junctions in PBC could contribute to elevated serum levels of conjugated bilirubin and bile salts in PBC.[25] The transient induction of MRP2 and (to a lesser extent) of BSEP in PBC III may be interpreted as a compensatory attempt to overcome the cholestatic injury before returning to normal range in more advanced stage IV PBC. Similar observations have been made in primary sclerosing cholangitis (PSC).[197] In analogy to the situation in experimental animals where Oatps are diversely regulated, mRNA expression of OATP1 also known as OATPA (*SLC21A3*) is up-regulated in PSC (in contrast to down-regulation of OATP2), which has been interpreted as an attempt to facilitate efflux of cholephiles from cholestatic hepatocytes into the sinusoidal blood.[202] This down-regulation of basolateral uptake systems (NTCP, OATP2) and overexpression of potentially compensatory efflux pumps (MRP3) and organic anion exchanges (OATP1) may protect the hepatocyte from further intracytoplasmatic accumulation of potentially toxic bile acids and other biliary constituents in chronic cholestatic disorders such as PBC and PSC. Moreover, strong expression of MRP3 in proliferating bile ducts may facilitate the return of bile salts from the injured/vanishing bile ducts.[200,201] Therefore, these changes represent adaptive rather than causative changes in chronic cholestasis. Whether the transcriptional and post-transcriptional mechanisms identified in experimental animal models of cholestasis (e.g., alterations of ligand-activated transcription factors; mistargeting and retrieval of transporters from the canalicular membrane, etc.) also apply to the human situation remains to be explored. Future studies will have to investigate the regulatory mechanisms of hepatobiliary transport systems in humans and to which extent experimental findings in mice and rats can be applied to human cholestatic liver diseases.[31]

Conclusions and Outlook

Exposure to cholestatic injury such as drugs, hormones, proinflammatory cytokines, and biliary obstruction, or the combination of hereditary and exogenous cholestatic factors result in altered expression and function of hepatobiliary transport systems. While some of these alterations may contribute to or maintain cholestasis, others (especially with prolonged duration of cholestasis) may be the result of adaptive processes which aim at limiting the extent of

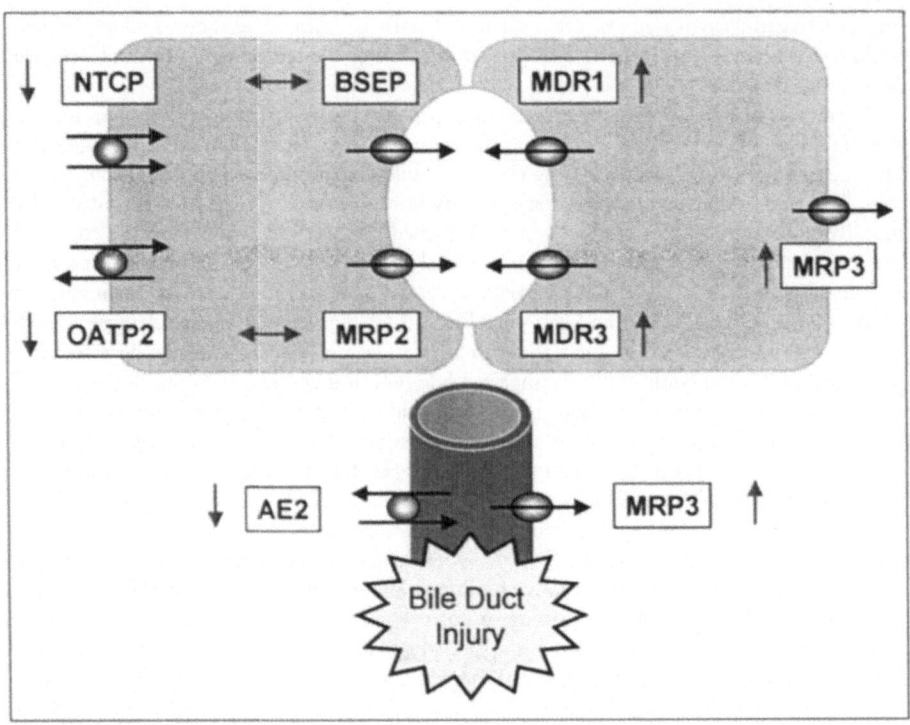

Figure 5. Changes in transporter expression in chronic cholestasis in humans (e.g., primary biliary cirrhosis). Down-regulation of basolateral uptake systems (NTCP, OATP2) and maintainance (BSEP, MRP2) or overexpression (MRP3) of efflux pumps may protect the hepatocyte from further intracytoplasmatic accumulation of potentially toxic bile acids and other biliary constituents in longstanding, chronic cholestasis. In addition, expression of the Cl^-/HCO_3^- anion exchanger (AE2) is reduced in hepatocytes (not shown) and cholangiocytes, which may contribute to impaired bile secretion in these patients. Increased expression of MRP3 in proliferating bile ducts may facilitate the return of bile salts from vanishing bile ducts. Therefore, these changes may represent secondary, adaptive changes in response to ongoing bile duct injury and progressive chronic cholestasis.

the cholestatic liver injury. These new insights may not only improve our understanding of the molecular mechanisms of cholestasis but may also be clinically relevant, since therapeutic strategies can target these molecular changes in transporter expression. Drugs used or proposed to treat cholestatic liver diseases (e.g., ursodeoxycholic acid, steroids, fibrates) may act via stimulation, restoration and/or reinsertion of defective hepatobiliary transport systems[93,173,203-208] (see related chapter by Rust & Beuers in this book). In addition, therapeutic strategies could be aimed at supporting and stimulating adaptive rescue pathways (e.g., alternative elimination routes for bile salts) in cholestasis. The increasing information on the molecular regulation of hepatobiliary transport systems should lead to the identification of novel therapies for cholestatic liver diseases.

Acknowledgements
This work was supported by the Jubilee Funds of the Austrian National Bank (grants 7171 and 8522), the Joseph Skoda Prize of the Austrian Society of Internal Medicine, and the Austrian Science Foundation (grant P15502).

References

1. Kullak-Ublick GA, Meier PJ. Mechanisms of cholestasis. Clinics in Liver Disease 2000; 4:357-385.
2. Trauner M, Meier PJ, Boyer JL. Molecular pathogenesis of cholestasis. N Engl J Med 1998; 339:1217-1227.
3. Phillips MJ, Poucell S, Oda M. Biology of disease: mechanisms of cholestasis. Lab Invest 1986; 54:593-608.
4. Trauner M, Meier PJ, Boyer JL. Molecular regulation of hepatocellular transport systems in cholestasis. J Hepatol 1999; 31:165-178.
5. Lee J, Boyer JL. Molecular alterations in hepatocyte transport mechanisms in acquired cholestatic liver disorders. Semin Liver Dis 2000; 20:373-384.
6. Whitehead MW, Hainsworth I, Kingham JG. The causes of obvious jaundice in South West Wales: perceptions versus reality. Gut 2001; 48:409-413.
7. Feldmann G. The cytoskeleton of the hepatocyte. Structure and functions. J Hepatol 1989; 8:380-386.
8. McNiven MA, Marlowe KJ. Contributions of molecular motor enzymes to vesicle-based protein transport in gastrointestinal epithelial cells. Gastroenterology 1999; 116:438-451.
9. Dranoff JA, McClure M, Burgstahler AD et al. Short-termregulation of bile acid uptake by microfilament-dependent translocation of rat ntcp to the plasma membrane. Hepatology 1999; 30:223-229.
10. Omary MB, Ku NO. Intermediate filament proteins of the liver: emerging disease associations and functions. Hepatology 1997; 25:1043-1048.
11. Kawahara H, Cadrin M, Perry G et al. Role of cytokeratin intermediate filaments in transhepatic transport and bile secretion. Hepatology 1990; 11:435-338.
12. Van Eyken P, Desmet VJ. Cytokeratins and the liver. Liver 1993; 13:113-122.
13. Fickert P, Trauner M, Fuchsbichler A et al. Cytokeratins as targets for bile acid-induced toxicity. Am J Pathol 2002; 160:491-9.
14. Song JY, Van Noorden CJF, Frederiks WM. Alterations of hepatocellular intermediate filaments during extrahepatic cholestasis in rat liver. Virchows Arch1997; 430:253-260.
15. Anderson JM. Leaky junctions and cholestasis: a tight correlation. Gastroenterology 1996; 110:1662-1674.
16. Mitic LL, Anderson JM. Molecular architecture of tight junctions. Annu Rev Physiol 1998; 60:121-124.
17. Mitic LL, Van Italie CM, Anderson JM. Molecular physiology and pathophysiology oftight junctions. I. Tight junction structure and function: lessons from mutant animals and proteins. Am J Physiol 2000; 279:G250-G254.
18. Anderson JM, Glade JL, Stevenson BR et al. Hepatic immunohistochemical localization of the tight junction protein ZO-1 in rat models of cholestasis. Am J Pathol 1989; 134:1055-1062.
19. Fallon MB, Mennone A, Anderson JM. Altered expression and localization of the tight junction protein ZO-1 after common bile duct ligation. Am J Physiol 1993; 264:C1439-C1447.
20. Fallon MB, Brecher AR, Balda MS et al. Altered hepatic localization and expression of occludin after common bile duct ligation. Am J Physiol 1995; 269:C1057-1062.
21. Rahner C, Stieger B, Landmann L. Structure-function correlation of tight junctional impairment following intrahepaticand extrahepatic cholestasis in rat liver. Gastroenterology 1996; 110:1564-1578.
22. Kawaguchi T, Sakisaka S, Sata M et al. Different lobular distributions of altered hepatocyte tight junctions in rat models of intrahepatic and extrahepatic cholestasis. Hepatology 1999; 29:205-216.
23. Lora L, Mazzon E, Martines D et al. Hepatocyte tight junctional permeability is increased in rat experimental colitis. Gastroenterology 1997; 113:1347-1354.
24. Kawaguchi T, Sakisaka S, Mitsuyama K et al. Cholestasis with altered structure and function of hepatocyte tight junction and decreased expression of canalicular multispecific organic anion transporter in a rat model of colitis. Hepatology 2000; 31:1285-1295.
25. Sakisaka S, Kawaguchi T, Taniguchi E et al. Alterations in tight junctions differ between primary biliary cirrhosis and primary sclerosing cholangitis. Hepatology 2001; 33:1460-8.
26. Fallon MB, Nathanson MH, Mennone A et al. Altered expression and function of hepatocyte gap junctions after common bile duct ligation in the rat. Am J Physiol 1995; 268:C1186-1194.
27. Nathanson MH, Schlosser SF. Calcium signaling mechanisms in liver in health and disease. In: Boyer JL, Ockner RK, eds. Progress in Liver Disease. Philadelphia: W.B. Saunders Company, 1996:1-27.
28. Bouscarel B, Kroll SD, Fromm H. Signal transduction and hepatocellular bile acid transport:cross talk between bile acids and second messengers. Gastroenterology 1999; 117:433-452.

29. Misra S, Ujhazy P, Gatmaitan Z et al. The role of phosphoinositide 3-kinase in taurocholate-induced trafficking of ATP-dependent canaliculat transporters in rat liver. J Biol Chem 1998; 273:26638-26644.
30. Misra S, Ujhazy P, Varticovski L et al. Phosphoinositide 3-kinase lipid products regulate ATP-dependent transport by sister of P-glycoprotein and multidrug resistance associated protein 2 in bile canalicular membrane vesicles. Proc Natl Acad Sci USA 1999; 96:5814-5819.
31. Boyer JL. Advancing the bile-ology of cholestatic liver disease. Hepatology 2001; 33:758-9.
32. Hofmann AF. Bile acids: the good, the bad, and the ugly. News Physiol Sci 1999; 14:24-29.
33. Kullak-Ublick GA, Stieger B, Hagenbuch B et al. Hepatic transport of bile salts. Semin Liver Dis 2000; 20:273-9.
34. Meier PJ, Stieger B. Bile salt transporters. Annu Rev Physiol 2002; 64:635-61.
35. Boyer JL. Nathanson: Bile formation. In: Schiff ER, Sorrell MF, Maddrey WC, eds. Schiff's Diseases of the Liver, Ed 8. Philadelphia: Lippincott-Raven Publishers, 1999:119-146.
36. Kanno N, LeSage G, Glaser S et al. Functional heterogeneity of the intrahepatic biliary epithelium. Hepatology 2000;31:555-61.
37. LeSage G, Glaser S, Alpini G. Regulation of cholangiocyte proliferation. Liver 2001; 21:73-80.
38. Baiocchi L, LeSage G, Glaser S et al. Regulation of cholangiocyte bile secretion. J Hepatol 1999; 31:179-191.
39. Arrese M, Ananthananarayanan M, Suchy FJ. Hepatobiliary transport: molecular mechanisms of development and cholestasis. Pediatr Res 1998; 44:141-7
40. Gartung C, Ananthanarayanan M, Rahman MA et al. Down-regulation of expression and function of the rat liver Na+/bile acid cotransporter in extrahepatic cholestasis. Gastroenterology 1996; 110:199-209.
41. Green RM, Beier D, Gollan JL. Regulation of hepatocyte bile salt transporters by endotoxin and inflammatory cytokines in rodents. Gastroenterology 1996, 111:193-198.
42. Moseley RH, Wang W, Takeda H et al. Effect of endotoxin on bile acid transport in rat liver:a potential model for sepsis-associated cholestasis. Am J Physiol 1996, 271:G137-G146.
43. Bolder U, Ton-Nu HT, Schteingart CD et al. Hepatocyte transport of bile acids and organic anions in endotoxemic rats:impaired uptake and secretion. Gastroenterology 1997;112:214-25.
44. Simon FR, Fortune J, Iwahashi M et al. Ethinyl estradiol cholestasis involves alterations in expression of liver sinusoidal transporters. Am J Physiol 1997, 271:G1043-G1052.
45. Trauner M, Arrese M, Lee H et al. Endotoxin downregulates rat hepatic *ntcp* gene expression via decreased activity of critical transcription factors. J Clin Invest 1998; 101:2092-2100 .
46. Dumont M, Jacquemin E, D'Hont C et al. Expression of the liver Na⁺-independent organic anion transporting polypeptide (Oatp-1) in rats with bile duct ligation. J Hepatol 1997; 27:1051-1056.
47. Vos TA, Hooiveld GJEJ, Koning H et al. Up-regulation of the multidrug resistance genes, mrp1 and mdr1b, and down-regulation of the organic anion transporter, mrp2, and the bile salt transporter, spgp, in endotoxemic rat liver. Hepatology 1998; 28:1637-1644.
48. Trauner M, Arrese M, Soroka CJ et al. The rat canalicular conjugate export pump (Mrp2) Is down-regulated in intrahepatic and obstructive cholestasis. Gastroenterology 1997; 113:255-264.
49. Lee JM, Trauner M, Soroka CJ et al. The molecular expression of the bile salt export pump, sister of P-glycoprotein, is selectively preserved in cholestatic liver injury. Gastroenterology 2000; 118:163-172.
50. Kakyo M, Unno M, Tokui T et al. Molecular characterization and functional regulation of a novel rat liver-specific organic anion transporter rlst-1. Gastroenterology 1999; 117:770-775.
51. Lund M, Kang L, Tygstrup N et al. P. Effects of LPS on transport of indocyanine green and alanine uptake in perfused rat liver. Am J Physiol 1999; 277:G91-100.
52. Buscher HP, Miltenberger C, MacNelly S et al. The histoautoradiographic localization of taurocholate in rat liver after bile duct ligation. evidence for ongoing secretion and reabsorption processes. J Hepatol 1989; 8:181-191.
53. Hirohashi T, Suzuki H, Ito K et al. Hepatic expresion of multidrug resistance-associated protein-like proteins maintained in Eisai hyperbilirubinemic rats. Molecular Pharmacol 1998; 53:1068-1075.
54. Ogawa K, Suzuki H, Hirohashi T et al. Characterization of inducible nature of MRP3 in rat liver. Am J Physiol 2000; 278:G438-446.
55. Schrenk D, Gant TW, Preisegger KH et al. Induction of multidrug restance gene expression during cholestasis in rats and nonhuman primates. Hepatology 1993; 17:854-860.
56. Soroka CJ, Lee JM, Azzaroli F et al. Cellular localization and up-regulation of multidrug resistance-associated protein 3 in hepatocytes and cholangiocytes during obstructive cholestasis in rat liver. Hepatology 2001; 33:783-91.
57. Donner MG, Keppler D. Up-regulation of basolateral multidrug resistance protein 3 (Mrp3) in cholestatic rat liver. Hepatology 2001; 34:351-9.

58. Pei QL, Kobayashi Y, Tanaka Y et al. Increased expression of multidrug resistance-associated protein 1 (mrp1) in hepatocyte basolateral membrane and renal tubular epithelia after bile duct ligation in rats. Hepatol Res 2002; 22:58-64.

59. Tanaka Y, Kobayashi Y, Gabazza EC et al.. Increased renal expression of bilirubin glucuronide transporters in a rat model of obstructive jaundice. Am J Physiol Gastrointest Liver Physiol 2002;282:G656-62.

60. Cao J, Vore M. Expression and functional characterization of hepatic multidrug resistance-associated proteins (Mrp) in pregnancy (Abstract). Hepatology 2001; 34:474A.

61. Raedsch R, Lauterburg BH, Hofmann AF. Altered bile acid metabolism in primary biliary cirrhosis. Dig Dis Sci 1981;26:394-401.

62. Hirohashi T, Suzuki H, Takikawa H et al. ATP-dependent transport of bile salts by rat multidrug resistance-associated protein 3 (Mrp3). J Biol Chem 2000; 275:2905-10.

63. Akita H, Suzuki H, Ito K et al. Characterization of bile acid transport mediated by multidrug resistance associated protein 2 and bile salt export pump. Biochim Biophys Acta 2001; 1511:7-16.

64. Akita H, Suzuki H, Hirohashi T et al.. Transport activity of human MRP3 expressed in Sf9 cells:comparative studies with rat MRP3. Pharm Res 2002; 19:34-41.

65. Akita H, Suzuki H, Sugiyama Y. Sinusoidal efflux of taurocholate is enhanced in Mrp2-deficient rat liver. Pharm Res 2001; 18:1119-25.

66. Kullak-Ublick GA. Regulation of organic anion and drug transporters of the sinusoidal membrane. J Hepatol 1999; 31:563-573.

67. Greim H, Trulzsch D, Czygan P et al. Mechanism of cholestasis. 6. Bile acids in human livers with or without biliary obstruction. Gastroenterology 1972;63:846-50.

68. Greim H, Trulzsch D, Roboz J et al. Mechanism of cholestasis. 5. Bile acids in normal rat livers and in those after bile duct ligation. Gastroenterology 1972; 63:837-45.

69. Eppen EF, van Mil SW, de Vree JM et al. FIC1, the protein affected in two forms of hereditary cholestasis, is localized in the cholangiocyte and the canalicular membrane of the hepatocyte. J Hepatol 2001; 35:436-43.

70. Zollner G, Fickert P, Zenz R et al. Hepatobiliary transporter expression in percutaneous liver biopsies of patients with cholestatic liver diseases. Hepatology 2001; 33:633-46.

71. Oude Elferink RPJ, Van Berge Henegouwen GP. Cracking the genetic code for benign recurrent and progressive familial intrahepatic cholestasis. J Hepatol 1998; 29:317-320.

72. Stieger B. FIC1. another bile salt carrier within the enterohepatic circulation? J Hepatol 2001; 35:522-4.

73. Roberts SK, Ludwig J, LaRusso NF. The pathobiology of biliary epithelia. Gastroenterology 1997; 112:269-279.

74. Alvaro D, Gigliozzi A, Attili AF. Regulation and deregulation of cholangiocyte proliferation. J Hepatol 2000; 33:333-40.

75. Theise ND, Saxena R, Portmann BC et al. The canals of Hering and hepatic stem cells in humans. Hepatology 1999; 30:1425-33.

76. Ros JE, Roskams T, Geuken M et al. The ABC transporter genes MDR1/Mdr1b, MRP1 and MRP3 are highly expressed in hepatic progenitor cells (Abstract). Hepatology 2001; 34:473A.

77. Alpini G, Glaser SS, Rodgers R et al. Functional expression of the apical Na⁺-dependent bile acid transporter in large but not small rat cholangiocytes. Gastroenterology 1997; 113:1734-1740.

78. Lazaridis KN, Pham L, Tietz P et al. Rat cholangiocytes absorb bile acids at their apical domain via the ileal sodium-dependent bile acid transporter. J Clin Invest 1997; 100:2714-2721.

79. Alpini G, Glaser S, Ueno Y et al. Bile acid feeding induces cholangiocyte proliferation and secretion; evidence for bile acid-regulated ductal secretion. Gastroenterology 1999; 116:179-186.

80. Alpini G, Ueno Y, Glaser SS et al. Bile acid feeding increased proliferative activity and apical bile acid transporter expression in both small and large rat cholangiocytes. Hepatology 2001; 34:868-76.

81. Slott PA, Liu MH, Tavoloni N. Origin, pattern, and mechanism of bile duct proliferation following biliary obstruction in the rat. Gastroenterology 1990; 99:466-77.

82. Elsing C, Fitscher BA, Boker C et al. Expression of a bile acid transporter in biliary epithelial cells from normal and cholestatic rat livers. Eur J Med Res 1999; 27:165-168.

83. Lee J, Azzaroli F, Wang L et al. Adaptive regulation of bile salt transporters in kidney and liver in obstructive cholestasis in the rat. Gastroenterology 2001; 121:1473-84.

84. Lazaridis KN, Tietz P, Wu T et al. Alternative splicing of the rat sodium/bile acid transporter changes its cellular localization and transport properties. Proc Natl Acad Sci USA 2000; 97:11092-11097.

85. Meerman L, Koopen NR, Bloks V et al. Biliary fibrosis associated with altered bile composition in a mouse model of erythropoietic protoporphyria. Gastroenterology 1999; 117:696-705.

86. Marinelli RA, Pham L, Agre P et al. Secretin promotes osmotic water transport in rat cholangiocytes by increasing aquaporin-1 water channels in plasma membrane. Evidence for a secretin-induced vesicular translocation of aquaporin-1. J Biol Chem 1997; 272:12984-12988.
87. Marinelli RA, Tietz PS, Pham LD et al. Secretin induces the apical insertion of aquaporin-1 water channels in rat cholangiocytes. Am J Physiol 1999; 276:G280-G286.
88. Tietz PS, Hadac EM, Miller LJ et al. Upregulation of secretin receptors on cholangiocytes after bile duct ligation. Regul Pept 2001; 97:1-6.
89. Kanno N, LeSage G, Glaser S et al. Regulation of cholangiocyte bicarbonate secretion. Am J Physiol Gastrointest Liver Physiol 2001; 281:G612-25.
90. Alpini G, Elias I, Glaser SS et al. gamma-Interferon inhibits secretin-induced choleresis and cholangiocyte proliferation in a murine model of cirrhosis. J Hepatol 1997; 27:371-80.
91. Spirli C, Nathanson MH, Fiorotto R et al. Proinflammatory cytokines inhibit secretion in rat bile duct epithelium. Gastroenterology 2001; 121:156-69.
92. McGill JM, Yen MS, Cummings OW et al. Interleukin-5 inhibition of biliary cell chloride currents and bile flow. Am J Physiol Gastrointest Liver Physiol 2001; 280:G738-45.
93. Alvaro D, Gigliozzi A, Marucci L et al. Corticosteroids modulate the secretory processes of the rat intrahepatic biliary epithelium. Gastroenterology 2002; 122:1058-1069.
94. Dumaswala R, Berkowitz D, Heubi JE. Adaptive response of the enterohepatic circulation of bile acids to extrahepatic cholestasis. Hepatology 1996; 23:623-9.
95. Sauer P, Stiehl A, Fitscher BA et al. Downregulation of ileal bile acid absorption in bile-duct-ligated rats. J Hepatol 2000; 33:2-8.
96. Arrese M, Trauner M, Sacchiero RJ et al. Neither intestinal sequestration of bile acids nor common bile duct ligation modulate the expression and function of the rat ileal bile acid transporter. Hepatology 1998; 28:1081-7.
97. Rost D, Mahner S, Sugiyama Y et al. Expression and localization of the multidrug resistance-associated protein 3 in rat small and large intestine. Am J Physiol Gastrointest Liver Physiol 2002; 282:G720-6.
98. St-Pierre MV, Kullak-Ublick GA, Hagenbuch B et al. Transport of bile acids in hepatic and non-hepatic tissues. J Exp Biol 2001; 204:1673-86.
99. Schaub TP, Kartenbeck J, Konig J et al. Expression of the conjugate export pump encoded by the mrp2 gene in the apical membrane of kidney proximal tubules. J Am Soc Nephrol 1997; 8:1213-21.
100. Arrese M, Trauner M, Ananthanarayanan M et al. Maternal cholestasis does not affect the ontogenic pattern of expression of the Na+/taurocholate cotransporting polypeptide (ntcp) in the fetal and neonatal rat liver. Hepatology 1998; 28:789-95.
101. Zinchuk VS, Okada T, Akimaru K et al. Asynchronous expression and colocalization of Bsep and Mrp2 during development of rat liver. Am J Physiol Gastrointest Liver Physiol 2002; 282:G540-8.
102. St-Pierre MV, Serrano MA, Macias RI et al. Expression of members of the multidrug resistance protein family in human term placenta. Am J Physiol Regul Integr Comp Physiol 2000; 279:R1495-503.
103. St-Pierre MV, Hagenbuch B, Ugele B et al. Characterization of an Organic anion transporting Polypeptide (OATP-B) in Human Placenta. J Clin Endocrinol Metab 2002; 87:1856-63.
104. Macias RI, Pascual MJ, Bravo A et al. Effect of maternal cholestasis on bile acid transfer across the rat placenta-maternal liver tandem. Hepatology 2000; 31:975-83.
105. Serrano MA, Brites D, Larena MG et al. Beneficial effect of ursodeoxycholic acid on alterations induced by cholestasis of pregnancy in bile acid transport across the human placenta. J Hepatol 1998; 28:829-39.
106. Muller M. Transcriptional control of hepatocanalicular transporter gene expression. Semin Liver Dis 2000; 20:323-337.
107. Haeussinger D, Schmitt M, Weiergraeber O et al. Short-term regulation of canalicular transport. Semin Liver Dis 2000; 20:307-321.
108. Schmitt M, Kubitz R, Wettstein M et al. Retrieval of the mrp2 gene encoded conjugate export pump from the canalicular membrane contributes to cholestasis induced by tert-butyl hydroperoxideand chloro-dinitrobenzene. Biol Chem 2000; 381:487-495.
109. Paulusma CC, Kothe MJ, Bakker CT et al. Zonal down-regulation and redistribution of the multidrug resistance protein 2 during bile duct ligation in rat liver. Hepatology 2000; 31:684-93.
110. Kubitz R, Wettstein M, Warskulat U et al. Regulation of the multidrug resistance protein 2 in the rat liver by lipopolysaccharide and dexamethasone. Gastroenterology 1999; 116:401-410.,
111. Kim PK, Chen J, Andrejko KM et al. Intraabdominal sepsis down-regulates transcription of sodium taurocholate cotransporter and multidrug resistance-associated protein in rats. Shock 2000; 14:176-81.
112. Chawla A, Saez E, Evans RM. Don't know much bile-ology. Cell 2000; 103:1-4.

113. Chawla A, Repa JJ, Evans RM et al. Nuclear receptors and lipid physiology: opening the X-files. Science 2001; 294:1866-70.
114. Lu TT, Repa JJ, Mangelsdorf DJ. Orphan nuclear receptors as eLiXiRs and FiXeRs of sterol metabolism. J Biol Chem 2001; 276:37735-8.
115. Mangelsdorf DJ, Evans RM. The RXR heterodimers and orphan receptors. Cell 1995; 83:841-50.
116. Mangelsdorf DJ, Thummel C, Beato M et al. The nuclear receptor superfamily:the second decade. Cell 1995; 83:835-9.
117. Denson LA, Auld KL, Schiek DS et al. Interleukin-1 beta suppresses retinoid transactivation of two hepatic transporter genes involved in bile formation. J Biol Chem 2000; 275:8835-43.
118. Ohata M, Lin M, Satre M et al. Diminished retinoic acid signaling in hepatic stellate cells in cholestatic liver fibrosis. Am J Physiol 1997; 272:G589-96.
119. Floreani A, Baragiotta A, Martines D et al. Plasma antioxidant levels in chronic cholestatic liver diseases. Aliment Pharmacol Ther 2000; 14:353-8.
120. Makishima M, Okamoto AY, Repa JJ et al. Identification of a nuclear receptor for bile acids. Science 1999; 284:1362-1365.
121. Parks DJ, Blanchard SG, Bledsoe RK et al. Bile acids:natural ligands for an orphan nuclear receptor. Science 1999; 284:1365-1368.
122. Denson LA, Sturm E, Echevarria W et al. The orphan nuclear receptor, shp, mediates bile acid-induced inhibition of the rat bile acid transporter, ntcp. Gastroenterology 2001; 121:140-7.
123. Sinal CJ, Tohkin M, Miyata M et al. Targeted disruption of the nuclear receptor FXR/BAR impairs bile acid and lipid homeostasis. Cell 2000; 102:731-44.
124. Zollner G, Fickert P, Silbert D et al. Induction of short heterodimer partner 1 precedes downregulation of Ntcp in bile duct-ligated mice. Am J Physiol Gastrointest Liver Physiol 2002; 282:G184-91.
125. Lee YK, Dell H, Dowhan DH et al. The orphan nuclearreceptor SHP inhibits hepatocyte nuclear factor 4 and retinoid X receptor transactivation: two mechanisms for repression. Mol Cell Biol 2000; 20:187-95.
126. Lu TT, Makishima M, Repa JJ et al. Molecular basis for feedback regulation of bile acid synthesis by nuclear receptors. Mol Cell 2000; 6:507-15.
127. Goodwin B, Jones SA, Price RR et al. A regulatory cascade of the nuclear receptors FXR, SHP-1, and LRH-1 represses bile acid biosynthesis. Mol Cell 2000; 6:517-26.
128. Ananthanarayanan M, Balasubramanian N, Makishima M et al. Human bile salt export pump promoter is transactivated by the farnesoid X receptor/bile acid receptor. J Biol Chem 2001; 276:28857-65.
129. Plass JR, Mol O, Heegsma J et al. Farnesoid X receptor and bile salts are involved in transcriptional regulation of the gene encoding the human bile salt export pump. Hepatology 2002; 35:589-96.
130. Kast HR, Goodwin B, Tarr PT et al. Regulation of multidrug resistance-associated protein 2 (ABCC2) by the nuclear receptors pregnane X receptor, farnesoid X-activated receptor, and constitutive androstane receptor. J Biol Chem 2002; 277:2908-15.
131. Xie W, Radominska-Pandya A, Shi Y et al. An essential role for nuclear receptors SXR/PXR in detoxification of cholestatic bile acids. Proc Natl Acad Sci USA 2001; 98:3375-80.
132. Staudinger JL, Goodwin B, Jones SA et al. The nuclear receptor PXR is a lithocholic acid sensor that protects against liver toxicity. Proc Natl Acad Sci USA 2001; 98:3369-74.
133. Inokuchi A, Hinoshita E, Iwamoto Y et al. Enhanced expression of the human multidrug resistance protein 3 by bile salt in human enterocytes. A transcriptional control of a plausible bile acid transporter. J Biol Chem 2001; 276:46822-9.
134. Cherrington NJ, Hartley DP, Li N et al. Organ distribution of multidrug resistance proteins 1, 2, and 3 (Mrp1, 2, and 3) mRNA and hepatic induction of Mrp3 by constitutive androstane receptor activators in rats. J Pharmacol Exp Ther 2002; 300:97-104.
135. Crawford JM, Boyer JL. Clinicopathology conferences: inflammation-induced cholestasis. Hepatology 1998; 28:253-260.
136. Trauner M, Fickert P, Stauber RE. Inflammation-induced cholestasis. J Gastroen Hepatol 1999; 14:946-959.
137. Nakamura J, Nishida T, Hayashi K et al. Kupffer cell-mediated down regulation of rat hepatic CMOAT/MRP2 gene expression. Biochem Biophys Res Commun 1999; 255:143-9.
138. Sturm E, Zimmerman TL, Crawford AR et al. Endotoxin-stimulated macrophages decrease bile acid uptake in WIF-B cells, a rat hepatoma hybrid cell line. Hepatology 2000; 31:124-30.
139. Green RM, Whiting JF, Rosenbluth AB et al. Interleukin-6 inhibits hepatocyte taurocholate uptake and sodium-potassium-adenosinetriphosphatase activity. Am J Physiol 1994; 267:G1094-100.

140. Beigneux AP, Moser AH, Shigenaga JK et al. The acute phase response is associated with retinoid X receptor repression in rodent liver. J Biol Chem 2000; 275:16390-9.
141. Lee HY, Suh YA, Robinson MJ et al. Stress pathway activation induces phosphorylation of retinoid X receptor. J Biol Chem 2000; 275:32193-9.
142. Miyake JH, Wang SL, Davis RA. Bile acid induction of cytokine expression by macrophages correlates with repression of hepatic cholesterol 7alpha-hydroxylase. J Biol Chem 2000; 275:21805-8.
143. Gupta S, Stravitz RT, Dent P et al. Down-regulation of cholesterol 7alpha-hydroxylase (CYP7A1) gene expression by bile acids in primary rat hepatocytes is mediated by the c-Jun N-terminal kinase pathway. J Biol Chem 2001; 276:15816-22.
144. Karpen SJ, Sun AQ, Kudish B et al. Multiple factors regulate the rat liver basolateral sodium-dependent bile acid cotransporter gene promoter. J Biol Chem 1996; 271:15211-21.
145. Jung D, Hagenbuch B, Gresh L et al. Characterization of the human OATP-C (SLC21A6) gene promoter and regulation of liver-specific OATP genes by hepatocyte nuclear factor 1 alpha. J Biol Chem 2001; 276:37206-14.
146. Shih DQ, Bussen M, Sehayek E et al. Hepatocyte nuclear factor-1alpha is an essential regulator of bile acid and plasma cholesterol metabolism. Nat Genet 2001; 27:375-82.
147. Kullak-Ublick GA, Jung D, Hagenbuch B et al. Organic anion transporting polypeptides, cholestasis, and nuclear receptors. Hepatology 2002; 35:732-3.
148. Wang B, Cai SR, Gao C et al. Lipopolysaccharide results in a marked decrease in hepatocyte nuclear factor 4 alpha in rat liver. Hepatology 2001; 34:979-89.
149. Geier A. Mertens RR, Gerloff T et al. Down-regulation of Mrp2 gene transcription in CBDL-treted rats is mediated by transcription factor EFI$_A$/YB-1 (Abstract). Hepatology 2001; 34:470A.
150. Ros JE, Schuetz JD, Geuken M et al. Induction of Mdr1b expression by tumor necrosis factor-alpha in rat liver cells is independent of p53 but requires NF-kappaB signaling. Hepatology 2001; 33:1425-31.
151. Stein U, Walther W, Laurencot CM et al. Tumor necrosis factor-alpha and expression of the multidrug resistance-associated genes LRP and MRP. J Natl Cancer Inst 1997; 89:807-13.
152. Kubitz R, D'Urso D, Keppler, D et al. Osmodependent dynamic localization of the multidrug resistance protein 2 in the rat hepatocyte canalicular membrane. Gastroenterology 1997; 113:1438-1442.
153. Kipp H, Arias IM. Newly synthesized canalicular ABC transporters are directly targeted from the Golgi to the hepatocyte apical domain in rat liver. J Biol Chem 2000; 275:15917-15925.
154. Soroka CJ, Pate MK, Boyer JL. Canalicular export pumps traffic with polymeric immunoglobulin A receptor on the same microtubule-associated vesicle in rat liver. J Biol Chem 1999; 274:26416-26424.
155. Webster CRL, Blanch CJ, Phillips J et al. Cell swelling-induced translocation of rat liver Na$^+$/ taurocholate cotransport polypeptide is mediated via the phosphoinositide 3-kinase signaling pathway. J Biol Chem 2000; 275:29754-29760.
156 Kipp H, Arias IM. Intracellular trafficking and regulation of canalicular ATP-binding cassette transporters. Semin Liver Dis 2000; 20:339-351.
157. Stieger B, Landmann L. Effects of cholestasis on membrane flow and surface polarity in hepatocytes. J Hepatol 1996; 24(Suppl 1):128-34.
158. Stieger B, Meier PJ, Landmann L. Effect of obstructive cholestasis on membrane traffic and domain-specific expression of plasma membrane proteins in rat liver parenchymal cells. Hepatology 1994; 20:201-12.
159. Marks DL, La Russo NF, McNiven MA. Isolation of the microtubile-vesicle motor kinesin from rat liver: selective inhibition by cholestatic bile acids. Gastroenterology 1995; 108:824-833.
160. Torok NJ, Larusso EM, McNiven MA. Alterations in vesicle transport and cell polarity in rat hepatocytes subjected to mechanical or chemical cholestasis. Gastroenterology 2001; 121:1176-84.
161. Rost D, Kartenbeck J, Keppler D. Changes in the localization of the rat canalicular conjugate export pump mrp2 in phalloidin-induced cholestasis. Hepatology 1999; 29:814-821.
162. Paulusma CC, Kothe MJ, Bakker CT et al. Zonal down-regulation and redistribution of the multidrug resistance protein 2 during bile duct ligation in rat liver. Hepatology 2000; 31:684-693.
163. Dombrowski F, Kubitz R, Chittattu A et al. Electron-microscopic demonstration of multidrug resistance protein 2 (Mrp2) retrieval from the canalicular membrane in response to hyperosmolarity and lipopolysaccharide. Biochem J 2000; 348:183-188.
164. Ortitz D, Moseley J, Li S et al. The cortactin binding protein Hax-1 specifically interacts with canalicular transporters MDR2 and BSEP (Abstract). Hepatology 2001; 34:255A.
165. Kubitz R, Huth C, Schmitt M, et al. Protein kinase C-dependent distribution of the multidrug resistance protein 2 from the canalicular to the basolateral membrane in human HepG2 cells. Hepatology 2001; 34:340-50.

166. Milkiewicz P, Elias E, Williams A et al. Changes in the distribution of hepatocellular transporters in human end-stage cholestatic liver disease (Abstract). Hepatology 2001; 34:471A.

167. Harada M, Sakisaka S, Terada K et al. A mutation of the Wilson disease protein, ATP7B, is degraded in the proteasomes and forms protein aggregates. Gastroenterology 2001; 120:967-74.

168. Keitel V, Kartenbeck J, Nies AT et al. Impaired protein maturation of the conjugate export pump multidrug resistance protein 2 as a consequence of a deletion mutation in Dubin Johnson syndrome. Hepatology 2000; 32:1317-28.

169. Bolder U, Schmidt A, Landmann L et al. Heat stress prevents impairment of bile acid transport in endotoxemic rats by a posttranscriptional mechanism. Gastroenterology 2002; 122:963-73.

170. Noe J, Hagenbuch B, Meier PJ et al. Characterization of the mouse bile salt export pump overexpressed in the baculovirus system. Hepatology 2001; 33:1223-31.

171. Beuers U, Nathanson MH, Isales CM et al. Tauroursodeoxycholic acid stimulates hepatocellular exocytosis and mobilizes extracellular Ca⁺⁺mechanisms defectice in cholestasis. J Clin Invest 1992; 2984-2993.

172. Beuers U, Throckmorton DC, Anderson MS et al. Tauroursodeoxycholic acid activates protein kinase C in isolated rat hepatocytes. Gastroenterology 1996; 110:1553-1563.

173. Beuers U, Bilzer M, Chittattu A et al. Tauroursodeoxycholic acid inserts the apical conjugate export pump, Mrp2, into canalicular membranes and stimulates organic anion secretion by protein kinase C-dependent mechanisms in cholestatic rat liver. Hepatology 2001; 33:1206-16.

174. Beuers U, Probst I, Soroka C et al. Modulation of protein kinase C by taurolithocholic acid in isolated rat hepatocytes. Hepatology 1999; 29:477-82.

175. Kipp H, Arias IM. Trafficking of canalicular ABC transporters in hepatocytes. Annu Rev Physiol 2002; 64:595-608.

176. Kipp H, Pichetshote N, Arias IM. Transporters on demand: intrahepatic pools of canalicular ATP binding cassette transporters in rat liver. J Biol Chem 2001; 276:7218-24.

177. Glavy JS, Wu SM, Wang PJ et al. Down-regulation by extracellular ATP of rat hepatocyte organic anion transport is mediated by serine phosphorylation of oatp1. J Biol Chem 2000; 275:1479-84.

178. Simon FR. The role of sex hormones and hepatic plasma membranes in the pathogenesis of cholstasis. In: Reyes HB, Leuschner U, Arias IM, eds. Pregnancy, sex hormones and the liver. Dordrecht: Kluwer Academic Publishers Press, 1996:51-59

179. Hyogo H, Tazuma S, Nishioka T et al. Phospholipid alterations in hepatocyte membranes and transporter protein changes in cholestatic rat model. Dig Dis Sci 2001; 46:2089-97.

180. Stieger B, Fattinger K, Madon J et al. Drug- and estrogen-induced cholestasis through inhibition of the hepatocellular bile dsalt export pump (Bsep) of rat liver. Gastroenterology 2000; 118:422-430.

181. Funk C, Pantze M, Jehle L et al. Troglitazone-induced intrahepatic cholestasis by an interference with thehepatobiliary export of bile acids in male and female rats. Correlation with thegender difference in troglitazone sulfate formation and the inhibition of thecanalicular bile salt export pump (Bsep) by troglitazone and troglitazonesulfate. Toxicology 2001; 167:83-98.

182. Fattinger K, Funk C, Pantze M et al. The endothelin antagonist bosentan inhibits the canalicular bile salt exportpump:a potential mechanism for hepatic adverse reactions. Clin Pharmacol Ther 2001; 69:223-31.

183. Funk C, Ponelle C, Scheuermann G et al. Cholestatic potential of troglitazone as a possible factor contributing to troglitazone-induced hepatotoxicity: in vivo and in vitro interaction at thecanalicular bile salt export pump (Bsep) in the rat. Mol Pharmacol 2001; 59:627-35.

184. Fattinger K, Cattori V, Hagenbuch B et al. Rifamycin SV and rifampicin exhibit differential inhibition of the hepatic rat organic anion transporting polypeptides, Oatp1 and Oatp2. Hepatology 2000; 32:82-6.

185. Arias IM. New genetics of inheritable jaundice and cholestatic liver disease. Lancet 1998; 352:82-83.

186. Jansen PL, Muller M. Progressive familial intrahepatic cholestasis types 1,2, and 3. Gut 1998; 42:766-767.

187. Jansen PLM, Muller M. The molecular genetics of familial intrahepatic cholestasis. Gut 2000; 47:1-5.,

188. Jansen PL, Muller M, Sturm E. Genes and cholestasis. Hepatology 2001; 34:1067-74.

189. Hinoshita E, Taguchi K, Inokuchi A et al. Decreased expression of an ATP-binding cassette transporter, MRP2, in human livers with hepatitis C virus infection. J Hepatol 2001; 35:765-73.

190. Dumoulin FL, Reichel C, Sauerbruch T et al. Semiquantitation of intrahepatic MDR3 mRNA levels by reverse transcription/competitive polymerase chain reaction. J Hepatol 1997; 26:852-6.

191. Oswald M, Kullak-Ublick GA, Paumgartner G et al. Expression of hepatic transporters OATP-C and MRP2 in primary sclerosing cholangitis. Liver 2001; 21:247-53.

192. Shneider BL, Fox VL, Schwarz KB et al. Hepatic basolateral sodium-dependent-bile acid transporter expression in two unusual cases of hypercholanemia and in exrahepatic biliary atresia. Hepatology 1997; 25:1176-1183.
193. Kogan D, Anathanarayanan M, Emre S et al. The bile salt export pump (BSEP/SPGP) is not down-regulated in human cholestasis associated with extrahepatic biliary atresia (Abstract). Hepatology 1999; 30:468A.
194. Nozawa S, Miyazaki M, Tou HI et al. Human MDR1 and MDR3 gene expression in the liver with obstructive jaundice (Abstract). Gastroenterology 1997; 112:A1349.
195. Shoda J, Kano M, Oda K et al. The expression levels of plasma membrane transporters in the cholestatic liver of patients undergoing biliary drainage and their association with the impairment of biliary secretory function. Am J Gastroenterol 2001; 96:3368-78.
196. Prieto J, Qian C, Garcia N et al. Abnormal expression of anion exchanger genes in primary biliary cirrhosis. Gastroenterology 1993; 105:572-578.
197. Medina JF, Martinez-Anso E, Vazquez JJ et al. Decreased anion exchanger 2 immunoreactivity in the liver of patients with primary biliary cirrhosis. Hepatology 1997; 25:12-17.
198. Prieto J, Garcia N, Marti-Climent JM et al. Assessment of biliary bicarbonate secretion in humans by positron emission tomography. Gastroenterology 1999; 117:167-172.
199. Vazquez JJ, Vazquez M, Idoate MA et al. Anion exchanger immunotreactivity in human slivary glands in health and Sjogren"s syndrome. Am J Pathol 1995; 146:1422-1432.
200. Zollner G, Fickert P, Silbert D et al. Alterations of hepatobiliary transporter expression in primary biliary cirrhosis (PBC) (Abstract). J Hepatol 2002; in press.
201. Roskams T, Ros JE, Libbrecht L et al. Multidrug resistance-associated protein 3 and multidrug resistance P-glycoprotein MDR1 are strongly expressed in reactive ductules in human liver disease (Abstract). Hepatology 2001; 34:479A.
202. Kullak-Ublick GA, Beuers U, Fahney C et al. Identification and functional characterization of the promoter region of the human organic anion transporting polypeptide gene. Hepatology 1997; 26:991-7.
203. Fickert P, Zollner G, Fuchsbichler A et al. Effects of ursodeoxycholic and cholic acid feeding on hepatocellular transporter expression in mouse liver. Gastroenterology 2001; 121:170-83.
204. Warskulat U, Kubitz R, Wettstein M et al. Regulation of bile salt export pump mRNA levels by dexamethasone and osmolarity in cultured rat hepatocytes. Biol Chem 1999; 380:1273-1279.
205. Leuschner M, Maier KP, Schlichting J et al. Oral budesonide and ursodeoxycholic acid for treatment of primary biliary cirrhosis: results of a prospective double-blind trial. Gastroenterology 1999; 117:918-925.
206. Kurihara T, Niimi A, Maeda M et al. Bezafibrate in the treatment of primary biliary cirrhosis:comparison with ursodeoxycholic acid. Am J Gastroenterol 2000; 95:2990-2992.
207. Miranda S, Vollrath V, Wielandt AM, et al. Overexpression of mdr2 gene by peroxisome proliferators in mouse liver. J Hepatol 1997; 26:1331-1339.
208. Trauner M, Graziadei IW. Mechanisms of action and therapeutic applications of ursodeoxycholic acid in chronic liver diseases. Aliment Pharm Therap 1999; 13:979-995.

CHAPTER 21

Pathophysiological Basis of Pruritus and Fatigue in Cholestasis

Mark G. Swain

Summary

Fatigue and pruritus are subjective complaints which are extremely common amongst patients with cholestasis, significantly impairing the quality of life of these patients. The genesis of these complaints appears to be complex and likely multi-factorial. Moreover, the central nervous system appears to play a direct role in the clinical manifestation and propagation of these disabling symptoms. However, recent developments are increasing our understanding of possible mechanisms which may underlie the genesis of these symptoms in cholestasis and which hopefully will provide novel future therapeutic modalities for clinical intervention.

Introduction

Pruritus and fatigue constitute the most common symptomatic complaints in patients with chronic cholestatic diseases such as primary biliary cirrhosis (PBC) and primary sclerosing cholangitis (PSC), occurring in the majority of patients.[1,2,3] These symptoms can be mild and unobtrusive or so severe that they profoundly impact on the quality of life of this group of patients.[3] Intractable pruritus in cholestatic patients can be so severe as to be an indication for liver transplantation.[4] Fatigue and pruritus are subjective complaints, and therefore are difficult to measure or quantify. Moreover, the subjective nature of these complaints has made the study of fatigue and pruritus extremely challenging and has hampered our understanding of the pathophysiological mechanism(s) underlying the development of these symptoms and has inhibited the development of specific therapies to treat fatigue and pruritus.

Pruritus in Cholestasis

Epidemiology

Pruritus is one of the classical symptoms associated with cholestatic syndromes, occurring in approximately 60—70% of patients with PBC and PSC.[2,3] Pruritus constitutes the second most common complaint in these patients after fatigue[2,3] and can range in severity from being trivial to being so severe as to make a patient suicidal. Pruritus can be divided into two components, itch and scratch.[5] Itch can be defined as an unpleasant sensation localized to the skin or mucosa which provokes the desire to scratch. Scratching is the physical act which may, or may not, occur in response to the itch and which constitutes an attempt to remove the cause of the itch. Cholestasis associated pruritus can be generalized or localized, often occurring on the soles of the feet or the palms. Moreover, pruritus in cholestatic patients can be constant or intermittent and typically is not relieved by scratching or skin emollients. Pruritus is often worse at night, disrupting sleep, and during winter months. Pruritus does not correlate with disease stage in PBC and PSC and can be the presenting symptom, even in early disease.[1-3,6,7]

Molecular Pathogenesis of Cholestasis, edited by Michael Trauner and Peter L.M. Jansen. ©2004 Eurekah.com and Kluwer Academic / Plenum Publishers.

Cholestasis associated pruritus is often exacerbated by pregnancy or by estrogen replacement therapy.[6,7]

Pathophysiology

Given that itch is a subjective sensation, it must relate to nervous activity. However, which specific components of the nervous system are involved in the genesis of pruritus are incompletely defined, but appear to include both the peripheral and central nervous systems (Fig. 1).

The superficial skin contains numerous afferent nerves which can perceive mechanical, thermal, and chemical stimuli. However, no specific receptors for pruritus have been identified in the skin and the current view is that activation of a plexus of free polymodal nocieptor nerves at the junction of the epidermis and dermis produces the itch sensation (reviewed in 8). These activated polymodal nociceptor nerves conduct impulses to the central nervous system through unmyelinated C-fibres, synapsing in the spinal cord where they interact with spinal interneurons and descending pathways.[8] Inhibitory spinal interneurons link C-fibres carrying pruritogenic signals with both afferent A-fibres and descending spinal pathways. Although the exact role of afferent myelinated A-fibres in the sensation of pruritus are unkown, they may be stimulated by scratching and, by synapsing with inhibitory interneurons situated in the dorsal horn of the spine, subsequently inhibit C-fibres carrying the pruritic signal and thereby reduce the pruritic sensation.[8] Pruritogenic stimuli are then carried by secondary neurons through the anterior commissure of the spinal cord and ascend in the anterolateral spinothalamic tract to the brainstem and ultimately to the thalamus and cerebral cortex. Although the role of the cerebral cortex in itch sensation is poorly understood,[9,10] the importance of central control of itch is highlighted by the association of itch with a number of primary central nervous system diseases. In addition, central itch can be induced pharmacologically by the intrathecal or intraspinal administration of opiates (reviewed in 11).

Pruritus in cholestatic patients has traditionally been suggested as being due to stimulation of cutaneous C-fibre afferents by substances retained in plasma as part of the cholestatic syndrome, with particular attention being given to retained bile acids.[12,13] Serum levels of bile acids are elevated in cholestasis and these bile acids accumulate in the skin.[12,13]. Moreover, bile acids injected into the skin induce pruritus in normal volunteers[14] and pruritus in cholestasis is often relieved by the administration of bile acid sequestrants or by biliary diversion.[15,16] However, pruritus in cholestatic patients does not correlate with skin concentrations of bile acids.[17] In addition, many cholestatic patients experience no pruritus despite the presence of elevated serum bile acid levels[18] and cholestatic patients with end stage liver failure can experience a resolution of their itching despite the persistence of very high serum bile acid levels.[18] These findings suggest that cholestatic pruritus is not simply due to elevated tissue bile acids. A recently proposed theory suggests that retained bile acids in cholestasis act within the liver to release an as yet unidentified substance which is the active pruritogen.[19] Interestingly, the bile acid sequestrant cholestyramine can relieve itch in disease states such as renal failure and polycythemia rubra vera which are not associated with increased serum bile acid levels.[20,21]

Histamine is a classical mediator of itch (reviewed in 22). Direct injection of histamine into the skin induces itch which is accompanied by a wheal and flare response. In cholestatic patients serum histamine levels are increased and pruritic patients have higher serum histamine levels than non-pruritic patients.[23] In addition, rats with experimental cholestasis demonstrate increased skin mast cell counts in association with increased plasma histamine levels.[24] These observations suggest a possible role of histamine in the genesis of cholestatic itching. However, pruritic cholestatic patients do not demonstrate cutaneous wheal and flare reactions in pruritic skin and respond poorly to anti-histamine treatment. However, histamine at concentrations that do not produce a wheal and flare response can enhance C-fibre activation induced by other stimuli.[25] Therefore, increased plasma and skin histamine levels in cholestatic patients may increase their susceptibility to the cutaneous pruritogenic effects of other unidentified compounds.

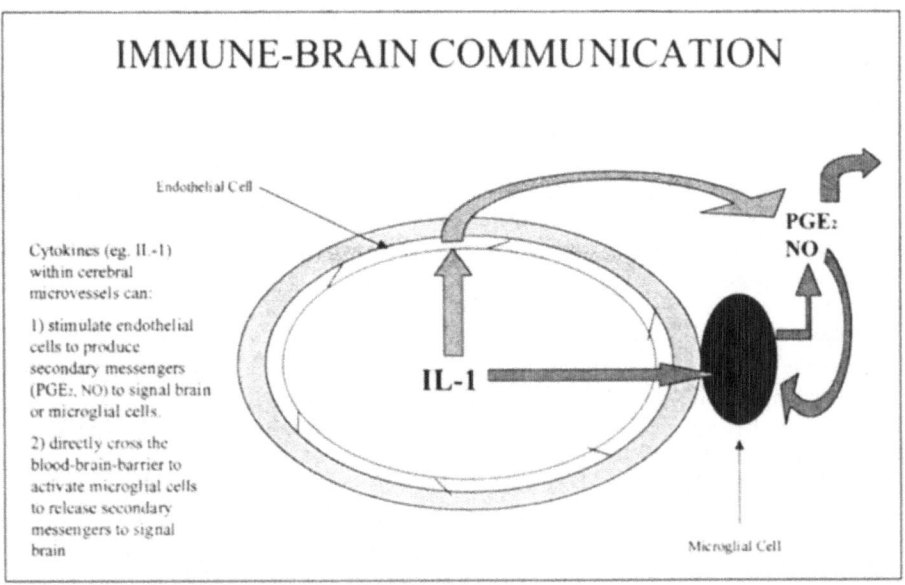

Figure 1. Potential mechanism(s) for blood-borne cytokines to communicate with central neural pathways. PGE$_2$, prostaglandin E$_2$; NO, nitric oxide; IL-1, interleukin 1.

A number of other compounds have been identified which have the capacity to enhance the pruritogenic effects of known pruritogens. These pruritus-augmenting compounds include endogenous opioids and the prostaglandin, PGE$_2$.[26,27] Patients with the cholestatic liver disease PBC exhibit increased plasma levels of endogenous opioids as well as enhanced cellular production of PGE$_2$.[28,29] Similar findings have been reported in rats with experimental obstructive cholestasis.[30] In addition, opioid receptor blockade appears to ameliorate itch in up to 50% of pruritic cholestatic patients.[31,32] However, pruritus does not appear to correlate with circulating levels of endogenous opioids in PBC patients, and opioid receptor blockade can ameliorate itch in patients who have pruritus not associated with liver disease.[38] These findings suggest that increased circulating levels of endogenous opioids are unlikely to play a direct role in cholestatic pruritus.

It is likely that pruritus in cholestasis is induced by a gut-derived pruritogen[5] which activates cutaneous afferent C-fibres, either directly or indirectly, possibly in concert with other pruritus-augmenting factors such as histamine or endogenous opioids. However, the perception of itch can be profoundly influenced through activation of higher central neural pathways which may suppress itch through descending inhibitory neural pathways or by modulating central itch control centers. Opioids are capable of eliciting pruritus when injected spinally or intrathecally, apparently by activating central μ opioid receptors.[34,35] Typically central opioid induced itch is most commonly isolated to the face,[34] a finding uncommonly encountered in cholestatic patients.[3] However, endogenous opioids acting at sites within the central nervous system have been implicated in the genesis of cholestatic pruritus. Specifically, treating pruritic cholestatic patients with the μ opioid receptor blockers naloxone, nalmefene, or naltrexone results in a reduction in pruritus in about half of treated patients.[31,32,36] In addition, opioid receptor antagonist-treated pruritic cholestatic patients can experience classical opiate withdrawal type symptoms and signs, highly suggestive of enhanced central opioidergic tone in these patients.[31,36] Interestingly, rats with experimental cholestatic liver disease demonstrate naloxone reversible analgesia as well as a downregulation of brain μ opioid receptors; findings consistent with enhanced activation of central μ opioid receptors in cholestatic rats.[37,38] These observations are consistent with an itch modulatory role of endogenous opioids within the central nervous system in cholestasis. Similarly, μ opioid receptor blockade can also ameliorate pruritus in patients with eczema and atopic dermatitis.[33]

Other potential pruritogens, which may be active in cholestatic patients, have been postulated but have not been adequately studied. These potential pruritogens include substance P, proteases and kinins, serotonin and the cytokine IL-2.[22] The serotonergic neurotransmitter system has recently been implicated in cholestatic pruritus by the demonstration that the 5-HT$_3$ receptor antagonist ondansetron ameliorated pruritus in a group of cholestatic patients.[39] Interestingly, 5-HT$_3$ receptor activation can modulate opiate-mediated itch, suggesting a possible synergistic activity of these two systems in the genesis of cholestatic itch.[40]

Treatment

Since the specific cause of pruritus in cholestasis has not been identified, treatments to date have been empiric, although in general they are reasonably effective. Treatment strategies have focused on potential peripheral sources of pruritogens, peripheral actions of pruritogens, and central pruritus modulating pathways.

The gut has traditionally been felt to be the likely peripheral source of the pruritogen(s) in cholestasis, and most attention has focused on gut derived bile acids. Oral administration of the non-absorbable bile acid binding resin cholestyramine is effective in reducing cholestatic pruritus in about 90% of patients, making it the first line of therapy in these patients.[15] However, cholestyramine binds many compounds in the gut preventing their absorption, and therefore the anti-pruritic effect of cholestyramine might be due to its binding of a non-bile acid, gut-derived pruritogen. The observation that the antibiotics rifampicin and metronidazole can ameliorate cholestatic itch suggest that gut flora may be important in the generation of the pruritogen(s) in cholestasis.[41,42] Rifampicin has a number of other actions distinct from its antibiotic function (eg. hepatic enzyme induction, hepatic bile acid uptake modulation) suggesting that the anti-pruritic effect of rifampicin may be multifactorial. Certainly treatment with rifampicin is an effective alternative treatment in the majority of patients who fail to respond to cholestyramine therapy.[41]

Other therapeutic modalities have targeted potential peripheral itch augmenting compounds, such as histamine and endogenous opioids; levels of which are elevated in plasma in cholestatic patients.[23,28] Both antihistamines (eg. hydroxyzine) and opioid receptor blockers (eg. naloxone, naltrexone) can improve itching in cholestatic patients, supporting a possible peripheral itch augmenting role of histamine and endogenous opioids in these patients. However, the anti-pruritic effects of anti-histamines have been suggested as being secondary to their central sedating properties. Interestingly, PGE$_2$ can augment itch induced by histamine, and cellular PGE$_2$ release is increased in cholestasis.[26,29] However, the anti-pruritic potential of prostaglandin synthesis inhibition has not been examined in cholestatic patients.

Recent clinical attention in the treatment of cholestatic itch has focused on manipulation of central itch modulation pathways; specifically, the opioid and serotonin systems. Certainly opioid receptor blockade appears to decrease itch in cholestatic patients as well as in a number of other conditions associated with pruritus.[31-33,36] In addition, the central inhibition or modulation of serotonergic neurotransmission may also be of benefit in the treatment of cholestatic pruritus.[39] Identifying and targeting central itch controlling pathways in cholestasis appears to represent an exciting potential area for research and therapeutic intervention in cholestatic pruritus.

Fatigue in Cholestasis

Epidemiology

In patients with cholestatic liver diseases fatigue is the most common complaint, occurring in the majority of patients with PSC and PBC.[1-3] Fatigue can be mild and relatively unobtrusive, or totally incapacitating. Typically the onset of fatigue is gradual and insidious, often

Table 1. Treatments for cholestatic pruritus

First Line Therapies

1) Cholestyramine (4 gm—16 gm/day)
2) Antihistamines (hydroxyzine 25-50 mg every 6-8 hrs., especially at night)

Second Line Therapies

1) Rifampicin (up to 10 mg/kg/day)
2) Naltrexone (25-50 mg/day)

Third Line Therapies

1) Plasmapheresis
2) External biliary drainage
3) Phenobarbitol (2-4 mg/kg/day)
4) Corticosteroids (10-20 mg/day)
5) Liver transplantation

leading to subtle changes in lifestyle to compensate, which may not be noticed by a patient for years. Therefore, patients with cholestatic liver disease should be questioned directly about fatigue. However, given its subjective nature fatigue is often overlooked by clinicians. Fatigue has a significant impact on the quality of life of cholestatic patients.[43,44] In patients with PBC, fatigue is the worst symptom in approximately 50% of patients, and causes severe disability with respect to activities of daily living in about 25% of patients.[44]

Given its subjective nature fatigue is difficult to quantify. However the complaint of fatigue does not correlate with disease severity parameters in cholestatic patients.[1,2,44] Moreover, no specific therapy exists to treat fatigue, making management of this symptom extremely difficult.

Pathophysiology

To understand and discuss the concept of fatigue in cholestasis it is crucial that peripheral fatigue be differentiated from central fatigue. Peripheral fatigue relates to dysfunction of muscles or nerves outside of the central nervous system and therefore is not relevant to the typical cholestatic patient in the absence of muscle wasting associated with liver failure. Central fatigue, on the other hand, relates to dysfunction occurring in neurotransmitter pathways within the central nervous system which give rise to feelings of malaise, lethargy, lassitude, or exhaustion which fit under the broad symptom termed fatigue (reviewed in 45). Patients with the cholestatic liver disease PBC have been reported as having normal electromyography studies but demonstrate high fatigue scores on questionnaires.[46] These findings are suggestive of altered central neurotransmission as being the main cause of fatigue in PBC patients. A recent study examined grip strength in fatigued and non-fatigued PBC patients compared to normal controls.[47] Baseline grip strength was similar in all three groups; however, grip strength in the fatigued PBC patient group declined to a greater extent upon repeated testing than did grip strength in the other 2 groups. Moreover, grip strength decline per test repeat correlated closely with scores obtained in these patients from fatigue questionnaires.[47] Given previous observations of normal EMG studies in PBC patients with fatigue,[46] the documented decline in muscle grip strength in fatigued PBC patients likely reflects a failure in the central drive stimulus which is needed for sustained or repeated muscle contraction.[48] However, grip strength decline may be of potential use in studies of cholestasis associated fatigue as an objective measure to supplement fatigue questionnaire results.

The concept of central fatigue implies altered neurotransmission within the brain; however, alterations in specific central neurotransmitter pathways that might subserve the genesis of central fatigue are poorly defined. Scientific attention has focused primarily upon the corticotropin-releasing hormone (CRH) and serotonergic neurotransmitter systems within the brain, (reviewed in 49) as abnormalities in these 2 systems have been implicated in the development of fatigue in a number of clinical scenarios associated with fatigue.

CRH-containing nerves have been localized within a number of brain areas and CRH is involved in coordinating a number of behavioral responses.[50,51] Moreover, infusion of CRH into the brains of rodents induces behavioral activation,[52] which has led to the postulate that defective central CRH release might result in behavioral depression and fatigue.[53,54] Clinical observation has suggested that defective central CRH release may contribute to fatigue in a variety of conditions, including the chronic fatigue syndrome and atypical depression.[53,54] Impaired central CRH release has been suggested in patients with PBC by the observation that infusion of exogenous CRH results in enhanced ACTH release from the pituitary gland.[55] CRH is the main central regulator of ACTH release from the anterior pituitary and enhanced responsiveness to exogenous CRH administration is suggestive of an upregulation of pituitary CRH receptors, possibly due to decreased central CRH release in these PBC patients. This hypothesis is supported by the observation of decreased central CRH levels and release in rats with experimental cholestasis,[56] coupled with increased expression of CRH type 1 receptors (main CRH receptor type mediating behavioral and ACTH responses to CRH);[57] within the brains of these animals.[58] Moreover, cholestatic rats demonstrate a marked increase in the behavior/locomotor activating effects of centrally infused CRH, compared to non-cholestatic controls, mediated by these upregulated central CRH type 1 receptors.[58] Therefore, defective central CRH release in cholestasis likely plays a central role in the genesis of cholestasis associated fatigue.

Altered central serotonergic neurotransmission has also been implicated in the genesis of central fatigue (reviewed in 59). In addition, serotonin containing nerves are intimately involved in the regulation of central CRH release. Specifically, serotonin containing nerves project from the dorsal raphè nucleus within the midbrain to areas throughout the brain, synapsing with CRH-containing neurons within the paraventricular nucleus of the hypothalamus.[60] Moreover, activation of these hypothalamic CRH-containing nerves during stress is mediated, at least in part, via these serotonergic inputs.[61] However, both increased and decreased central serotonergic neurotransmission have been implicated in fatigue states. Patients with the chronic fatigue syndrome demonstrate increased sensitivity to hypothalamic neuronal activation by serotonin receptor agonists, suggesting the presence of defective central serotonin release in these patients.[62] In addition, pharmaceutical augmentation of central serotonergic neurotransmission has been utilized with some effect in treating fatigue a number of conditions strongly associated with fatigue.[63] Alternatively, rats which have been exercised to exhaustion demonstrate increased central serotonin levels,[64] and treatment of athletes with serotonin-reuptake inhibitors decreases their time to exercise-induced exhaustion.[65] These observations suggest a role for enhanced central serotonergic neurotransmission in the genesis of fatigue. Altered central serotonergic neurotransmission has been documented in rats with experimental cholestasis. Specifically, cholestatic rats demonstrate increased midbrain 5-HT_{1A} receptor sensitivity and responsiveness compared to non-cholestatic controls.[66] Activation of midbrain 5-HT_{1A} receptors decreases serotonin release from serotonergic nerves which project throughout the brain.[61] Desensitization of these midbrain 5-HT_{1A} receptors in cholestatic rats results in an amelioration of fatigue-like behavior as determined in a swim-paradigm of fatigue assessment.[67] Interestingly, serotonin involvement in fatigue in hepatitis C-induced cirrhosis was suggested by the observation of an amelioration of fatigue in a hepatitis C-infected patient treated with the 5-HT_3 receptor antagonist ondansatron.[68] Obviously, serotonergic modulation of fatigue is likely complex but represents a potentially fruitful area for future investigation.

Other neurotransmitter systems, including the norepinephrine and dopamine systems, have been implicated in arousal, locomotion, and fatigue states.[69,70] However, investigation of these other pathways in the genesis of central fatigue awaits further study.

The question that obviously remains is; How do abnormalities in central neurotransmission which give rise to fatigue come about in the cholestatic patient?. It seems apparent that the diseased liver must communicate with the brain to bring about these changes. Communication between the damaged liver and brain may involve neural and/or immune pathways. The liver is a richly innervated organ and afferent signaling from the liver to the brain occurs via both vagal afferents and spinal nerve projections, although vagal signaling has been best characterized (Fig. 2).[71,72] Intraperitoneal inflammation induced by the injection of endotoxin causes increased expression of the neural activation marker Fos within the brain of rodents,[73] and also induces lethargy/malaise in these animals.[74] These effects can be abolished by subdiaphragmatic vagotomy.[74] In addition, intraperitoneal administration of endotoxin or the cytokine IL-1β to rodents causes increased CRH expression within the hypothalamus; an effect blocked by subdiaphragmatic vagotomy.[75,76] Moreover, administration of IL-1β directly into the portal vein of rats produces increased electrical activity in the hepatic branch of the vagus nerve.[77] IL-1β is one of a number of proinflammatory cytokines including IL-6 and TNFα which are synthesized within the liver and are likely released within the liver during cholestatic liver injury,[78,79] potentially activating hepatic vagal afferents to signal the brain.

Systemic release of inflammatory mediators (eg. cytokines) from the diseased liver in cholestatic liver disease may also directly signal the brain. Kupffer cells within the liver constitute the largest population of fixed tissue macrophages within the body and represent a rich potential source of inflammatory mediators, including cytokines. A number of cytokines have been identified as potentially carrying signals from damaged tissues to the brain. The cytokines which have received the greatest scientific attention in this regard have been IL-1β, IL-6 and TNF (reviewed in 80,81). Increased plasma levels of all of these cytokines have been identified in patients with cholestatic liver diseases[79] and plasma endotoxin levels, which can stimulate cytokine release, are elevated in patients with PBC.[82] Moreover, elevated circulating endotoxin, IL-6 and TNF levels have been documented in cholestatic rats and mice.[83-85] Lethargy and decreased locomotion are well characterized behaviors observed in rodents treated intravenously with cytokines and endotoxin.[80,86] Circulating endotoxins and/or cytokines may communicate with the brain to bring about changes in neurotransmitter systems either directly, by entering the central nervous system through areas devoid of an intact blood-brain-barrier, or indirectly through the induction of the synthesis and release of secondary mediators (eg. nitric oxide, PGE_2) from vascular or peri-vascular cells within the brain (reviewed in refs. 80,81)(Fig. 3). In addition, circulating endotoxins and/or cytokines can induce the de novo synthesis and release of cytokines within the brain.[80,81,86] Interestingly, cholestatic rats demonstrate enhanced suppression of locomotor activity in response to centrally administered IL-1β compared to non-cholestatic controls.[87] In addition, systemic endotoxin administration to rats results in increased hypothalamic CRH mRNA expression which can be blocked by the intracerebroventricular pre-administration of a specific IL-1 receptor antagonist.[88] These observations are consistent with the existence of a potential humoral-immune communication pathway between the liver and the brain in the setting of cholestatic liver disease.

Chronic stress has a profound effect on central neurotransmitter pathways. Chronic stress is felt to be an etiological factor in the development of depression and anxiety disorders. Moreover, as a patient, being diagnosed with a progressive cholestatic liver disease such as PBC or PSC coupled with all of the uncertainties with respect to disease manifestations, lack of a cure, potential need for liver transplantation, and fears of death constitute an ongoing significant stress. In patients with PBC fatigue scores on questionnaires closely correlate with measures of depression and anxiety.[1] A loss of the sense of control about ones life has been suggested as being closely linked to the development of fatigue in patients with multiple sclerosis.[90] Interestingly, chronic stress has profound effects on central CRH levels in rodents and abnormal responses to exogenous CRH have been documented in depressed patients and in patients with

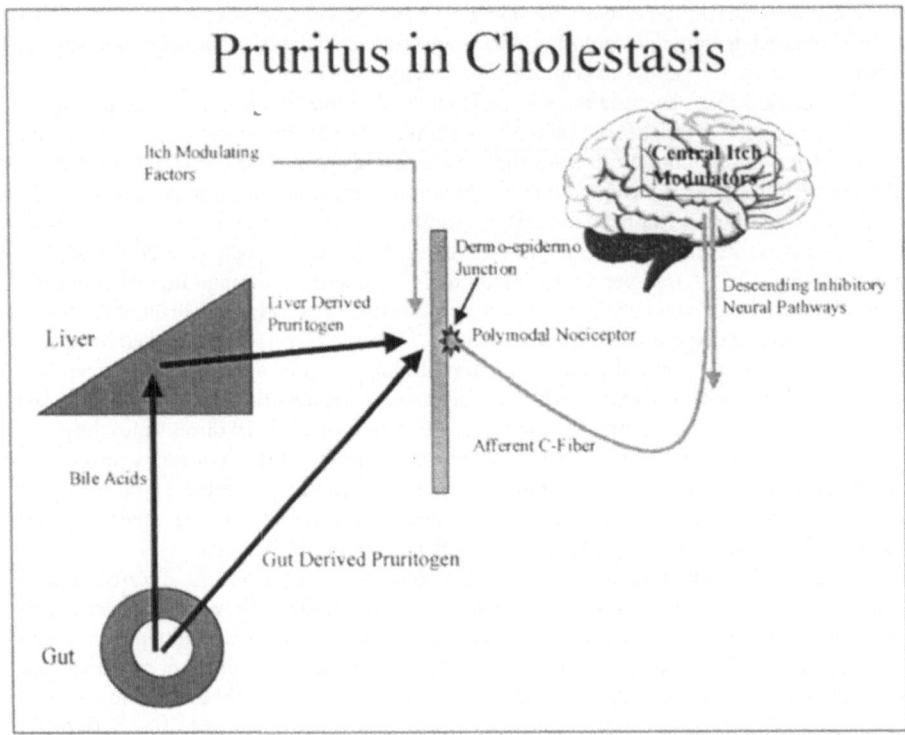

Figure 2. Potential peripheral and central pathways regulating pruritus in cholestatic patients.

post-traumatic stress disorder.[91] It therefore seems likely that fatigue may be exacerbated in PBC and PSC patients after they have been given their diagnosis. However, the fact that the majority of cholestatic patients with fatigue experience this symptom for a period of time before they were diagnosed with a specific cholestatic disease argues that, although chronic stress related to their diagnosis may exacerbate fatigue, it is unlikely the precipitating event in most patients.

Treatment

Since the cause of fatigue in cholestatic liver disease is unknown, there are currently no specific therapies available to treat it. However, general measures can be employed to help the patient deal with their fatigue. Specifically, identifiable causes of fatigue including hypothyroidism, anemia, diabetes, renal failure or electrolyte disturbances, should be ruled out. Medications linked with fatigue should be discontinued if possible (eg. benzodiazepines, β blockers). Patients should be counseled on good sleep hygiene and limiting caffeine and alcohol ingestion. Moreover, patients should be instructed on the importance of maintenance of an adequate aerobic exercise program and stress reduction.

The neuroendocrine and neurotransmitter systems postulated to be involved in the induction of fatigue are closely linked to those implicated in the genesis of mood disorders such as depression.[45] In addition, depression and fatigue often co-exist in cholestatic patients.[1,44] In an animal model of cholestatic liver injury behaviours consistent with depression (ie. anhedonia, or loss of interest in pleasureful activities, and a loss in social interest) have been identified and correlate with changes in brain CRH levels and serotonergic neurotransmission.[66,67,92,93] These findings strongly suggest a link between mood disorder and fatigue in cholestatic liver disease

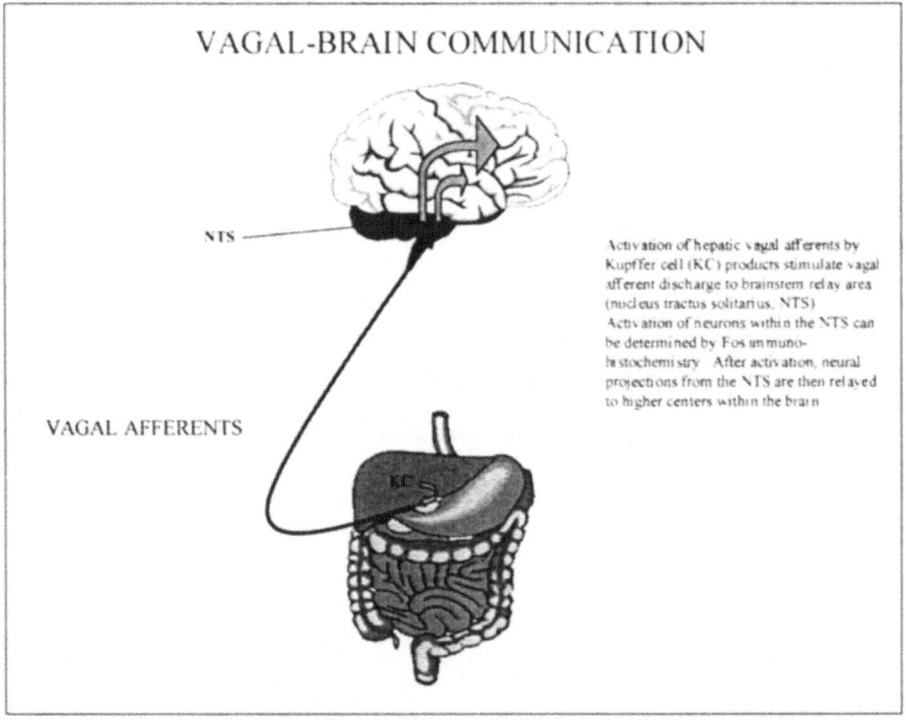

VAGAL-BRAIN COMMUNICATION

NTS

VAGAL AFFERENTS

Activation of hepatic vagal afferents by Kupffer cell (KC) products stimulate vagal afferent discharge to brainstem relay area (nucleus tractus solitarius, NTS). Activation of neurons within the NTS can be determined by Fos immuno-histochemistry. After activation, neural projections from the NTS are then relayed to higher centers within the brain.

Figure 3. Potential liver-brain signaling pathway via hepatic vagal nerve afferents in cholestasis.

based on defects in common central pathways. Furthermore, these findings suggest that treatments effective for depression may also be useful in treating fatigue in cholestatic patients. Therefore, in a given fatigued cholestatic patient symptoms or signs of significant depression should be directly sought, and if identified they should be treated. A trial of antidepressant therapy should be considered in cholestatic fatigued patients with signific...cant depressive symptoms.

Recently the potential use of cognitive-behavioural therapy has been suggested for patients with fatigue syndromes and has achieved some success in patients with the chronic fatigue syndrome[94] and in patients with other chronic diseases.[95] Cognitive behavioural therapy is based on a patients thoughts and beliefs about their illness (cognition) and the way they cope with it (behaviour).[95] The aim of the approach is to identify the cognitions and behaviours in a given patient that may be contributing to their disease related behaviour, namely fatigue. Cognitions and behaviours identified as contributing to fatigue can then be modified using self-help techniques.

References

1. Cauch-Dudek K, Abbey S, Stewart DE et al. Fatigue in primary biliary cirrhosis. Gut 1998; 43:705-10.
2. Lindor K, Wiesner RH, MacCarty RC et al. Advances in primary sclerosing cholangitis. Am J Med 1990; 89:73-80.
3. Witt-Sullivan H, Heathcote J, Cauch K et al. The demography of primary biliary cirrhosis in Ontario, Canada. Hepatology 1990; 12:98-105.
4. Maddrey W, van Thiel DH. Liver transplantation: an overview. Hepatology 1988; 8:948-59.
5. Ekblom A, Some neurophysiological aspects of itch. Sem Dermatol 1995; 14:262-70.
6. Raiford DS. Pruritus of chronic cholestasis. QJM 1995; 88:603-7.
7. Bergasa NV. The pruritus of cholestasis. Sem Dermatol 1995; 14:302-12.

8. Kam PCA, Tan KH. Pruritus—itching for a cause and relief? Anesthesia 1996; 51:1133-38.
9. Hagabarth KE, Kerr DIB. Central influence on spinal afferent conduction. J Neurophysiol 1954; 17:295-307.
10. Bradford FK. Ablations of frontal cortex in cats with special reference to enhancement of the scratch reflex. J Neurophysiol 1939; 2:192-201.
11. Jones EA, Bergasa NV. The pruritus of cholestasis: from bile acids to opiate agonists. Hepatology 1990; 11:884-887.
12. Schoenfield LJ, Sjovall J, Perman E. Bile acids on the skin of patients with pruritic hepatobiliary disease. Nature 1967; 213:93-94.
13. Stiehl A. Bile acids and bile acid sulfates in the skin of patients with cholestasis and pruritus. Z Gastroenterol 1974; 12:121-124.
14. Kirby J, Heaton KW, Burton JL. Pruritic effect of bile salts. Br Med J 1974; 4: 693-695.
15. Datta DV, Sherlock S. Cholestyramine for long term relief of the pruritus complicating intrahepatic cholestasis. Gastroenterology 1966; 50:323-332.
16. Whitington PF, Whitington GL. Partial external diversion of bile for the treatment of intractable pruritus associated with intrahepatic cholestasis. Gastroenterology 1988; 95:130-136.
17. Bartholomew TC, Summerfield JA, Billing BH et al. Bile acid profiles of human serum and skin interstitial fluid and their relationship to pruritus studied by gas chromatography—mass spectrometry. Clin Sci 1982; 63:65-73.
18. Murphy GM, Ross A, Billing BH. Serum bile acids in primary biliary cirrhosis. Gut 1972; 13:201-206.
19. Ghent CN. Pruritus of choestasis is related to effects of bile salts on the liver, not the skin. Am J Gastro 1987; 82:117-118.
20. Silverberg DS, Iaina A, Reisin E et al. Cholestyramine in uremic pruritus. Br Med J 1977; 1:752-753.
21. Chanarin I, Szur I. Relief of intractable pruritus in polycythemia rubra vera with cholestyramine. Br J Hematol 1970; 29:669-670.
22. Hagermark O, Itch mediators. Sem Dermatol 1995; 14:271-276.
23. Gittlen SC, Schulman ES, Maddrey WV. Raised histamine concentrations in chronic cholestatic liver disease. Gut 1990; 31:96-99.
24. Clements WD, O'Rourke DM, Rowlands BJ et al. The role of mast cell activation in cholestatic pruritus. Agents Actions 1994; 41:C30-31.
25. Kress M, Koltzenburg M, Reeh PW et al. Responsiveness and functional attributes of electrically localized terminals of cutaneous C-fibers in vivo and in vitro. J Neurophysiol 1992; 68:581-595.
26. Greaves MW, McDonald-Gibson W. Itch: Role of prostaglandins. BMJ 1973; 3:608-609.
27. Fjellner B, Hagermark O. Potentiation of histamine-induced itch and flare response in human skin by the enkephalin analogue FK-33-824, bendorphin and morphine. Arch Dermatol Res 1982; 274:29-37.
28. Thornton JR, Losowksy MS. Plasma methionine enbephalin concentration and prognosis in primary biliary cirrhosis. BMJ 1988; 297:1241-1242.
29. Chiricolo M, Lenzi M, Bianchi F et al. Immune dysfunction in primary biliary cirrhosis. II. Increased production of prostaglandin E. Scand J Immunol 1989; 30:363-367.
30. Swain MG, Rohman RB, Xu H et al. Endogenous opioids accumulate in plasma in a rat model of acute cholestasis. Gastroenterology 1992; 103:630-635.
31. Bergasa NV, Alling DW, Talbot T et al. Effects of raloxone infusions in patients with the pruritus of cholestasis. Ann Int Med 1995; 123:161-167.
32. Wolfhagen FMJ, Sternier E, Hop WLJ et al. Oral raltrexone treatment for cholestatic pruritus: A double-blind placebo-controlled study. Gastroenterology 1997; 113:1264-1269.
33. Monroe EW. Efficacy and safety of nalmefene in patients with severe pruritus caused by chronic urticaria and atopic dermatitis. J Am Acad Dermatol 1989; 21:135-136.
34. Scott PV, Fischer HBJ. Spinal opiate analgesia and facial pruritus: a neural theory. Postgrad Med J 1982; 58:531-535.
35. Berstein JE, Grinzi RA. Butorphanol-induced pruritus antagonized by naloxone. J Am Acad Dermatol 1981; 5:227-228.
36. Bergasa NV, Alling DW, Talbot TL et al. Oral nalmefene therapy reduces scratching activity due to the pruritus of cholestasis: a controlled study. J Am Acad Dermatol 1999; 41:431-434.
37. Bergasa NV, Alling DW, Vergalla J et al. Cholestasis in the male rat is associated with naloxone-reversible antinociception. J Hepatol 1994; 20:85-90.
38. Bergasa NV, Rothman RB, Vergalla J et al. Central mu-opioid receptors are down-regulated in a rat model of acute cholestasis. J Hepatol 1992; 15:220-224.
39. Muller C, Pongratz S. Pidlich J et al. Treatment of pruritus in chronic liver disease with the 5-hydroxytryptamine receptor type 3 antagonist ondansetron: a randomized, placebo-controlled, double-blind cross-over study. Eur J Gastroenterol Hepatol 1998; 10:865-870.
40. Kyriadkides K, Hussain SK, Hobbs GJ. Management of opioid-induced pruritus: a role for 5-HT3 antagonists. Br J Anesth 1999; 82:439-441.

41. Backs L, Pars A, Elena M et al. Effects of long-term rifampicin administration in primary biliary cirrhosis. Gastroenterology 1992; 102:2077-2080.
42. Berg CL, Gollan JL. Primary biliary cirrhosis: new therapeutic directions. Scand J Gastroenterol 1992; 192:543-549.
43. Younossi ZM, Kiwi ML, Boparai N et al. Cholestatic liver diseases and health-related quality of life. Am J Gastroenterol 1000; 95:497-502.
44. Huet PM, Deslauriers J, Tran A et al. Impact of fatigue on the quality of life of patients with primary biliary cirrhosis. Am J Gastroenterol 2000; 95:760-767.
45. Swain MG. Fatigue in chronic disease. Clin Sci 2000;99:1-8.
46. Jalan R, Gibson H, Lombard MG. Patients with PBC have central but no peripheral fatigue. Hepatology 1996; 24:A162.
47. Goldblatt J, James OFW, Jones DEJ. Grip strength and subjective fatigue in patients with primary biliary cirrhosis. JAMA 2001; 285:2196-2197.
48. Enoka RM, Stuart DG. Neurobiology of muscle fatigue. J Appl Physiol 1992; 72:1631-1648.
49. Beam J, Wessely S. Neurobiological aspects of the chronic fatigue syndrome. Eur J Clin Invest 1994; 24:79-90.
50. Swanson LW, Sawchenko PE, Rivier J et al. Organization of ovine corticotropin-releasing factor immunoreactive cells and fibres in the rat brain: an immunohistochemical study. Neuroendocrinology 1983; 36:165-186.
51. Koob GF, Heinrichs SC, Pich EM et al. The role of corticotropin-releasing factor in behavioural responses to stress. Ciba Found Symp 1993; 172:277-289.
52. Sutton RE, Koob GF, LeMoal M et al. Corticotropin-releasing factor produces behavioural activation in rats. Nature 1982; 297:331-333.
53. Clauw DJ, Chrousos GP. Chronic pain and fatigue syndromes: overlapping clinical and neuroendocrine features and potential pathogenic mechanisms. Neuroimmunomodulation 1997; 4:134-153.
54. Gold PW, Chrousos GP. The endocrinology of melancholic and atypical depression: relation to neurocircuitry and somatic consequences. Proc Assoc Am Phys 1998; 111:22-34.
55. Swain MG, Mogiakou MA, Bergasa NV et al. Facilitation of ACTH and cortisol responses to corticotropin-releasing hormone (CRH) in patients with primary biliary cirrhosis. Hepatology 1994; 20:A197.
56. Swain MG, Patchev V, Vergalla J et al. Suppression of hypothalamic-pituitary-adrenal axis responsiveness to stress in a rat model of acute cholestasis. J Clin Invest 1993; 91:1903-1908.
57. Grigoriadis DE, Lovenberg TW, Chalmers DT et al. Characterization of corticotropin-releasing factor receptor subtypes. Ann NY Acad Sci 1996; 780:60-80.
58. Burak K, Le T, Swain MG. Increased sensitivity to the locomotor activating effects of corticotropin-releasing hormone (CRH) in cholestatic rats: Implications for cholestasis associated fatigue. Gastroenterology (in press).
59. Neeck G, Crofford LJ. Neuroendocrine perturbations in fibromyalgia and chronic fatigue syndrome. Rheum Dis Clin North Am 2000; 26:989-1002.
60. Liposits Z, Phelix C, Paull W. Synaptic interaction of serotonergic axons and corticotropin-releasing factor (CRF) synthesizing neurons in the hypothalamic paraventricular nucleus of the rat. Histochemistry 1987; 86:541-549.
61. Roth BL. Multiple serotonin receptors: clinical experimental aspects. Ann Clin Psychiatry 1994; 6:67-78.
62. Bakheit AMO, Behan PO, Dinan TG et al. Possible upregulation of hypothalamic 5-hydroxytryptamine receptors in patients with postviral fatigue syndrome. Br Med J 1992; 304:1010-1012.
63. Goldenberg D, Mayskiy M, Mossey C et al. A randomized, double-blind cross-over trial of fluoxetine and amitryptiline in the treatment of fibromyalgia. Arthritis Rheum 1996; 39:1852-1859.
64. Bailey SP, Davis JM, Ahlborn EN. Neuroendocrine and substrate responses to altered brain 5-HT activity during prolonged exercise to fatigue. J Appl Physiol 1993; 74:3006-3012.
65. Wilson WM, Maughan RJ. Evidence for a possible role of 5-hydroxytryptamine in the genesis of fatigue in man: administration of paroxetine, a 5-HT re-uptake inhibitor reduces the capacity to perform prolonged exercise. Exp Physiol 1992; 77:921-924.
66. Burak K, Le T, Swain MG. Increased midbrain 5-HT1A receptor number and responsiveness in cholestatic rats. Brain Res 2001; 892:376-379.
67. Swain MG, Maric M. Improvement in cholestasis-associated fatigue with a serotonin receptor antagonist using a novel rat model of fatigue assessment. Hepatology 1997; 25:291-294.
68. Jones EA. Relief from profound fatigue associated with chronic liver disease by long term andansetron therapy. Lancet 1999; 357:397.
69. Crofford LJ, Engleberg NC, Demitrack MA. Neurohormonal perturbations in fibromyalgia. Baillieres Clin Rheumatol 1996; 10:365-378.

70. Brunello N, Akiskal H, Boyer P et al, Langer SZ, Mendlewicz J, Paes de Souza M, Placidi GF, Racagori G, Wessley S. Dysthymia: Clinical picture, extent of overlap with chronic fatigue syndrome, neuropharmacological considerations, and new therapeutic vistas. J Affect Disord 1999; 52:275-290.
71. Adachi A. Projection of the hepatic vagal nerve in the medulla oblengata. J Auton New Syst 1984; 10:287-293.
72. Magni F, Carobi C. The afferent and preganglionic parosympathetic innervation of the rat liver, demonstrated by the retrograde transport of horseradish peroxidase. J Auton Nerv Syst 1983; 8: 237-260.
73. Wan W, Wetmore L, Sorensen CM et al. Neural and biochemical mediators of endotoxin and stress-induced c-fos expression in the rat brain. Brain Res Bull 1994; 34:7-14.
74. Konsman JP, Luheshi GN, Bluthe RM et al. The vagus nerve mediates behavioural depression, but not fever, in response to peripheral immune signals; a functional anatomical analysis. Eur J Neurosci 2000; 12:4434-4446.
75. Suda T, Tozawa F, Ushiyama T et al. Interleukin-1 stimulates corticotropin-releasing factor gene expression in rat hypothalamus. Endocrinology 1990; 126:1223-1228.
76. Gaykema RP, Dijkstra I, Tilders FJ. Subdiaphragmatic vagotomy suppresses endotoxin-induced activation of hypothalamic corticotropin-releasing hormone neurons and ACTH secretion. Endocrinology 1995; 136:4717-4720.
77. Niijma A. The afferent discharges from sensors for interleukin-1 beta in the hepatoportal system in the anesthetized rat. J Auton Nerv Syst 1996; 61:287-291.
78. Swain MG. Cytokines and neuroendocrine dysregulation in obstructive cholestasis: pathophysiological implications. J Hepatol 2001; 35:416-418.
79. Tilg H, Wilmer A, Vogel W et al. Serum levels of cytokines in chronic liver diseases. Gastroenterology 1991; 103:264-74.
80. Turnbull AV, Rivier CL. Regulation of the hypothalamic-pituitary-adrenal axis by cytokines: Actions and mechanisms of action. Physiol Rev 1999; 79:1-71.
81. Licinio J, Wong M-L. Cytokines and the brain: Pathways and mechanisms for cytokine signaling of the central nervous system. J Clin Invest 1997; 100:2941-2947.
82. Yamamoto Y, Sezai S, Sakurabayashi S et al. A study fo endotoxemia in patients with primary biliary cirrhosis. J Int Med Res 1994; 22:95-99.
83. Bemelmans MH, Gouma DJ, Greve JW et al. Cytokines tumour necroses factor and interleukin-6 in experimental biliary obstruction in mice. Hepatology 1992; 15:1132-1136.
84. Swain MG, Appleyard CB, Wallace JL et al. TNFα facilitates inflammation-induced glucocorticoid secretion in rats with biliary obstruction. J Hepatol 1997; 26:361-368.
85. Clements WD, Erwin P, McCaigue MD et al. Conclusive evidence of endotoxemia in biliary obstruction. Gut 1998; 42:293-299.
86. Dantzer R. Cytokine-induced sickness behavior: Where do we stand? Brain Behav Immun 2001; 15:7-24.
87. Swain MG, Beck P, Rioux K et al. Augmented interleukin-1 beta-induced depression of locomotor activity in cholestatic rats. Hepatology 1998; 28:1561-1565.
88. Kakucska I, Qi Y, Clark BD et al. Endotoxin-induced corticotropin-releasing hormone gene expression in the hypothalamic paraventricular nucleus is mediated centrally by interleukin-1. Endocrinology 1993; 133:815-821.
89. Ehlert U, Gaab J, Heinrichs M. Psychoneuroendocrinological contributions to the etiology of depression, posttraumatic stress disorder, and stress-related bodily disorders: the role of the hypothalamic-pituitary-adrenal axis. Biol Psych 2001; 57:141-152.
90. Schwartz CE, Coulthard-Morris L, Zeng Q. Psychosocial correlates of fatigue in multiple sclerosis. Arch Phys Med Rehabil 1996; 77:165-170.
91. Yehuda R. Biology of posttraumatic stress disorder. J Clin Psychiatry 2001; 62 Suppl 17:41-46.
92. Swain MG, Le T. Chronic cholestasis in rats induces anhedenia and a loss of social interest. Hepatology 1998; 28:6-10.
93. Swain MG, Maric M. Defective corticotropin-releasing hormone mediated neuroendocrine and behavioural responses in cholestatic rats: implications for cholestatic liver disease-related sickness behaviours. Hepatology 1995; 22:1560-1564.
94. Whiting P, Bagnall AM, Sowden AJ et al. Interventions for the treatment and management of chronic fatigue syndrome: a systematic review. JAMA 2001; 286:1360-1368.
95. Kroenke K, Swindle R. Cognitive-behavioural therapy for somatization and symptom syndromes: a critical review of controlled clinical trials. Psychother Psychosom 2000; 69:205-215.

CHAPTER 22

Bone Disease in Chronic Cholestatic Liver Disease

Harald Dobnig and Astrid Fahrleitner

Summary

P atients with chronic cholestatic liver disease are especially prone to osteoporosis. Hepatic osteodystrophy typically presents as low turnover osteoporosis with initially normal bone resorption that becomes elevated as disease progresses. The mechanisms leading to low bone formation remain largely unclear. Histological studies suggest that impaired osteoblastic proliferation rather than osteoid synthesis per osteoblast is the main contributor to decreased bone formation rate. A decrease in hepatic IGF1 production or an increase in bilirubin or a growth inhibitor retained in plasma through impaired hepatic function may eventually be responsible for impaired osteoblastic proliferation. Similary to postmenopausal osteoporosis age, body weight, and onset of menopause constitute important clinical determinants of bone mass. Less clear is the role of vitamin D or other fat soluble vitamins, genetic polymorphisms of candidate genes modulating bone modeling and remodeling as well as the impact of severity and length of cholestatic liver disease on bone metabolism.

Because clinical signs of osteoporosis may develop late in the course of cholestatic liver disease and biochemical markers of calcium and vitamin D metabolism are typically not remarkable, early bone densitometry is recommended for all patients. Depending on dietary preferences, all patients should have an adequate intake of calcium and vitamin D and most are well advised to take calcium and vitamin supplements. Patients fulfilling the WHO criteria for osteoporosis at either the femoral neck or lumbar spine should be treated preferentially with bisphosphonates (in the absence of esophageal varices). Based on current literature, female patients with osteoporosis may also be given estrogens but here, beyond the standard precautions, regular monitoring of liver function is still recommended.

All patients being considered for OLT should be given calcium and vitamin D. For these patients, pretreatment with bisphosphonates may be considered even with T-scores below −1, since many patients who sustain fractures in the early posttransplantation period do not fulfill classical osteoporosis criteria prior to OLT. There is a definite lack of well-designed, randomized, controlled trials involving patients with cholestatic liver disease and newer substances that have been proven effective in the treatment of postmenopausal osteoporosis. Such trials would lead to better prophylaxis and treatment of osteoporosis in the pre-, peri- and posttransplantation period.

Introduction

It is well established that osteoporosis is the predominant histological feature when osteopenia is present in patients with chronic cholestatic liver disease. Osteomalacia is only rarely found and is usually restricted to end-stage liver disease. A variety of liver disease entities are associated with osteopenia, but it most often accompanies chronic cholestatic liver diseases such as

Molecular Pathogenesis of Cholestasis, edited by Michael Trauner and Peter L.M. Jansen. ©2004 Eurekah.com and Kluwer Academic / Plenum Publishers.

primary biliary cirrhosis (PBC) and primary sclerosing cholangitis (PSC). This may be due to the cholestasis itself, or in the case of PBC osteopenia may be more evident because osteoporosis is more likely to be encountered in postmenopausal women[1,2]

Although the reported percentages of patients with osteoporosis vary considerably in published studies, it is nonetheless obvious that there are young patients with a degree of osteopenia that is uncommon for their age and warrants prophylactic measures and further observation of their bone mass development. Little is known about the most important consequence of osteoporosis, namely fractures, in patients with chronic cholestatic liver disease. Current literature suggests a prevalence of vertebral fractures in PBC patients of 10% to 30%. This is high given the average age of the studied populations (42 to 56 years) and when compared to a woman's life-time risk of 15% for vertebral fractures.

Bone status becomes particularly important when a patient is considered for orthotopic liver transplantation (OLT) because fractures may develop rapidly under immunosuppressive therapy, even when bone density values are above the WHO criteria for osteoporosis. It would seem advisable to evaluate bone status in all liver-transplantation patients on the basis of clinical, biochemical and radiological findings and at all disease stages since therapeutic options are available either to conserve bone mass or to minimize bone loss in the posttransplantation period.

Physiology of Bone Metabolism

When discussing metabolic bone disease it is important to consider the biology of bone remodeling. Bone is made up of two histologically different subtypes, cortical and trabecular bone, which account for approximately 20% and 80% of bone mass, respectively. Trabecular bone has a larger bone surface than cortical bone and is turned over more rapidly because of a greater number of remodeling sites per unit of bone. It is estimated that within a year roughly 30% of trabecular and 2-3% of cortical bone is replaced by approximately one million remodeling sites that are actively involved in resorption and formation processes at a given time. Consequently, the trabecular bone compartment accounts for 80% of the turnover, although it represents only 20% of the skeleton mass.

The first step in the remodeling cycle is initiated by bone lining cells that cover the trabecular surface and that begin to retreat in response to poorly understood signals that probably originate within the bone. Osteoclasts arise from precursor cells in the adjacent haematopoietic marrow and either migrate to the bone surface or, as has been recently suggested burrow underneath the bone lining cells.[3] They then resorb mineralized bone matrix over a period of approximately two weeks.[4] When the resorption phase is over, osteoblasts migrate to the resorption pit and synthesize unmineralized bone matrix, which starts to become mineralized within 1-2 weeks. The time required for the completion of a new bone structural unit is approximately 3 to 5 months.[4]

Bone turnover is generally balanced, with the amount of bone formed in the remodeling process equalling the amount resorbed. The mechanism guiding this delicate balance between resorption and formation is thought to be primarily communicated by cytokines that are locally released from bone matrix during the resorption process. In postmenopausal women net bone loss occurs when resorption exceeds formation or a normal resorption rate is followed by decreased bone formation. When bone turnover is elevated the number of active remodeling sites determines the extent of net bone loss, since an increase multiplies net bone loss present at each bone structural unit. In principle, women are more prone to trabecular perforations and men to trabecular thinning; this may partly account for the higher fracture incidence rates seen in the former.

Presentation of Hepatic Osteodystrophy

Clinical and Radiological Findings

Patients with PBC and PSC usually suffer from symptoms derived from their underlying cholestatic liver disease and osteopenia will typically not be recognized until a symptomatic fracture occurs. Since bone turnover is greatest at sites rich in trabecular bone, vertebral and rib fractures are usually among the first to occur. Vertebral fracture typically involves minimal or no trauma and only some 30% of patients have severe, acute pain at the fracture site, with little paravertebral radiation of pain. Interestingly, the majority of patients may lose height without any acute bone pain, probably because the fracture evolves at a very slow pace.

In the few and small cross-sectional studies that looked at fractures in patients with chronic cholestatic disease (mean ages 42-56 years), 5-30% of the patients were reported to have one or more prevalent vertebral fractures.[1,5-11] The largest series comprised 78 patients with PBC and PSC and found vertebral fractures prior to orthotopic liver transplantation in 5% and 7%, respectively.[1]

Spinal radiographs are little sensitive to pick up osteopenia and only bone mass decrements of 30% and more can be diagnosed reproducibly as osteopenic. Similar conclusions were drawn from a study that reported spinal osteopenia diagnosed by radiological criteria to be present in only 23% of the PBC patients who had bone mineral density (BMD) measurements below the threshold value.[12]

Bone Densitometry

Clinical recognition of osteoporosis increasingly relies on non-invasive imaging techniques such as bone densitometry, which allows assessment of bone mass and estimation of fracture risk. Only densitometry can provide timely identification of patients who require more specific diagnostic procedures and therapy. The method most commonly used is dual-energy x-ray densitometry (DEXA), usually performed at the lumbar spine and femoral neck. Attention must be paid to artefacts (e.g., compression fractures or osteoarthritis) that may falsely increase BMD values and lead to underdiagnosis of osteoporosis.

Results of bone mass measurements are given in standard-deviation differences compared to the mean values of either age- and gender-matched controls (Z score), or a gender-matched young normal population (T Score). The World Health Organization defines osteoporosis as a BMD value that is 2.5 SD below the mean of young normals (T score).[13]

Most studies using bone densitometry report an osteoporosis prevalence rate in patients with cholestatic liver disease of 20% to 56% at the lumbar spine,[1,5,14-18] and 15% to 32% at the femoral neck.[16,17,19] In one small trial that included only patients with advanced cholestatic liver disease, prevalence rates were as high as 80% for the spine and 60% for the femoral neck region.[20] When results of these studies are normalized for age they translate into mean Z-score values of –0.7 for the spine (mean of 6 studies),[2,5,11,17,18,21] -0.7 for the femoral neck (3 studies)[5,17,18] and –0.6 (2 studies) for the radius.[11,22]

If the young mean age of the above cholestatic patient populations (45-56 years) and the reported osteoporosis prevalence of 20% to 56% is contrasted to a prevalence of roughly 2-5% in a matched healthy population in that age group, the true dimension of osteopenia becomes evident. Springer et al[17] calculated from their own data that the relative risk of severe bone loss at the spine (defined as a Z score of < –2) was 5 times greater in patients with PBC when compared to controls.

Histomorphometric Findings

Dynamic histomorphometry using double tetracycline labeling followed by iliac crest bone biopsy allows tissue level rates of bone formation and resorption to be calculated. In most studies, a reduced bone formation rate could be demonstrated as the primary histological abnormality.[6,11,23,24] Since the majority of studies have found the mineral apposition rate, which reflects the amount of osteoid produced by osteoblasts, to be unchanged,[6,25-28] low bone formation rates are likely to be the result of a decrease in mineralizing surfaces.[6,11] These results strongly suggest that a decrease in osteoblast proliferation rates rather than a reduction in the actual synthesis rate of osteoid by the individual osteoblast is responsible for low bone formation rates.

Results on bone resorption are more difficult to obtain due to inherent methodological problems and here the results are much more inconsistent: bone resorption rates have been found to be either unchanged[11,23,24] or increased.[6,25,26,29] It is likely, however, that results on bone resorption are influenced by the severity of the underlying liver disease, with normal rates at earlier stages of disease and increased rates later on.[6,25] In a homogeneous population of premenopausal women with PBC, a careful histomorphometric study reported a three-fold increase in the birth rate of remodeling units (activation frequency) and a strong correlation to serum concentrations of bilirubin and bile acids as well as stool fat excretion,[25] thus suggesting a linkage between severity of cholestatic liver disease and extent of bone resorption.

Pathogenetic Factors

Although other chronic liver disorders, including chronic autoimmune and viral hepatitis, haemochromatosis and alcohol-related liver disease, are also associated with osteoporosis, many studies point out that patients with chronic cholestatic syndromes such as PBC and PSC seem to be more vulnerable.[1,20,30] Since there are no large prospective studies that could potentially elucidate important pathogenetic factors responsible for the development or aggravation of bone disease, one has to rely on cross-sectional studies of rather small patient populations that suggest a variety of responsible factors (Table 1).

Even though it may seem obvious that patients with more advanced stages of liver disease generally suffer from more severe forms of osteoporosis, the literature does not provide a uniform picture. Some studies propose an inverse correlation between BMD and severity of cholestasis, defined either as histological grading[31-33] or the Mayo risk score.[19,21,34] Similarly, the presence of endoscopic and/or ultrasonographic signs of portal hypertension was associated with lower bone mass and increased fracture prevalence.[2] Other groups, however, were unable to relate the Mayo risk score[2,23,35,36] or histological staging indices[2,5,11] to the results of BMD measurements. Attempts to correlate serum levels of bilirubin or albumin have, with the exception of one study in which weak associations were reported,[21] failed to yield any results in that respect.[2,11,17,33,37] The reported discrepancies may be due to heterogeneities among small patient cohorts, differences in menopausal status, varying percentages of patients in histological subgroups, liver biopsy sampling errors or different lengths of disease duration.

Weighing the available evidence, it seems reasonable to assume a causal relationship between both severity and duration of liver disease and the degree of bone involvement. In a well documented cohort of PSC patients, Hay et al support this belief with their finding that osteoporosis was limited to patients with advanced disease stages (8-9-fold elevated serum bilirubin, steatorrhea, histological grades III or IV) and was not detectable in newly diagnosed cases (normal serum bilirubin, no steatorrhea, histological grades mainly II or III).

Not surprisingly, in patient populations with PBC, more general demographic data such as age or menopausal status were also found to correlate significantly with BMD.[2,17,23,35] Since these patients, unlike those with other chronic liver diseases, are almost exclusively female, it cannot be ruled out that this gender-effect may have led to an overestimation of osteoporosis prevalence in the past. The vast majority of studies did not correct for gender differences or for the onset of menopause in their cohorts.

Chronic liver disease accelerates the development of hypogonadism due to both reduced hypothalamic release of gonadotrophins and primary gonadal failure.[38] There is also one report suggesting a lower mean age at onset of menopause (46 years) specifically for PBC patients.[5] In this study, BMD measurements were 1 SD lower for these women when compared to age-matched women with regular menstrual cycles.

Table 1. Potential pathophysiological factors implicated in PBC- and PSC-associated bone loss

General Factors
- Age
- Menopausal status
- Age at menopause
- Low body weight

Specific Hepatic Factors
- Severity of disease
- Length of disease
- Hypogonadism due to liver disease
- Hyperbilirubinemia
- Low vitamin D levels
- Decreased calcium absorption
- Low vitamin K levels
- Growth factors   – decreased hepatic IGF1 production
                 – retention of mitogenic inhibitors of osteoblastogenesis
- Genetic factors (VDR- and COLIA1- gen polymorphisms)
- Associated disorders (e.g., collagenoses, hyperthyroidism, coelic disease)

Another important predictor of osteoporosis in patients with PBC is low body weight, which is correlated to BMD.[17,24,25,35] Compared to patients with non-cholestatic liver disease, those with chronic cholestatic syndromes weighed an average of 10 kg less prior to OLT,[39] although body composition itself seems unaltered.[5]

Bile salts are essential for the absorption of vitamin D and early studies considered osteomalacia to be the likely cause of osteopenia in patients with chronic cholestatic syndromes.[40] Indeed, 60% of patients with PBC[29] and 80% of those with accompanying osteoporosis show decreased calcium absorption and some of these patients have low 25-hydroxyvitamin D levels at the same time.[23,29] It is difficult to predict to what degree serum levels of vitamin D or the presence of steatorrhea modulate calcium malabsorption in an individual patient but it is likely that both mechanisms play a role.

Serum 25-hydroxyvitamin D levels are frequently found to be low to low-normal[11,23,25,29,33] or normal.[5,11,24,26] Hepatic 25-hydroxylation has not been found to be the rate-limiting step, given adequate oral vitamin D supply and/or sunlight exposure.[41,42] A normal "free" fraction of serum 25-hydroxyvitamin D[43] and adequate 1,25-dihydroxyvitamin D levels despite low total serum levels further argue against a significant role of vitamin D in the pathogenesis of osteoporosis.[21,44] Finally, none of the studies to date demonstrated a relationship between serum vitamin D levels and BMD.[33,45] Although the latter correlation may not readily be expected for methodological reasons, clinical trials evaluating supplementation with vitamin D or 25-hydroxyvitamin D have generally failed to reverse or halt further progression of bone loss as evaluated by histomorphometric, densitometric or fracture criteria.[27,28,36] It thus seems unlikely that the vitamin D system plays an important pathogenetic role in advancing bone loss.

In recent years a number of studies have looked at potential physiological changes associated with various vitamin D receptor (VDR) polymorphisms. Tissue specific differences in sensitivity to 1,25-dihydroxyvitamin D by different VDR's have been proposed as possible mechanism of action.[46] In two very recent studies evaluating 72[17] and 61[47] women with PBC, one group found a significant correlation between lumbar spine (and not femoral neck) BMD and VDR genotype,[17] whereas results of the other group were negative regarding baseline BMD and prospective bone loss.[47] The latter group, however, described another gene encoding the collagen type I alpha1 (COLIA1) polymorphism that yielded a Z-score difference of 0.6 SD in lumbar spine BMD between the SS and ss genotype. The validity of such results needs to be confirmed, however, due to the comparatively small number of subjects included and the 5% prevalence of the ss-genotype.

Fairly normal calcium homeostasis in patients with cholestatic liver syndromes is also reflected by the consistent finding of normal parathyroid function in all[25,48-51] but one study.[5] Although there could be a number of reasons for intestinal hyperparathyroidism to develop in patients with chronic cholestasis, normal levels of ionized calcium[11,29] and free 25-hydroxyvitamin D[43] make significant secondary hyperparathyroidism unlikely.

Vitamins K and A are other fat-soluble vitamins that have been found to be decreased to varying degrees in PBC patients.[52,53] Recent studies suggest that vitamin K may influence the degree of osteocalcin carboxylation and thus bone quality[54] and that supplementation with this vitamin may reduce vertebral and hip fractures without changes in bone mass.[55] Although vitamin A deficiency seems to be a common finding in PBC patients,[53] relatively high intake of retinol rather than vitamin A deficiency seems to increase fracture risk.[56] Thus, though fat-soluble vitamins, especially vitamin D and K, may be decreased in a small percentage of patients with cholestatic liver disease, the clinical importance of this finding is probably slight.

As mentioned above, histomorphometric studies point toward low bone formation as the primary abnormality in hepatic osteodystrophy. This is further substantiated by the finding of consistently low serum osteocalcin levels.[11,48,57-59] Osteocalcin is thought to reflect bone formation and has also been shown to correlate significantly to bone formation rate in PBC patients.[11] Histologic findings suggest that a decrease in osteoblast proliferation accounts for low bone formation rather than a decrease in osteoblast synthesis rate. Theoretically, this may be due to reduced trophic factors produced by the liver, such as IGF-1 or increased concentrations of possible growth inhibitors such as bilirubin, which increase with impaired hepatic function.

IGF-1 is an anabolic hormone that is produced locally in bone; dependent on growth hormone and insulin, it is also synthesized in the liver. In bone, IGF-1 stimulates osteoblast proliferation and differentiation. In animal models of rats with experimentally induced liver cirrhosis or chronic non-cirrhotic liver injury, IGF-1 treatment partially restored bone mass.[60-61] In patients with alcoholic liver cirrhosis, IGF-1 levels were decreased by more than 50% compared to healthy controls.[62] Low IGF-1 levels may simply be due to a difference in nutritional status or may be an epiphenomenon. In any case, a causal relationship between bone metabolism and liver disease has not yet been established.

Osteoblast dysfunction in cholestasis may also be due to retained substances in plasma. In a bioassay using normal human osteoblast-like cells, plasma from jaundiced patients caused a 44% decrease in the mitogenic activity which was independent of the cause of jaundice.[63] Removal of bilirubin by photobleaching offset this inhibiting effect in vitro and unconjugated bilirubin rather than bile acids led to a dose-dependent decrease in osteoblast proliferative capacity. It is, however, unlikely that bilirubin is the only factor responsible for decreased bone formation, since a significant relationship to bilirubin serum levels has never been firmly established.

Bone Metabolism in the Posttransplantation Period

With OLT being performed more often, bone disease at the time of referral to a transplantation center may be less severe than it was a few years ago.

Fractures in the post-transplantation period commonly occur in patients with low BMD values prior to OLT.[1] In one study, irrespective of underlying liver disease, patients lost considerable amounts of bone during the first three months.[64] Due to lower mean pre-transplant BMD values, fractures occurred 4-5 times more often in PBC and PSC patients.

Most of the pre- and posttransplantation fracture data refer to heterogeneous patient populations. According to these studies new vertebral fractures occur in 14% to 31% during the first months post OLT.[21,65-71] In a small longitudinal study of 20 PBC patients, vertebral fractures occurred in 35% and any other clinical fracture in 65% in the posttransplantation period.[21]

Similarly high overall fracture rates of 31% to 43% were reported for 78 patients with cholestatic liver disease.[1] Approximately three months after OLT, bone mass again starts to increase and usually reaches pretransplantation levels after one year.[21] After 2 years, BMD is approximately 2-5% above pretransplantation values.[1,21]

Bone loss and elevated bone turnover as well as the dramatic increase in fracture risk weeks to months after OLT probably are the result of immunosuppressive therapy and postoperative immobilisation, which initiate bone resorption processes.[48,71] Corticosteroids affect bone metabolism in several ways. They are well known to decrease osteoblast function and proliferation and thus bone formation, to increase osteoclast activity, renal calcium-, and parathyroid hormone secretion, and to reduce calcium absorption and gonadal function.[31] The consequence is marked bone loss with preferential involvement of the trabecular bone compartment, especially during the first 3-6 months after transplantation.[64,72] Recently, glucocorticoids have been shown in vitro[73,74] and in humans as well[75] to decrease osteoprotegerin, a member of the tumor necrosis factor receptor superfamily and a key regulator of bone turnover. In a prospective study of cardiac transplant patients, there was a strong correlation between BMD decrease at the lumbar spine and femoral neck and the decrease in serum osteoprotegerin levels.[76] A similar mechanism may also apply to patients after OLT.

Little is known about long-term consequences of posttransplantation bone disease but there are data to suggest that the majority of patients will continue to be osteopenic or osteoporotic and that fracture risk may also be increased after more than one year following OLT.[67,77-80]

Recognition of Hepatic Osteodystrophy

Attention should focus on bone densitometry for early recognition of bone disease. DEXA measurements should be performed both at the lumbar spine and femoral neck and interpretation should take into account findings from lumbar spine x-rays (anterior-posterior, lateral) to exclude any artefacts that may lead to falsely elevated BMD values. Although osteoporosis is by far the predominant diagnosis when low BMD is present, this does not exclude osteomalacia, which may be found in the occasional patient, especially at later disease stages. Results on BMD measurements should follow the guidelines given by the World Health Organization, which propose the presence of osteoporosis when BMD measurements are more than 2.5 SD below the mean value of young, healthy controls.

As far as evaluation of bone status is concerned, an active case-finding strategy may be the minimum approach for patients with chronic cholestatic liver disease and it might be advisable, given the prevalence of osteoporosis in this patient population and the general lack of specific clinical symptoms, that all patients should undergo osteodensitometry.

Since patients with PBC may have had a premature menarche, a gynaecological history should be obtained. Past fractures, signs of osteopenia on x-rays, a family history of osteoporosis, height loss, low body weight, and/or a history of glucocorticoid treatment, should all alert the physician and lead to early evaluation of bone mineral density.

When there is frank osteoporosis, or a low Z-score in younger patients, the following tests can exclude secondary causes of osteoporosis: serum calcium, albumin (to normalize serum calcium levels), serum phosphate, serum-creatinine, serum electrophoresis, 25-hydroxyvitamin D, intact parathyroid hormone, sex hormones including luteinizing hormone, and thyroid function test. Typically, osteocalcin values will be low-normal or below the normal range, indicating decreased bone formation. Importantly, many collagen-related markers of bone turnover including carboxy-terminal (PICP)—and amino-terminal propeptides of type I collagen (PINP), cross-linked carboxy-terminal telopeptides of type I collagen (ICTP) and urinary markers such as N-telopeptides (NTX) are correlated to the histological stage of liver disease and do not reflect bone turnover because of altered liver collagen metabolism.[81]

If osteoporosis is present, a thoracolumbar spinal radiograph should be obtained. The majority of patients will only have abnormal bone densitometry values and should be treated if they fulfill the WHO criteria for osteopenia, osteoporosis, or manifest osteoporosis.

Management of Osteoporosis in Cholestasis

The only long-term treatment options for patients with chronic cholestatic liver disease are either treatment with ursodeoxycholic acid (UDCA) or, with more advanced disease, OLT. A randomized, double-blind three-year treatment study with UDCA, however, showed no difference in bone density values,[34] while after OLT, bone status deteriorates at least temporarily and usually does not improve noticeably for two or more years. Ideally, prevention of osteoporosis should start at early stages of the disease.

General precautions for prevention of bone loss are an adequate calcium intake of 1500 mg daily, sufficient sunlight exposure, maintenance of an optimal level of mobility, avoidance of smoking and excessive alcohol consumption, and treatment of hypogonadism and vitamin D deficiency if present.

Numerous osteoporosis treatment trials in patients with postmenopausal osteoporosis have shown the efficacy of calcium and vitamin D supplementation in maintaining BMD or halting age-related bone loss for several years. A reduction in clinical fractures, including hip fractures, has been reported in elderly and institutionalized populations.[82,83]

Since small trials using 25-hydroxvitamin D in PBC patients did not show improvements in histological or radial densitometric follow-up studies, it was concluded that there is little evidence to support the routine administration of vitamin D beyond the recommended daily allowance of 400-800 IU for all patients.[27,28,84] Recently, Shiomi et al reported prevention of bone loss at the lumbar spine in a randomized study of 34 PBC patients with 1,25 dihydroxyvitamin D (calcitriol) and proposed it as effective treatment for osteoporosis in this patient population.[85] Despite these somewhat negative results with active vitamin D metabolites, patients with chronic liver diseases have the potential for abnormalities in calcium and vitamin D metabolism and it would seem prudent to ensure that their intake of calcium and vitamin D is adequate.

As to other pharmacological treatments, there is no specific reason to assume that drugs tested successfully in postmenopausal osteoporosis will not work in patients with hepatic osteodystrophy. For each new drug, however, there should be small trials to look specifically for any potential side effects that could influence liver function in patients with chronic liver disease.

Currently, alendronate, risedronate, raloxifen, and to some extent calcitonin have been studied in large, double-blind placebo-controlled trials and shown to reduce vertebral and to a certain degree non-vertebral fractures in postmenopausal osteoporosis.[86-90]

Estrogens reduce bone turnover and increase or stabilize BMD; however, except for two small prospective trials,[91-92] evidence for a reduction in clinical fractures is limited to observational studies only.[93-94] Estrogens have thus far been used reluctantly in patients with PBC, since they impair bile flow.[95,96] Two small studies (one with oral, one with transdermal estrogen) of 1 and 2 years' duration reported increases in lumbar spine BMD and no significant changes in liver function tests and concluded that estrogens are safe and effective in this patient population.[97,98] Clearly, randomized trials with the inclusion of proper controls are now warranted. Raloxifen, a newly available drug for the treatment of osteoporosis with partial estrogen-agonistic and –antagonistic properties has not been tested at all in patients with hepatic osteodystrophy.

Bisphosphonates, such as alendronate and risedronate are powerful anti-resorptive drugs that decrease bone turnover and thus the birth rate of bone remodeling units by approximately 50%. These compounds increase bone density not through a truly anabolic effect on bone formation but increase secondary mineralization of newly formed bone matrix by decreasing bone turnover. With the exception of intermittently administered parathyroid hormone, only bisphosphonates have thus far also proven to decrease non-vertebral fractures and are thus ideally suited for treatment of elderly patients with an increased risk of hip fractures. Bisphosphonates should not, however, be used in patients with advanced liver disease and esophageal varices due to their potential for precipitating variceal hemorrhage. Esophageal irritation in animal studies, and most likely also in humans may be caused by prolonged contact with the tablet, reflux of acidic gastric contents and exacerbation of preexisting esophageal damage.[99] Since severe esophagitis and esophageal ulcers have been described in otherwise healthy patients with postmenopausal osteoporosis[100] and an ulcer on a varix could cause a massive bleed bisphosphonate treatment should be avoided in patients with esophageal varices. Intravenous application of bisphosphonates would circumvent such a problem, however we still lack data proving that intermittent parenteral treatment leads to a favourable reduction in fracture occurence in postmenopausal osteoporotic women.

Unfortunately, clinical results for liver patients with osteoporosis are only available for the first-generation bisphosphonate, etidronate. In two studies, etidronate did not change lumbar spine BMD but prevented bone loss in one trial in which the placebo group lost bone due to concomitant prednisone and azathioprine treatment.[101,102]

In postmenopausal women with established osteoporosis, intranasal calcitonin (200 IU daily) was shown to decrease vertebral fractures over a five-year treatment period in a double-blind placebo-controlled trial.[90] Subcutaneous calcitonin every other day (40 IU) for 6 months could not produce any measurable effect on bone density in PBC patients;[103] however, due to the low dosage and the short study period, clear conclusions certainly cannot be drawn from this study.

There are two more studies with rather complicated treatment designs that were considered positive by their respective authors. One was a trial using 1,25 dihydroxyvitamin D, calcitonin and calcium within an ADFR dosing scheme;[104] the study revealed a positive BMD response at the spine but lacked appropriate controls. The second study compared the effects of etidronate and fluoride and showed an increase in lumbar spine BMD only for etidronate, which was also better tolerated than fluoride.[105]

Prevention of Bone Loss in the Early and Later Posttransplantation Period

The most vulnerable period is the first 3-6 months after transplantation when bone loss is greatest and patients are very susceptible to fractures. Two recent studies investigated the influence of calcitonin[106] or intravenous pamidronate[107] in this important peri- and early posttransplantation period.

Calcitonin at a dose of 100 IU daily for the first 6 months following OLT was unable to prevent or reduce accelerated bone loss or spontaneous fractures,[106] whereas 3-monthly intravenous infusions of pamidronate that were started before transplantation and continued up to 9 months after OLT effectively prevented symptomatic vertebral fractures after liver transplantation.[107] Thirty-eight percent of untreated patients acquired fractures as compared to zero percent in the group treated with pamidronate. This patient population was heterogeneous as to underlying liver disease, though the majority had PBC. These results are certainly very encouraging and point for the first time to a possibly successful prevention of transplantation-associated osteoporosis, but they definitely need to be confirmed in a randomized double-blind trial.

Several studies looked at possible improvements in BMD or fracture rates within a stable clinical period after OLT. By far the largest trial included 283 patients at least 6 months after OLT who were assigned to 5 treatment groups.[108] They received either 0.25 or 0.5ug of 1,25 dihydroxyvitamin D daily with or without calcium and one treatment arm received additional fluoride. Mean BMD increases were roughly 5% and 10% at the spine and 4% and 5% at the femoral neck for the calcitriol/calcium combination therapy. The calcitriol/fluoride arm gained the most, with an increase of 11% at the lumbar spine and 13% at the femoral neck. Unfortunately, neither BMD results for the untreated control group nor levels of significance for BMD analyses were given. The overall fracture incidence in the control group, however, was 2.8% and higher than the 0.8% of the pooled treatment arms.

Isoniemi et al reported improvements of 6% and 4% at the lumbar spine and femoral neck with transdermal estrogen in an open study of 33 patients with PBC after 2 years of treatment.[109] Finally, Valero et al compared the effects of intramuscular calcitonin (40 IU per day) with a cyclical regime of etidronate in patients who were 1-17 months post transplantation and found 6% and 8% increases after one year of treatment.[110] This study, however, also lacked an appropriate control group, so it is difficult to say to what extent this increase was related to the study drug or to spontaneous BMD recovery after OLT.

Conclusions

Given the high prevalence of osteoporosis and the large number of fractures following liver transplantation evaluation of bone status is justified and should be performed more often, ideally in all patients with hepatic osteodystrophy. Specific trials should focus on drugs that have proven effective in the treatment of postmenopausal osteoporosis in patients with hepatic dysfunction primarily to address safety issues. Preclinical studies are needed to better characterize factors involved in the process of impaired bone formation.

References

1. Porayko MK, Wiesner RH, Hay JE et al. Bone disease in liver transplant recipients: Incidence, timing, and risk factors. Transplant Proc 1991; 23:1462-1465.
2. Isaia G, Di Stefano M, Roggia C et al. Bone disorders in cholestatic liver diseases. Forum (Genova) 1998; 8:28-38.
3. Parfitt AM. The bone remodeling compartment: a circulatory function for bone lining cells. J Bone Miner Res 2001; 16:1575-1582.
4. Bone HG, Adami S, Rizzoli R et al. Weekly administration of alendronate: Rationale and plan for clinical assessment. Clin Ther 2000; 22: 15-28.
5. Bagur A, Mautalen C, Findor J et al. Risk factors for the development of vertebral and total skeleton osteoporosis in patients with primary biliary cirrhosis. Calcif Tissue Int 1998; 63:385-390.
6. Stellon AJ, Webb A, Compston J et al. Low bone turnover state in primary biliary cirrhosis. Hepatology 1987; 7:137-142.
7. Shih MS, Anderson C. Does "hepatic osteodystrophy" differ from peri- and postmenopausal osteoporosis? A histomorphometric study. 1987; 41:187-191.
8. Compston JE. Hepatic osteodystrophy: vitamin D metabolism in patients with liver disease. Gut 1986; 27:1073-90.
9. Stellon AJ, Webb A, Compston J et al. Lack of osteomalacia in chronic cholestatic liver disease. Bone 1986; 7:181-185.
10. Stellon AJ, Davies A, Compston J et al. Osteoporosis in chronic cholestatic liver disease. Q J Med 1985; 57:783-790.
11. Hodgson SF, Dickson ER, Wahner HW et al. Bone loss and reduced osteoblast function in primary biliary cirrhosis. Ann Intern Med 1985; 103:855-860.
12. Hay JE, Lindor KD, Wiesner RH et al. The metabolic bone disease of primary sclerosing cholangitis. Hepatology 1991; 14:257-261.
13. Kanis JA, Melton LJ, Christiansen C et al. The diagnosis of osteoporosis. J Bone Miner Res 1994; 9:1137-1141.
14. Lakatos PL, Firneisz G, Lakatos P et al. Follow-up study of bone mineral density in postmenopausal patients with primary biliary cirrhosis. Orv Hetil 2001; 142:503-508.
15. Hamburg SM, Piers DA, van den Berg AP et al. Bone mineral density in the long term after liver transplantation. Osteoporos Int 2000; 11:600-606.
16. Bonkovsky HL, Hawkins M, Steinberg K et al. Prevalence and prediction of osteopenia in chronic liver disease. Hepatology 1990; 12:273-280.
17. Springer JE, Cole DEC, Rubin LA et al. Vitamin D-Receptor genotypes as independent predictors of decreased bone mineral density in primary biliary cirrhosis. Gastroenterology 2000; 118:145-151.
18. Newton J, Francis R, Prince M et al. Osteoporosis in primary biliary cirrhosis revisited. Gut 2001; 49:282-287.
19. Pereira SP, Bray GP, Pitt PI et al. Non-invasive assessment of bone density in primary biliary cirrhosis. Eur J Gastroenterol Hepatol 1999; 11:323-328.
20. Crosbie OM, Freaney R, McKenna MJ et al. Bone density, vitamin D status, and disordered bone remodeling in end-stage chronic liver disease. Calcif Tissue Int 1999; 64:295-300.
21. Eastell R, Dickson ER, Hodgson SF et al. Rates of vertebral bone loss before and after liver transplantation in women with primary biliary cirrhosis. Hepatology 1991; 14:296-300.
22. Trautwein C, Possienke M, Schlitt HJ et al. Bone density and metabolism in patients with viral hepatitis and cholestatic liver diseases before and after liver transplantation. Am J Gastroenterol 2000; 95:2343-2351.

23. Maddrey WC. Bone disease in patients with primary biliary cirrhosis. Prog Liver Dis 1990; 9:537-554.
24. Diamond TH, Stiel D, Lunzer M et al. Hepatic osteodystrophy. Gastroenterology 1989; 96:213-221.
25. Hodgson SF, Dickson ER, Eastell R et al. Rates of cancellous bone remodeling and turnover in osteopenia associated with primary biliary cirrhosis. Bone 1993; 14:818-827.
26. Cuthbert JA, Pak CYC, Zerwekh JE et al. Bone disease in primary biliary cirrhosis: Increased bone resorption and turnover in the absence of osteoporosis or osteomalacia. Hepatology 1984; 4:1-8.
27. Herlong FH, Recker RR, Maddrey WC. Bone disease in primary biliary cirrhosis: Histologic features and response to 25-hydroxyvitamin D. Gastroenterology 1982; 83:103-108.
28. Matloff DS, Kaplan MM, Neer RM et al. Osteoporosis in primary biliary cirrhosis: effects of 25-hydroxyvitamin D3 treatment. Gastroenterology 1982; 83:97-102.
29. Mitchison HC, Malcolm AJ, Bassendine MF et al. Metabolic bone disease in primary biliary cirrhosis at presentation. Gastroenterology 1988; 94:463-470.
30. McDonald JA, Dunstan CR, Dilworth P et al. Bone loss after liver transplantation. Hepatology 1991; 14:613-619.
31. Wolfhagen FHJ, van Buuren HR, Vleggaar FP, Schalm SW. Management of osteoporosis in primary biliary cirrhosis. Bailliere's Clin Gastroenterology 2000; 14:629-641.
32. van Berkum FN, Beukers R, Birkenhagen JC et al. Bone mass in women with primary biliary cirrhosis: the relation with histological stage and use of glucocorticoids. Gastroenterology 1990; 4:1134-1139.
33. Hay JE, Lindor KD, Russell H et al. The metabolic bone disease of primary sclerosing cholangitis. Hepatology 1991; 14:257-261.
34. Lindor KD, Janes CH, Crippin JS et al. Bone disease in primary biliary cirrhosis: does ursodeoxycholic acid make a difference? Hepatology 1995; 21:389-392.
35. Ninkovic M, Love SA, Tom B et al. High prevalence of osteoporosis in patients with chronic liver disease prior to liver transplantation. 2001; 69:321-326.
36. Hay JE. Bone disease in cholestatic liver disease. Gastroenterology 1995; 108:276-283.
37. Cemborain A, Castilla-Cortazar I, Garcia M et al. Osteopenia in rats with liver cirrhosis:beneficial effects of IGF-I treatment. J Hepatol 1998; 28:122-131.
38. Bell H, Raknerud N, Falch JA et al. Inappropriately low levels of gonadotrophins in amenorrhoeic women with alcoholic and non-alcoholic cirrhosis. Eur J Endocrinol 1995; 132:444-449.
39. Keogh JB, Tsalamandris C, Sewell RB et al. Bone loss at the proximal femur and reduced lean body mass following liver transplantation: A longitudinal study. Nutrition 1999; 15:661-664.
40. Atkinson M, Nordin BEC, Snenock S. Malabsorption and bone disease in prolonged obstructive jaundice. Q J Med 1966; 99:299-312.
41. Krawitt EL, Grundman MJ, Mawer EB. Absorption, hydroxylation and excretion of vitamin D3 in primary biliary cirrhosis. Lancet 1977; 2:1246-1249.
42. Skinner RK, Long RG, Sherlock S et al. 25-hydroxylation of vitamin D in primary biliary cirrhosis. Lancet 1977; 1:720-721.
43. Diamond TH, Stiel D, Mason R et al. Serum vitamin D metabolites are not responsible for low turnover osteoporosis in chronic liver disease. J Clin Endocrinol Metab 1989; 69:1234.
44. Kaplan MM, Goldberg MJ, Mation DS et al. Effect of 25-hydroxyvitamin D3 on vitamin D metabolites in primary biliary cirrhosis. Gastroenterology 1981; 81:681-685.
45. Aguilar JI, Radick J, Kodali V et al. Bone disease and plasma vitamin K levels in PBC patients. Gastroenterology 1993; 104:A868 (abstract).
46. Howard H, Nguyen T, Morrison N et al. Genetic influences on bone density: Physiological correlates of vitamin D receptor gene alleles in premenopausal osteoporosis. J Clin Endocrinol Metab 1995; 80:2800-2805.
47. Pares A, Guanabens N, Alvarez L et al. Collagen type I alpha 1 and vitamin D receptor gene polymorphisms and bone mass in primary biliary cirrhosis. Hepatology 2001; 33:554-560.
48. Vedi S, Greer S, Skingle SJ et al. Mechanism of bone loss after liver transplantation: A histomorphometric analysis. J Bone Miner Res 1999; 14:281-286.
49. Pauletzki J, Paumgartner G et al. Lebertransplantation bei akutem Leberversagen: Wer? Wann? Wie? Wien Klin Wochenschr 1998; 110:547-550.
50. Yeganehfar W, Wamser P, Rockenschraub S et al. Lebertransplantation bei akutem Leberversagen. Wien Klin Wochenschr 1998; 110:570-578.
51. Hawkins FG, Leon M, Lopez MB et al. Bone loss and turnover in patients with liver transplantation. Hepato-Gastroenterol 1994; 41:158-161.
52. Kowdley KV, Emond MJ, Sadowski JA et al. Plasma vitamin K1 level is decreased in primary biliary cirrhosis. Am J Gastroenterol 1997; 92:2059-2061.
53. Phillips JR, Angulo P, Petterson T et al. Fat-soluble vitamin levels in patients with primary biliary cirrhosis. Am J Gastroenterol 2001; 96:2745-2750.

54. Sugiyama T, Kawai S. Carboxylation of osteocalcin may be related to bone quality: a possible mechanism of bone fracture prevention by vitamin K. J Bone Miner Metab 2001; 19:146-149.
55. Booth SL, Tucker KL, Chen H et al. Dietary vitamin K intakes are associated with hip fracture but not with bone mineral density in elderly men and women. Am J Clin Nutr 2000; 71:1201-1208.
56. Freskanich D, Singh V, Willett WC et al. Vitamin A intake and hip fractures among postmenopausal women. JAMA 2002; 287:47-54 (abstract).
57. Giannini S, Nobile M, Ciuffreda M et al. Long-term persistence of low bone density in orthotopic liver transplantation. Osteoporos Int 2000; 11:417-424.
58. Delmas PD. Osteoporosis in patients with organ transplants: a neglected problem. Lancet 2001; 357:325-326.
59. Guardiola J, Xiol X, Sallie R et al. Ann Intern Med 1999; 131:752-755.
60. Cemborain A, Castilla-Cortazar I, Garcia M et al. Effects of IGF-I treatment on osteopenia in rats with advanced liver cirrhosis. J Physiol Biochem 2000; 56:91-99.
61. Chemborain A, Castilla-Cortazar I, Garcia M et al. Osteopenia in rats with liver cirrhosis: beneficial effects of IGF-I treatment. J Hepatol 1998; 28:122-131.
62. Schimpff RM, Lebrec D, Donnadieu M. Somatomedin production in normal adults and cirrhotic patients. Acta endocrinol 1977; 86:355-362.
63. Janes CH, Dickson ER, Okazaki R et al. Role of hyperbilirubinemia in the impairment of osteoblast proliferation associated with cholestatic jaundice. J Clin Inv 1995; 95:2581-2586.
64. Abdelhadi M, Eriksson SA, Eriksson S et al. Bone mineral status in end-stage liver disease and the effect of liver transplantation. Scand J Gastroenterol 1995; 30:1210-1215.
65. Arnold JC, Hauser D, Ziegler R et al. Bone disease after liver transplantation. Transplant Proc 1992; 24:2709-2710.
66. Meys E, Fontages E, Fourcade N et al. Bone loss after orthotopic liver transplantation. Am J Med 1994; 97:445-450.
67. Leidig-Bruckner G, Hosch S, Dodidou P et al. Frequency and predictors of osteoporotic fractures after cardiac or liver transplantation: a follow-up study. Lancet 2001; 357:342-347.
68. Ninkovic M, Skingle SJ, Bearcroft PW et al. Incidence of vertebral fractures in the first three months after orthotopic liver transplantation. Eur J Gastroenterol Hepatol 2000; 12:931-935.
69. Compston JE, Greer S, Skingle SJ et al. Early increases in plasma parathyroid levels following transplantation. J Hepatol 1996; 25:715-718.
70. Navasa M, Monegal A, Guanabens N et al. Bone fractures in liver transplant patients. Brit J Rheumatol 1994; 33:52-53.
71. McDonald JA, Dunstan CR, Dilworth P et al. Bone loss after liver transplantation. J Hepatol 1991; 14:613-619.
72. Feller RB, McDonald JA, Sherbon KJ et al. Evidence of continuing bone recovery at a mean of 7 years after liver transplantation. Liver Transplant Surg 1999; 407-413.
73. Vidal NOA, Braendstroem H, Jonsson KB et al. Osteoprotegerin mRNA is expressed in primary human osteoblast-like cells: down-regulation by glucocorticoids. J Endocrinol 1998; 159:191-195.
74. Hofbauer LC, Gori F, Riggs BL et al. Stimulation of osteoprotegerin ligand and inhibition of osteoprotegerin production by glucocorticoids in human osteoblastic lineage cells: potential paracrine mechanism of glucocorticoid-induced osteoporosis. Endocrinology 1999; 140:4382-4389.
75. Bornefalk E, Dahlen I, Johannson G et al. Serum levels of osteoprotegerin: effects of glucocorticoid and growth hormone. Bone 1998; 23(Suppl 1):S136.
76. Dobnig H, Fahrleitner A, Prenner G et al. Serum osteoprotegerin is a major determinant of bone density and vertebral fracture status in patients after cardiac transplantation. Bone 2001; 28:S96-97. (Abstract).
77. Crosbie OM, Freaney R, McKenna MJ et al. Predicting bone loss following orthotopic liver transplantation. Gut 1999; 44:430-434.
78. Floreani A, Fries W, Luisetto G et al. Bone metabolism in orthotopic liver transplantation: a prospective study. Liver Transpl Surg 1998; 4:311-319.
79. Hyder Hussaini S, Oldroyd B, Stewart SP et al. Regional bone mineral density after orthotopic liver transplantation. Eur J Gastroenterol Hepatol 1999; 11:157-163.
80. Compston JE, Greer S, Skingle SJ et al. Early increase in plasma parathyroid hormone levels following liver transplantation. J Hepatol 1996; 25:715-718.
81. Guanabens N, Pares A, Alvarez L et al. Collagen-related markers of bone turnover reflect the severity of liver fibrosis in patients with primary biliary cirrhosis. J Bone Miner Res 1998; 13:731-738.
82. Chapuy MC, Arlot ME, Duboeuf F et al. Vitamin D3 and calcium to prevent hip fractures in the elderly women. N Engl J Med 1992; 327:1637-1642.
83. Dawson-Hughes B, Harris SS, Krall EA et al. Effect of calcium and vitamin D supplementation on bone density in men and women 65 years of age or older. N Engl J Med 1997; 337:670-676.

84. Reichel H, Koeffler HP, Norman AW. The role of vitamin D endocrine system in health and disease. N Engl J Med 1989; 320:980-991.
85. Shiomi S, Masaki K, Habu D et al. Calcitriol for bone loss in patients with primary biliary cirrhosis. J Gastroenterol 1999; 34:241-245.
86. Black DM, Cummings SR, Karpf DB et al. Randomised trial of effect of alendronate on risk of fracture in women with existing vertebral fractures. Fracture Intervention Trial Research Group. Lancet 1996; 348:1535-1541.
87. Reginster J, Minne HW, Sorensen OH et al. Randomized trial of the effects of risedronate on vertebral fractures in women with established postmenopausal osteoporosis. Vertebral efficacy with risedronate therapy (VERT) Study Group. Osteoporos Int 2000; 11:83-91.
88. Harris ST, Watts NB, Genant HK et al. Effects of risedronate treatment on vertebral and nonvertebral fractures in women with postmenopausal osteoporosis: a randomized controlled trial. Vertebral efficacy with risedronate therapy (VERT) Study Group. JAMA 1999; 282:1344-1352.
89. Ettinger B, Black DM, Mitlak BH et al. Reduction of vertebral fracture risk in postmenopausal women with osteoporosis treated with raloxifene: results from a 3-year randomized clinical trial. Multiple outcomes of raloxifene evaluation (MORE) investigators. JAMA 1999; 282:637-645.
90. Chestnut CH3rd, Silverman S, Andriano K et al. A randomized trial of nasal spray salmon calcitonin in postmenopausal women with established osteoporosis: the prevent recurrence of osteoporotic fractures study. PROOF Study Group. Am J Med 2000:109; 267-276.
91. Lufkin EG, Wahner HW, O'Fallon WM et al. Treatment of postmenopausal osteoporosis with transdermal estrogen. Ann Intern Med 1992; 117:1-9.
92. Lindsay R, Hart DM, Forrest C et al. Prevention of spinal osteoporosis in oophorectomised women. Lancet 1980; 2:1151-1154.
93. Kanis JA, Johnell O, Gullberg B et al. Evidence for efficacy of drugs affecting bone metabolism in preventing hip fracture. BMJ 1992; 305:1124-1128.
94. Michaelsson K, Baron JA, Farahmand BY et al. Hormone replacement therapy and risk of hip fracture: population based case-control study. The Swedish Hip Fracture Study Group. Osteoporos Int 1998; 8:540-546.
95. Schreiber AJ, Simon FR. Estrogen-induced cholestasis: clues to pathogenesis and treatment. Hepatology 1983; 3:607-613.
96. Kontturi M, Sottaniemi E. Effect of estrogen on liver function of prostatic cancer patients. Brit Med J 1969; 4:204-205.
97. Olsson R, Mattsson LA, Obrant K et al. Estrogen-progesteron therapy for low bone mineral density in primary biliary cirrhosis. Liver 1999; 19:188-192.
98. Crippin JS, Jorgensen RA, Dickson ER et al. Hepatic osteodystrophy in primary biliary cirrhosis: effects of medical treatment. Am J Gastroenterol 1994; 89:47-50.
99. Peter CP, Handt LK, Smith SM. Esophageal irritation due to alendronate sodium tablets: possible mechanisms. Dig Dis Sci 1998; 43:1998-2002.
100. Donahue JG, Chan KA, Andrade SE et al. Gastric and duodenal safety of daily alendronate. Arch Intern Med 2002; 162:936-942.
101. Wolfhagen FHJ, van Buuren HR, den Ouden JW et al. Cyclical etidronate in the prevention of bone loss in corticosteroid-treated primary biliary cirrhosis. J Hepatol 1997; 26:325-330.
102. Lindor KD, Jorgensen RA, Tiegs RD et al. Etidronate for osteoporosis in primary biliary cirrhosis: a randomized trial. J Hepatol 2000; 33:878-882.
103. Camisasca M, Crosignani A, Battezzati PM et al. Parenteral calcitonin for metabolic bone disease associated with primary biliary cirrhosis. Hepatology 1994; 633-637.
104. Floreani A, Zappala F, Fries W et al. A 3-year pilot study with 1,25-Dihydroxyvitamin D, calcium, and calcitonin for severe osteodystrophy in primary biliary cirrhosis. J Clin Gastroenterol 1997; 24:239-244.
105. Guanabens N, Pares A, Monegal A et al. Etidronate versus fluoride for treatment of osteopenia in primary biliary cirrhosis: Preliminary results after 2 years. Gastroenterol 1997; 113:219-224.
106. Hay JE, Malinchoc M, Dickson ER. A controlled trial of calcitonin therapy for the prevention of post-liver transplantation atraumatic fractures in patients with primary biliary cirrhosis and primary sclerosing cholangitis. J Hepatol 2001; 34:292-298.
107. Reeves HL, Francis RM, Manas DM et al. Intravenous bisphosphonate prevents symptomatic osteoporotic vertebral collapse in patients after liver transplantation. Liver Transpl Surg 1998; 4:404-409.
108. Neuhaus R, Kubo A, Lohmann R et al. Calcitriol in prevention and therapy of osteoporosis after liver transplantation. Transplant Proc 1999; 31:472-473.
109. Isoniemi H, Appelberg J, Nilsson CG et al. Transdermal oestrogen therapy protects postmenopausal liver transplant women from osteoporosis. A 2-year follow-up study. J Hepatol 2001; 34:299-305.
110. Valero MA, Loinaz C, Larrodera L et al. Calcitonin and bisphosphonates treatment in bone loss after liver transplantation. Calcif Tissue Int 1995; 57:15-19.

CHAPTER 23

Fat Absorption and Lipid Metabolism in Cholestasis

Anniek Werner, Folkert Kuipers and Henkjan J. Verkade

Introducton

The liver has a central role in control of various aspects of lipid metabolism. Primarily, the liver produces bile, constituents of which are required for efficient intestinal fat absorption. Additionally, biliary secretion of cholesterol (either as such, or after metabolism in the form of bile salts) and phospholipids from the liver into the intestine is of major importance in body lipid homeostasis. The liver is the major source of plasma lipoproteins: it synthesizes apoproteins (i.e., apo A-I, apo B, apo E) that regulate many complex metabolic interconversions between lipoprotein classes, as well as lipoprotein lipid constituents as cholesterol, triglycerides and phospholipids. The liver is also the major site of clearance of circulating lipoproteins, which are subsequently catabolized in the hepatocytes. Additionally, the liver synthesizes enzymes (e.g., LCAT, CETP, PLTP, LPL) which are involved in lipoprotein metabolism in the plasma compartment. Finally, the liver is the site of active synthesis, metabolism and/or oxidation of various lipid classes, including long-chain polyunsaturated fatty acids (Fig. 1).

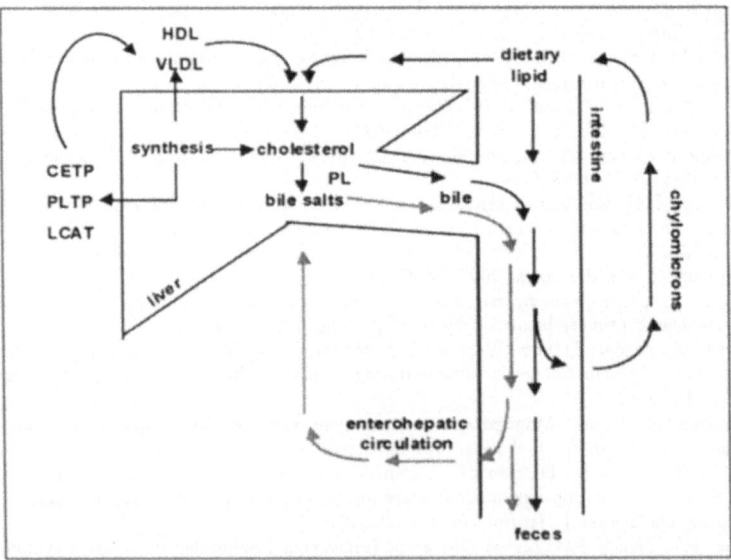

Figure 1. The liver as a central organ in lipid metabolism.

Molecular Pathogenesis of Cholestasis, edited by Michael Trauner and Peter L.M. Jansen.
©2004 Eurekah.com and Kluwer Academic / Plenum Publishers.

In view of this multitude of essential functions that are in part strongly interrelated, it is evident that disturbances in bile formation in cholestatic liver disease will have a strong impact on various aspects of lipid metabolism in the body. Consequences of cholestasis, which is functionally defined as decreased or absent bile flow from the liver into the intestine, may be related to:

1. the absence of specific bile components at their sites of action, particularly in the intestine
2. disruption of the continuous flux of lipids from the liver into bile and intestine, resulting in accumulation of toxic and non-toxic bile components in the body, most notably in hepatocytes, with concomitant alterations in hepatocyte function.
3. characteristic alterations in plasma lipoprotein composition associated with cholestatic liver diseases such as decreased HDL levels and the appearance of lipoprotein X.

Intestinal Lipid Absorption

Dietary Lipid Classification

Dietary fat comprises a wide array of lipid classes, which have been categorized according to the nature of their interactions with water into polar and non-polar lipids.[1-3] Polar lipids, which are insoluble in water, are cholesteryl esters, hydrocarbons and carotene. Polar lipids are divided into 3 subclasses; firstly the insoluble non-swelling amphiphiles which form a thin stable monolayer in water; secondly the insoluble swelling amphiphiles which form both stable monolayers in water as well as laminated lipid-water structures called liquid crystals; and finally the soluble amphiphiles, which possess strong polar groups that render these molecules soluble in water at low concentrations, forming both unstable monolayers and micelles.[4,5] Examples of class 1 polar lipids are triacylglycerols (TAG), diacylglycerols (DAG), non-ionized long chain fatty acids (LCFA), unesterified cholesterol and the fat soluble vitamins A, D, E and K. Class 2 insoluble swelling amphiphiles are monoacylglycerols (MAG), ionized fatty acids (FA) and phospholipids (PL). Examples of class 3 soluble amphiphiles are sodium salts of long chain fatty acids and bile salts. The absence of biliary components during cholestasis will differentially affect solubilization and absorption of lipid classes due to their different interactions with water. In this chapter we will concentrate on digestion, absorption and metabolism of the main dietary lipids TAG, PC, cholesterol and fat-soluble vitamins, under physiological and cholestatic conditions.

TAG is the major fat in human diet, contributing 90 to 95% of energy provided by dietary fat. The majority of luminal phospholipid is phosphatidylcholine (PC), which is mostly of biliary origin (10-20 g daily in humans), with a dietary contribution of 1-2 g per day.[1,4]

The predominant dietary sterol is cholesterol (0.5 g/day,[1,4] which is mostly of animal origin although small amounts are also present in vegetables. Beta-sitosterol is the most important plant sterol (which account for 25% of dietary sterols), but it is virtually not absorbed by humans under physiological conditions due to the activity of the intestinal half-transporters ABCG5/G8, which have recently been postulated to play a major role in efficient efflux of absorbed dietary sterols from the enterocyte into the intestinal lumen, and from the liver into the bile ducts.[6,7]

A specific lipid group is formed by the essential fatty acids (EFA), which are long-chain poly-unsaturated fatty acids that cannot be synthesized de novo by mammalian cells and therefore must be provided by the diet. EFA's are mostly present in the diet in the form of TAG and PC. Additionally, biliary PC is an important source of intestinal EFA's since it contains up to 40 mol% of EFA's.

The fat-soluble vitamins A, D, E and K, compounds that are required in small quantities for maintenance of normal cell and organ function,[8-11] are class 1 polar lipids and depend upon micellar solubilization for intestinal uptake. Absorption rates differ between vitamin species, averaging 50 to 80% for vitamins A, D and K but only 20-30% for vitamin E.[12] Additionally, there may be competition between vitamin species for intestinal absorption- or transport-sites,[13] although minimal information is available regarding the nature and function of these sites.

The sequence of processes involved in intestinal lipid absorption can be divided into intraluminal and intracellular events.

Intraluminal Phase of Lipid Absorption

Before translocation from the intestinal lumen into the enterocytes can occur, dietary lipids must undergo a number of physicochemical alterations. This is achieved in a sequence of events called the intraluminal phase of lipid digestion and absorption, including:

- emulsification of dietary lipid
- lipolysis
- solubilization (micelles, vesicles)
- translocation of lipolytic products across the enterocyte membrane

Emulsification and Lipolysis

In humans, the first step in dietary fat digestion starts in the stomach with mechanical emulsification and partial TAG hydrolysis by gastric lipase, resulting in the lipolytic products DAG and free fatty acids. Gastric lipase does not hydrolyze PL or cholesterol ester, but its activity in the stomach accounts for 10 to 30% of TAG-lipolysis.[1,2,14,15] The remaining part of TAG digestion is brought about in the duodenal lumen by pancreatic lipase, which acts mainly on the sn-1 and sn-3 position of TAG-molecules, releasing 2-MAG and free fatty acids.[14,16] Pancreatic lipase is abundantly present in pancreatic juice, in accordance with the clinical observation that only severe pancreatic insufficiency results in lipid malabsorption. In the presence of bile salts, pancreatic lipase requires the cofactor pancreatic co-lipase[1] for adequate TAG-hydrolysis, since TAG-droplets covered with bile salts are not accessible to pancreatic lipase. Binding of pancreatic co-lipase to the TAG / water interface facilitates binding of pancreatic lipase.

Digestion of phospholipids occurs entirely in the duodenal lumen, predominantly by pancreatic phospholipase A2. Phospholipase A2, requiring calcium and the presence of bile salts for activation, hydrolyzes phospholipids at the sn-2 position resulting in free fatty acids and lyso-phosphatidylcholine (lyso-PC).

Dietary cholesterol is mainly present as free cholesterol, and only 10-15% as cholesterol ester. Cholesterol esters must be hydrolyzed in the duodenum by pancreatic cholesterol esterase (CE) before absorption can take place. Human cholesterol esterase (also known as carboxyl ester lipase, bile salt-stimulated lipase, monoglyceride lipase, pancreatic non-specific lipase or human milk lipase) does not only hydrolyze cholesterol esters, but also acts on TAG (sn-1, sn-2, sn-3), PL (sn-1, sn-2) and lipidic vitamin esters [17] and its activity is greatly enhanced by the presence of bile salts.

Solubilization of Lipolytic Products

For diffusion through the unstirred water layer, which separates the brush border membrane of enterocytes from the liquid luminal contents of the intestine, solubilization of lipolytic products is required. The most important function of biliary bile salts, phospholipids and cholesterol in the intestinal lumen appears to be their ability to increase the solubility of lipolytic products in the luminal aqueous phase by formation of mixed micelles. Mixed micelles were first described by Hoffman and Borgstrom[18] as disc-like aggregates of amphiphilic biliary and dietary components, which orient themselves with their hydrophobic parts to the inside of the micelles and their hydrophilic polar headgroups towards the aqueous outside. This conformation increases the solubility of FFA and MAG 100-1000 fold. [4] Mixed micelles contain bile salts (class 3 polar lipids), hydrogenated fatty acids (class 1 polar lipids), fatty acid ions (class 2 polar lipids), MAG (class 2 polar lipid), phospholipids (class 2 polar lipids) and cholesterol (class 1 polar lipid), and are about 4 nm in diameter.[1-5]

Carey and Patton[19] described the co-existence of these mixed micelles in the intestinal lumen with unilamellar liquid crystalline vesicles or liposomes. They demonstrated that only when intraluminal bile salt concentrations exceed a critical micellar concentration, mixed bile salt / lipid micelles are formed. However, when bile salt concentrations are decreased, large (20-60 nm) unilamellar liquid crystalline vesicles or liposomes are formed.[20,21] All classes of lipolytic products can be incorporated into disc-shaped micelles as well as liquid crystalline vesicles. Since both phases co-exist, quantification of the relative contribution of the two phases remains difficult, especially since continuous exchange of 2-MAG and FA between both structures occurs. The dissociation rate of lipolytic products from vesicles and their subsequent translocation across the enterocyte membrane appeared to be slower than dissociation rates from mixed micelles, as demonstrated by Narayan and Storch.[22]

The existence of liquid crystalline vesicles is thought to have specific pathophysiological consequences for lipid absorption in conditions where intraluminal bile salt concentrations are diminished, as in cholestasis. It has been demonstrated that although lipid absorption rates are slower in bile salt-deficient states, fat uptake can still occur rather efficiently. Porter et al reported on a bile fistula patient who continued to absorb up to 80% of dietary lipid, despite the obvious bile salt deficiency and the hundred-fold decreased free fatty acid concentration in the aqueous phase of the small intestinal lumen.[23] Mansbach et al found similar results in patients with bile salt malabsorption, where the strong decrease in solubilized fatty acid concentration led to an only mild degree of lipid malabsorption.[24] Solubilization of lipolytic products into liquid crystalline vesicles during intestinal bile-salt deficiency could explain the slower but preserved rate of lipid absorption in cholestasis.

Nishioka et al studied the importance of phospholipid/cholesterol vesicles for lipid absorption during bile deficiency (Fig. 2). Intraduodenal administration of ^{13}C-labeled linoleic acid (LA) or palmitic acid (PA) to chronically bile-diverted rats was associated with strongly decreased plasma concentrations of ^{13}C-LA and ^{13}C-PA. Subsequent intraduodenal supplementation with PC-cholesterol vesicles significantly reconstituted plasma concentrations of labeled palmitic acid. However, there appeared to be a delay in plasma appearance of both lipids, since at 5h after lipid administration plasma concentrations were still increasing. These observations are in concordance with the slower dissociation and translocation rates of lipolytic products from vesicles compared to mixed micelles, as proposed by Narayan and Storch.[22]

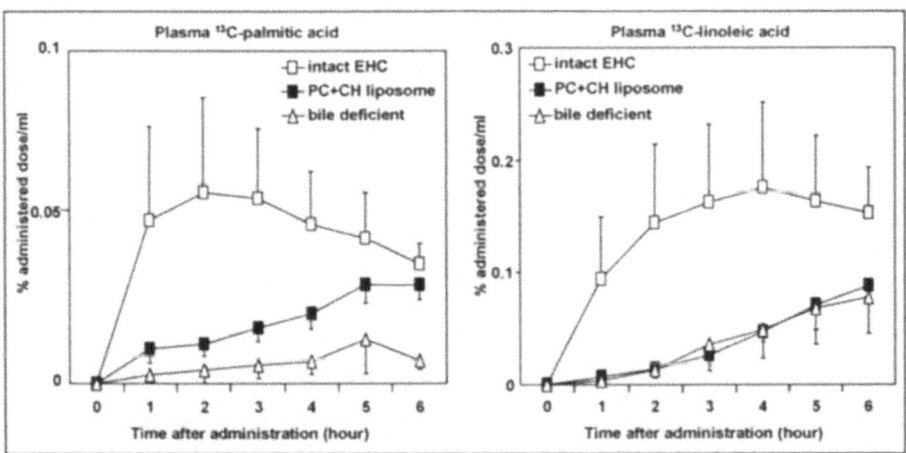

Figure 2. Palmitic acid and linoleic acid absorption in bile diverted rats.

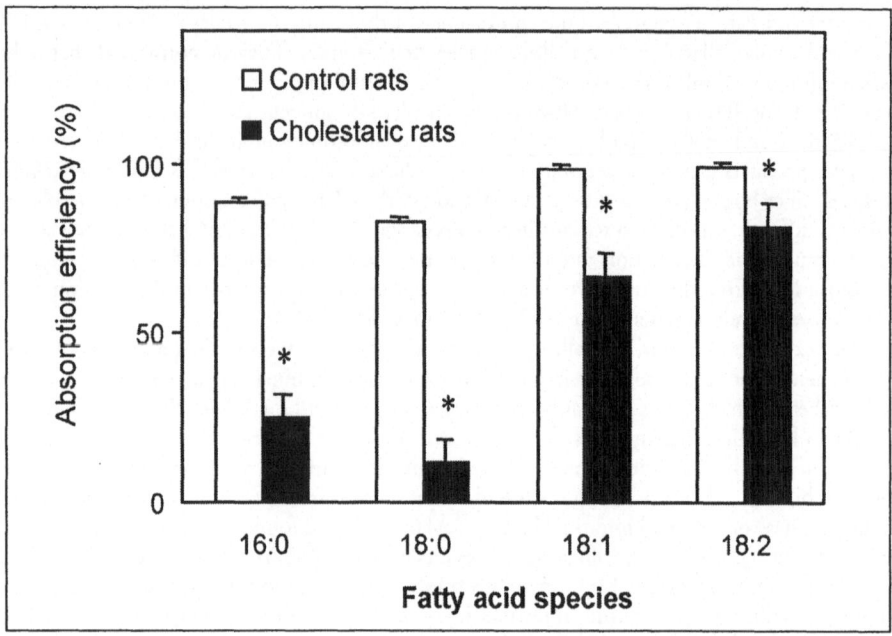

Figure 3. Saturated and unsaturated fatty acid absorption in cholestatic rats.

It is important to note the lipid-class difference in the dependence on bile for solubilization and consecutive uptake of fats (Fig. 3). Their less hydrophobic nature renders (poly-)unsaturated fatty acids (PUFA's) less dependent upon bile for solubilization than the more hydrophobic saturated fatty acids, despite longer chain length. Even in complete absence of intestinal bile salts, absorption of unsaturated fatty acids has been demonstrated to be relatively well preserved (80%) compared to that of saturated shorter chain fatty acids (50%), although absorption remained significantly lower than in the presence of bile (96-98%).[25]

In contrast to long acyl-chain lipids, short and medium acyl-chain lipids do not depend upon the luminal presence of bile salts for adequate uptake, and can directly be transferred from the intestinal lumen through the enterocytes. Medium-chain triglyceride-based formulas are therefore widely used as energy providers in conditions where intestinal solubilization is impaired. For absorption of cholesterol and fat-soluble vitamins however, micellar solubilization by bile salts is crucial.[26]

Translocation

For many years, translocation of lipolytic products across the unstirred water layer and the enterocyte membrane has been assumed to occur through passive diffusion. In the past decade, reports from Stremmel, Weber, Hauser et al[27,28] suggested that an active carrier-mediated process is involved in transport of FA across the intestinal brush border membrane, where a family of fatty acid binding proteins (FABP) was discovered. Many members of this family appeared to have both transport and fatty acid esterifying capacities.[29,30] Stahl et al identified FATP4, abundantly present in the apical membrane of mature enterocytes, as the principal intestinal transporter of long-chain fatty acids.[31]

A fatty acid-transporter not specific for the enterocyte was identified by Harmon et al[32] who isolated a 88-kDa membrane protein termed FAT (fatty acid transporter) which appeared to be the rat homologue of human CD36. CD36 is expressed in platelets, macrophages and endothelial cells as well as in intestine, adipose tissue, heart and muscle, where it mediates long-chain fatty acid uptake. Impaired CD36 function is associated with a large (60-80%)

defect in fatty acid uptake in these tissues.[33] Some authors have suggested interactions between different fatty acid transporters to regulate intestinal fatty acid uptake,[34] however, the exact molecular mechanism by which translocation of lipolytic products occurs is still a matter of debate.

Uptake of cholesterol is highly specific, since the plant sterol β-sitosterol is poorly absorbed under physiological conditions despite its structural similarity to cholesterol. In healthy individuals, 50-60% of dietary cholesterol is taken up, whereas absorption of plant sterols is less than 1%.[4,35] The half-transporters ABCG5/G8, which have been implied in the autosomal recessive disorder sitosterolemia, are thought to be responsible for efficient efflux of absorbed dietary sterols from enterocytes into the intestinal lumen, and from liver into bile,[6,7] possibly with different affinities for sterol species. Sitosterolemia patients appear to have 30-fold increased plasma levels of plant sterols, as well as moderately increased cholesterol absorption and decreased biliary sterol secretion, resulting in sterol accumulation and atherosclerosis.[36,37]

Intracellular Phase of Lipid Absorption

After translocation across the apical membrane of the enterocyte, dietary lipids migrate to the endoplasmic reticulum. Intestinal fatty acid binding protein (I-FABP) has been implied in intracellular transport of fatty acids inside the enterocyte. However, Agellon et al demonstrated in *Fabpi -/-* mice that I-FABP is not essential for dietary fat absorption,[38] which leaves the molecular mechanism of cellular fatty acid transfer to be elucidated.

At the cytosolic membrane of the smooth endoplasmic reticulum, re-esterification of absorbed fatty acids into TAG takes place. Two different biochemical pathways are involved in TAG resynthesis, of which the MAG-pathway is the most important under physiological conditions. In the MAG-pathway, 2-MAG is re-acylated to DAG and subsequently to TAG by mono-acylglycerol acyltransferase (MGAT) and diacylglycerol acyltransferase enzymes (DGAT1 and 2),[39,40] respectively. The alternative route of re-esterification is the alpha-glycerophosphate pathway, which involves conversion of glycerol-3-phosphate via phosphatidic acid to DAG and subsequently to TAG, also mediated by DGAT. Under physiological conditions there is an abundant supply of 2-MAG and FFA during lipid absorption, and the 2-MAG route will predominate over the alpha-glycerophosphate route.

Newly synthesized triglycerides from both pathways are thought to be metabolically distinct: TAG from the 2-MAG route is secreted more rapidly across the basolateral membrane than TAG originated from the alpha-glycerophosphate route. It has been suggested that the DAG from each pathway enter into separate intracellular pools. DAG from the alpha-glycerophosphate route is preferentially used for de novo PC synthesis.

Absorbed cholesterol, either from biliary or dietary origin, enters the free cholesterol pool inside the enterocyte, which also contains cholesterol that originates from absorption of shed intestinal mucosal membranes and from de novo synthesis. Cholesterol is transported into the lymphatic system mainly as cholesterol ester (CE) in the neutral lipid core of chylomicrons. The enzymes involved in cholesterol esterification are the acyl CoA cholesterol acyltransferases ACAT-2 and ACAT-1. Inhibition of ACAT activity has been found to decrease absorption of dietary cholesterol, associated with lymphatic release of aberrant apo-B containing lipoproteins devoid of cholesterol esters, containing mostly TAG in their cores.[39,41]

Newly synthesized TAG and CE form lipid droplets in the smooth endoplasmic reticulum (SER), where packaging occurs, mainly into lipoprotein particles called chylomicrons (CM). In the SER, the surface of these nascent chylomicrons is covered with phospholipids, cholesterol and apolipoproteins apo A-I, apo A-IV and apo B-48. Under physiological conditions, surface coat PL of lymph chylomicrons are predominantly of biliary origin rather than of dietary sources,[42] whereas chylomicron TAG FA-composition closely resembles that of dietary TAG. As fat absorption and TAG resynthesis proceeds, lipoprotein particles increase in size and in number and eventually end up in vesicles filled with pre-chylomicrons which are transported towards the Golgi apparatus. Here, modification of pre-chylomicrons into mature CM, including terminal glycosylation, occurs, followed by translocation to the lateral surface of the enterocyte where CM are exocytosed into the interstitium and into the lymph. Nascent chylomicrons have diameters between 100-1000 nm. Mesenteric lymph ducts drain into the thoracic duct, which enters the systemic circulation at the level of the jugular vein.

In recent years, it has become appreciated that biliary phospholipid secretion is necessary for proper intestinal chylomicron assembly and thus for secretion of lipid into lymph.

Studies in rats with interruption of the enterohepatic circulation by cholestyramine feeding[43] or by manipulation of bile composition by dietary means[44] revealed an accumulation of lipid in enterocytes. This phenomenon was also seen by Tso in bile diverted rats,[45] where subsequent administration of bile acids could only partially reinstate lipid transport into lymph. Only after administrating biliary PC, lymphatic lipid transport was fully restored.

Voshol et al demonstrated a delayed plasma triglyceride appearance after an oral lipid bolus in *Mdr2-/-* mice lacking biliary phospholipid secretion. This aberrant postprandial plasma TAG response was accompanied by normal fecal fat excretion and accumulation of lipid droplets in the intestinal wall, suggesting a relatively well preserved intestinal lipid uptake into enterocytes in the absence of biliary phospholipids, but a delay in subsequent chylomicron secretion.[46] Intestinal PC requirements for CM production might be comparable to that of the liver for VLDL secretion. In choline deficiency,[47] the decreased hepatic PC synthesis results in impaired production of VLDL. Similarly, enterocytes might require biliary PC for appropriate intestinal chylomicron assembly and secretion into lymph.

Alterations in Lipid Homeostasis during Cholestasis

Lipid Malabsorption

Fat malabsorption as a consequence of disturbed bile secretion is associated with weight loss due to energy deficiency (in children additionally complicated by impaired growth and development), with fat-soluble vitamin deficiencies and with essential fatty acid deficiency.

Several compensatory mechanisms for fat malabsorption during bile deficiency have been described in animal models for cholestasis. Minich et al reported on lipid malabsorption in rats with chronic bile diversion, in which bile is absent from the intestinal lumen but no (toxic) biliary components accumulate in the body. These rats appeared to substantially compensate for their markedly decreased dietary fat absorption by strongly increasing their food ingestion.[25] Subsequent morphological examination of the small intestine revealed that both villus and crypt height were significantly increased in bile-diverted rats compared to controls.

Porter and Knoebel et al described a compensatory mechanism designated as "the absorptive reserve of the small intestine".[23,48] Under physiological conditions, only the proximal part of the intestine is involved in fat absorption. In situations where proximal fat absorption appears impaired, more distal parts can also contribute.

Minich et al also demonstrated that the amount of dietary lipid strongly affects the efficacy of lipid absorption in bile deficient states. In rats with chronic bile diversion, absorption of dietary lipid remained highly efficient when regular low-fat chow was fed (84% of ingested lipid). However, when rats were fed a high-fat diet, lipid absorption coefficients decreased to around 50%, indicating that compensatory mechanisms for lipid absorption in bile deficiency have limited capacity.[5] Detailed knowledge of different compensation mechanisms and alternative routes of lipid absorption during bile deficiency are important for developing dietary treatment strategies for nutritional deficiencies in cholestatic patients.

Fat-Soluble Vitamin Deficiency

Although great variability exists between studies in defining biochemical vitamin deficiency, many reports have indicated the existence of significantly decreased fat-soluble vitamin levels under cholestatic conditions.[49-51] Fat-soluble vitamins (class I polar lipids, see above) are highly dependent on intraluminal solubilization by bile acids, and lack of bile flow inevitably results in malabsorption and depletion of fat-soluble vitamin stores. The extent of deficiency appears to be highly vitamin-species specific; for example, Phillips et al[49] reported biochemical deficiencies of vitamin A, D, E and K in 34%, 13%, 2% and 8% of PBC patients, respectively. Similar values were found by Kaplan and Kowdley et al.[12,50,51]

Vitamin A deficiency in chronic cholestasis can not only result from intestinal malabsorption, but hepatic secretion of retinol binding protein (RBP) may also be diminished, leading to low plasma levels of retinol and impaired delivery to target tissues such as retina and epithelial cells.[8] Deficiency of vitamin K can lead to life-threatening hemorrhages due to the vitamin K-dependence of clotting factors II, VII, IX, X and proteins S and C. Biliary atresia patients have been reported to present with stroke as their first consequence of cholestasis-induced lipid malabsorption.[52] Although vitamin D deficiency during lipid malabsorption can partially be circumvented by endogenous photosynthesis of vitamin D3 in the skin, many chronically ill patients are not adequately exposed to sunlight resulting in low vitamin D and calcium levels and impaired bone mineralization.[8,9] Prolonged vitamin E deficiency in cholestatic children leads to a degenerative neuromyopathy eventually resulting in peripheral neuropathy, muscular weakness, ophthalmoplegia and retinal dysfunction which appears irreversible to a significant degree. The irreversibility and severity of many of the symptoms associated with fat-soluble vitamin deficiencies mandate strict monitoring and correction of vitamin status in cholestatic patients.

Essential Fatty Acid (EFA)-Deficiency

EFA-deficiency as a consequence of overall lipid malabsorption in cholestasis is well recognized. However, in the last decade it has become apparent that EFA-deficiency in itself can impair the efficiency of lipid absorption. Levy et al observed decreased biliary bile salt secretion rates in EFA-deficient rats,[53] implying impaired bile formation as a probable cause for EFAD-induced fat malabsorption. However, in a mouse model for EFA-deficiency, we recently reported an increased bile flow and biliary bile salt secretion compared to EFA-sufficient controls. Additionally, lipid absorption in *Mdr2-/-* mice, secreting phospholipid-free bile, was equally affected by EFA-deficiency as that in control mice. Thus, altered biliary phospholipid secretion as major contributor to the pathophysiological mechanism behind fat malabsorption during EFA-deficiency in mice is excluded.[54] Apart from an apparent species specificity in the effects of EFAD, it is concluded that fat malabsorption in EFAD mice is not due to impaired bile formation, and it is suggested that EFA-deficiency affects intracellular events of dietary fat absorption occurring in the enterocyte. If true, EFA-deficiency during cholestasis may further compromise dietary fat absorption.

Lipid Metabolism

During lipid absorption, the intestine releases large amounts of TAG-rich chylomicrons into the circulation. During fasting however, the liver is the major source of TAG-rich lipoproteins by secreting VLDL. Both liver and intestine are capable of synthesizing HDL, which are secreted as nascent particles containing predominantly phospholipids and unesterified cholesterol. Another major lipoprotein, LDL, is formed in the plasma compartment as a product of VLDL catabolism. Additionally, the liver synthesizes apoproteins that are essential structural and enzymatic components of lipoproteins. Apoproteins act as cofactors for enzymes crucial for cholesterol esterification or triglyceride lipolysis. Apo A-I activates the cholesterol esterifying enzyme LCAT, which is also synthesized in the liver. Apo C-II is required for lipoprotein lipase activation, which hydrolyzes lipoprotein triglycerides, thus converting chylomicrons into chylomicron remnants and VLDL into IDL and ultimately LDL. Apo E and apo B are crucial for receptor-mediated uptake of lipoproteins by peripheral cells, as well as for hepatic uptake of end products of lipoprotein catabolism.[55,56] Lipoprotein remnant uptake by the liver, mediated by SR-B1 and hepatic lipase, provides a feedback inhibition mechanism for cholesterol homeostasis by regulating activity of HMG-CoA reductase, the key enzyme in hepatic cholesterol neosynthesis. Hepatocyte damage due to toxic accumulation of bile acids in cholestasis may disrupt synthesis of apoproteins and of other enzymes involved in lipoprotein formation and metabolism such as LCAT, CETP and PLTP, with concomitant derangements in plasma and hepatic lipid homeostasis which will be discussed below.

Lipoprotein X in Cholestasis

Biliary excretion is the principal route for cholesterol disposal from the body (either direct or after conversion into bile acids), and cholestasis thoroughly deranges the whole body sterol balance. The well-recognized increase in plasma free cholesterol observed in cholestasis is accompanied by an equimolar elevation of plasma phospholipid.[57,58] Hypercholesterolemia in cholestasis, particularly in extrahepatic forms, is accompanied by plasma appearance of an aberrant lipoprotein, LpX.[59,60] Kostner and Laggner et al described LpX as a 40-100 nm bilamellar vesicle with an aqueous lumen, predominantly composed of phospholipids and free cholesterol in equimolar amounts and containing only minor amounts of TAG (3%) and cholesteryl ester (2%). [61, 62] Gradient ultracentrifugation revealed that LpX is isolated in the LDL-fraction,[63] and contains apo C as well as albumin. Manzato et al hypothesized that LpX particles represent biliary vesicles regurgitated from liver into plasma of cholestatic subjects,[58] since both LpX and bile vesicles are composed of PC and free cholesterol. The presence of apo C and albumin, as well as the observation that the cholesterol/PC ratio in LpX differs from that in bile, can be explained by plasma interactions of LpX with other lipoproteins. LpX is not readily taken up by the liver, thus LpX-cholesterol does not participate in feedback inhibition of hepatic cholesterol synthesis. This could contribute to the paradox of increased hepatic cholesterol neosynthesis in hypercholesterolemia during cholestasis. Felker et al observed LpX-like vesicles within bile canaliculi of bile duct ligated rats, indicating a biliary origin of the particle.[63,64] Oude Elferink et al demonstrated the biliary origin of LpX in bile duct ligated *Mdr2-/-* mice, which secrete phospholipid-free bile. In contrast to controls, bile duct ligation in *Mdr2-/-* mice resulted in decreased plasma cholesterol and PC concentrations and a complete absence of LpX particles, suggesting that during cholestasis, biliary lipid reflux occurs from bile into the plasma compartment.[65] Bloks et al described the presence of LpX in ferrochelatase-deficient mice whose livers are not uniformly cholestatic but, instead, show enhanced bile flow and biliary bile salt secretion rates. Since LCAT activity was not impaired in these animals, the authors propose that formation of LpX is related to relative undersecretion of biliary PC and cholesterol that is observed in these animals.[66] Although the exact nature of LpX formation in ferrochelatase-deficient mice remains to be established, it is evident that this aberrant particle can be formed in situations in which bile formation per se is not impaired.

HDL in Cholestasis

Apart from the increased lipid content in the LDL fraction in the form of LpX, the appearance of TAG-rich LDL and the decreased plasma VLDL-concentrations, chronic cholestasis is associated with strongly decreased plasma HDL concentrations (<10%).[67] The mechanism behind this may involve either an increased HDL clearance rate during cholestasis, or a decreased HDL synthesis. Recent work indicates that bile salts, accumulating in hepatocytes during cholestasis, are able to suppress apo A-I gene transcription via a negative farnesoid X receptor (FXR) response element mapped to the C-site of the apo A-I promoter.[68] As HDL is partly derived from CM surface remnants, low plasma HDL levels could also result from decreased CM formation during intestinal bile deficiency, or from defective HDL formation from CM surface remnants by PLTP. [69] Upon ultracentrifugation, the HDL observed in cholestasis is in the density range of bilamellar discoidal particles, enriched in free cholesterol and phospholipid with decreased apo A-I and apo A-II contents and increased apo E, resembling so-called "nascent" HDL particles. Such particles are normally not found in the plasma compartment in considerable amounts because of rapid transformation by concerted actions of LCAT, CETP and PLTP.

Lipoprotein-Metabolizing Enzymes in Cholestasis

LCAT

Lecithin cholesterol acyl transferase (LCAT) and hepatic lipase (HL) are two of the key enzymes in lipoprotein metabolism. Both proteins are produced in the liver, but LCAT is active in the circulation at the surface of HDL whereas hepatic lipase resides at the hepatic endothelial cell lining. LCAT, a 60 kDa glycoprotein that converts cholesterol and phosphatidylcholines into cholesteryl esters and lyso-PC, is activated by apo A-I. Its cholesterol esterifying activity not only moves cholesterol from the HDL surface into the core and thereby promotes the flux of cholesterol from cell membranes into HDL, but it also leads to morphological changes of the HDL particle. Nascent disc-shaped HDL becomes spherical as cholesteryl esters accumulate in the HDL core. Hepatic lipase and LCAT hydrolytic activities together account for over 80% of disappearance of PC from plasma.[70] Impaired hepatic synthesis of these enzymes in cholestasis may thus contribute to increased plasma PC concentrations. Both plasma cholesteryl ester as well as LCAT concentrations are decreased in cholestatic subjects, and the plasma appearance of "nascent" discoidal HDL particles is assumed to be a direct result from defective LCAT functioning. Furthermore, discoidal HDL has also been described in association with primary familial LCAT-deficiency.[57]

CETP

Cholesteryl ester transfer protein (CETP) transfers excess cholesteryl esters from HDL to VLDL and LDL in exchange for TAG,[71,72] thus participating in the so-called reverse cholesterol transport. CETP-activity results in homogenous fatty acid species distribution between lipoprotein fractions. Activity of CETP is decreased 25% in cholestasis, associated with a decreased LA-content of VLDL-TG and cholesteryl esters compared to those of HDL.[71] Faust et al demonstrated that fatty acid absorption regulates CETP secretion in CaCo-2 cells,[73,74] and several animal and human studies[75-78] revealed that high-fat diets can increase CETP activity. Freeman et al suggested that serum TAG levels above 1.4 mmol/l are required for significant CETP-mediated lipid exchange between LDL and VLDL.[79] The mechanism for the decreased CETP-activity in cholestatic subjects is not obvious because of the multiple origin of CETP synthesis (liver, intestine, adipose tissue, macrophages).[80,81] However, since hepatocytes are the predominant source of CETP, impaired hepatic CETP-synthesis remains a likely contributor to the decreased CETP activity in cholestasis.[71]

PLTP

Plasma phospholipid transfer protein (PLTP) circulates bound to HDL and mediates transfer of phospholipids from apo B-containing lipoproteins into HDL, thus modulating HDL size and lipid composition. PLTP activity generates pre-beta HDL, the major acceptor of cholesterol in the reverse-cholesterol transport route. PLTP-knockout mice have been shown to have markedly reduced HDL-levels due to defective transfer of phospholipids from triglyceride-rich lipoproteins into HDL.[82] Liver, adipose tissue and lung are presumably the major sources of circulating PLTP. Impaired hepatic synthesis of PLTP in cholestatic conditions can markedly reduce circulating HDL levels and additionally, due to the stimulatory effect of PLTP on CETP activity,[83] it can further deteriorate the already impaired CETP function. Recently, PLTP has been identified as an FXR target gene,[84] providing a molecular basis for reduced PLTP gene expression under cholestatic conditions.

Essential Fatty Acid (EFA)-Metabolism in Cholestasis

Socha et al reported on decreased plasma arachidonic acid levels in pediatric cholestatic patients, which was attributed to impaired hepatic microsomal desaturase and/or elongase activity.[85,86] However, Minich et al recently demonstrated that conversion of [^{13}C]-linoleic acid to [^{13}C]-arachidonic acid was not significantly different in short-term bile duct ligated rats compared to controls. Accordingly, Δ6-desaturase activity as determined in hepatic microsomes was not altered.[87] These results are in agreement with observations of de Vriese et al, who found no differences in Δ9-, Δ6- and Δ3-desaturase activities in liver microsomes of cholestatic and non-cholestatic rats.[88] Decreased net uptake of the parent EFA linoleic acid observed in cholestatic subjects appears the predominant cause of low plasma AA-levels, rather than post-absorptive EFA-metabolism. Yet, in rats with long-term bile duct ligation impaired hepatic β-oxidative capacity has been reported.[89]

Nutritional Therapy in Cholestasis

Chronic cholestasis is frequently accompanied by nutritional deficiencies due to inadequate dietary intake, maldigestion, malabsorption and/or defective metabolism of nutrients. Additionally, requirements of energy and/or specific nutrients may be increased during cholestasis. Generally the recommended caloric intake for patients with chronic cholestasis is 130% of recommended daily allowance, usually accomplished by dietary supplementation with glucose polymers and/or MCT oil enriched with essential fatty acid-rich oils.[90,91] The enteral route is preferred but in severe chronic malabsorption, nasogastric and nocturnal feedings are often required.[71]

Fat-soluble vitamin deficiency is frequently present, particularly in cholestatic children, for which adequate and rapid correction is required. For treatment of vitamin D deficiency a regimen of oral 25-OHD is recommended, at a dose of 2-4 μg/kg/d in children and 50-100 _g in adults, with regular measurements of cholecalciferol levels in plasma to exclude development of toxicity. Vitamin K supplements of 2.5-5 mg 2-7 times a week are currently recommended as prophylaxis for all children with chronic cholestasis.[8,9] Most cholestatic children absorb the phylloquinone form of vitamin K adequately. In adults, vitamin K supplements are only recommended when blood tests suggest deficiency. For correction of vitamin E deficiency, standard oral forms of vitamin E (α-tocopherol, α-tocopheryl acetate, α-tocopheryl succinate) are recommended at doses starting from 10-25 IU/kg/d increasing to 100-200 IU/kg/d.

In situations where normalization of plasma vitamin E levels is not reached, a water-soluble form of vitamin E called d-α-tocopheryl polyethylene glycol-1000 succinate (TPGS) has been shown to markedly improve vitamin E status in cholestatic patients.[92] Argao et al demonstrated that absorption of other fat-soluble vitamins is greatly enhanced by simultaneous administration with TPGS.[93] Recommended dosage of vitamin A in chronic cholestasis is 10000 IU if given with TPGS.[92] Irrespective of the form of vitamin supplementation that is chosen, plasma vitamin levels should be carefully monitored to avoid excessive serum levels and toxicity.

Conclusions

Cholestatic liver disease can disturb many aspects of lipid absorption and metabolism.

Accumulation of potentially toxic bile components in hepatocytes due to disruption of the flux of bile from the liver into the gut can damage hepatocytes, resulting in impaired synthetic function and decreased production of enzymes involved in lipoprotein metabolism. Also, lipoprotein secretion appears to be disturbed during cholestasis, reflected by decreased HDL-levels and appearance of the aberrant lipoprotein X in plasma. The absence of biliary components from the intestinal lumen during cholestasis can strongly impair uptake of dietary fat and fat-soluble vitamins, resulting in a variety of nutritional deficiencies. Prolonged survival of patients with chronic cholestasis in the past decades will require critical evaluation of nutrient deficiencies and adequate treatment strategies in order to prevent permanent sequelae and to improve quality of life.

References

1. Carey MC, Hernell O. Digestion and absorption of fat. Seminars in gastrointestinal disease.1992; 3:189-208.
2. Carey MC, Small DM, Bliss CM. Lipid digestion and absorption. Ann Rev Physiol 1983; 45:651-77.
3. Carey MC, Small DM. The characteristics of mixed micellar solutions with particular reference to bile. Am J Med 1970; 49:590-608.
4. Tso P. Intestinal lipid absorption. In: Tso P, ed. Physiology of the gastrointestinal tract. 3 ed. New York: Raven Press, 1994:1867-907.
5. Verkade HJ, Tso P. Biophysics of intestinal luminal lipids. In: Mansbach CM, Tso P, Kuksis A, eds. Intestinal Lipid metabolism, 2000:1-18
6. Berge KE, Tian H, Graf GA et al. Accumulation of dietary cholesterol in sitosterolemia caused by mutations in adjacent ABC transporters. Science 2000; 290:1771-75.
7. Lee MH, Lu K, Hazard S, Yu H et al. Identification of a gene, ABCG5, important in the regulation of dietary cholesterol absorption. Nat Genet 2001; 27:79-83.
8. Sokol RJ. Fat-soluble vitamins and their importance in patients with cholestatic liver diseases. Gastroenterol Clin North Am 1994; 23:673-705.
9. Sokol RJ. vitamin deficiency and replacement in childhood cholestasis. In: Lentze M, Reichen J, eds. pediatric cholestasis. Dordrecht/Boston/London: Kluwer Academic Publishers, 1991:289-303.
10. Argao EA, Heubi JE. Fat-soluble vitamin deficiency in infants and children. Curr Opin Pediatr 1993; 5:562-66.
11. Jeppesen PB, Hoy CE, Mortensen PB. Deficiencies of essential fatty acids, vitamin A and E and changes in plasma lipoproteins in patients with reduced fat absorption or intestinal failure. European Journal of Clinical Nutrition 2000; 54:632-42.
12. Meydani M, Martin KR. Intestinal absorption of fat-soluble vitamins. In: Mansbach CM, Tso P, Kuksis A, eds. Intestinal lipid metabolism. Dordrecht/Boston/London: Kluwer academic publishers, 2000:367-78.
13. Hollander D. Intestinal absorption of vitamins A, E, D, and K. J Lab Clin Med 1981; 97:449-62.
14. Borgstrom B. Digestion and absorption of lipids. Int Rev Physiol 1977; 12:305-23.
15. Liao TH, Hamosh P, Hamosh M. Fat digestion by lingual lipase: mechanism of lipolysis in the stomach and upper small intestine. Pediatr Res 1984; 18:402-9.
16. Mattson FH, Volpenhein RA. Carboxylic ester hydrolases of rat pancreatic juice. J Lipid Res 1966; 7:536-43.
17. Lombardo D, Guy O. Studies on the substrate specificity of a carboxyl ester hydrolase from human pancreatic juice. II. Action on cholesterol esters and lipid-soluble vitamin esters. Biochim Biophys Acta 1980; 611:147-55.
18. Hofmann AF, Borgstrom B. The intraluminal phase of fat digestion in man: the lipid content of the micellar and oil phases of intestinal content obtained during fat digestion and absorption. J Clin Invest 1964; 43:247-57.
19. Patton JS, Carey MC. Watching fat digestion. Science 1979; 204:145-48.
20. Staggers JE, Hernell O, Stafford RJ et al. Physical-chemical behavior of dietary and biliary lipids during intestinal digestion and absorption 1. Phase behavior and aggregation states of model lipid systems patterned after aqueous duodenal contents of healthy adult human beings. Biochemistry 1990; 29:2028-40.
21. Hernell O, Staggers JE, Carey MC. Physical-chemical behavior of dietary and biliary lipids during intestinal digestion and absorption 2. Phase analysis and aggregation states of luminal lipids during duodenal fat digestion in healthy adult human beings. Biochemistry 1990; 29:2041-56.
22. Narayanan VS, Storch J. Fatty acid transfer in taurodeoxycholate mixed micelles. Biochemistry 1996; 35:7466-73.
23. Porter HP, Saunders DR, Tytgat G et al. Fat absorption in bile fistula man. A morphological and biochemical study. Gastroenterology 1971; 60:1008-19.
24. Mansbach CM, Newton D, Stevens RD. Fat digestion in patients with bile acid malabsorption but minimal steatorrhea. Dig Dis Sci 1980; 25:353-62.
25. Minich DM, Kalivianakis M, Havinga R et al. Bile diversion in rats leads to a decreased plasma concentration of linoleic acid which is not due to decreased net intestinal absorption of dietary linoleic acid. Biochim Biophys Acta 1999; 1438:111-19.
26. Voshol PJ, Havinga R, Wolters H et al. Reduced plasma cholesterol and increased fecal sterol loss in multidrug resistance gene 2 P-glycoprotein-deficient mice. Gastroenterology 1998; 114:1024-34.
27. Stremmel W, Lotz G, Strohmeyer G et al. Identification, isolation, and partial characterization of a fatty acid binding protein from rat jejunal microvillous membranes. J Clin Invest 1985; 75:1068-76.

28. Schulthess G, Lipka G, Compassi S et al. Absorption of monoacylglycerols by small intestinal brush border membrane. Biochemistry 1994; 33:4500-4508.
29. Coe NR, Smith AJ, Frohnert BI et al. The fatty acid transport protein (FATP1) is a very long chain acyl-CoA synthetase. J Biol Chem 1999; 274:36300-36304.
30. Watkins PA, Pevsner J, Steinberg SJ. Human very long-chain acyl-CoA synthetase and two human homologs: initial characterization and relationship to fatty acid transport protein. Prostaglandins Leukot Essent Fatty Acids 1999; 60:323-28.
31. Stahl A, Hirsch DJ, Gimeno RE et al. Identification of the major intestinal fatty acid transport protein. Mol Cell 1999; 4:299-308.
32. Harmon CM, Abumrad NA. Binding of sulfosuccinimidyl fatty acids to adipocyte membrane proteins: isolation and amino-terminal sequence of an 88-kD protein implicated in transport of long-chain fatty acids. J Membr Biol 1993; 133:43-49.
33. Ibrahimi A, Abumrad NA. Role of CD36 in membrane transport of long-chain fatty acids. Curr Opin Clin Nutr Metab Care 2002; 5:139-145.
34. Abumrad NA, Sfeir Z, Connelly MA et al. Lipid transporters: membrane transport systems for cholesterol and fatty acids. Curr Opin Clin Nutr Metab Care 2000; 3:255-262.
35. Salen G, Ahrens EH Jr, Grundy SM. Metabolism of beta-sitosterol in man. J Clin Invest 1970; 49:952-67.
36. Gregg RE, Connor WE, Lin DS et al. Abnormal metabolism of shellfish sterols in a patient with sitosterolemia and xanthomatosis. J Clin Invest 1986; 77:1864-72.
37. Bhattacharyya AK, Connor WE. Beta-sitosterolemia and xanthomatosis. A newly described lipid storage disease in two sisters. J Clin Invest 1974; 53:1033-43.
38. Vassileva G, Huwyler L, Poirier K et al. The intestinal fatty acid binding protein is not essential for dietary fat absorption in mice: FASEB J 2000; 14:2040-2046.
39. Buhman KF, Accad M, Farese RV. Mammalian acyl-CoA:cholesterol acyltransferases. Biochim Biophys Acta 2000; 1529:142-54.
40. Chang CC, Sakashita N, Ornvold K et al. Immunological quantitation and localization of ACAT-1 and ACAT-2 in human liver and small intestine. J Biol Chem 2000; 275:28083-92.
41. Martins IJ, Mortimer BC, Redgrave TG. Effect of the ACAT inhibitor CL 277,082 on apolipoprotein B48 transport in mesenteric lymph and on plasma clearance of chylomicrons and remnants. Arterioscler Thromb Vasc Biol 1997; 17:211-16.
42. Patton GM, Clark SB, Fasulo JM et al. Utilization of individual lecithins in intestinal lipoprotein formation in the rat. J Clin Invest 1984; 73:231-40.
43. Cassidy MM, Lightfoot FG, Grau L et al. Lipid accumulation in jejunal and colonic mucosa following chronic cholestyramine (Questran) feeding. Dig Dis Sci 1985; 30:468-76.
44. Arjmandi BH, Ahn J, Nathani S et al. Dietary soluble fiber and cholesterol affect serum cholesterol concentration, hepatic portal venous short-chain fatty acid concentrations and fecal sterol excretion in rats. J Nutr 1992; 122:246-53.
45. Tso P, Kendrick H, Balint JA et al. Role of biliary phosphatidylcholine in the absorption and transport of dietary triolein in the rat. Gastroenterology 1981; 80:60-65.
46. Voshol PJ, Minich DM, Havinga R et al. Postprandial chylomicron formation and fat absorption in multidrug resistance gene 2 p-glycoprotein-deficient mice. Gastroenterology 2000; 118:173-82.
47. Verkade HJ, Fast DG, Rusinol AE et al. Impaired biosynthesis of phosphatidylcholine causes a decrease in the number of very low density lipoprotein particles in the Golgi but not in the endoplasmic reticulum of rat liver. J Biol Chem 1993; 268:24990-24996.
48. Knoebel LK. Intestinal absorption in vivo of micellar and nonmicellar lipid. Am J Physiol 1972; 223:255-61.
49. Philips JR, Angulo P, Petterson T et al. Fat-soluble vitamin levels in patients with primary biliary cirrhosis. Am J Gastroenterology 2001; 96:2745-50.
50. Kowdley KV, Emond MJ, Sadowski JA et al. Plasma vitamin K1 level is decreased in primary biliary cirrhosis. Am J Gastroenterol 1997; 92:2059-61.
51. Kaplan MM, Elta GH, Furie B et al. Fat-soluble vitamin nutriture in primary biliary cirrhosis. Gastroenterology 1988; 95:787-92.
52. van den Anker JN, Sinaasappel M. Bleeding as presenting symptom of cholestasis. J Perinatol 1993; 13:322-24.
53. Levy E, Garofalo C, Rouleau T et al. Impact of essential fatty acid deficiency on hepatic sterol metabolism in rats. Hepatology 1996; 23:848-57.
54. Werner A, Minich DM, Havinga R et al. Fat malabsorption in essential fatty acid-deficient mice is not due to impaired bile formation. Am J Physiol Gastrointest Liver Physiol 2002; 283:G900-G908.
55. Sabesin SM, Bertram PD, Freeman MR. Lipoprotein metabolism in liver disease. Adv Intern Med 1980; 25:117-46.

56. Brown MS, Goldstein JL. Lipoprotein receptors in the liver. Control signals for plasma cholesterol traffic. J Clin Invest 1983; 72:743-47.
57. Miller JP. Dyslipoproteinaemia of liver disease. Baillieres Clin Endocrinol Metab 1990; 4:807-32.
58. Manzato E, Fellin R, Baggio G et al. Formation of lipoprotein-X. Its relationship to bile compounds. J Clin Invest 1976; 57:1248-60.
59. Soros P, Bottcher J, Maschek H et al. Lipoprotein-X in patients with cirrhosis: its relationship to cholestasis and hypercholesterolemia. Hepatology 1998; 28:1199-205.
60. Seidel D, Alaupovic P, Furman RH. A lipoprotein characterizing obstructive jaundice. I. Method for quantitative separation and identification of lipoproteins in jaundiced subjects. J Clin Invest 1969; 48:1211-23.
61. Kostner GM, Laggner P, Prexl HJ et al. Investigation of the abnormal low-density lipoproteins occurring in patients with obstructive jaundice. Biochem J 1976; 157:401-7.
62. Laggner P, Glatter O, Muller K et al. The lipid bilayer structure of the abnormal human plasma lipoprotein X. An X-ray small-angle-scattering study. Eur J Biochem 1977; 77:165-71.
63. Felker TE, Hamilton RL, Havel RJ. Secretion of lipoprotein-X by perfused livers of rats with cholestasis. Proc Natl Acad Sci USA 1978; 75:3459-63.
64. Crawford AR, Smith AJ, Hatch VC et al. Hepatic secretion of phospholipid vesicles in the mouse critically depends on mdr2 or MDR3 P-glycoprotein expression. Visualization by electron microscopy. J Clin Invest 1997; 100:2562-67.
65. Oude Elferink RP, Ottenhoff R, van Marle J et al. Class III P-glycoproteins mediate the formation of lipoprotein X in the mouse. J Clin Invest 1998; 102:1749-57.
66. Bloks VW, Plosch T, Van Goor H et al. Hyperlipidemia and atherosclerosis associated with liver disease in ferrochelatase-deficient mice. J Lipid Res 2001; 42:41-50.
67. Tallet F, Vasson MP, Couderc R et al. Characterization of lipoproteins during human cholestasis. Clin Chim Acta 1996; 244:1-15.
68. Claudel T, Sturm E, Duez H et al. Bile acid-activated nuclear receptor FXR suppresses apolipoprotein A-I transcription via a negative FXR response element. J Clin Invest 2002; 109:961-71.
69. Huuskonen J, Olkkonen VM, Jauhiainen M et al. The impact of phospholipid transfer protein (PLTP) on HDL metabolism. Atherosclerosis 2001; 155:269-81.
70. Shamburek RD, Zech LA, Cooper PS et al. Disappearance of two major phosphatidylcholines from plasma is predominantly via LCAT and hepatic lipase. Am J Physiol 1996; 271:E1073-E1082.
71. Korsten MA, Lieber CS. Nutrition in pancreatic and liver disorders. In: Shils ME, Olson JA, Shike M et al, eds. Modern nutrition in health and disease. 9 ed. Philadelphia, Baltimore, New York, London, Buenos Aires, Hong Kong, Sydney, Tokyo: Lippincott Williams & Wilkins, 1999:1066-79.
72. Swenson TL, Brocia RW, Tall AR. Plasma cholesteryl ester transfer protein has binding sites for neutral lipids and phospholipids . J Biol Chem 1988; 263:5150-5157.
73. Faust RA, Albers JJ. Synthesis and secretion of plasma cholesteryl ester transfer protein by human hepatocarcinoma cell line, HepG2. Arteriosclerosis 1987; 7:267-75.
74. Faust RA, Albers JJ. Regulated vectorial secretion of cholesteryl ester transfer protein (LTP-I) by the CaCo-2 model of human enterocyte epithelium. J Biol Chem 1988; 263:8786-89.
75. Quinet EM, Agellon LB, Kroon PA et al. Atherogenic diet increases cholesteryl ester transfer protein messenger RNA levels in rabbit liver. J Clin Invest 1990; 85:357-63.
76. Quig DW, Zilversmit DB. Plasma lipid transfer activity in rabbits: effects of dietary hyperlipidemias. Atherosclerosis 1988; 70:263-71.
77. Groener J, van Ramshorst E, Katan M et al. Diet modulates plasma neutral lipid transfer protein activity in normolipidemic human subjects. Klin Wochenschr 1990; 68(Suppl 22):106-7.:106-7.
78. Son YS, Zilversmit DB. Increased lipid transfer activities in hyperlipidemic rabbit plasma. Arteriosclerosis 1986; 6:345-51.
79. Freeman DJ, Caslake MJ, Griffin BA et al. The effect of smoking on post-heparin lipoprotein and hepatic lipase, cholesteryl ester transfer protein and lecithin:cholesterol acyl transferase activities in human plasma. Eur J Clin Invest 1998; 28:584-91.
80. Nagashima M, McLean JW, Lawn RM. Cloning and mRNA tissue distribution of rabbit cholesteryl ester transfer protein. J Lipid Res 1988; 29:1643-49.
81. Jiang XC, Masucci-Magoulas L, Mar J et al. Down-regulation of mRNA for the low density lipoprotein receptor in transgenic mice containing the gene for human cholesteryl ester transfer protein. Mechanism to explain accumulation of lipoprotein B particles. J Biol Chem 1993; 268:27406-12.
82. Kawano K, Qin S, Vieu C et al. Role of hepatic lipase and scavenger receptor BI in clearing phospholipid/free cholesterol-rich lipoproteins in PLTP-deficient mice. Biochim Biophys Acta 2002; 1583:133-40.

83. Lagrost L, Athias A, Herbeth B et al. Opposite effects of cholesteryl ester transfer protein and phospholipid transfer protein on the size distribution of plasma high density lipoproteins. Physiological relevance in alcoholic patients. J Biol Chem 1996; 271:19058-65.
84. Urizar NL, Dowhan DH, Moore DD. The farnesoid X-activated receptor mediates bile acid activation of phospholipid transfer protein gene expression. J Biol Chem 2000; 275:39313-17.
85. Socha P, Koletzko B, Pawlowska J et al. Essential fatty acid status in children with cholestasis, in relation to serum bilirubin concentration. J Pediatr 1997; 131:700-706.
86. Socha P, Koletzko B, Swiatkowska E et al. Essential fatty acid metabolism in infants with cholestasis. Acta Paediatr 1998; 87:278-83.
87. Minich DM, Havinga R, Stellaard F et al. Intestinal absorption and postabsorptive metabolism of linoleic acid in rats with short-term bile duct ligation. Am J Physiol Gastrointest Liver Physiol 2000; 279(6):G1242-G1248.
88. Vriese De SR, Savelli JL, Poisson JP et al. Fat absorption and metbolism in bile duct ligated rats. Nutrition and Metabolism. 1 A.D.; 45:209-16.
89. Krahenbuhl S, Talos C, Reichen J. Mechanisms of impaired hepatic fatty acid metabolism in rats with long-term bile duct ligation . Hepatology 1994; 19:1272-81.
90. Moreno LA, Gottrand F, Hoden S et al. Improvement of nutritional status in cholestatic children with supplemental nocturnal enteral nutrition. J Pediatr Gastroenterol Nutr 1991; 12:213-16.
91. Bavdekar A, Bhave S, Pandit A. Nutrition management in chronic liver disease. Indian J Pediatr 2002; 69:427-31.
92. Sokol RJ, Butler-Simon N, Conner C et al. Multicenter trial of d-alpha-tocopheryl polyethylene glycol 1000 succinate for treatment of vitamin E deficiency in children with chronic cholestasis. Gastroenterology 1993; 104:1727-35.
93. Argao EA, Heubi JE, Hollis BW et al. d-Alpha-tocopheryl polyethylene glycol-1000 succinate enhances the absorption of vitamin D in chronic cholestatic liver disease of infancy and childhood. Pediatr Res 1992; 31:146-50.

CHAPTER 24

Medical Therapy of Cholestatic Liver Diseases

Christian Rust and Ulrich Beuers

Summary

Cholestasis and its sequelae are the hallmark of chronic cholestatic liver diseases and can be a feature of virtually all liver diseases at some point. Ursodeoxycholic acid (UDCA), a dihydroxy bile acid, is the only drug approved for the treatment of patients with primary biliary cirrhosis (PBC), the most common chronic cholestatic liver disease, and has been studied for several other cholestatic syndromes. Although the effectiveness of UDCA has been shown most convincingly in PBC, beneficial effects have also been observed in patients with primary sclerosing cholangitis (PSC), intrahepatic cholestasis of pregnancy (ICP), some pediatric liver diseases and chronic graft-versus-host-disease (GVHD). Several potential mechanisms of action of UDCA have been proposed including stimulation of impaired hepatocellular secretion, cytoprotective and anti-apoptotic effects, anti-oxidative and immunomodulating mechanisms and alteration of the bile acid pool. Except for UDCA, other medical approaches for the treatment of cholestatic liver diseases have not been clearly proven to be beneficial. Although immunosuppressive substances have been disappointing as single agents in the treatment of chronic cholestatic liver diseases such as PBC and PSC, a combination of these substances with UDCA is currently under study. However, it still remains unclear whether combination therapy is indeed superior to monotherapy with UDCA.

Introduction

Cholestasis is observed in a variety of clinical syndromes and is the main feature of a number of chronic progressive liver diseases, which finally lead to cirrhosis, liver failure and death. Hepatic retention of hydrophobic, potentially toxic bile acids has long been implicated as a major cause of liver damage.[1] Accumulation of these bile acids within the hepatocyte is thought to play a key role in liver injury during cholestasis. Indeed, hepatic levels of the toxic bile acids chenodeoxycholate and deoxycholate correlate with the degree of liver damage.[2] Cholestasis can be caused by obstruction of extra- or intrahepatic bile ducts or by impairment of hepatocellular bile secretion. Therapy of extrahepatic bile duct obstruction is usually not medical and is thus not in the focus of this chapter. Numerous cholestatic disorders are characterized by a decreased expression of hepatocellular transport proteins resulting in marked impairment of hepatocellular uptake and canalicular secretion of bile acids and other cholephils. The molecular mechanisms of cholestasis have been reviewed in detail in this book.

The majority of adult patients with chronic cholestasis suffer from primary biliary cirrhosis (PBC) and primary sclerosing cholangitis (PSC); both diseases are described in this book. The most common cholestatic liver diseases in children are progressive familial intrahepatic cholestasis (Byler's disease) and biliary atresia. However, these diseases are rare.

Molecular Pathogenesis of Cholestasis, edited by Michael Trauner and Peter L.M. Jansen.
©2004 Eurekah.com and Kluwer Academic / Plenum Publishers.

Ursodeoxycholic acid (UDCA) is a physiological bile acid that has been introduced as a choleretic in the 1950s, is used for the medical therapy of cholesterol gallstones since 1975 and was introduced for the treatment of chronic cholestatic liver diseases in the late 1980s.[3,4] In this chapter, we provide an overview on the use of UDCA in cholestatic liver diseases and discuss the proposed mechanisms of action of this bile acid. Other medical approaches for the treatment of cholestatic liver diseases are also briefly discussed.

Role of UDCA in Cholestatic Liver Diseases

Primary Biliary Cirrhosis

Primary biliary cirrhosis (PBC) is a progressive liver disease which eventually proceeds to cirrhosis and death unless liver transplantation is performed. UDCA is the only drug currently approved specifically for the treatment of patients with PBC. UDCA consistently improves biochemical parameters including serum bilirubin levels. Serum bilirubin is a powerful prognostic marker for PBC in the absence as well as in the presence of UDCA therapy[5] and patients with two consecutive serum bilirubin levels higher than 6 mg/dl have a mean survival of only 2.1 years.[6] Several randomized double-blind placebo-controlled studies have shown that UDCA improves biochemical and histological features.[7-9] UDCA also delays histological progression of early-stage disease[10] and development of severe fibrosis or cirrhosis in patients with PBC.[11] In an analysis combining the raw data of three large clinical trials, 548 patients were followed for up to 4 years and treated with a dose of 13-15 mg/kg/day UDCA or placebo;[12] UDCA therapy significantly improved survival free of liver transplantation. A long-term observational study also indicated that the 10-year survival of UDCA-treated patients is better than that of untreated patients, as predicted by the Mayo risk score for PBC.[13] It is noteworthy that a recent meta-analysis of randomized controlled studies of UDCA in PBC did not find a beneficial effect of UDCA on survival free of liver transplantation.[14] However, this study included trials of different length of follow-up, mostly less than two years, which were performed with various doses of UDCA, subobtimal in part – a daily dose of 13-15 mg/kg is currently regarded as optimal for PBC. Thus, this meta-analysis has to be interpreted with caution, because survival analyses in a disease with a very long natural history over decades are ideally based on longer follow-up periods. A second meta-analysis included five studies with a follow-up of at least 4 years and application of therapeutic doses of 13-15 mg/kg/day. This analysis found a 32% reduction in the risk of death or need of liver transplantation in patients treated with UDCA.[15] Nonetheless, these conflicting meta-analyses may ask for additional data in order to prove the efficacy of UDCA in the long-term treatment of PBC. At present, we recommend to treat PBC patients with UDCA at a dose of 13-15mg/kg/day. Treatment should be initiated as early as possible and should be continued for the entire life. In patients who suffer from pruritus as well as in late stage disease, UDCA should be started at low doses and should be slowly increased to 13-15 mg/kg/day to avoid transient aggravation of pruritus in the early treatment phase.

Primary Sclerosing Cholangitis

Primary sclerosing cholangitis (PSC) is a rare progressive cholestatic liver disease of unknown etiology. As in PBC, immunologic abnormalities have been described. PSC is characterized by intrahepatic and extrahepatic bile duct fibrosis as well as associated inflammatory changes resulting in strictures and dilatations of bile ducts. Approximately 75% of patients have concomitant inflammatory bowel disease.[16] Based on the benefit of UDCA in patients with PBC, UDCA was also tested in a number of small trials for patients with PSC.[17-19] In these studies, a significant improvement of biochemical as well as histologic parameters was demonstrated for patients receiving UDCA and no major side effects were reported. However, the largest randomized placebo-controlled study in patients with PSC (n=105) could not find a clinical benefit of UDCA treatment when administered at a dose of 13-15 mg/kg/day during a mean follow-up period of 2.2 years.[20] Some limitations of this study must be considered when interpreting its results. A large percentage of the patients had advanced disease and these patients might not respond well to medical treatment; for the remainder of the patients, the follow-up period might have been too short to show a clinical benefit. In addition, bile duct strictures typical for PSC will not respond to UDCA therapy alone, but may need mechanical intervention. Indeed, bile duct strictures developed during UDCA treatment.[21] When UDCA treatment was combined with endoscopic dilatation of bile duct strictures, the survival appeared significantly improved in comparison to calculated survival without treatment.[21] Recently, it has also been shown that the dosage of UDCA may be important for patients with PSC. In a pilot study, regular-dose (10-15 mg/kg/day) and high-dose (25-30 mg/kg/day) UDCA were compared in thirty patients with PSC.[22] Changes in the Mayo risk score after 1 year of treatment were significant only in the group treated with high-dose UDCA; the high-dose treatment was well tolerated. Another pilot study compared high-dose UDCA (20 mg/kg/day) to placebo in 26 patients with PSC and found an improvement in liver biochemistry as well as a reduction in progression in cholangiographic appearances and liver fibrosis in patients receiving high-dose treatment.[23] Thus, medical therapy of PSC should include UDCA ($\geq$ 15 mg/kg/d) in combination with regular endoscopic dilatation of major strictures. UDCA therapy should be started early in the course of the disease. In addition, patients should be included in trials to determine the optimal dosage and assess the true long-term benefit of UDCA in PSC.

Intrahepatic Cholestasis of Pregnancy

Intrahepatic cholestasis of pregnancy (ICP) is a rare disorder occurring mainly in the third trimester of pregnancy. It is characterized by pruritus in the mother as well as by an increased risk of premature delivery and stillbirth of the child. Serum levels of unconjugated bile acids and progesterone metabolites are increased in ICP. In addition, the normal fetal-to-maternal transfer of bile acids across the placenta is impaired, resulting in accumulation of toxic bile acids in the fetus.[24] Treatment with UDCA restores the ability of the placenta to transfer bile acids and reduces serum bile acids and progesterone levels in the mother.[25] UDCA also appears to be safe for mother and fetus, even at higher concentrations.[26] One small randomized, double-blind, placebo-controlled study demonstrated an improvement of pruritus and serum liver tests in ICP patients.[27] In mothers receiving UDCA, all babies were born near term as compared to 70% preterm deliveries and one stillbirth in the placebo group. However, with only 15 patients included in this study, further data have to be awaited. Thus, UDCA appears to be beneficial and safe in the treatment of ICP, but large controlled trials are required before UDCA can be generally recommended for the treatment of ICP.

Other Cholestatic Disorders

UDCA has been evaluated for cystic fibrosis (CF), because liver disease contributes to the morbidity of CF. A small randomized, placebo-controlled study showed clinical, biochemical and nutritional improvement in CF patients treated with UDCA for one year.[28] Liver histology also improved after 2 years of UDCA treatment in CF patients.[29] However, it remains unclear if UDCA treatment affects the natural history of the disease.

Since cholestatic liver disease caused by chronic graft-versus-host-disease (GVHD) is similar to PBC, UDCA was used in combination with immunosuppressive agents in patients after bone marrow transplantation.[30] Therapy with UDCA is safe and may exert beneficial effects in addition to those induced by immunosuppressive therapy of liver GVHD. In the same study, the development of veno-occlusive disease was also prevented in patients treated with UDCA. However, larger trials are needed to confirm the positive preliminary results of UDCA treatment. Use of UDCA has also been reported in a number of pediatric disorders.[31]

Mechanisms of Action of UDCA

The clinical use of UDCA for cholestatic diseases is supported by increasing experimental evidence for its potential mechanisms of action including stimulation of impaired hepatocellular secretion, cytoprotective and anti-apoptotic effects, and alteration of the bile acid pool.[32] UDCA is a hydrophilic, non-toxic bile acid, that accounts for not more than 3 % of the bile acid pool in man. However, during therapeutic application of 13-15mg/kg/day, UDCA is the predominant bile acid in serum and bile.[33] Experimental studies to date favor three main mechanisms of action that will be discussed in detail.

Stimulation of Hepatocellular Secretion

Hepatocellular retention of toxic bile acids leads to liver damage under cholestatic conditions.[34] Thus, stimulation of secretion of bile acids and other potentially toxic compounds is a therapeutic goal for these liver diseases. UDCA seems to exert its beneficial effects in cholestatic diseases at least in part by this mechanism. Indeed, UDCA amides stimulate secretion of bile acids and other cholephils in isolated hepatocytes,[35] in the perfused rat liver[36,37] as well as in the model of the bile fistula rat in vivo.[38] More importantly, stimulation of secretion of bile acids and phospholipids was also documented in patients with PBC and PSC during UDCA treatment.[39,40] In experimental cholestasis, vesicle-mediated targeting of proteins to the canalicular membrane is impaired.[36, 41] Experimental evidence suggested that UDCA may stimulate hepatobiliary exocytosis and, thereby, insertion of carrier proteins into the apical hepatocellular membrane.[36, 42-45] The capacity of UDCA to stimulate hepatocellular secretion under cholestatic conditions is probably due to insertion of vesicles containing transport proteins into the canalicular membrane.[36,43,44,46] A proof for this assumption was provided recently in a series of experiments which showed that the taurine conjugate of UDCA (TUDCA) significantly enhances the density of the conjugate export pump, Mrp2, in canalicular membranes and thereby stimulates secretion of potentially toxic compounds by cholestatic hepatocytes.[46] This process of vesicular exocytosis is regulated by a complex network of signals, that have been partially elucidated in recent years.

Cytosolic free calcium $[Ca^{++}]_i$ seems to be critical for TUDCA-induced exocytosis in the model of the perfused rat liver.[36,47] TUDCA, but not the trihydroxy bile acid taurocholic acid (TCA) acid, induces a sustained elevation of $[Ca^{++}]_i$ in isolated hepatocytes.[47,48] TUDCA also selectively induces translocation of the Ca^{++}-sensitive α-isoform of PKC, a key mediator of regulated exocytosis, to hepatocellular membranes and activation of membrane-bound PKC.[43,49,50] Inhibition of α-PKC by the PKC inhibitor bisindolylmaleimide-I markedly impairs TUDCA-induced secretion of the model Mrp2 substrate dinitrophenyl-S-glutathione (GS-DNP) in experimental cholestasis strongly supporting the concept that TUDCA exerts anticholestatic effects, at least in part, by Ca^{++}- and α-PKC-dependent mechanisms.[46] In contrast to TUDCA, the monohydroxy bile acid TLCA markedly impairs bile flow and hepatobiliary exocytosis and decreases the density of key transport proteins in canalicular membranes.[46] TLCA may impair calcium influx in hepatocytes,[47,51] a phenomenon observed also in other models of cholestasis.[36] In addition, TLCA may activate the Ca^{++}-independent ϵ-isoform of PKC at the level of the canalicular membrane.[52] ϵ-PKC has been associated with impairment of vesicle-mediated targeting and insertion of membrane proteins in secretory cells. Preliminary evidence suggests that TLCA-induced activation of ϵ-PKC and cholestasis are mediated by phosphoinositide 3-kinase (PI3-K)-dependent mechanisms.[53] Interestingly, ϵ-PKC binding to canalicular membranes is reversed by TUDCA in TLCA-induced cholestasis.[46]

Canalicular bile acid secretion may be increased by TUDCA via alternative signaling pathways independent of PKC in normal liver.[44] Activation of the small GTP-binding protein Ras and the mitogen-activated protein kinases (MAPK), extracellular signal-regulated kinase (Erk)-1 and Erk-2 on one hand and the Ras/Raf-independent MAPK p38 on the other hand mediate TUDCA-induced bile acid secretion in the perfused rat liver. TUDCA-induced Ras-activation appears to be PI3-K-dependent.[44,54,55] In the experimental model chosen, TUDCA induced a transient and concentration-dependent activation of p38(MAPK) and of Erk-2; TUDCA-induced stimulation of taurocholic acid excretion was accompanied by a p38(MAPK)-dependent insertion of the bile salt export pump (Bsep) into the canalicular membrane.[55]

Recently, the transcriptional regulation of canalicular transport systems by UDCA and the potentially more toxic cholic acid (CA) has also been addressed.[56] In hepatocytes of mice fed a UDCA or CA supplemented diet, both UDCA and CA up-regulated Bsep and Mrp2 mRNA whereas only CA up-regulated Mdr2 expression. However, the effects of UDCA on expression of hepatocellular transporter systems under cholestatic conditions remain unclear.

Thus, UDCA and its conjugates may enhance the impaired secretory capacity of the cholestatic liver by modulating complex intracellular signaling cascades including calcium, α-PKC and different MAP kinases (Fig. 1).

Protection Against Bile Duct Injury

Hydrophobic bile acids may damage membranes of hepatocytes and cholangiocytes. Even under physiological conditions, levels of hydrophobic bile acids are three orders of magnitude higher in bile than in serum. Protection of bile ducts is thus necessary and is partly accomplished by hepatocellular secretion of phospholipids into bile. An impressive example of the protective capacity of phospholipids is the Mdr2-knockout mouse that lacks the ability to secrete phospholipids. These mice develop a chronic nonsuppurative cholangitis resembling changes in chronic cholestasis.[57] However, when Mdr-2 knockout mice were fed UDCA, ductular proliferation and portal inflammation were markedly decreased.[58] Indeed, the inflammatory reaction around altered bile ducts seen in both PBC and PSC is less severe in patients treated with UDCA.[7,8,10,18,59]

Hydrophobic bile acids damage cell membranes at high micromolar or millimolar concentrations in vitro whereas UDCA amides counteract these damaging effects of hydrophobic bile acids.[60-62] This observation may be explained by the formation and structure of simple and mixed micelles rather than a direct membrane interaction.[61] However, the relevance of these in vitro findings may be restricted to the biliary tree, because millimolar bile acid concentrations are only observed within the biliary lumen.[63]

These studies support the impression that UDCA partly exerts its beneficial effects by protecting against the consequences of bile duct destruction, but not by preventing ongoing bile duct destruction itself.

Protection Against Hepatocyte Apoptosis

Since widespread necrosis is not prominent in cholestatic liver disease, it was proposed that hepatocyte cell death during cholestasis occurs by apoptosis rather than by necrosis.[64] For example, in liver tissue from patients with PBC, apoptotic features occur more frequently than in normal controls.[65] The mechanisms of apoptosis by toxic bile acids have been partially elucidated in recent years. Toxic bile acids like glycochenodeoxycholic acid (GCDCA) directly cause apoptosis in rat hepatocytes by ligand-independent activation of Fas.[66] Subsequently, caspase 8 is activated, followed by activation of Bid which leads to mitochondrial dysfunction by chaperoning Bax to the mitochondrial membrane.[67] Bid and Bax are pro-apoptotic members of the Bcl-2 family.[68] Bax is thought to induce the mitochondrial membrane permeability transition (MMPT), a phenomenon characterized by a sudden permeability of the inner mitochondrial membrane to ions, followed by mitochondrial swelling.[69] Mitochondrial swelling may induce the release of cytochrome c from the intermembrane space to the cytosol, where it activates caspase 9 through interaction with the apoptotic protease-activating factor 1 (APAF-1) and subsequently causes the apoptotic demise of the cell.[70]

UDCA exerts antiapoptotic effects in vitro as well as in vivo in the model of the bile acid-fed rat.[71,72] Coadministration of UDCA with a toxic bile acid was shown to reduce apoptosis in human hepatocytes by at least 50%.[72] It was also shown that UDCA inhibits the induction of apoptosis not only by toxic bile acids, but also by other substances such as ethanol, transforming growth factor-b1 (TGF-β1), anti-Fas antibodies and okadaic acid.[72] The antiapoptotic effect of UDCA was associated with a reduction of the MMPT.[72,73] It was suggested that UDCA may reduce MMPT by binding to Bax and prevention of Bax translocation from cytosol

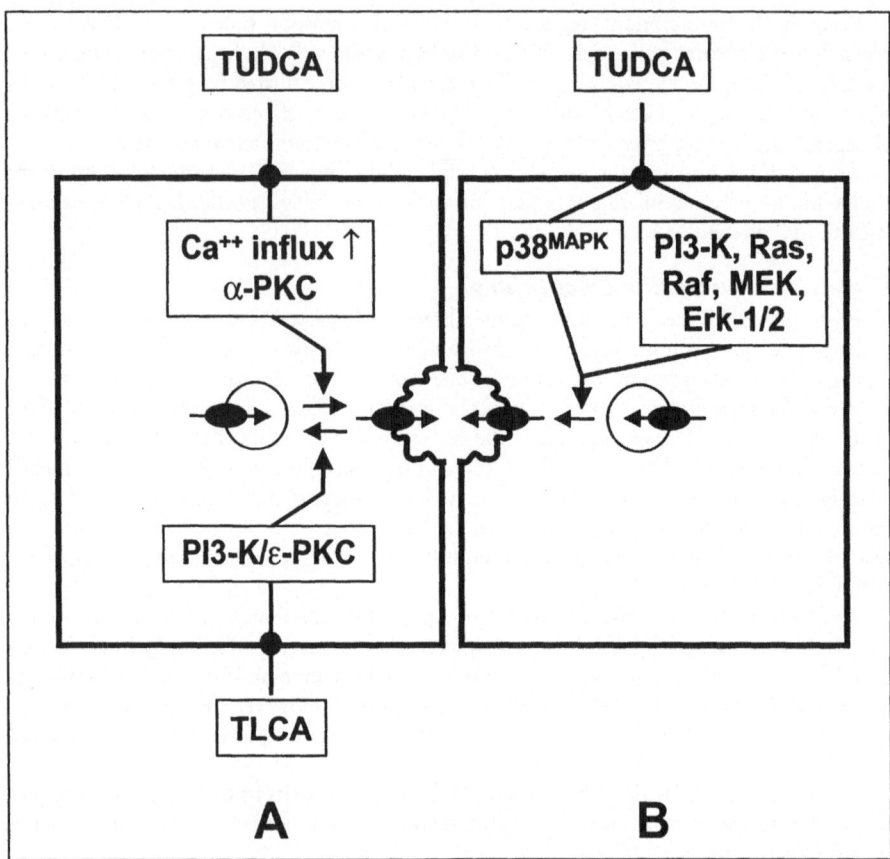

Figure 1. Stimulation of hepatocellular secretion by UDCA. Model A: TUDCA stimulates apical vesicular exocytosis and insertion of key canalicular transporters such as the conjugate export pump, Mrp2, via Ca^{++}-and α-PKC-dependent mechanisms in experimental cholestasis. In contrast, the cholestatic bile acid TLCA reduces density of membrane carriers and bile secretory capacity of the liver cell by PI3-K and putatively ε-PKC-dependent mechanisms at the canalicular membrane.[36,53] Model B: TUDCA induces bile acid secretion by activation of p38(MAPK) or by PI3-K-, Ras-, Raf-, MEK- and Erk-1/2-dependent mechanisms and by translocation of the bile salt export pump Bsep to the canalicular membrane in normal liver.[44,55]

to mitochondria.[74] In addition, UDCA impairs mitochondrial cytochrome c release by diminishing mitochondrial membrane alterations.[72] The anti-apoptotic properties of UDCA appear to be greatly enhanced by adding an NO-releasing moiety to UDCA.[75] Interestingly, an adaptive phenomenon to resist cell death by the MMPT occurs during cholestasis through an increase in mitochondria cardiolipin content.[76] Other anti-apoptotic compounds, e.g., specific caspase inhibitors, are currently developed, but have not yet been tested in clinical studies. Thus, the impact of antiapoptotic mechanisms for the beneficial effects of UDCA in cholestatic liver disease are unclear at present. A model of antiapoptotic mechanisms of UDCA is shown in Figure 2.

Role of Other Medical Therapies in Cholestatic Liver Diseases

The development of specific therapies for PBC or PSC is delayed by our incomplete understanding of these diseases. Although the pathomechanisms of chronic cholestatic liver diseases are still unclear, dysregulation of immunologic mechanisms seems to be crucial for the destruction

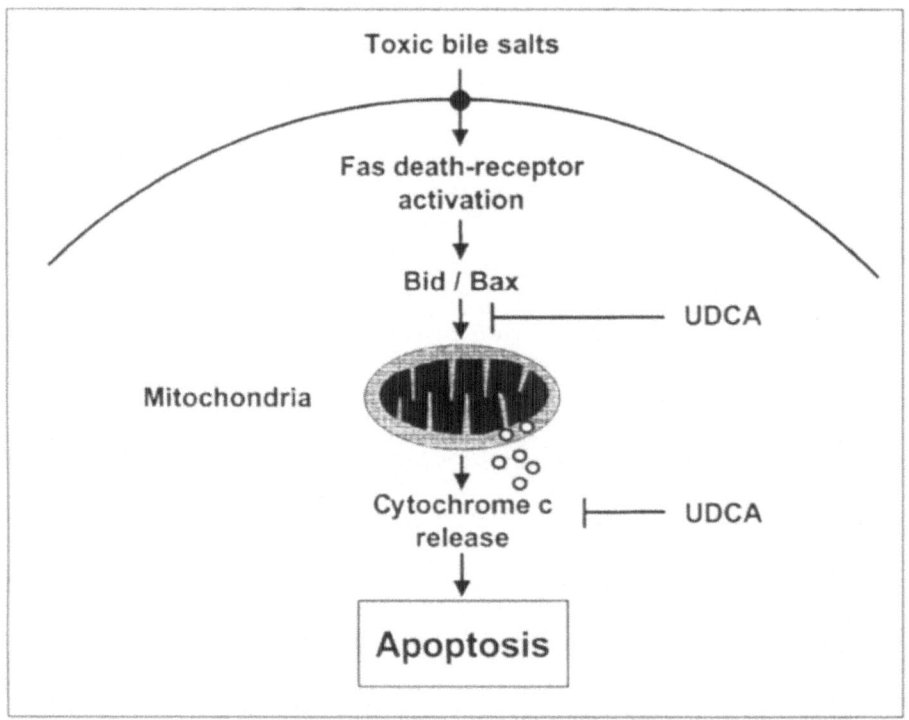

Figure 2. Anti-apoptotic effects of UDCA. Toxic bile acids induce hepatocellular apoptosis via ligand-independent activation of the Fas death-receptor. Subsequently, caspase 8 followed by Bid are activated. Bid causes mitochondrial dysfunction by bringing Bax to the mitochondrial membrane which causes release of cytochrome C into the cytosol. UDCA prevents bile acid-induced apoptosis in hepatocytes by reducing (i) mitochondrial dysfunction and (ii) release of cytochrome c (adapted from[82] with permission).

of bile ducts observed in these diseases. Thus, a number of immunosuppressive substances have been evaluated in the treatment of PBC and PSC. These substances are attractive, because their point of action would be early in the development of the diseases. So far, azathioprin, chlorambucil, cyclosporin A, corticosteroids and methotrexate have been tested in randomized studies.[31,77] Unfortunately, these substances either showed no marked beneficial effects on the long-term course in patients with PBC or PSC, or serious side effects excluded long-term treatment. However, a number of small trials and pilot studies demonstrated that a combination of UDCA with immunosuppressive agents improved biochemical and histological parameters when compared to UDCA alone. The combination of UDCA with prednisone,[78] prednisone plus azathioprine[79] and budesonide[80] have been tested. Budesonide is of particular interest as this glucocorticosteroid has a high first-pass metabolism avoiding some of the adverse systemic effects of regular corticosteroids. Based on the promising results of these studies, several larger, controlled studies evaluating combinations of UDCA with budesonide, budesonide and azathioprine, methotrexate, and other immunosupresssive agents are currently under way. Results of these studies are eagerly awaited.

A novel approach is the combination of UDCA with a second choleretic drugs. In a pilot study, 23 PBC patients with incomplete response to UDCA therapy received UDCA in combination with sulindac, a nonsteroidal anti-inflammatory drug with strong choleretic properties, or UDCA alone.[81] Sulindac was well tolerated and combination therapy with UDCA significantly improved liver biochemistry as compared to UDCA monotherapy. The effect on the long-term outcome, however, is unclear, and sulindac may exert hepatotoxic effects. Bezafibrate has also been reported to improve biochemical markers of cholestasis in patients with PBC who did not respond completely to treatment with UDCA alone. Further studies are awaited.

Antifibrotic treatment has also been evaluated in PBC patients. Administration of colchicine or malotilate did not clearly improve symptoms, biochemical tests, histological features or survival.[77] D-penicillamine caused severe side effects, but did not beneficially affect markers of disease activity. Thus, antifibrotic drugs are not recommended for treatment of PBC at present.

References

1. Javitt J. Cholestasis in rats induced by taurolithocholate. Nature 1966; 210:1262-63.
2. Schmucker DL, Ohta M, Kanai S et al. Hepatic injury induced by bile salts: correlation between biochemical and morphological events. Hepatology 1990; 12(5):1216-21.
3. Makino I, Tanaka H. From a choleretic to an immunomodulator: historical review of ursodeoxycholic acid as a medicament. J Gastroenterol Hepatol 1998; 13(6):659-64.
4. Poupon R, Chretien Y, Poupon RE et al. Is ursodeoxycholic acid an effective treatment for primary biliary cirrhosis? Lancet 1987; 1(8537):834-6.
5. Bonnand AM, Heathcote EJ, Lindor KD et al. Clinical significance of serum bilirubin levels under ursodeoxycholic acid therapy in patients with primary biliary cirrhosis. Hepatology 1999; 29(1):39-43.
6. Shapiro JM, Smith H, Schaffner F. Serum bilirubin: a prognostic factor in primary biliary cirrhosis. Gut 1979; 20(2):137-40.
7. Heathcote EJ, Cauch-Dudek K, Walker V et al. The Canadian Multicenter Double-blind Randomized Controlled Trial of ursodeoxycholic acid in primary biliary cirrhosis. Hepatology 1994; 19(5):1149-56.
8. Poupon RE, Poupon R, Balkau B. Ursodiol for the long-term treatment of primary biliary cirrhosis. The UDCA-PBC Study Group. N Engl J Med 1994; 330(19):1342-7.
9. Lindor KD, Dickson ER, Baldus WP et al. Ursodeoxycholic acid in the treatment of primary biliary cirrhosis. Gastroenterology 1994; 106(5):1284-90.
10. Pares A, Caballeria L, Rodes J et al. Long-term effects of ursodeoxycholic acid in primary biliary cirrhosis: results of a double-blind controlled multicentric trial. UDCA-Cooperative Group from the Spanish Association for the Study of the Liver. J Hepatol 2000; 32(4):561-6.
11. Corpechot C, Carrat F, Bonnand AM et al. The effect of ursodeoxycholic acid therapy on liver fibrosis progression in primary biliary cirrhosis. Hepatology 2000; 32(6):1196-9.
12. Poupon RE, Lindor KD, Cauch-Dudek K et al. Combined analysis of randomized controlled trials of ursodeoxycholic acid in primary biliary cirrhosis. Gastroenterology 1997; 113(3):884-90.
13. Poupon RE, Bonnand AM, Chretien Y et al. Ten-year survival in ursodeoxycholic acid-treated patients with primary biliary cirrhosis. The UDCA-PBC Study Group. Hepatology 1999; 29(6):1668-71.
14. Goulis J, Leandro G, Burroughs A. Randomised controlled trials of ursodeoxycholic-acid therapy for primary biliary cirrhosis: a meta-analysis. Lancet 1999; 354:1053-60.
15. Lindor KD, Poupon R, Heathcote EJ et al. Ursodeoxycholic acid for primary biliary cirrhosis. Lancet 2000; 355(9204):657-8.
16. Lee YM, Kaplan MM. Primary sclerosing cholangitis. N Engl J Med 1995; 332(14):924-33.
17. Stiehl A, Walker S, Stiehl L et al. Effect of ursodeoxycholic acid on liver and bile duct disease in primary sclerosing cholangitis. A 3-year pilot study with a placebo- controlled study period. J Hepatol 1994; 20(1):57-64.
18. Beuers U, Spengler U, Kruis W et al. Ursodeoxycholic acid for treatment of primary sclerosing cholangitis: a placebo-controlled trial. Hepatology 1992; 16(3):707-14.
19. O'Brien CB, Senior JR, Arora-Mirchandani R et al. Ursodeoxycholic acid for the treatment of primary sclerosing cholangitis: a 30-month pilot study. Hepatology 1991; 14(5):838-47.
20. Lindor KD. Ursodiol for primary sclerosing cholangitis. Mayo Primary sclerosing cholangitis-Ursodeoxycholic acid Study Group. N Engl J Med 1997; 336(10):691-5.
21. Stiehl A, Rudolph G, Sauer P et al. Efficacy of ursodeoxycholic acid treatment and endoscopic dilation of major duct stenoses in primary sclerosing cholangitis. An 8-year prospective study. J Hepatol 1997; 26(3):560-6.
22. Harnois DM, Angulo P, Jorgensen RA et al. High-dose ursodeoxycholic acid as a therapy for patients with primary sclerosing cholangitis. Am J Gastroenterol 2001; 96(5):1558-62.
23. Mitchell SA, Bansi DS, Hunt N et al. A preliminary trial of high-dose ursodeoxycholic acid in primary sclerosing cholangitis. Gastroenterology 2001; 121:900-907.

24. Fagan EA. Intrahepatic cholestasis of pregnancy. Clin Liver Dis 1999; 3(3):603-32.
25. Serrano MA, Brites D, Larena MG et al. Beneficial effect of ursodeoxycholic acid on alterations induced by cholestasis of pregnancy in bile acid transport across the human placenta. J Hepatol 1998; 28(5):829-39.
26. Mazzella G, Nicola R, Francesco A et al. Ursodeoxycholic acid administration in patients with cholestasis of pregnancy: effects on primary bile acids in babies and mothers. Hepatology 2001; 33(3):504-8.
27. Palma J, Reyes H, Ribalta J et al. Ursodeoxycholic acid in the treatment of cholestasis of pregnancy: a randomized, double-blind study controlled with placebo. J Hepatol 1997; 27(6):1022-8.
28. Colombo C, Battezzati PM, Podda M et al. Ursodeoxycholic acid for liver disease associated with cystic fibrosis: a double-blind multicenter trial. The Italian Group for the Study of Ursodeoxycholic acid in Cystic fibrosis. Hepatology 1996; 23(6):1484-90.
29. Lindblad A, Glaumann H, Strandvik B. A two-year prospective study of the effect of ursodeoxycholic acid on urinary bile acid excretion and liver morphology in cystic fibrosis- associated liver disease. Hepatology 1998; 27(1):166-74.
30. Essell JH, Schroeder MT, Harman GS et al. Ursodiol prophylaxis against hepatic complications of allogeneic bone marrow transplantation. A randomized, double-blind, placebo-controlled trial. Ann Intern Med 1998; 128(12 Pt 1):975-81.
31. Poupon R, Chazouilleres O, Poupon RE. Chronic cholestatic diseases. J Hepatol 2000; 32(1 (Suppl)):129-40.
32a. Beuers U, Boyer JL, Paumgartner G. Ursodeoxycholic acid in cholestasis: potential mechanisms of action and therapeutic applications. Hepatology 1998; 28(6):1449-53.
   32b Trauner M, Graziadei IW. Mechanisms of action and therapeutic applications of ursodeoxycholic acid in chronic liver diseases. Aliment Pharmacol Ther 1999;13:979-95.
33. Hofmann AF. Pharmacology of ursodeoxycholic acid, an enterohepatic drug. Scand J Gastroenterol Suppl 1994; 204:1-15.
34. Greim H, Trulzsch D, Czygan P et al. Mechanism of cholestasis. 6. Bile acids in human livers with or without biliary obstruction. Gastroenterology 1972; 63(5):846-50.
35. Kitani K, Ohta M, Kanai S. Tauroursodeoxycholate prevents biliary protein excretion induced by other bile salts in the rat. Am J Physiol 1985; 248(4 Pt 1):G407-17.
36. Beuers U, Nathanson MH, Isales CM et al. Tauroursodeoxycholic acid stimulates hepatocellular exocytosis and mobilizes extracellular Ca++ mechanisms defective in cholestasis. J Clin Invest 1993; 92(6):2984-93.
37. Heuman DM, Mills AS, McCall J et al. Conjugates of ursodeoxycholate protect against cholestasis and hepatocellular necrosis caused by more hydrophobic bile salts. In vivo studies in the rat. Gastroenterology 1991; 100(1):203-11.
38. Kitani K, Kanai S, Sato Y, Ohta M. Tauro alpha-muricholate is as effective as tauro beta-muricholate and tauroursodeoxycholate in preventing taurochenodeoxycholate-induced liver damage in the rat. Hepatology 1994; 19(4):1007-12.
39. Jazrawi RP, de Caestecker JS, Goggin PM et al. Kinetics of hepatic bile acid handling in cholestatic liver disease: effect of ursodeoxycholic acid. Gastroenterology 1994; 106(1):134-42.
40. Stiehl A, Rudolph G, Sauer P et al. Biliary secretion of bile acids and lipids in primary sclerosing cholangitis. Influence of cholestasis and effect of ursodeoxycholic acid treatment. J Hepatol 1995; 23(3):283-9.
41. Larkin JM, Palade GE. Transcytotic vesicular carriers for polymeric IgA receptors accumulate in rat hepatocytes after bile duct ligation. J Cell Sci 1991; 98(Pt 2):205-16.
42. Crawford JM, Strahs DC, Crawford AR et al. Role of bile salt hydrophobicity in hepatic microtubule-dependent bile salt secretion. J Lipid Res 1994; 35(10):1738-48.
43. Beuers U, Throckmorton DC, Anderson MS et al. Tauroursodeoxycholic acid activates protein kinase C in isolated rat hepatocytes. Gastroenterology 1996; 110(5):1553-63.
44. Schliess F, Kurz AK, vom Dahl S et al. Mitogen-activated protein kinases mediate the stimulation of bile acid secretion by tauroursodeoxycholate in rat liver. Gastroenterology 1997; 113(4):1306-14.
45. Haussinger D, Saha N, Hallbrucker C et al. Involvement of microtubules in the swelling-induced stimulation of transcellular taurocholate transport in perfused rat liver. Biochem J 1993; 291(Pt 2):355-60.
46. Beuers U, Bilzer M, Chittattu A et al. Tauroursodeoxycholic acid inserts the apical conjugate export pump, Mrp2, into canalicular membranes and stimulates organic anion secretion by protein kinase C-dependent mechanisms in cholestatic rat liver. Hepatology 2001; 33(5):1206-16.
47. Beuers U, Nathanson MH, Boyer JL. Effects of tauroursodeoxycholic acid on cytosolic Ca2+ signals in isolated rat hepatocytes. Gastroenterology 1993; 104(2):604-12.

48. Bouscarel B, Fromm H, Nussbaum R. Ursodeoxycholate mobilizes intracellular Ca2+ and activates phosphorylase a in isolated hepatocytes. Am J Physiol 1993; 264(2 Pt 1):G243-51.
49. Bouscarel B, Kroll SD, Fromm H. Signal transduction and hepatocellular bile acid transport: cross talk between bile acids and second messengers. Gastroenterology 1999; 117(2):433-52.
50. Stravitz RT, Rao YP, Vlahcevic ZR et al. Hepatocellular protein kinase C activation by bile acids: implications for regulation of cholesterol 7 alpha-hydroxylase. Am J Physiol 1996; 271(2 Pt 1):G293-303.
51. Combettes L, Berthon B, Doucet E et al. Bile acids mobilise internal Ca2+ independently of external Ca2+ in rat hepatocytes. Eur J Biochem 1990; 190(3):619-23.
52. Beuers U, Probst I, Soroka C et al. Modulation of protein kinase C by taurolithocholic acid in isolated rat hepatocytes. Hepatology 1999; 29(2):477-82.
53. Beuers U, Soroka C, Denk GU et al. Taurolithocholic acid impairs hepatocellular bile acid and organic anion secretion by a phosphatidylinositol 3-kinase-dependent mechanism. Hepatology 2001; 34: in press.
54. Kurz AK, Block C, Graf D et al. Phosphoinositide 3-kinase-dependent Ras activation by tauroursodesoxycholate in rat liver. Biochem J 2000; 350 Pt 1:207-13.
55. Kurz AK, Graf D, Schmitt M et al. Tauroursodesoxycholate-induced choleresis involves p38(MAPK) activation and translocation of the bile salt export pump in rats. Gastroenterology 2001; 121(2):407-19.
56. Fickert P, Zollner G, Fuchsbichler A et al. Effects of ursodeoxycholic and cholic acid feeding on hepatocellular transporter expression in mouse liver. Gastroenterology 2001; 121(1):170-83.
57. Smit JJ, Schinkel AH, Oude Elferink RP et al. Homozygous disruption of the murine mdr2 P-glycoprotein gene leads to a complete absence of phospholipid from bile and to liver disease. Cell 1993; 75(3):451-62.
58. Van Nieuwkerk CM, Elferink RP, Groen AK et al. Effects of Ursodeoxycholate and cholate feeding on liver disease in FVB mice with a disrupted mdr2 P-glycoprotein gene. Gastroenterology 1996; 111(1):165-71.
59. Combes B, Markin RS, Wheeler DE et al. The effect of ursodeoxycholic acid on the florid duct lesion of primary biliary cirrhosis. Hepatology 1999; 30(3):602-5.
60. Heuman DM, Bajaj R. Ursodeoxycholate conjugates protect against disruption of cholesterol- rich membranes by bile salts. Gastroenterology 1994; 106(5):1333-41.
61. Heuman DM, Bajaj RS, Lin Q. Adsorption of mixtures of bile salt taurine conjugates to lecithin-cholesterol membranes: implications for bile salt toxicity and cytoprotection. J Lipid Res 1996; 37(3):562-73.
62. Guldutuna S, Zimmer G, Imhof M et al. Molecular aspects of membrane stabilization by ursodeoxycholate. Gastroenterology 1993; 104(6):1736-44.
63. Heuman DM. Hepatoprotective properties of ursodeoxycholic acid. Gastroenterology 1993; 104(6):1865-70.
64. Patel T, Gores GJ. Apoptosis and hepatobiliary disease. Hepatology 1995; 21(6):1725-41.
65. Koga H, Sakisaka S, Ohishi M et al. Nuclear DNA fragmentation and expression of Bcl-2 in primary biliary cirrhosis. Hepatology 1997; 25(5):1077-84.
66. Faubion W, Guicciardi M, Miyoshi H et al. Toxic bile salts induce rodent hepatocyte apoptosis via direct activation of Fas. J Clin Invest 1999; 103:137-145.
67. Eskes R, Desagher S, Antonsson B et al. Bid induces the oligomerization and insertion of Bax into the outer mitochondrial membrane. Mol Cell Biol 2000; 20(3):929-35.
68. Adams J, Cory S. The Bcl-2 protein familiy: arbiters of cell survival. Science 1998; 281(5381):1322-6.
69. Kluck RM, Esposti MD, Perkins G et al. The pro-apoptotic proteins, Bid and Bax, cause a limited permeabilization of the mitochondrial outer membrane that is enhanced by cytosol. J Cell Biol 1999; 147(4):809-22.
70. Martinou JC, Desagher S, Antonsson B. Cytochrome c release from mitochondria: all or nothing. Nat Cell Biol 2000; 2(3):E41-3.
71. Benz C, Angermuller S, Tox U et al. Effect of tauroursodeoxycholic acid on bile-acid-induced apoptosis and cytolysis in rat hepatocytes. J Hepatol 1998; 28(1):99-106.
72. Rodrigues CM, Fan G, Ma X et al. A novel role for ursodeoxycholic acid in inhibiting apoptosis by modulating mitochondrial membrane perturbation. J Clin Invest 1998; 101(12):2790-9.
73. Botla R, Spivey JR, Aguilar H et al. Ursodeoxycholate (UDCA) inhibits the mitochondrial membrane permeability transition induced by glycochenodeoxycholate: a mechanism of UDCA cytoprotection. J Pharm Exp Ther 1995; 272(2):930-8.
74. Guicciardi ME, Gores GJ. Is ursodeoxycholate an antiapoptotic drug? Hepatology 1998; 28(6):1721-3.

75. Fiorucci S, Mencarelli A, Palazzetti B et al. An NO derivative of ursodeoxycholic acid protects against Fas-mediated liver injury by inhibiting caspase activity. Proc Natl Acad Sci USA 2001; 98(5):2652-7.
76. Lieser MJ, Park J, Natori S et al. Cholestasis confers resistance to the rat liver mitochondrial permeability transition. Gastroenterology 1998; 115(3):693-701.
77. Beuers U. Primary biliary cirrhosis: treatments other than ursodeoxycholic acid. Primary biliary cirrhosis. Eastbourne, UK: West End Studios, Ltd, 1999; 115-118.
78. Leuschner M, Guldutuna S, You T et al. Ursodeoxycholic acid and prednisolone versus ursodeoxycholic acid and placebo in the treatment of early stages of primary biliary cirrhosis. J Hepatol 1996; 25(1):49-57.
79. Wolfhagen FH, van Hoogstraten HJ, van Buuren HR et al. Triple therapy with ursodeoxycholic acid, prednisone and azathioprine in primary biliary cirrhosis: a 1-year randomized, placebo-controlled study. J Hepatol 1998; 29(5):736-42.
80. Leuschner M, Maier KP, Schlichting J et al. Oral budesonide and ursodeoxycholic acid for treatment of primary biliary cirrhosis: results of a prospective double-blind trial. Gastroenterology 1999; 117(4):918-25.
81. Leuschner MS, Schlichting JK, Ackermann H et al. Sulindac and ursodeoxycholic acid (UDCA) improve primary biliary cirrhosis (PBC) in patients not responding to UDCA. A prospective controlled pilot study. Hepatology 2000; 32(4):309A.
82. Lazaridis KN, Gores GJ, Lindor KD. Ursodeoxycholic acid 'mechanisms of action and clinical use in hepatobiliary disorders'. J Hepatol 2001; 35(1):134-46.

CHAPTER 25

Hepatocyte Transplantation and Liver-Directed Gene Therapy

Chandan Guha, Siddhartha S. Ghosh, Sung W. Lee,
Namita Roy-Chowdhury and Jayanta Roy-Chowdhury

Introduction

The research on liver-directed gene therapy and hepatocyte transplantation has progressed in parallel. Hepatocytes, with or without genetic modification have been used to introduce normal genes into patients or animal models with inherited disorders, while gene transfer can be used to expand hepatocytes in culture or abrogate allograft rejection. In this article, we discuss the current state of the two areas of research, as well as the hurdles that remain to be crossed for widespread application of these advances in the treatment of liver diseases.

Hepatocyte Transplantation

Although liver transplantation has dramatically improved the prognosis for patients with acute or end-stage liver failure, and inherited metabolic disorders, because of the complexity and associated morbidity and mortality of this procedure, hepatocyte transplantation is being explored as a convenient and safer alternative. Transplanting isolated hepatocytes by percutaneous or transjugular infusion into the portal vein, or injecting into the splenic pulp or the peritoneal cavity, is a less invasive procedure compared with liver transplantation. As the host liver is not removed or resected, the loss of graft function should not worsen liver function. Furthermore, isolated hepatocytes could be, potentially, cryopreserved for ready access.[1] After extensive evaluation in experimental animals, clinical trials of hepatocyte transplantation have been initiated at several institutions for acute or chronic liver failure and inherited metabolic disorders. Hepatocytes transplantation has been explored as a vehicle for ex vivo gene therapy and is being considered for rescuing patients from radiation-induced liver damage resulting from radiotherapy for liver tumors.

Hepatocyte Transplantation for Liver-Based Metabolic Diseases

Hepatocyte transplantation has been tested on a number of animal models with liver-based metabolic disorders. The Gunn rat is an animal model of Crigler-Najjar syndrome type 1 (CN1). This mutant strain of Wistar rats lacks bilirubin-UDP-glucuronosyltranferase (UGT1A1) activity and consequently, accumulate toxic plasma levels of unconjugated bilirubin.[2] Hepatocyte transplantation has been tested most extensively in Gunn rats and Nagase genetically analbuminemic (NAR) rats. In both models, transplantation of normal donor hepatocytes ameliorated the metabolic deficit.[3] When the liver is structurally normal, which is the case in many inherited liver based disorders, hepatocytes injected into the splenic pulp or infused into the portal vein migrate to the liver and become integrated into the hepatic cords within days. These cells become morphologically identical to the host cells and function throughout life, and can be recognized only by virtue of biochemical or genetic markers.[4,5] Hepatocyte transplantation has also resulted in partial correction of metabolic disorders in low density lipoprotein receptor-deficient[6] Watanabe heritable hyperlipidemic rabbits (an animal model of familial hypercholesterolemia), and the Long-Evans Cinnamon rat (an animal model for Wilson's disease).[7] Most functional liver proteins are present in great excess, therefore, the initial assumption was that the transplantation of a small fraction of the total hepatocyte mass (~1-5%) should correct many metabolic disorders. However, most experimental and clinical studies have not borne out this expectation. Although repeated hepatocyte infusions could improve the response to transplantation to some extent,[8] the most dramatic results have been obtained by massive preferential repopulation of the host liver by engrafted hepatocytes.

Massive Repopulation of the Liver

In inherited disorders which result in death of host hepatocytes, such as fumarylacetoacetate hydrolase (FAH) deficient mutant mice (a model of hereditary tyrosinemia type I) or transgenic mice overexpressing the plasminogen activator (uPA), transplanted normal hepatocytes can spontaneously repopulate the liver, eventually replacing almost all host hepatocytes.[9] However, the transplant recipient mice remain susceptible to the development of hepatocellular carcinomas. Therefore, continuing cancer risk or recurrent disease should be considered in determining the clinical indications for hepatocyte transplantation for specific diseases.

In most metabolic diseases, however, the host hepatocytes do not have a rapid turnover. In these cases, the host hepatocytes respond to proliferative stimuli to the same extent, as do the engrafted cells, precluding preferential growth of transplanted cells. Therefore, for a more general application, strategies are being developed to inhibit host hepatocyte regeneration, while providing a proliferative stimulus to the transplanted cells. In experimental model, the proliferative pressure is exerted by performing partial hepatectomy, causing reperfusion injury, inducing Fas-mediated apoptosis of host hepatocytes or administering pharmacological doses of thyroid hormone.[10,11] For preventing the proliferation of host hepatocytes, a plant alkaloid, retrorsine, which blocks the hepatocyte cell cycle has been utilized.[12] However, as retrorsine is potentially carcinogenic, X-irradiation of the liver has been evaluated as a part of the preparative regimen.[13,14] In initial studies, the livers of recipient dipeptidylpeptidase IV-deficient rats were irradiated (50 Gy) and the rats were subjected to partial hepatectomy. Following transplantation of hepatocytes from congeneic normal rats (~0.1% of the rat liver hepatocyte mass), the engrafted cells proliferated preferentially, forming small clusters in 3 weeks, and replacing nearly all host hepatocytes in 3 months (Fig. 1). Transplantation of normal hepatocytes from congeneic donor rats into Gunn rats that had undergone partial hepatectomy and preparative hepatic irradiation resulted in massive repopulation of the liver by the normal cells (Fig. 2), resulting in complete normalization of serum bilirubin levels.[15] Interestingly, hepatocyte transplantation completely prevented radiation-induced morphological injury of the liver and the synthetic function of the liver was fully retained in the massively repopulated liver.[14] Efforts to stimulate the proliferation of the transplanted hepatocytes by non-surgical means, such as controlled expression of FasL in the liver by gene transfer are underway.[16]

Clinical Experience with Hepatocyte Transplantation for Inherited Metabolic Disorders

Patients with CN1, ornithine transcarbamylase (OTC) deficiency and α_1-antitrypsin deficiency have been transplanted with allogeneic hepatocytes. Two children with OTC deficiency showed significant evidence of the enzyme activity. However, one of these patients died shortly from hyperammonemia. In the other, hyperammonemia was corrected for a 10-day period, but recurred after that time, and liver transplantation was required for correction of the metabolic abnormality (I.J. Fox, personal communication). Direct evidence of function of transplanted hepatocytes was obtained in a CN1 patient (UGT1A1 deficiency).[17,18] Allograft rejection was prevented with standard Tacrolimus and prednisone therapy. After transplantation,

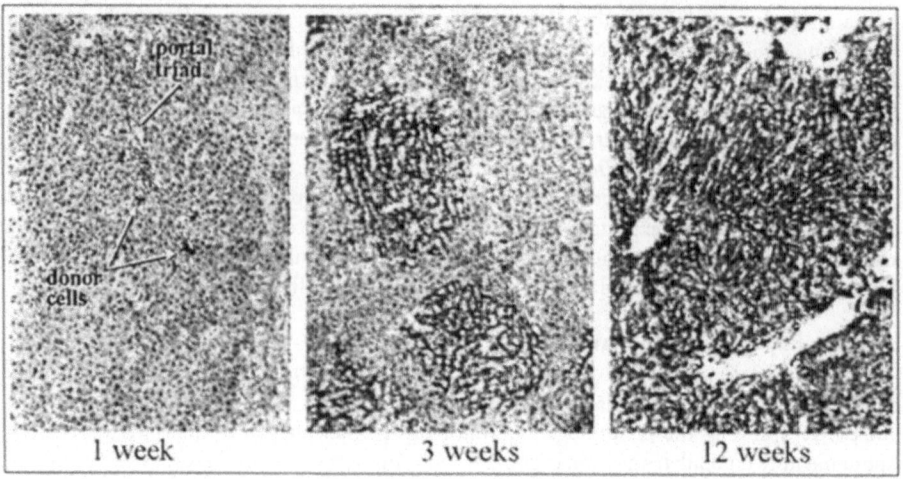

Figure 1. Massive repopulation of the liver by preparative irradiation: DPPIV-deficient F344 rats received preparative hepatic irradiation (50 Gy) and 66% hepatectomy, followed by transplantation of 2 x 10^6 hepatocytes from congeneic DPPIV-positive F344 rats. Liver biopsies performed 1, 3 and 12 weeks after the transplantation show progressive and massive replacement of the host hepatocytes with the transplanted cells, the plasma membranes of which are positively stained for DPPIV activity.

the serum bilirubin concentration declined to half of the pretransplantation level and UGT1A1 activity in a liver biopsy specimen increased to ~5% of normal. Although the procedure did not cause portal hypertension or other significant complications, but the extent of replacement of hepatic UGT1A1 activity was not sufficient to eliminate the need for phototherapy.

Hepatocyte Transplantation for Acute Liver Failure

As the liver architecture usually remains normal in acute liver failure and the liver can, potentially, regenerate, hepatocyte transplantation provides a logical approach to provide temporary metabolic support to buy time for recovery of the liver. Hepatocyte transplantation dramatically improves the survival of rodents with acute liver failure induced by hepatotoxins, liver ischemia, or 90% hepatectomy.[19,20] In pigs with ischemic acute liver failure, hepatocyte transplantation prevented the development of intracranial hypertension.[21] However, hepatocyte transplantation in patients with acute liver failure has provided beneficial results only when used as a "bridge" while awaiting liver transplantation.

In initial clinical studies on patients with fulminant liver failure, human fetal hepatocytes were injected into the peritoneal cavity, resulting in improved survival in patients with grade III encephalopathy.[22] However, since hepatocytes injected into the peritoneal cavity with a supporting scaffold have not been found to survive long-term, in subsequent studies, hepatocytes have been transplanted at other sites. Most trials have used adult hepatocytes. In one study, five out of seven transplant candidates who received hepatocyte infusion through the splenic artery survived to undergo subsequent liver transplantation.[23] In clinical trials on patients who were not candidates for liver transplantation, hepatocytes infusion into the portal vein complications were not serious, but included pulmonary embolization with the hepatocytes, transient hemodynamic instability and sepsis.[24-26] In some recipients, engrafted hepatocytes were demonstrated within the spleens. Improvement in ammonia, prothrombin time, encephalopathy, cerebral perfusion pressure and cardiovascular stability has been reported anecdotally. Although these studies suggested that the hepatocyte transplantation had provided some benefit, unequivocal evidence of function of the engrafted cells has been difficult to obtain, probably because relatively small numbers of hepatocytes had been transplanted.

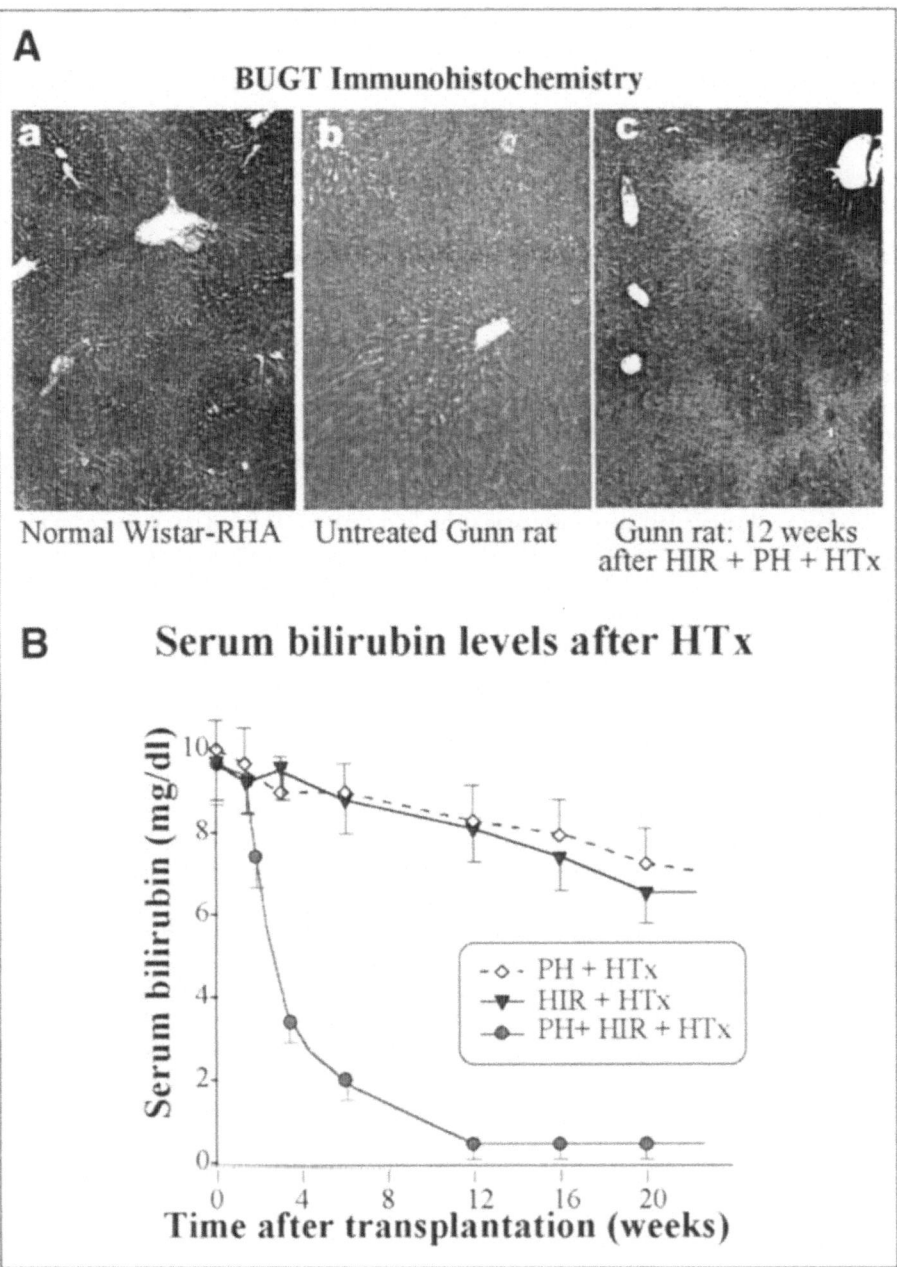

Figure 2. Repolulation of the Gunn rat liver by engrafted normal hepatocytes: Gunn rats were subjected to preparative irradiation of the liver and partial hepatectomy as in Figure 1 and were transplanted 2 x 10[6] hepatocytes from congeneic normal Wistar rats. 2A. (Reproduced from ref. 15 with permission). Immunohistochemistry of a normal Wistar RHA rat liver, an untreated Gunn rat liver and a Gunn rat liver 12 weeks after hepatocyte transplantation. 2B, Serum bilirubin levels (means + standard deviation) at various time points after transplantation subsequent to the following preparative regimens as marked in the legend box: 70% hepatectomy alone; hepatic irradiation alone; hepatic irradiation plus partial hepatectomy.

A part of the difficulty in translating the results of animal studies on acute liver failure to clinical application is that in contrast to the situation in animal models, liver regeneration is slower in patients with acute liver failure.[27] The improvement in survival in experimental acute liver failure does not always imply engraftment, since injection of hepatocyte lysates or bone marrow cells also improve the animals. Recent studies on a mouse model of acute liver failure that more closely resembles the clinical situation suggest that repeated hepatocyte infusions may be required for improved patient survival.[28]

Hepatocyte Transplantation for Chronic Liver Failure

Animal Studies

Efficacy of hepatocyte transplantation has been evaluated mostly in animal models of hepatic encephalopathy induced by portacaval shunting. In rats with end-to-side portacaval shunts, intrasplenic transplantation of hepatocytes improved behavioral score and amino acid imbalance,[29] and prevented ammonium chloride-induced hepatic coma.[30] The engrafted hepatocytes proliferated in the spleen, forming structures resembling hepatic chords.

Hepatocyte transplantation has also been shown to improve metabolic abnormalities and to prolong survival in rats with decompensated stable liver cirrhosis, induced by chronic administration of phenobarbital and carbon tetrachloride.[31] Hepatocyte transplantation into cirrhotic livers causes a severe and prolonged increase in portal pressure.[32] Although some transplanted hepatocytes migrate into cirrhotic nodules after intraportal infusion, it unlikely that enough cells can engraft to significantly improve liver function. It is also uncertain whether the transplanted cells could function within the cirrhotic nodules. However, hepatocytes engrafted in the spleen have been shown to improve metabolic function in rats with stable end-stage cirrhosis.

Clinical Experience

In an early clinical trial on ten patients with liver cirrhosis, 10-600 million hepatocytes were harvested from the left lateral liver segments of the patients and were injected into the spleen, or infused into the splenic artery or the portal vein. In one patient, ascites and encephalopathy resolved and hepatocytes were detected in the spleen by ^{99m}PMT-radioisotope uptake 11 months after transplantation.[33] These studies suggested that hepatocytes within cirrhotic nodules could function when present in a different environment. In subsequent studies, up to ten billion allogeneic hepatocytes were infused into the splenic artery of patients with decompensated chronic liver disease. Nuclear scanning demonstrated the presence of engrafted hepatocytes in the spleen. The procedure was well tolerated and was associated with improvement in encephalopathy, hepatic protein synthesis and renal function.[26] Unfortunately, the improvement of liver function following hepatocyte transplantation in chronic liver failure patients has been relatively poor, compared with that observed in animal models. The difference could have resulted from the fact that in animal experiments, hepatocytes were injected into the splenic pulps, whereas in most of the clinical studies the cells were infused into the splenic artery.[26] Hepatocytes infused into arteries of rats are lost rapidly.[34]

Current Issues

Currently, less than 30% of the hepatocytes transplanted into the liver engraft. Since the number of hepatocytes that can be transplanted at one time is limited, better elucidation of the mechanism of hepatocyte engraftment is needed to augment the survival of the transplanted cells, thereby reducing the need for repeated cell transplantation. As most of the high quality donor livers are utilized for whole organ transplantation, it has been difficult to obtain well preserved livers for harvesting hepatocytes. Cryopreservation of high quality hepatocytes could provide ready access to the cells and could permit repeated hepatocyte transplantation from a single donor. At this time, however, cryopreserved hepatocytes have not been shown to engraft as well as fresh hepatocytes.[35]

Current technology does not permit long-term expansion of primary hepatocytes in vitro. Only the hepatocytes that have been immortalized by gene transfer are capable of long-term growth and correcting metabolic abnormalities in liver failure after transplantation.[30,36] Xenogeneic cells could offer a soluion to the scarcity of human donor hepatocyts, but such cells face immunological processes that are different from that following allotransplantation.[37,38] So far, the short survival of xenografts has precluded the determination as to whether organs or tissues from another species will functionally substitute for human organs.[39] Transmission of infectious agents through xenogeneic liver cells also remains a potential concern.[40] On the other hand, it is possible, that engrafted xenogeneic hepatocytes could be resistant to recurrence of infection by human-specific viruses.

Management of immunosuppression remains a special problem for hepatocyte transplantation, because in many cases detection of early rejection can be difficult. In some cases, continued functioning of the engrafted hepatocytes can be determined by direct biochemical analysis, e.g., analysis of pigments excreted in bile in the case of CN1. In other situations, viability of the transplanted cells may be difficult to determine despite repeated liver biopsies, because of the non-uniformity of the distribution of the transplanted cells in the liver. New non-invasive approaches to the detection of the transplanted hepatocytes are needed.

Liver-Directed Gene Therapy

Potential Indications

Liver-directed gene therapy is being contemplated for both inherited disorders and acquired conditions, such as infectious and neoplastic diseases, cirrhosis of the liver and immune rejection of transplants. Missing gene products causing inherited diseases could be replaced by transferring genes expressing those proteins. In other cases, specific genes could be overexpressed for therapeutic purposes, such as the overexpression of metalloproteases for the treatment of cirrhosis. In some situations, such as CN1 or low-density lipoprotein receptor deficiency (familial hypercholesterolemia), the missing gene product must be expressed in the liver for appropriate metabolic effect. In other cases, the large size of liver and secretory characteristics of the liver could be exploited to generate "biodrugs" for export out of the hepatocytes. Such proteins include coagulation factors, hormones or vaccines. In some cases, proteins that are normally expressed at extrahepatic sites could be expressed in the liver for specific purposes. For example, the catalytic subunit of the apolipoprotein B mRNA editing enzyme (APOBEC-1), which is normally expressed in the intestinal epithelial cells, could be expressed ectopically in the liver to switch the hepatic apolipoprotein production from apo B100 to apo B48, thereby reducing the production of low density lipoproteins.[41] Similarly, hepatic expression of PDX, a homeobox protein that is responsible for pancreatic differentiation, could result in the secretion of insulin from hepatocytes.[42] In other cases, nucleic acids may be used to inhibit the expression of deleterious proteins, such as viral proteins or mutant α_1-antitrypsin. This could be accomplished by transfecting synthetic antisense RNAs, ribozymes[43] or inhibitory double-stranded RNAi.[44,45] Genes expressing antisense RNAs, RNAi, ribozymes or dominant negative proteins[46] could be transferred into the liver. Finally, technologies are being developed to correct mutations within the endogenous genes in intact organisms. This could be accomplished by targeted replacement of the defective gene,[47] or by site-directed correction of a target genomic sequence.[48] These newer strategies permit the repaired gene to remain under the control of the endogenous promoter, whereby physiological regulation is retained. A partial list of inherited and acquired disorders targeted for liver-directed gene therapy is provided in Table 1. Gene therapies for cancer and infectious diseases of the liver pose some special problems, and have been discussed separately.

Gene Therapy for Cancer

Nucleic acid transfer approaches are being designed to kill the tumor cells or inhibiting their growth, reduce tumor blood supply, evoke anti-tumor immune response, or to augment

Table 1. Some liver-based disorders that are current targets for gene therapy

I. Inherited Liver-Based Disorders

- Disorders causing damage to hepatocytes or to the liver architecture:
  α_1-antitrypsin deficiency
  Glycogen storage diseases, e.g., von Gierke's disease and Pompe's disease
  Wilson's disease
  Progressive familial intrahepatic cholestasis
  Tyrosinemia
- Disorders that do not affect hepatocyte longevity or liver architecture:
  Crigler-Najjar syndrome type I
  Familial hypercholesterolemia and other lipid metabolic disorders
  Mucopolysaccharidosis VII
  Ornithine transcarbamylase deficiency
  Maple syrup urine disease
  Phenylketonuria
  Hemophilia A and B
  Oxalosis

II. Acquired Diseases

- Infectious diseases:
  Hepatitis B and C (prophylaxis and treatment), and malaria (prophylaxis).
- Liver tumors:
  Hepatomas, cholangiocarcinomas, metastatic tumors
  Extrahepatic tumors (inhibition of neovascularization)
- Cirrhosis of the liver
- Allograft or xenograft rejection

the effect of chemotherapy and radiotherapy. P53, a sentinel gene of the cell cycle, has been transferred to induce apoptosis of tumor cells.[49] The herpes simplex virus thymidine kinase (HSV-TK), which converts a prodrug, ganciclovir, to toxic ganciclovir phosphate,[50] or cytosine deaminase and purine nucleoside phosphorylase (which converts fludarabine to a diffusible toxic metabolite)[51] have been expressed as a "suicide gene" in an attempt to ablate tumors. Ganciclovir phosphate may diffuse to neighboring cells, killing them by "by-stander effect", extending number of tumor cells killed. To reduce the toxic effect on normal cells, the suicide genes have been targeted to tumor cells by tagging the DNA to monoclonal antibody directed at cell surface proteins that are preferentially expressed in tumor cells, such as AF-20, a 180-kDa tumor-specific cell surface glycoprotein, expressed in hepatoma cell lines.[52] Another strategy to increase the therapeutic ration is to use tumor-specific promoters (e.g., alpha-fetoprotein or carcinoembryonic antigen) to drive the transgene expression. A different approach uses E1B-mutant adenoviruses that are replicate preferentially in P53-deficient tumor cells,[53] although P53 deficiency may not be always required for the replication of these mutant adenoviruses.

Since killing the tumor cells by topical expression of toxic gene products does not eliminate tumors at metastatic sites, efforts are being made to inhibit neovascularization, that is needed for both primary and metastatic tumor growth, by the expressing angiostatin or endostatin genes.[54,55]

Host immune response against tumor antigens is important in eliminating tumor cells that are often present as micrometastasis. Expression of tumor-specific "neoantigens" may evoke this necessary immune response, provided they are presented properly to the immune system. Tumor antigens, such as those from melanoma cells, have been used for DNA-based tumor vaccination. Genetic manipulation of antigen-presenting cells may augment the host immune response against tumor cells. Since cytokines, such as TGFβ or IL10, secreted by large tumors, suppress immune response[56] "debulking" the tumor by surgery, radiotherapy, chemotherapy or gene therapy may enhance the immune response.

Currently, the greatest potential of gene therapy for cancer is an adjunct to chemotherapy or radiotherapy. Irradiation of tumors augments tumor cell transduction using recombinant viruses. Transgene expression could be driven by radiation inducible promoters.[57] On the other hand, inhibition of the expression in the tumor cells of radioprotective proteins, such as ATM ("mutated in ataxia telangiectasia"), could increase radiosensitivity of the tumors.[58]

Gene Therapy for Infectious Diseases

Nucleic acid-based approaches are being explored both for prophylaxis and treatment of hepatic viral infections. DNA vaccination[59] can be safer and cheaper by eliminating the possibility of vaccine contamination infectious agents from tissue culture. The injected DNA itself can enhance the immunoreactivity of the expressed antigen. Expression of the antigenic peptides directly in antigen presenting cells could augment the immunoresponse by improving presentation. Expression of specific cytokines can be used to promote the proliferation of antigen presenting cells, as well as the maturation of T cells to helper T cells. Gene transfer is also being used to express single chain antibodies or antibody fragments in hepatocytes to make them resistant to viral infections.[60]

Gene therapy can also be used to interfere with the viral life-cycle by introducing synthetic antisense RNAs, ribozymes, or DNA ribonucleases (see below). Ribozymes, antisense RNAs or dominant negative proteins[61] can also be expressed within the target cells by gene transfer

Gene Therapy for Inherited Disorders

As many inherited disorders are caused by the abnormality of a single gene, the effect of gene therapy can be evaluated directly and precisely in these conditions. For this reason, rare single gene abnormalities continue to be important targets of liver-directed gene therapy. Table 1 contains a partial list of inherited disorders that are currently targets of gene therapy.

Methods of Gene Delivery to the Liver

Genes may be delivered to the liver by systemic administration, infusion into the portal vein, hepatic artery or bile duct, or direct injection into the liver. Alternatively, hepatocytes isolated from the liver, or immortalized liver cells can be transduced with a therapeutic gene and then transplanted into the liver as a part of an "ex vivo" gene therapy approach. Commonly used methods for delivering genes to the liver of intact organisms, such as those using replication-deficient recombinant viruses, or non-viral vehicles are briefly discussed below. Frequently used methods of gene transfer to the liver are listed in Table 2.

Vectors Based on Recombinant Viruses

The attainable infectious titer, ability of the virus to infect non-dividing cells, integration into the host genome, repeatability of administration and safety of the vector system are important considerations in selecting recombinant viruses for gene therapy for specific diseases.

Retrovirus-Based Vectors

Proviral DNA, complimentary for the RNA genome of retroviruses, is integrated into the host genome and are transmitted to the progeny of the transduced cells.[62] Tumor retroviruses, such as, Moloney's murine leukemia virus (MoMuLV), have been used extensively for generating recombinant vectors. The viral structural genes are replaced by target transgenes, leaving packaging signal (ψ) intact.[63] The viral proteins are provided in *trans* by "packaging cell lines", generating a replication-deficient recombinant virus. The envelope protein determines the range

Table 2. Advantages and limitations of liver-directed gene therapy methods

Method	Integration and Persistence	In vivo Gene Transfer Efficiency to Liver	Liver-Specificity	Immunological and Other Issues
VIRAL VECTORS:				
Leukemia type retroviruses	Integration is required.	Requires cell division. Low efficiency in quiescent cells, e.g., hepatocytes.	None	Non-immunogenic. Difficult to obtain very high titers. Envelope can be pseudotyped with other viral proteins to increase stability and broaden host range.
Immunodeficiency type retroviruses	Integration is required.	Can infect non-dividing cells, but cell cycling may be needed for hepatocytes. Intermediate to high efficiency for liver.	None	Non-immunogenic. Difficult to obtain very high titers. Envelope can be pseudotyped with other viral proteins to increase stability and broaden host range.
Adeno-associated virus	Both episomal and integrated forms persist in the nucleus.	Can infect both quiescent and dividing cells. Low to intermediate efficiency for liver.	None	Causes humoral immune response. Can be grown at high titers. Site-specificity of integration of the wild type virus is lost in the absence of *rep*. Can undergo lytic cycle in the presence of helper proteins. Limited packaging space.
Simian virus 40	Integrates into the host genome	Infects both quiescent and dividing cells. High efficiency.	None	No significant immune response. Can be grown at high titers. Limited packaging space.
Adenovirus	Episomal. Transgene is expressed for several months in the absence of host immune response.	Efficiently infects both dividing and non-dividing cells. Very high efficiency for liver upon systemic administration.	Liver targeted in rodents and mice, but not in humans	Evokes both antibody and cell-mediated immune response. Deletion of viral genes reduces primary immunogenicity, but does not permit repeated injection. Host tolerization or coexpression of immunomodulatory genes permits repeated gene transfer.
Hybrid viruses	Combines the advantages of different viruses.	Efficiency of adenoviral vectors is combined with the integrating proprties of other viruses.	Can be liver targeted	Adenoviral capsid proteins may evoke immune response. Site-specific integration or episomal replication may be possible.

continued on next page

Table 2. Advantages and limitations of liver-directed gene therapy methods (continued)

Method	Integration and Persistence	In vivo Gene Transfer Efficiency to Liver	Liver-Specificity	Immunological and Other Issues
NON-VIRAL VECTORS:				
Naked DNA or RNA injected into the liver or injected i.v. in a high volume to cause hepatic congestion	Transient	Intermediate efficiency	Can be liver targeted	Immune response to the transfected DNA is possible.
Transposon-based integration	Permanent	Low to intermediate efficiency	Can be liver targeted	Immune response to the transfected DNA is possible.
Receptor and/or liposome-mediated plasmid delivery	Transient	Intermediate efficiency	Can be liver targeted	Probably non-immunogenic
Oligonucleotides designed to correct mutations: triplex forming oligonucleotides, RNA-DNA chimera, single-stranded oligonucleotides.	Permanent	Low efficiency	Liver targeted	Non-immunogenic

of species infectable by the recombinant retrovirus. The host range can be expanded by engineering proteins from other viruses, such as the G-protein of the vesicular stomatitis virus (VSV) into the retroviral envelope. After the recombinant virus enters the cell, the RNA genome is reverse-transcribed into double-stranded DNA provirus, which is transported to the nucleus and is integrated into the host genome. For the tumor viruses, this process requires resolution of the nuclear envelope that occurs during cell division. Therefore, the tumor retroviruses are inefficient in infecting primary hepatocytes, which undergo mitosis infrequently. Lentiviruses, which are retroviruses of the immunodeficiency group, form preintegration complexes that can be translocated into non-dividing nuclei. Therefore, recombinant lentiviruses are being developed for liver-directed gene therapy. However, it has been suggested that recombinant lentiviruses may need the hepatocytes to be in cell cycle for efficient gene transfer in vivo.[64]

Recombinant Adenovirus

Adenovirus types 5 and 2, are large linear double-stranded DNA viruses that are commonly used for preparing gene transfer vectors. Recombinant adenoviruses can transfer genes into both dividing and quiescent cells with high efficiency. Recombinant adenoviruses localize preferentially to the liver after intravenous administration in rodents[65]. However, clinical trials in human subjects indicate transduction of human liver cells may not occur with such high efficiency.[66] It is unclear whether this difference is based on the density of the adenoviral receptor, CAR, on the cell surface of hepatocytes of different species.

Adenoviral vectors are generated by disruption of the early region-1 (E1) by the insertion of the transgene. The E1 region encodes transcription factors required for the expression of adenoviral genes, and its disruption markedly inhibits the expression of the viral proteins. The recombinant virus is generated in packaging cells that provide the E1 gene products in trans. To abolish the possibility of viral gene expression and to increase the space available for insertion of the transgene, all structural genes of the virus have been deleted from the vectors.[67] These "gene-deleted" vectors require helper adenoviruses to provide the structural proteins.

Cellular and humoral immune responses to adenoviral proteins limit the clinical application of adenovectors. In humans, antibodies are memory cells often exist from previous infections by the adenoviruses, and naïve individuals readily develop humoral and cell-mediated immunity against the viral antigens after the initial injection of the recombinant adenovirus. Neutralizing antibodies block gene transfer following subsequent administrations of the vector. Adenovirus-specific cytotoxic lymphocytes attack the host cells that are infected by the recombinant virus, resulting in liver damage and rapid loss of the transgene after secondary gene transfer.[68] The helper-dependent, adenovectors, in which all viral genes deleted, retain their immunogenicity because of the viral proteins provided in trans by the packaging cells.[69] Although these vectors express transgenes for a longer duration than do the first generation adenoviral vectors, secondary or tertiary administration fails to transfer the transgenes.

Development of immune response requires presentation of antigenic peptides by antigen presenting cells. Following docking of the antigen presenting cells with uncommitted T cells, the antigen presenting cells and the T cells costimulate each other via the B7-CD28 and CD40-CD40 ligand interactions. Inhibition of the B7-CD28 costimulation prevents effective immune response. CTLA4-Ig, a soluble inhibitory protein inhibits this costimulation. Injecting CTLA4-Ig alone at the time of administration of adenovectors does not prevent antibody formation.[70] However, coexpression of CTLA4-Ig along with the target transgene permits multiple administration of the recombinant adenovirus[71] (Fig. 3). Host tolerance toward adenoviral proteins can be induced by injecting recombinant adenoviruses in utero or in newborn rats,[65] inoculating adenoviral proteins into the thymus of young adult rats,[72] or orally administering of small doses of adenoviral proteins.[73] Since wildtype adenoviruses are human pathogens, safety concerns persist regarding tolerization of humans to adenoviral antigens remain. Adenoviral proteins may have additional toxic effects. A patient with ornithine transcarbamylase deficiency died shortly after recombinant adenovirus administration. The death was thought to have resulted from massive cytokine release, rather than an immune response to the virus.

As discussed above, the hepatotropism of recombinant adenoviruses that is seen in rodents was not found in human trials.[74] Therefore, attempts are being made to modify the viral surface proteins to change their tropism, so that they can be internalized by hepatocytes through receptors that are normally present on the hepatocyte surface.[75]

Herpes Simplex Virus-1

Herpes simplex virus-1 (HSV-1) is a 150 kb double-stranded DNA virus with a broad host range[76] that can infect both dividing and non-dividing cells. However, long-term gene expression in the liver has not been achieved with the currently available HSV vectors.

Recombinant Baculovirus

The *Autographa californica* nuclear polyhedrosis virus (AcNPV), which is usually used to generate recombinant proteins in insect cells,[77] can infect hepatocytes, but not other mammalian cells and can express transgenes that are driven by mammalian or viral promoters.

Recombinant Adeno-Associated Virus

The wildtype adeno-associated virus-2 (AAV-2) is a small (4.7 kb) single-stranded DNA parvovirus that integrates preferentially on the q13.4-ter arm of human chromosome 19.[78] The AAV can remain latent for long periods, but can cause lytic infection following infection with a "helper virus," such as adeno- or herpes simplex virus, or when the cell is exposed to genotoxic stimuli, e.g., UV light or irradiation.[79] AAV type 2 is internalized following binding to a cell surface heparin sulfate proteoglycan. The 145-bp inverted terminal repeats (ITR) that flank the AAV genome are needed for integration into the host genome. Following internalization, AAV vectors may remain as episomes or may integrate into the host genome over several weeks. The viral Rep protein directs the site specificity of AAV integration into chromosome 19. Recombinant vectors that lack this gene, do not exhibit the site-specificity of integration and in fact may remain in the nucleus as episomes for prolonged periods.[80]

Infusion of high doses of the recombinant virus into the portal vein of factor IX-deficient dogs resulted in the appearance of 5% of normal levels of factor IX activity in plasma.[81] Efficiency of AAV-mediated gene transfer to tumors could be augmented when used in combination with tumor irradiation or chemotherapy.[82] Recombinant AAV causes a humoral immune response, which may be a problem if the vector needs to be readministered.

Simian Virus 40–Based Vectors

Simian virus 40 (SV40) is a non-enveloped 5.2 kb double-stranded DNA papova virus. The large (Tag) and small (tag) T antigens are transcription factors required for the expression of the viral structural genes, VP1, VP2 and VP3. The recombinant vector is generated by replacing the *Tag* genes of the viral genome by the target transgene (Fig. 4). All three structural genes can be deleted to make additional space for inserting transgenes. The recombinant virus is generated by transfecting the recombinant viral genome into a helper cells (e.g., COS cells) that provide Tag in trans.[83] Absence of the *Tag* gene makes the recombinant virus replication deficient and markedly reduces its immunogenicity. Up to 4.7 kb of exogenous DNA. can be inserted into SV40 vectors. The recombinant vector can be generated at 10^9 infectious units (IU)/ml and concentrated to 10^{12} IU/ml.[84,85] Recombinant SV40 integrates into the genome of both dividing and non-dividing cells. The range of target cells is broad and includes hepatocytes, neural cells and bone marrow cells.[83-85]

Non-Viral Vectors

Naked DNA

Injection of naked DNA into the muscle results in gene expression for several weeks. However, direct injection into the liver results in local inflammatory response and a shorter duration of expression. For experiments in mice or rats, the plasmid solution can be injected rapidly in

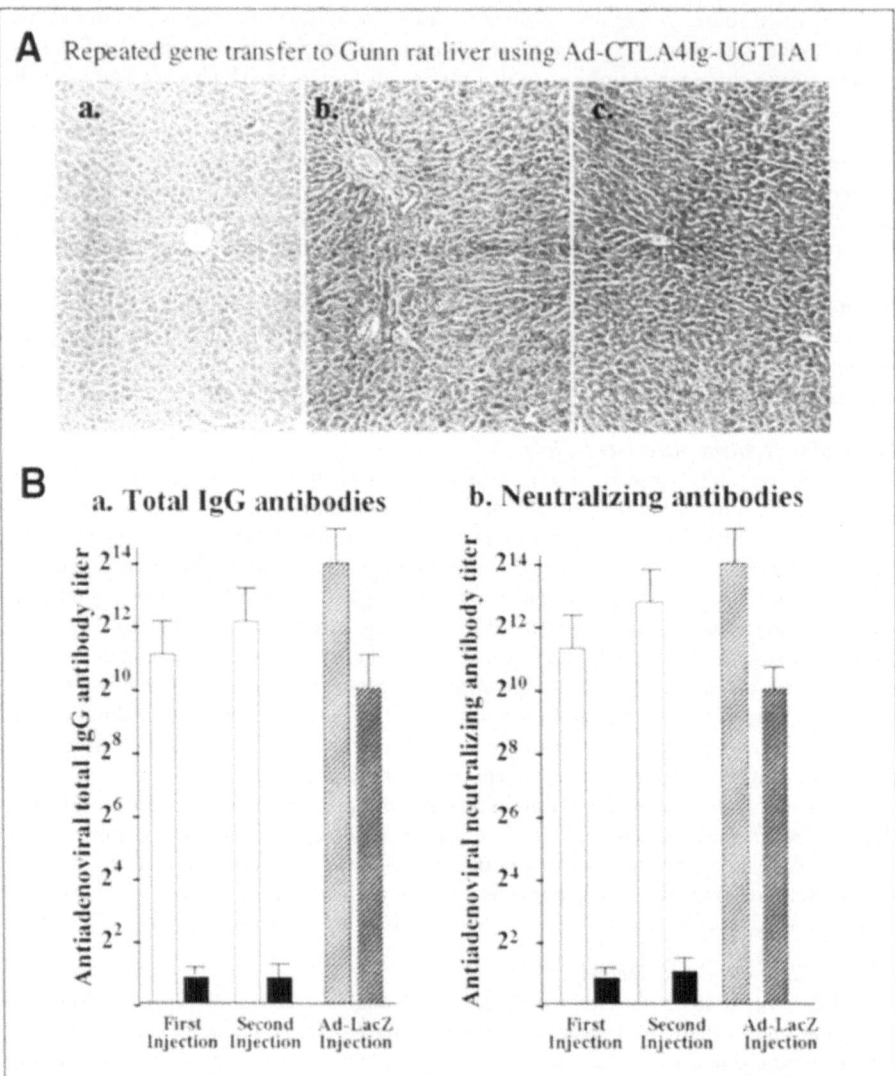

Figure 3. Adenoviral vector, coexpressing CTLA4Ig: A recombinant adenovirus, expressing both CTLA4-Ig and human UGT1A1 (Ad-hUGT1A1-CTLA4Ig) was injected intravenously into Gunn rats. 3A. Immunohistochemistry of liver sections: panel a, untreated Gunn rat; panel b, six months after the first injection; panel c, 2 weeks after the second injection (one year after the first injection). 3B. Antibody response: panel a, total IgG antibodies, panel b, neutralizing antibodies. Open bars, a first generation adenovector, expressing only UGT1A1 (Ad-hUGT1A1); solid bars, Ad-hUGT1A1-CTLA4Ig. After two injections of Ad-hUGT1A1-CTLA4Ig (black background, white hatch), or Ad-hUGT1A1 (white background, black hatch), a third injection of a first generation adenovector, expressing *E. coli* β-glucuronidase (Ad-LacZ) was administered. Note that Ad-LacZ caused an immune response in both groups, indicating that CTLA4-Ig caused immune ignorance, but did not tolerize the host permanently to adenoviral proteins. Cytotoxic lymphocyte response: the rat groups and injections were identical to those in 3B. Cytotoxicity of lymphocytes of the treated rats against adenovirally infected hepatocytes was measured by ALT release from the hepatocytes in the media (panel a) or the secretion of interferon-gamma into the media by the lymphocytes. 3D. Serum bilirubin levels after injection of the first generation virus, Ad-hUGT1A1 (interrupted line) or Ad-hUGT1A1-CTLA4Ig (solid line).

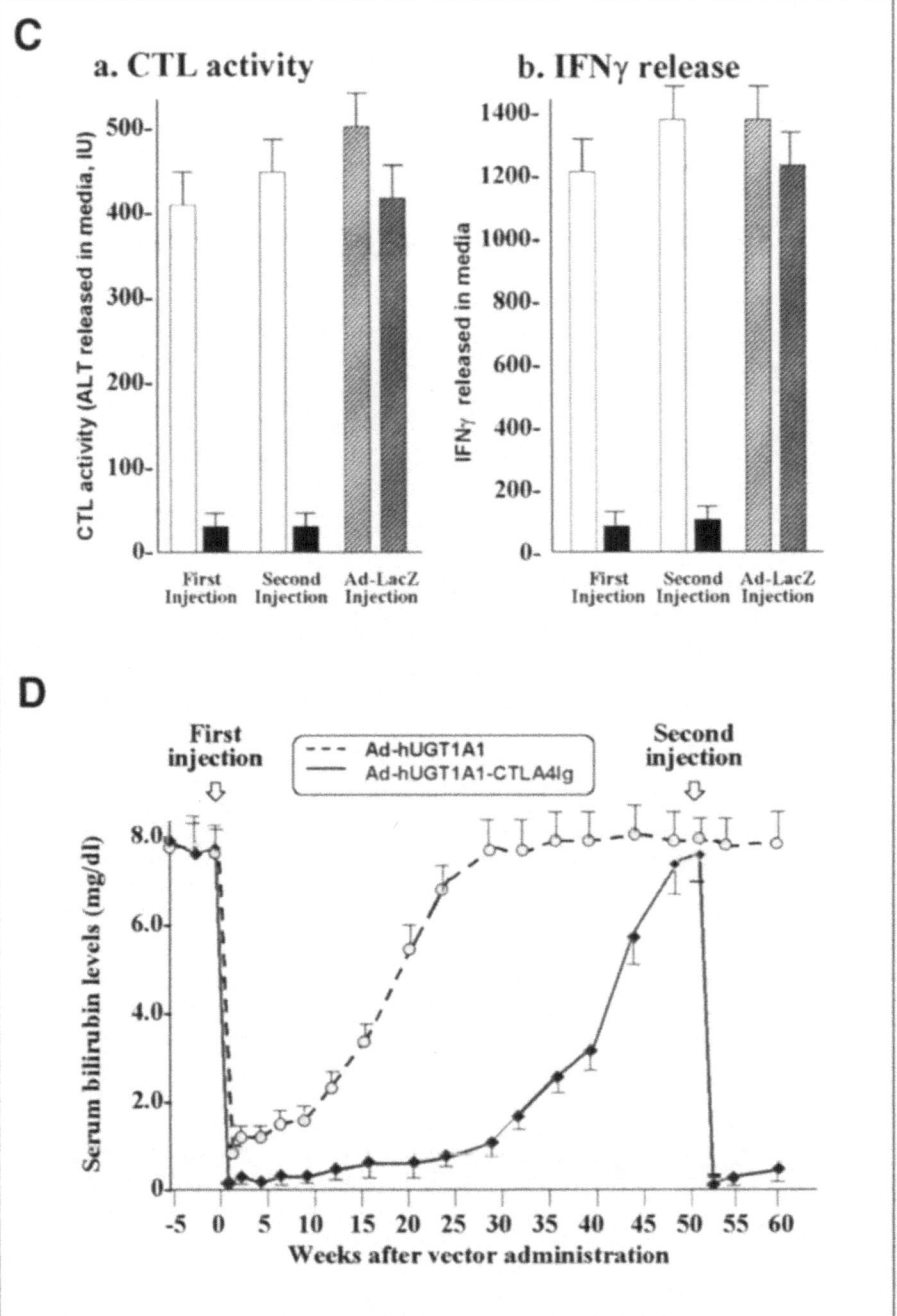

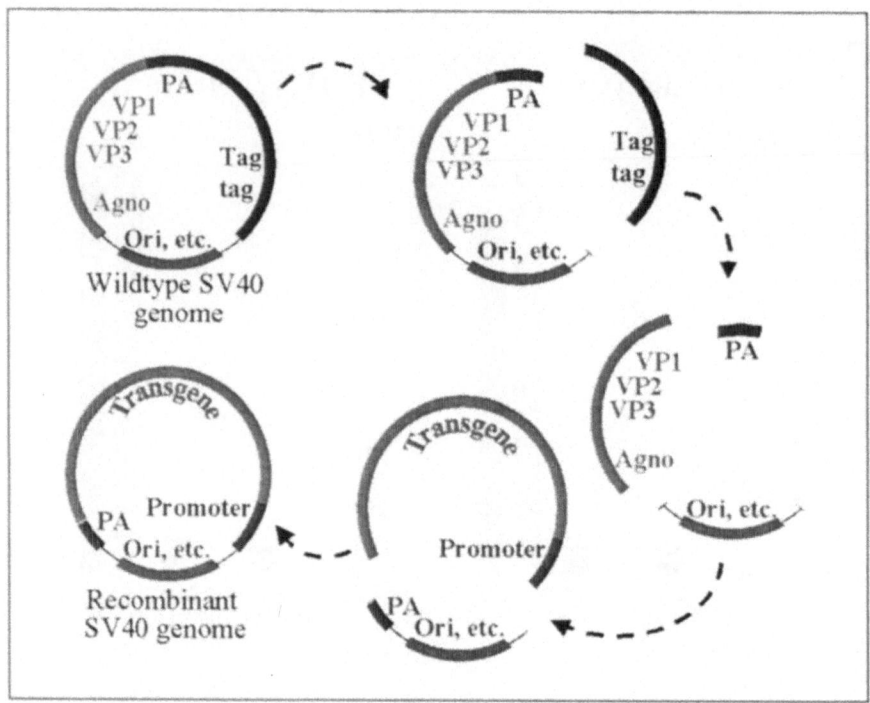

Figure 4. Generation of a recombinant SV40 vector. The wildtype viral genome is a circular DNA, containing the origin, promoter and the viral assembly signal, the coding region for the transcription factors large (Tag) and small (tag) antigens, the coding regions for the regulatory protein, Agno and the capsid proteins, VP1, VP2 and VP3 and the polyadenylation site (PA). The coding sequences for all viral proteins are removed and replaced by the transgene, driven by an internal promoter. The SV40 promoter is inactivated by placing a transcription termination signal downstream to it. The recombinant SV40 genome is transfected into COS cells that provide the viral proteins to generate the recombinant, replication defective virus.

a large volume, causing transient hepatic vascular congestion, which increases the transfection efficiency.[86] As the injection of simple plasmids results in only short-term gene expression, this approach has little potential for clinical use. Modification of the plasmid for enhancing integration using a transposon-based mechanism has been discussed later.

Non-Viral Vectors

Non-viral vectors can be composed of lipid microcapsules (liposomes) or lipid-nucleic acid complexes ("lipoplex"), polycations, e.g., poly-L-lysine (PLL), polyethylenimine (PEI), polyglucosamines, lipopolyamines or cationic peptides, or polycation-lipid hybrids ("Lipopolyplex"). Lipopolyplexes are formed by compacting DNA constructs by addition of polycations, which may be then encapsulated into liposomes, forming particles that are smaller than liposomes and are better protected from nuclease degradation. Cationic lipid transfecting agents (e.g., lipofectamine, DOTAP), and cationic polymers, (e.g., poly-L-lysine and PEI) are used commonly for in vitro gene transfer,[87] but are toxic in vivo or inactivated by plasma. For targeted gene delivery to hepatocytes, transfer vehicles have been modified to include ligands for the hepatocyte-preferred asialoglycoprotein receptor (ASGPr). Galactose-terminated peptides (e.g., Asialoorosomucoid or asialofetuin) or galactose can be conjugated to polylysine or PEI for targeted delivery to the liver.[88] Liposomes composed of galactocerebrosides have also been used for receptor-mediated nucleic acid delivery to the liver.[89]

Inhibition of Gene Expression: Ribozymes, Antisense RNA, RANi and DNA Ribonucleases

Ribozymes are RNA enzymes that hybridize to complementary RNA sequences and catalyze endoribonucleolytic cleavage, causing rapid degradation of the RNA molecule.[90] For hairpin ribozymes, the presence of a guanosine residue immediately downstream to the cleavage site is essential, but the effect is enhanced by a GUC sequence. Hammerhead ribozymes require only a UA, UC or UU dinucleotide for cleavage. Ribozymes can be synthesized and packaged for cellular uptake, or expressed in cells from transfected DNA. Recently, double stranded small ribonucleic acid molecules have been used effectively to degrade mRNA molecules and dramatically down-regulated the expression of specific genes. These inhibitory RNA molecules can be synthesized in vitro or expressed as hairpin molecules by transferring DNAs.

Hammerhead[90] and hairpin[91] ribozymes have been used to inhibit the production of hepatitis B and hepatitis C viruses in cultured cells. Following transfection of HuH7 cells (a differentiated human hepatoma line) with antisense oligonucleotides using ASGPr-polylysine complexes, the cells became resistant to HCV-directed protein synthesis.[91] Multiple sites within a viral RNA genome can be targeted simultaneously using a series of ribozymes expressed from a single vector, thereby inhibiting the development of drug-resistant mutants.[92] Since Hepatitis B virus replicates via a pregenomic RNA intermediate, it could also be a target for ribozyme-based therapy.[90]

DNA ribonucleases are synthetic single-stranded DNAs consisting of a 15-nucleotide catalytic domain, flanked by two RNA-binding domains. DNA ribonucleases cleave RNA with even more efficiency than do ribozymes.[93] DNA ribonucleases have been reported to specifically cleave the hepatitis C viral genome[93,94] and the RNA intermediates of hepatitis B.[95]

Homologous Recombination

In theory, site-specific recombination of DNA could be used to repair mutated or damaged DNA. Several proteins, including homologs of yeast recombination proteins rad51 and/or rad52, are important in homologous recombination in higher eukaryotes[96]. Homologous recombination, which has been used extensively in cultured cells, has been rarely successful in vivo.[96] Since homologous recombination is cell cycle-regulated, it occurs rarely in quiescent cells, such as hepatocytes.[97]

Triplex DNA

This method is based on generating a triple-chain DNA by the binding of a single stranded nucleic acid to the major groove of a homopurine region of the double stranded chromosomal DNA.[98] For efficient triplex formation, the polypurine regions needs to 12-14 nucleotide long and guanine-rich. Cross-linking agents, such as psoralen or other mutagens can be attached covalently to the oligonucleotide.[98] After intercalation of the psoralen at the target $5'ApT^{3'}$ site, UV-irradiation causes cross-linking of the thymines in the two strands. This substrate is then repaired by endogenous cellular DNA repair mechanisms, producing the characteristic T:A to A:T transversions. Single strand DNAs with cross-linking agents may also promote insertions and deletions via the excision repair pathway.[99] Bifunctional oligonucleotides containing triple-helix-forming sequences, as well as conventional Watson and Crick base pairs have been used in culltured cells to cover a broader range of genomic sequences than is possible with conventional triple-helix forming oligonucleotides.[100]

Single Nucleotide Modification

This method utilizes the cellular mismatch repair machinery to correct genomic mutations. A synthetic oligonucleotide, complimentary to the targeted genomic DNA, except for a single mismatched base is synthesized. Most successful studies have utilized a DNA/RNA chimera. The RNA component, consisting of 2'-O-methyl ribonucleic acid residues, is included to increase the strength of hybridization with the targeted genomic DNA sequence. The mismatch between the oligonucleotide and the genomic DNA triggers the mismatch repair enzymes of the cell, permanently correcting the mutation. Plasmid mutation studies in *E. coli* strains deficient in specific repair proteins,[101] indicated that this process required both RecA (an enzyme needed for homologous recombination) and MutS (a mismatch repair protein). Recombination and mismatch repair pathways are evolutionarily conserved and HuH-7 cell extracts, containing the mismatch repair protein hMSH2 can complement MutS-deficient E. coli in converting the mutant aminoglycoside resistance gene. Wild-type P53 may inhibit the initial pairing step in this repair process by inhibiting RecA and its human homolog Rad51.[102]

In intact rats, RNA-DNA chimeric oligonucleotides were delivered to the liver by receptor-mediated endocytosis, using lactosylated PEI or glactocerebroside-containing liposomes. The method has been shown to result in a Ser365Arg conversion of the rat factor IX gene[103] and insertion of a guanosine base, which is deleted in the bilirubin-UDP-glucuronosyltranferase (*UGT1A1*) gene in jaundiced Gunn rats.[104] In the latter case, expression of the wildtype functional enzyme was demonstrated and serum bilirubin levels were reduced significantly.[104] Tagalakis and associates have demonstrated a high level of conversion of the gene expressing human apolipoprotein E2 in a transgenic mouse. The RNA-DNA chimera has also been used successfully for site-directed gene conversion of the apolipoprotein A2 gene in human lymphocytes,[105] the nonsense mutation of carbonic anhydrase II in nude mouse primary kidney tubular cells[106] and the missense mutation in the tyrosinase gene in albino mouse melanocytes.[107]

Transposon-Based Gene Delivery

Transposons are ubiquitous elements in eukaryotic genomes that can move from one location to another by a "cut-and-paste" mechanism. Emmons and associates identified transposable elements in fish genomes that had homology with the Tc1 transposon of the nematode *Caenorhabditis elegans*.[108] These transposons of the Tc1/*mariner* superfamily were found to have short inverted repeats, flanking the coding sequences of a transposase. Hackett and associates generated a transposable element from an ancestral Tc1-like fish transposon that had acquired multiple mutations during evolution. This transposon was termed *"Sleeping Beauty"*, because it had remained dormant for millions of years and was made functional by correcting mutations in vitro. It is a 1.6-kb element, flanked by 250-bp terminal inverted repeats. DNA sequences, flanked by the terminal inverted repeats can efficiently integrate into mammalian cellular genome, in the presence of the transposase protein. The integration occurs at TA dinucleotide sites, which are duplicated upon insertion of the transposable element.[109] The Sleeping Beauty transposon system has been used to effect integration of the coding region of human factor IX into the chromosomes of factor IX-deficient hemophilic mouse hepatocytes.[110]

Acknowledgments

This work was upported in part by NIH grants: RO1-DK 46057 (to JRC), RO1-DK 39137 (to NRC), the Liver Research Core Center grant DK-P30-41296 (Director, David A. Shafritz), The Gene therapy Core of the Institute of Human Genetics of Albert Einstein College of Medicine and an American Cancer Society grant, RPG-00-066-01-CCE (to CG). The authors acknowledge the important contributions by many investigators in the fields of hepatocyte transplantation and liver-directed gene therapy that were not cited in this limited review.

References

1. Moshage HJ, Rijntjes PJ, Hafkenscheid JC, et al. Primary culture of cryopreserved adult human hepatocytes on homologous extracellular matrix and the influence of monocytic products on albumin synthesis. J Hepatol 1988; 7:34-44.
2. Demetriou A, Levenson SM, Whiting J et al. Replacement of hepatic functions in rats by transplantation of microcarrier-attached hepatocytes. Science 1986; 233:1190-1192.
3. Demetriou A, Levenson SW, Whiting J et al. Organization, morphology and function of microcarrier-attached transplanted hepatocytes in rats. Proc Natl Acad Sci USA 1986; 83:7475-7479.
4. Gupta S et al. Permanent engraftment and function of hepatocytes delivered to the liver: Implications for gene therapy and liver repopulation. Hepatol 1991; 14:144-148.

5. Gupta S, Rajvanshi P, Lee C-D. Integration of transplanted hepatocytes in host liver plates demonstrated with dipeptidylpeptidase IV deficient rats. Proc Natl Acad Sci (USA) 92:5860-5864 (1995;

6. Eguchi S et al. Treatment of hypercholesterolemia in the Watanabe rabbit using allogeneic hepatocellular transplantation under a regeneration stimulus. Transplantation 1996; 62:588-593.

7. Irani A et al. Correction of liver disease following transplantation of normal rat hepatocytes into Long-Evans Cinnamon rats modeling Wilson's disease. Mol Ther 2001; 3:302-309.

8. Rozga J et al. Repeated intraportal hepatocyte transplantation in analbuminemic rats. Cell Transplant 1995; 4:237-43.

9. Overturf K, Al-Dhalimy M, Ou C et al. Serial transplantation reveals the stem-cell-like regenerative potential of adult mouse hepatocytes. Am J Pathology 1997; 151:1273-1280.

10. Ilan Y et al. Massive repopulation of rat liver by transplantation of hepatocytes into specific lobes of the liver and ligation of portal vein branches to other lobes. Transplantation 1997; 64:8-13.

11. Oren R et al. Role of thyroid hormones in stimulating liver repopulation in the rat by transplanted hepatocytes. Hepatology 1999; 30:903-913.

12. Laconi E et al. Long-term, near-total liver replacement by transplantation of isolated hepatocytes in rats treated with retrorsine. Am J Pathol 1998; 153:319-29.

13. Guha C et al. Liver irradiation: a potential preparative regimen for hepatocyte transplantation. Int J Radiat Oncol Biol Phys 2001; 49:451-457.

14. Guha C, Sharma A, Gupta S et al. Amelioration of radiation induced liver damage in partially hepatectomized rats by hepatocyte transplantation. Cancer Research 199959:5871-5874.

15. Guha C, parashar B, Deb NJ et al. normal hepatocyte correct serum bilirubin after repopulation of Gunn rat liver subjected to irradiation/partial resection. Hepatology 2002; 36:354-362.

16. Takahashi M, Deb NJ, Kawashita Y et al. A novel strategy for in vivo expansion of transplanted hepatocytes using preparative hepatic irradiation and FasL-induced hepatocellular apoptosis. Gene therapy 2002; In press.

17. Fox IJ et al. Treatment of Crigler-Najjar syndrome type I with hepatocyte transplantation. New Eng J Med 1998; 338:1422-1426.

18. Roy Chowdhury J, Strom SC, Roy Chowdhury N et al. Hepatocyte transplantation in humans: Gene therapy and more. Pediatrics 1998; 102:647-648.

19. Makowka L et al. Reversal of toxic and anoxic induced hepatic failure by syngeneic, allogeneic, and xenogeneic hepatocyte transplantation. Surgery 1980; 88:244-53.

20. Demetriou AA, Reisner A, Sanchez J et al. Transplantation of microcarrier-attached hepatocytes into 90% partially hepatectomized rats. Hepatology 1988; 8:1006-1009.

21. Arkadopoulos N et al. Transplantation of hepatocytes for prevention of intracranial hypertension in pigs with ischemic liver failure. Cell Transplant 1998; 7:357-63.

22. Habibullah CM, Syed IH, Qamar A et al. Human fetal cell transplantation in patients with fulminant hepatic failure. Transplantation 1994; 58:951-952.

23. Strom SC et al. Hepatocyte transplantation as a bridge to orthotopic liver transplantation in terminal liver failure. Transplantation 1997; 63:559-69.

24. Bilir B et al. Hepatocyte transplantation in acute liver failure. Liver Transpl 6:32-40, 41-43 (2000;

25. Soriano H et al. Hepatocellular transplantation (HCT) in children wiht fulminant liver failure (FLF). Hepatology 1997; 26:239A(443).

26. Strom S, Roy Chowdhury J, Fox I. Hepatocyte transplantation for the treatment of clinical disese. Seminars in Liver Disease 1999; 19:39-48.

27. Miyazaki M et al. Reversal of lethal, chemotherapeutically induced acute hepatic necrosis in rats by regenerating liver cytosol. Surgery 1983; 94:142-50.

28. Braun K, Degen J, Sandgren E. Hepatocyte transplantation in a model of toxin-induced liver disease: variable therapeutic effect during replacement of damaged parenchyma by donor cells. Nat Med 2000; 6:320-326.

29. Ribeiro J et al. Intrasplenic hepatocellular transplantation corrects hepatic encephalopathy in portacaval-shunted rats. Hepatology 1992; 15:12-18.

30. Schumacher IK et al. Transplantation of conditionally immortalized hepatocytes to treat hepatic encephalopathy. Hepatology 1996; 24:337-43.

31. Kobayashi N et al. Hepatocyte transplantation in rats with decompensated cirrhosis. Hepatology 2000; 31:851-7.

32. Gagandeep S et al. Transplanted hepatocytes engraft, survive, and proliferate in the liver of rats with carbon tetrachloride-induced cirrhosis. J Pathol 2000; 191:78-85.

33. Mito M, Kusano M. Hepatocyte transplantation in man. Cell Transplantation 1993; 2:65-74.

34. Mito M, Kusano M, Ohnishi T et al. Hepatocellular transplantation. Gastroenterol Jpn 1978; 13:480-90.

35. David P et al. Engraftment and albumin production of intrasplenically transplanted rat hepatocytes (Sprague-Dawley), freshly isolated versus cryopreserved, into Nagase analbuminemic rats (NAR). Cell Transplant 2001; 10:67-80.
36. Runge D, Michalopoulos G, Strom S et al. Recent advances in human hepatocyte culture systems. Biochem Biophys Res Commun 2000; 274:1.
37. Platt J et al. Transplantation of discordant xenografts: a review of progress. Immunol Today 1990; 11:450-456.
38. Ramirez P et al. Life-supporting human complement regulatory decay accelerating factor transgenic pig liver xenograft maintains the metabolic function and coagulation in the nonhuman primate for up to 8 days. Transplantation 2000; 70:989-998.
39. Kanazawa A, Platt J. Prospects for xenotransplantation of the liver. Semin Liver Dis 2000; 20:511-522.
40. Platt J. New directions for organ transplantation. Nature 1998; 392:11-17.
41. Greeve J, Jona VK, Roy-Chowdhury N et al. Hepatic gene transfer of the catalytic subunit of the apolipoprotein B mRNA editing enzyme, APOBEC-1, leads to reduction of LDL in normal and Watanobe heritable hyperlipidemic rabbits. J Lipid Res 1996; 37:2001-2017.
42. Ferber S, Halkin A, Cohen H et al. panceatic and duodenalhomeobox gene 1 induces expression of insulin gene in liver and ameliorates streptozotocin-induced hyperglycemia. Nat Med 2000; 6:568-572.
43. Ozaki I, Zern MA, Liu S et al. Ribozyme-mediated specific gene replacement of the α1- antitrypsin gene in human hepatoma cells. J Hepatology 1999; 31:53-60.
44. Novina CD, Murray MF, Dykxhoorn DM et al. siRNA-directed inhibition of HIV-1 infection. Nature Medicine 2002; 8:681-686.
45. Brummelkamp TR, Bernards R, Agami R. A system for stable expression of short interfering RNAs in mammalian cells. Science 2002; 296:550-553.
46. Scaglioni P, Malegari M, Takahashi M et al. Use of dominant negative mutants of the hepadnaviral core protein as antiviral agents. Hepatology 1996; 24:1010-1017.
47. Roy Chowdhury J. Prospects of liver cell transplantation and liver directed gene therapy. In: Seminars. In: Berk PD, ed. Liver Disease. Thieme: 1999:1-6.
48. Kmiec EB, Kren BT, Steer CJ. Targeted gene repair in mammalian cells using chimeric RNA/DNA oligonucleotides. In: Friedman T, ed. Development of Human Gene therapy. Cold Spring Harbork: Cold Spring Harbor Laboratory Press, 1999:643-670.
49. Roth JA, Nguyen D, Lawrence DD et al. Retrovirus-mediated wild-type p53 gene transfer to tumors of patients with lung cancer. Nature Med 1996; 2:985-991.
50. Kokoris MS, Sabo P, Adman ET et al. Enhancement of tumor ablation by a selected HSV-1 thymidine kinase mutant. Gene Ther 1999; 6:1415-1426.
51. Mohr L, Shankara S, Yoon S-K et al. Gene therapy of hepatocellular carcinoma in vitro and in vivo in nude mice by adenoviral transfer of the Escherichia coli purine nucleoside phosphorylase gene. Hepatology 2000; 31:606-614.
52. Mohr L, Schauer JI, Boutin RH et al. Targeted gene transfer to hepatocellular carcinoma cells in vitro using a novel monoclonal antibody-based gene delivery. Hepatology 1999; 29:82-89.
53. Harada JN, Berk AJ. p53-Independent and -dependent requirements for E1B-55K in adenovirus type 5 replication. J Virol 1999; 73:5333-5344.
54. Tanaka T, Cao Y, Folkman J et al. Viral vector-targeted anti-angiogenic gene therapy utilizing an angiostatin complementary DNA. Cancer Res 1998; 58:3362-3369.
55. Blezinger P, Wang J, Gondo M et al. Systemic inhibition of tumor growth and tumor metastases by intramuscular administration of the endostatin gene. Nature Biotech 1999; 17:343-348.
56. Fakhrai H, Dorigo O, Shawler DL et al. Eradication of established intracranial rat gliomas by transforming growth factor b antisense gene therapy. Proc Natl Acad Sci USA 1996; 93:2909-2914.
57. Kawashita Y, Ohtsuru A, Kaneda Y et al. Regression of hepatocellular carcinoma in vitro and in vivo by radiosensitizing suicide gene therapy under the inducible and spatial control of radiation. Human Gene Ther 1999; 10:1509-1519.
58. Fan Z, Chakravarty P, Alfieri A et al. Adenovirus mediated antisense ATM gene transfer sensitizes prostrate cancer cell to radiation. Cancer Gene therapy 2000; 7:1307-1314.
59. Encke J, zu Pulitz J, Geissler M et al. Genetic immunization generates cellular and humoral immune responses against the nonstructural proteins of the hepatitis C virus in a murine model. J Immunol 1998; 161:4917-4923.
60. zu Putlitz J, Skerra A, Wands JR. Intracellular expression of a cloned antibody fragment interferes with hepatitis B virus surface antigen secretion. Biochem Biophys Res Comm 1999; 255:785-791.
61. Scaglioni P, Malegari M, Takahashi M et al. Use of dominant negative mutants of the hepadnaviral core protein as antiviral agents. Hepatology 1996; 24:1010-1017.

62. Verma IM, Somia N. Gene therapy—promises, problems and prospects. Nature 1997; 389:239-242.
63. Kalpana GV. Retroviral vectors for liver-directed gene therapy. Semin Liver Dis 1999; 19:27-37.
64. Park F, Ohashi K, Chiu W et al. Efficient lentiviral transduction of liver requires cell cycling in vivo. Nature Genet 2000; 24:49-52.
65. Takahashi M, Ilan Y, Sengupta K et al. Induction of tolerance to recombinant adenoviruses by injection into newborn rats: Long term amelioration of hyperbilirubinemia in Gunn rats. J Biol Chem 1996; 271:26536-26542.
66. Roy-Chowdhury J, Horwitz MS. Evolution of adenoviruses as gene therapy vectors. Mol Ther 2002; 5:340-344.
67. Mitani K, Graham FL, Caskey CT et al. Rescue, propagation, and partial purification of a helper virus-dependent adenovirus vector. Proc Natl Acad Sci USA 1995; 92:3854-3858.
68. Yang Y, Li Q, Ertl HCJ et al. Cellular and humoral immune responses to viral antigen create barriers to lung-directed gene therapy with recombinant adenoviruses. J Virol 1995; 67:2004-2015.
69. Kafri T et al. Cellular Immune response to adenoviral vector infected cells does not require de novo viral gene expression: implecation for gene therapy. Proc Natl Acad Sci USA 1998; 95:11377-11382.
70. Kay MA, Meuse L, Gown AM et al. Transient immunomodulation with anti-CD40 ligand antibody and CTLA4Ig enhances persistence and secondary adenovirus-mediated gene transfer into mouse liver. Proc Natl Acad Sci USA 1997; 94:4686-4691.
71. Thummala NR, Ghosh SS, Lee SW et al. A non-immunogenic adenoviral vector, coexpressing CTLA4Ig and bilirubin-uridinediphosphoglucuronateglucuronosyltransferase permits long-term, repeatable transgene expression in the Gunn rat model of Crigler-Najjar syndrome. Gene therapy 2002; 9:981-990.
72. Ilan Y, Attavar P, Takahashi M et al. Induction of central tolerance by intrathymic inoculation of adenoviral antigens into the host thymus permits long term gene therapy in Gunn rats. J Clin Invest 1996; 98:2640-2647.
73. Ilan Y, Prakash R, Davidson A et al. Oral tolerization to adenoviral antigens permits long term gene expression using recombinant adenoviral vectors. J Clin Invest 1997; 99:1098-1106.
74. Raper SE et al. A pilot study of in vivo liver-directed gene transfer with an adenoviral vector in partial ornithinetranscarbamylase deficiency. Hum Gene Ther 2002; 13:163-175.
75. Einfeld DA et al. Reducing the native tropism of adenovirus vectors requires removal of both CAR and integrin interaction. J Virol 2000; 75:11284-11291.
76. Fung Y, Federoff HG, Brownlee M et al. Rapid and efficient gene transfer in human hepatocytes by herpes viral vectors. Hepatology 1995; 22:723-729.
77. Hoffmann C, Sandig V, Jennings G. Efficient gene transfer into human hepatocytes by baculovirus vectors. Proc Natl Acad Sci USA 1995; 92:10099-10103.
78. Samulski RJ, Zhu X, Xiao X et al. Targeted integration of adeno-associated virus (AAV) into human chromosome 19. EMBO J 1991; 10:3941-3950.
79. Walz C, Schlehofer JR, Flentje M et al. Adeno-associated virus sensitizes HeLa cell tumors to gamma rays. J Virol 1992; 66:5651-5657.
80. Song S, Laipis PJ, Berns KI et al. Effect of DNA-dependent protein kinase on the molecular fate of the rAAV2 genome in skeletal muscle. Proc Natl Acad Sci USA 2001; 98:4084-4088.
81. Wang L, Nichols TC, Read MS et al. Sustained expression of therapeutic level of factor IX in hemophilia B dogs by AAV-mediated gene therapy in liver. Mol Ther 2000; 1:154-158.
82. Alexander IE, Russell DW, Miller AD. DNA-damaging agents greatly increase the transduction of nondividing cells by adeno-associated virus vectors. J Virol 1994; 68:8282-8287.
83. Strayer DS, Zern MA, Roy Chowdhury J. What can SV40-derived vectors do for gene therapy? Curr Opin Mol Ther 2002; 4:313-323.
84. Sauter BV, Parashar B, Roy Chowdhury N et al. A replication deficient rSV40 mediates liver-directed gene transfer and a long-term amelioration of jaundice in Gunn rats. Gastroenterology 2000; 119:1348-1357.
85. Strayer DS, Branco F, Zern MA et al. Durability of Transgene Expression and Vector Integration: Recombinant SV40-Derived Gene therapy Vectors. Mol Ther 2002; In press.
86. Liu F, Song YK, Liu D. Hydrodynamics-based transfectionin animals by systemic administrationof plasmid DNA. Gene Ther 1999; 61258-1266.
87. Zabner J, Fasbender AJ, Moninger T et al. Cellular and molecular barriers to gene transfer by a cationic lipid. J Biol Chem 1995; 270:18997-19007.
88. Findeis MA, Wu CH, Wu GY. Ligand-based carrier systems for delivery of DNA to hepatocytes. 1994; 247:341-351.
89. Kren BT, Bandyopadhyay P, Roy Chowdhury N et al. Oligonucleotide-mediated site-directed gene repair. Meths Enzymol 2002; 346:14-35.

90. Sakamoto N, Wu CH, Wu GY. Intracellular cleavage of hepatitis C virus RNA and inhibition of viral protein translation by hammerhead ribozymes. J Clin Invest 1996; 98:2720-2728.
91. Welch PJ, Tritz R, Yei S et al. Intracellular application of hairpin ribozyme genes against hepatitis B virus. Gene Ther 1997; 4:736-743.
92. Welch PJ, Yei S, Barber JR. Ribozyme gene therapy for hepatitis C virus infection. Clin Diag Virol 1998; 10:163-171.
93. Asahina Y, Ito Y, Wu CH et al. DNA ribonucleases that are active against intracellular hepatitis B viral RNA targets. Hepatology 1998; 28:547-554.
94. Wu CH, Wu GY. Targeted inhibition of hepatitis C virus-directed gene expression in human hepatoma cell lines. Gastroenterology 1998; 114:1304-1312.
95. Pan WH, Devlin HF, Kelly C et al. A selection system for identifying accessible sites in target RNAs. Rna-A Publication of the Rna Society 2001; 7(4):610-621.
96. Aravind L, Walker DR, Koonin EV. Conserved domains in DNA repair proteins and evolution of repair systems. Nucleic Acids Res 1997; 27:1223-1242.
97. Yamamoto A, Taki T, Yagi H et al. Cell cycle-dependent expression of the mouse Rad51 gene in proliferating cells. Mol Gen Genet 1996; 251:1-12.
98. Chan PP, Glazer PM. Triplex DNA: fundamentals, advances, and potential applications for gene therapy. J Mol Med 1997; 75:267-282.
99. Faruqi AF, Datta HJ, Carroll D et al. Triple-helix formation induces recombination in mammalian cells via a nucleotide excision repair-dependent pathway. Mol Cell Biol 2000; 20:990-1000.
100. Culver KW, Hsieh W-T, Huyen Y et al. Correction of chromosomal point mutations in human cells with bifunctional oligonucleotides. Nature Biotechnol 1999; 17:989-993.
101. Kren BT, Metz R, Kumar R et al. Gene repair using RNA/DNA oligonucleotides. Sem Liver Dis 1999; 19:93-104.
102. Stürzbecher H-W, Donzelmann B, Henning W et al. p53 is linked directly to homologous recombination processes via RAD51/RecA protein interaction. EMBO J 1996; 15:1992-2002.
103. Kren BT, Bandyopadhyay P, Steer CJ. In vivo site-directed mutagenesis of the factor IX gene by chimeric RNA/DNA oligonucleotides. Nature Med 1998; 4:285-290.
104. Kren BT, Parashar B, Bandyopadhyay P et al. Correction of the UDP-glucuronosyl-transferase gene defect in the Gunn rat model of Crigler-Najjar syndrome type I with a chimeric oligonucleotide. Proc Natl Acad Sci USA 1999; 96:10349-10354.
105. Tagalakis AD, Graham IR, Riddell DR et al. Gene correction of the apolipoprotein (Apo) E2 phenotype to wild-type ApoE3 by in situ chimeroplasty. J Biol Chem 2001; 276:13226-13230.
106. Lai L-W, Chau B, Lien Y-H. In vivo Gene targeting in carbonic anhydrous II deficient mice by chimeric RNA/DNA oligonucleotides. Conference Proceedings: 2nd Annual Meeting of the American Society of Gene therapy, Washington DC: 1999:236a.101.
107. Alexeev V, Igoucheva O, Domashenko A et al. Localized in vivo genotypic and phenotypic correction of the albino mutation in skin by RNA-DNA oligonucleotide. Nature Biotechnol 2000; 18:43-47.
108. Radice AD, Bugaj B, Fitch DHA et al. Wdespread occurance of TC1 transposon family: TC1-like transposons from teleost fish. Mol Gen Genet 1994; 244:606-612.
109. Ivics Z, Hackett PB, Plasterk RH et al. Molecular reconstruction of Sleeping Beauty, a Tc1-like transposon from fish, and its transposition in human cells. Cell 1997; 91:501-510.
110. Yant SR, Meuse L, Chiu W et al. Somatic integration and long-term transgene expression in normal and hemophilic mice using a DNA transposon system. Nature Genetics 2000; 25:35-41.

Index

A

α1-antitrypsin deficiency 341, 347
ABC transporter 4, 22, 23, 25, 48, 49, 52-56, 58, 59, 78, 187, 199, 203, 270, 288
ABC-A family 23
ABC-B family 25
ABCB4 4, 24-26, 80, 102, 172, 178-180, 278
ABCB11 4, 24, 25, 27, 28, 78-80, 91, 102, 171-173, 178-180, 278
ABC-C family 28
ABCC2 4, 24, 25, 28, 80, 100, 101, 172, 195, 196, 198-201, 203-205, 278
ABCC3 100, 172, 174, 196, 199, 278
ABCC7 80, 91, 172
ABC-G family 29
Acinar gradient 157
Adeno-associated virus-2 (AAV-2) 351
Adenovirus 348, 350, 352
Alagille syndrome (AGS) 80, 83, 91
Alendronate 308
Alkaline phosphatase (ALP) 62, 65, 149, 176, 247, 249, 256, 257, 259
Allograft rejection 161-163, 340, 341
ALT 127, 132, 247, 352
ANCA 252
Anion exchanger isoform (AE2) 5, 271, 279, 281
Anticipation 83
Antihistamine 11, 292
Antimitochondrial antibodies (AMA) 71, 190, 221, 223, 225-228, 230-232, 234-236, 258, 259
Antineutrophilic antibodies 247, 252
Antinuclear antibodies (ANA) 221, 230, 231, 247, 258
Apoptosis 23, 73, 98, 104, 119, 126-132, 154, 156, 223, 224, 236, 333, 334, 341, 346
AST 247
Atherosclerosis 29, 191, 319
Atox1 215-217
ATP7b (also ATP7B) 90, 172, 177, 211, 213, 214, 215, 216, 217, 218
ATP8B1 78-80, 83, 88, 89, 91, 172, 178, 188, 270
Autographa californica nuclear polyhedrosis virus (AcNPV) 351

B

B7-1 222
Bacterial antigens 228, 248
Bax/Bak 130
Bcl-2 126-129, 131-133
Benign recurrent intrahepatic cholestasis (BRIC) 22, 78, 79, 80, 83, 84, 86, 88, 91, 170, 172, 175, 177, 188, 189
Bile acids 2, 6, 11, 16, 21, 29, 40, 48, 50, 57, 59, 62-64, 67-69, 71, 72, 78, 96-105, 112-114, 116-121, 126-132, 139, 141, 145, 170-176, 178-181, 223, 234, 248, 271-275, 279, 281, 290, 292, 304, 306, 320-322, 329-335
Bile acid synthesis 103, 145, 179
Bile duct epithelial cell (BEC) 5, 151, 161, 162, 186, 188, 222-225, 230, 235, 248, 249
Bile flow 2-5, 21, 23, 39-41, 62, 63, 65, 68, 72, 101, 149, 153, 160, 164, 171, 175, 188, 203, 204, 213, 256, 259, 279, 308, 315, 320-322, 332
Bile salts 1-6, 9-13, 15, 16, 21, 23, 24, 26-28, 52, 78, 98, 101, 102, 104, 113, 126, 132, 136, 143, 144, 153-155, 157, 170-178, 180, 186-191, 200, 203, 223, 224, 256, 259-262, 266, 268-281, 305, 314-318, 321, 322, 332, 335
Bile salt export pump (BSEP) 4, 16, 23-25, 27, 48, 78, 80, 98, 102, 104, 113, 126, 170-176, 178, 179, 189, 200, 224, 256, 259, 260, 262, 266, 268, 272, 275, 278, 279, 281, 332, 335
Biliary atresia 72, 73, 154, 158, 163-165, 278, 321, 329
Bilirubin metabolism 80, 196
Bone densitometry 301, 303, 307
Bone metabolism 301, 302, 306, 307
Bone turnover 302, 303, 307, 308
Bromosulfophthalein (BSP) 3, 11, 13-16, 171, 196, 198, 199
Byler's disease 176, 188, 329